Y0-BQR-861

Hazardous Waste Management

Hazardous Waste Management

GAYNOR W. DAWSON
BASIL W. MERCER

A Wiley-Interscience Publication
JOHN WILEY & SONS
New York • Chichester • Brisbane • Toronto • Singapore

Library of Congress Cataloging in Publication Data:

Dawson, Gaynor W.
  Hazardous waste management.

  "A Wiley-interscience publication."
  Includes index.
1. Hazardous wastes—United States—Handbooks, manuals,
etc. I. Mercer, Basil W. II. Title.

TD811.5.D39 1985   363.7'28   85-9313
ISBN 0-471-82268-X

Printed in the United States of America

10 9 8 7 6 5 4 3 2 1

Preface

This handbook is intended for a broad spectrum of engineers and administrators responsible for the management of hazardous wastes. It addresses issues of concern for the regulator, the waste generator, the operator, and the local planners. As such, the text deals with waste management issues at several levels: (1) philosophical, with respect to constructing an overall framework for management; (2) technical, with respect to the selection and design of specific approaches; (3) economic, with respect to the relative cost of alternatives; and (4) sociological, with respect to the growing need for public interface and approval.

The material presented here incorporates both the empirical results of widespread research and development efforts and the opinions of the authors based on their observations and experience at all levels of the management network. The problem of how to deal with hazardous wastes is a relatively new one. Its meteoric rise in public awareness has fostered real growth in funding for development and investigative programs in the public and private sectors. Consequently, the field is quite dynamic, subject to rapid and significant changes. By their nature, then, certain elements of this text will be temporal and must be viewed relative to the context in which they are provided. Other elements are virtually timeless and can be applied to waste problems regardless of location or source. Discussions in each of the major topic areas are constructed from a historical perspective, with major emphasis on technology in the United States. Complementary information on foreign programs, especially those in Europe, is provided to broaden the focus and facilitate the integration of diverse elements into a more holistic view. It is the authors' hope that this will serve to stimulate creative thought on the part of the reader and thereby assist in the advancement of the state of knowledge while it educates.

GAYNOR W. DAWSON
BASIL W. MERCER

Richland, Washington
Harrison, Idaho
January 1986

Contents

1 INTRODUCTION   **1**

The Stage Is Set   1
Content   4
Terminology   8

2 THE REGULATION OF HAZARDOUS WASTE MANAGEMENT   **11**

Historical Perspective   11
RCRA   15
Related Legislation   22
State Programs   32
Foreign Regulatory Programs   37

3 DEFINING HAZARDOUS WASTES   **43**

Differentiating Hazardous Wastes   43
Defining Hazardous Wastes   46
Designating Hazardous Wastes   52
Extraction Procedures   73
Criteria Selection Rationales   78
Prioritizing Concern   91
Epilogue   103

4 QUANTIFYING HAZARDOUS WASTES   **107**

Introduction   107
Surveying Hazardous Wastes   107
Current Estimates of Hazardous Waste Generation Rates   119

5 FACILITY SITING   **133**

Introduction   133
Siting Technology   134
Public Acceptance   165
Financial Considerations   180

6 THE HAZARDOUS WASTE MANAGEMENT INDUSTRY 187

Introduction 187
A Fledgling Industry 188
Coming of Age 189
A Glimpse into the Future 199
Waste Exchange 203
Industry Outside the United States 207
The Prototypical Processing Plant 208
On-Site and Off-Site Trade-Offs 215

7 ABANDONED DISPOSAL SITES 227

Introduction 227
The Problem Set 228
Site Assessment 234
Site Mitigation 262
No-Action Alternative 272

8 HAZARDOUS WASTE TRANSPORTATION 279

Introduction 279
The Industry 281
Regulation via Manifests 289
Regulation Outside the United States 303

9 TREATMENT PROCESSES 307

Introduction 307
Wastewater Treatment 309
Cost of Wastewater Treatment 356

10 INCINERATION 365

Introduction 365
Process Description 366
Types of Incinerators 375
Miscellaneous Types of Incinerators 380
Cost of Incineration 383

11 LANDFILL DISPOSAL 387

Introduction 387
Landfill Design 389

Overview   410
Fixation and Encapsulation Methods   411
Landfill Disposal Costs   414

12   OCEAN DUMPING AND UNDERGROUND INJECTION   417

Introduction   417
Ocean Dumping Regulations   418
Ocean Dumping Operations   420
Underground Injection Regulations   423
Salt Formation Disposal   431
Cost of Ocean Dumping and Deep-Well Injection   433

APPENDIXES   437

A   Summary and Conclusions of the 1973 Report to Congress
     on Hazardous Waste Management   437
B   EPA Listing of Hazardous Waste as a Part of the
     RCRA Regulations, Section 3001   441
C   Proposed California Hazardous Waste Listing,
     February 1975   449
D   Minnesota State Survey Forms   451
E   Model Industrial Survey Forms   459
F   Site Selection Criteria Areas as Delineated in the
     1973 Report to Congress   467
G   Criteria for PCB Disposal: Excerpts from 40 CFR 761   469
H   Texas Site Qualification Guidelines   475
Ia Priority Abandoned Sites November 1981 List of
     114 EPA Priority Superfund Sites   481
Ib Expanded 1982 List of Proposed National
     Priorities Sites and Status   487
J   Unit Costs for Remedial Actions   497
K   Combustion Properties of Hazardous Materials   513

INDEX   523

Hazardous Waste Management

1

Introduction

The Stage Is Set

On August 2, 1978, Dr. Robert P. Whalen, commissioner of health for the state of New York, declared that a medical emergency existed in the city of Niagara Falls. This action was the result of a series of intensive chemical and medical examinations throughout the Love Canal area of the city that revealed the presence of over 82 organic chemicals in air and leachate, as well as an apparent rate of miscarriage among women nearly 1.5 times as high as that in the general populace. One of the chemicals identified was a known human carcinogen, and 11 were known or presumed animal carcinogens. The resultant state action was immediate evacuation of children under the age of two and all pregnant women. This was followed by evacuation of all 97 families from the emergency zone.

The series of events leading to what has been described by Dr. Whalen as a "major human and environmental tragedy without precedent and unparalleled in New York State's history" had their beginning in the 1920s, when the unfinished barge canal was first employed for burial of wastes. The ramifications of that action and subsequent events that allowed those chemicals to come into contact with the citizenry will be felt through the end of the century. This tragic episode, more than any other single occurrence, has brought the issue of chemical waste management squarely into the American consciousness. In the highly charged emotional atmosphere that accompanies such a disaster, the public has come to question the burdens that appear encumbent on a society of a complex industrial nature. The depth and content of associated public concerns must be put in perspective if society is to select an optimum course between the benefits to be derived from expanding technology and the concomitant costs. Hence, it is desirable that a comprehensive review and evaluation be made of chemical waste management. It is with that objective in mind that the present work has been undertaken.

1

It is important to note at the outset that although the drama of Love Canal awakened the public to the potential consequences of hazardous waste mismanagement, recognition of the problem set preceded the summer of 1978 by many years. The Atomic Energy Act of 1954 addressed the special hazards associated with radioactive waste materials and mandated the use of special care in their management. In October 1970, Congress enacted Public Law 91-512, the Resource Recovery Act of 1970. Section 212 of that act mandated a feasibility study for a system of national hazardous waste disposal sites. Specifically, Congress directed that

> The Secretary[1] shall submit to the Congress no later than two years after the date of enactment of the Resource Recovery Act of 1970, a comprehensive report and plan for the creation of a system of national disposal sites for the storage and disposal of hazardous wastes, including radioactive, toxic chemical, biological, and other wastes which may endanger public health or welfare. Such report shall include: 1) a list of materials which should be subject to disposal in any such site; 2) current methods of disposal of such materials; 3) recommended methods of reduction, neutralization, recovery, or disposal of such materials; 4) an inventory of possible sites including existing land or water disposal sites operated or licensed by Federal agencies; 5) an estimate of the cost of developing and maintaining sites including consideration of means for distributing the short- and long-term costs of operating such sites among the users thereof; and 6) such other information as may be appropriate.

During the fulfillment of this mandate, the newly formed U.S. Environmental Protection Agency developed the first quantitative description of hazardous waste management in the United States and characterized the nature of the threat posed by improper management. Both points are important to achieve an understanding of subsequent legislation and to allow proper evaluation of management technology.

Although the Love Canal incident has become a byword for the impacts associated with hazardous wastes, it is but one example of the kinds of damage and circumstances that may accompany improper management. In a report to Congress, the Environmental Protection Agency documented a series of incidents displaying the full range of hazards with which one must be concerned. A brief review will be instructive.

Drinking Water Contamination

Employees of a building contractor on the outskirts of Perham, Minnesota, were hospitalized for gastrointestinal complications in 1972. It was determined that the symptoms stemmed from arsenic poisoning, which was ultimately traced to a newly constructed well. The water was found to contain up to 21 ppm arsenic as a result of contact with grasshopper bait that had been buried nearly 40 years previously and left unmarked.

Toxic Vapors

Attempts to process ponded wastes from alkyl lead production in the San Francisco Bay area resulted in the intoxication of employees at the site and in surrounding areas. When these wastes were trucked to other locations, toll collectors working on the shipment route were also affected.

Environmental Damage

Over 216,000 fish were killed and a 4-mile stretch of the Clinch River virtually sterilized in 1967 when a containment dike of an alkaline waste lagoon burst and released nearly 400 acre-ft of flyash wastes. An additional 77-mile stretch of river was impacted. Over two years later, once-abundant molluscs had yet to reestablish themselves.

Property Damage

Dioxin-contaminated waste oil sprayed on three Missouri horse arenas and a farm road for dust control left toxic residuals that were ultimately responsible for the deaths of over 63 Appaloosa and quarter horses, 6 dogs, and 12 cats. Closure of the arenas resulted in loss of business on termination of a sale agreement for one of the arenas. Final damages were estimated at $500,000 in subsequent lawsuits.

Explosive Hazard

A bulldozer operator was killed and his $91,000 piece of equipment destroyed when a drum of unidentified chemical waste exploded during compaction operations at the Kin-Buc landfill in New Jersey. Ultimately, six citations covering 36 items were issued under the Occupational Safety and Health Act of 1970.

Reactive Materials

A young tank-truck driver was killed near Iberville, Louisiana, when the waste liquid he was discharging in a disposal lagoon reacted with the ponded liquors to produce a toxic plume of hydrogen sulfide gas. Irate townspeople in the area effectively closed down the disposal site soon after by burning the access bridge.

Flammability Hazard

Over $200,000 in property damage was sustained at a hazardous waste facility in Shakopee, Minnesota, when a major conflagration broke out in a diked drum storage area. The contents of nearly 4000 55-gal drums were consumed, including solvents, paint sludges, and waste oils.

Direct-Contact Poisoning

A small child in Arkansas was hospitalized for organophosphate poisoning after playing in a pile of abandoned 55-gal drums. The spent containers had been employed for various pesticides, including methyl parathion, ethyl parathion, toxaphene, and DDT. Residual concentrations were sufficient for intoxication.

Food-Chain Concentration

Alkyl mercury poisoning of three children in New Mexico was traced to contaminated pork. The hog had been raised by the family on floor sweepings from a seed company that employed methyl mercury seed dressing. A fourth child sustained congenital mercury poisoning after the mother consumed the same pork during her first trimester of pregnancy.

Major Economic Dislocation

Over 75 miles of the James River in Virginia were closed to the taking of sport and commercial fish when it was determined that the illegal dumping of Kepone production wastes had contaminated the water, sediments, and biota of the area. Subsequent litigation has indicated loss claims in the tens to hundreds of millions of dollars, and the cost of restoring the James has been estimated to exceed $3 billion. Current projections suggest that more than 100 years may be required for the river to cleanse itself to the point that biota will maintain tissue levels below the FDA action level.

Numerous other incidents have been documented in the past decade demonstrating the far-reaching implications of hazardous waste mismanagement. The remainder of this text is directed to exploring the breadth of those implications and the diversity of the activities required for proper management.

Content

This book is divided into chapters of two types: The first type deals with overall concepts at the heart of waste management, and the second deals with the practical aspects of the technology. Both types are technical in nature, but they also include discussions of the nontechnical environment that will greatly influence the character of any hazardous waste management program. In this way, we hope to provide decision makers with the inputs required to challenge preconceived notions and, in so doing, attempt to construct optimal approaches. Even the definitive and most technical portions of the text should not be taken as a recipe for decision making. Environmental problems do not neatly fit into packages that can be addressed with preselected prescriptions. Each has a uniqueness that must be viewed from all sides before a course of action is selected. Hence, the reader is encouraged to store the technical content of the book as one might amass a set of tools, while working with the conceptual matter until he has developed a consistent operational philosophy. With that accomplished, the reader will be ready to design and build programs to meet the needs of any given situation.

Chapter 2 begins with a brief history of legislation and regulations addressing the management of hazardous wastes. The narrative reveals the evolution of current law as the Congresses and administrations have responded to growing public pressures. It is current law that forms the backdrop for the technical discussion offered here and defines the policy issues that must be revealed to assure good management. As with many pieces of environmental law that seek to improve the quality of life through trade-offs between reduced emissions and the concomitant costs of that reduction, hazardous waste management regulations raise many questions with respect to the optimal approach toward control:

How clean is clean enough?
What are the practical limits of zero risk?
What costs are justified for approaching those limits?
When does the ease of administering design standards outweigh the merits of performance standards?
To what level must financial responsibility be required?
What is a reasonable limit to liability?

Although these issues will not be resolved solely as a result of the development of hazardous waste management programs, they will influence the direction of that development and in the process move closer toward resolution.

In the third chapter, this book turns to a discussion of hazardous waste classification systems and construction of a working definition for hazardous wastes. This issue is central to any program for management in that it delineates the boundaries by which materials will be included and others excluded. Selection of a single approach obviates a balance among technical judgment, ease of administration, and economics. The current proliferation of systems demonstrates the lack of consensus on where that balance lies. Classification systems can also be employed to provide prioritization among wastes and thus assist in directing scarce resources to the areas of greatest hazard. Although this concept is well understood, it is often neglected.

Given the general description of what hazardous wastes are in Chapter 3, Chapter 4 turns to an accounting of these wastes. Throughout the development of hazardous waste management programs in the United States, the inventory step has preceded all else. This has generated an inconsistent and often misleading data base, because those conducting the inventory have not been applying a common definition. As a consequence, it has been difficult to quantitatively describe hazardous waste generation in the past; but, good techniques have been developed to accomplish that end, and estimates are rapidly improving with promulgation of rules for mandatory industry reporting.

The fifth chapter addresses one of the most controversial and difficult of the facets of hazardous waste management—facility siting. It is currently thought that inability to site new facilities with sufficient capacity to meet demands is the single major obstacle to successful implementation and enforcement of the Resource Conservation and Recovery Act. The difficulties in siting new facilities go far beyond the technical issues related to the integrity of engineered and natural barriers. The issue of social acceptance of perceived risks becomes paramount. Hence, the technologists must ultimately rely heavily on the social sciences to determine the makeup of receptive populations, the optimal means of incorporating the public into the decision-making process, and the most effective programs for relaying technical information to the general public. To date, the body of knowledge available to the social scientist is heavily tied to recent efforts to site nuclear facilities and repositories for nuclear wastes. If private industry is to maintain its role in chemical waste management, institutional arrangements are also required to ensure adequate financing for potential liabilities and long-term care of facilities beyond the active life of a site.

Chapter 6 is directed to a discussion of the hazardous waste management industry. In the confusion surrounding the news media's coverage of the more spectacular cases of hazardous waste mismanagement, the public has not been apprised that at the same time there is a very capable industry quietly going about the business of providing proper management for these wastes. Many of the current operators had their origins in solid waste management or the secondary recovery industries. Operations designed to recover waste solvents and oils expanded to more exotic source materials, salvaging those commodities deemed profitable under the economics of the time, and treating and disposing of the residuals. As with all industries, waste management firms include those whose profits are derived from inadequate handling as well as those who are doing an exemplary job. In the penetrating light of current public scrutiny, the former are being identified. This, in turn, has shaken the credibility of the latter and set the stage for debate of a number of difficult problems, including

Siting
Disposal of destructible materials
Closure and ownership
Long-term liability and financial responsibility
Site restoration

The industry also includes a number of exchanges that provide a clearing-house through which waste chemicals can be transferred from the generator to a party interested in using them as raw materials.

Chapter 7 is directed to a discussion of the uncontrolled landfill problem. Since the revelation of Love Canal, subsequent investigations have identified numerous sites where improperly managed wastes have been abandoned. These sites are now sources of active or potential contamination of ground-water and surface waters and the atmosphere. Embryonic programs designed to address the uncontrolled site problem include three major phases—characterization, assessment, and remediation. The first two are designed to determine if the risks posed by a site are sufficient to warrant the costs associated with the third. Assessment and remediation are new and developing areas of engineering that in the end will require a much broader knowledge of how chemicals react with soil and minerals. Provided such knowledge is forthcoming, the classic approach of exhumation and disposal will be complemented by an array of options for in-place immobilization and in-situ destruction.

Chapter 8 addresses the transportation of hazardous wastes. Economies of scale and the geologic uniqueness of areas ideally suited to chemical disposal have provided ample incentive for the transportation of wastes over great distances. Indeed, some chemical wastes have been shipped across the continent for management at special facilities. Understanding waste transportation is also a key for future enforcement programs. The transgressions of the past have often stemmed from the transportation link. Unscrupulous haulers have diverted to improper sites or dumped loads in uncontrolled areas. Thus it is that a mainstay of current regulatory programs is the manifest form that tracks the movement of the waste and provides an audit trail for enforcement agents.

The ninth chapter addresses treatment technology. Although treatment facilities for hazardous wastes are relatively new, the unit processes employed have been developed and applied for many years to wastewater treatment and chemical production. As such, there is a good body of knowledge on many of the more broadly applicable techniques. The key to selection lies first in recognizing the limitations of individual processes for specific applications and then in estimation of potential costs. This evaluation process cannot be conducted in the abstract. After preliminary selection of one or more approaches, bench-scale and/or pilot-scale work with wastes is required to understand the operating parameters and thereby allow accurate analysis of both treatability and costs.

Chapter 10 addresses hazardous waste incineration technology. In many respects, this can be considered another means of treatment, because a residue is produced that ultimately requires disposal. Incineration is often viewed as a more desirable but also more expensive alternative to landfill disposal. This may be true for a large number of organic wastes, but certainly not for inorganic wastes. Therefore, heavy use of incineration will not eliminate the need for landfills. This chapter discusses both the design features of different incineration systems and the combustion properties of different waste constituents. The latter is important both for calculation of fuel requirements and for consideration of energy recovery through production of steam. Either is important with the increasing cost of energy.

Chapter 11 is directed to the workhorse of current hazardous waste management systems, the landfill or surface repository. An abundance of land and the low operating requirements have made landfills the preferred, apparent low-cost method of disposal. Recent changes in law and the realization that all repositories can and will fail are changing this attitude. As waste generators consider the liability associated with failure of a landfill, they begin to question the true cost of landfills. This is causing more and more generators to turn to incineration or other means of destruction for organic-based wastes. This will leave the landfill for inorganic wastes, which someday may be economically recoverable. Hence, there is more consideration being given to design of disposal sites as interim storage facilities, with retrievability an important factor. Alternatively, landfills are viewed as long-term facilities to disperse wastes back into the environment in concentrations that pose no significant impacts. This is a politically charged approach and will be a long time in achieving acceptance.

The final chapter addresses ocean disposal and deep-well injection. Both disposal methods were once considered highly undesirable but are once again being pursued. In the case of ocean disposal, it is recognized that the ocean is the ultimate repository of all wastes, regardless of disposal technique. Therefore, if properly designed, this approach can be safely implemented while skipping the intermediate steps wherein toxic constituents move through the air, groundwater, and surface water on their way to the sea. Deep-well disposal appears attractive where briny, unusable aquifers exist and can accept high-salt wastes.

Terminology

Common usage of various terms as well as apparently conflicting references in selected legislation and regulations has led to some very imprecise understandings of the terminology surrounding hazardous waste management and associated problems. For clarity, then, it is instructive to establish a basic understanding of the key terms to be employed in this work.

The term *waste* refers to any discarded material whose use potential has been diminished by physical or chemical modification or whose generation was a secondary or unintended consequence of some other primary activity. The definition is intentionally broad, such that it incorporates both products that have lost their primary value (off-specification or contaminated batches) and by-products that may ultimately be recycled. The operative principle here is one of economics. Materials are wasted when the economics of use are such that costs exceed benefits and the generator finds it prudent to discard the material. If a generator would not undertake an activity solely for the production of a given material, then that material is a waste associated with the activity. The boundaries are vague, because economics constantly change with temporal and spatial factors. For example, the markets for reclaimed oil and paper have gone through significant fluctuations in the past, with changes in supply and demand structures of raw materials. For the most part, however, generators of these recycled wastes would not continue to operate solely to produce waste oil or paper. These materials are by-products of other economically attractive activities.

The term "waste" is further complicated by its association with the verb "to waste," meaning "discard." Common practice often involves long-term storage of waste materials in anticipation of market changes that may raise the value of a material. This raises the question whether these materials are in fact raw materials stockpiled for the future or wastes placed in an interim facility. Usage, herein, focuses the definition of waste through use of the term "discarded" and assumes that storage beyond a typical inventory period of 180–380 days constitutes the first stage of a process of waste disposition.

The term *solid waste*, as employed here, goes beyond the confines of what the delimiter "solid" normally conveys. Common usage associates solid waste with refuse and garbage, as generated by households and commercial enterprises. Current legislative language, however, incorporates many non-solid waste forms, such as liquids, slurries, and contained gases. This convention is in part a result of the historical development of effluent regulation. Controls designed to protect air and water quality have focused on atmospheric emissions and continuous wastewater discharges. When it was recognized that this created incentives for disposing of residuals on the land, lawmakers associated the necessary controls with regulation of the major input to the land: solid wastes. Hence the definition of solid waste as presented in the Resource Conservation and Recovery Act (RCRA) of 1976:

(27) The term "solid waste" means any garbage, refuse, sludge from a waste treatment plant, water supply treatment plant, or air pollution

control facility and other discarded material, including solid, liquid, semisolid, or contained gaseous material resulting from industrial, commercial, mining, and agricultural operations, and from community activities, but does not include solid or dissolved material in domestic sewage, or solid or dissolved materials in irrigation return flows or industrial discharges which are point sources subject to permits under section 402 of the Federal Water Pollution Control Act, as amended (86 Stat. 880), or source, special nuclear, or by-product material as defined by the Atomic Energy Act of 1954, as amended (68 Stat. 923).

This definition is employed in the present work as well.

The term *hazardous waste* is one that has proved most difficult to define clearly. Indeed, Section 3001 of the guidelines and regulations for hazardous wastes as mandated by RCRA is devoted to a description of how one distinguishes between a hazardous waste and a nonhazardous waste. Similarly, an entire chapter of this work is dedicated to a discussion of what constitutes a hazardous waste. For clarity, however, it must be noted at this point that usage here is restricted to those materials that are intrinsically hazardous as a function of physical or chemical properties, but not including those that are hazardous because of biological (etiological, pathogenic, etc.) or radiological properties. This is not to imply that there is a distinction in the degree of hazard or concern between these two groups. Rather, it is intended as a means of narrowing the focus of this work to a specific subset of hazardous wastes: those that might more accurately be described as chemical wastes.

The term *waste management* is used here to describe the full range of activities that accompany custodianship and disposition of wastes from the point of generation to the point of final disposal. In this way, the definition embraces all aspects of waste handling, storage, treatment, transportation, and disposal, thereby setting the stage for "cradle-to-grave" control.

The term *waste disposal* is used in this text to define those activities that are conducted as a final act of control over a waste and constitute a release of the waste to the environment. Distinction is made with respect to control after disposal. If, for instance, the waste is merely put in a repository in a retrievable form, such an action constitutes long-term storage, not disposal. Similarly, incineration constitutes both treatment of solid and liquid materials and disposal of gaseous residuals. Disposal is not performed on the solid materials until the ash is discarded.

Numerous other terms used throughout the text are defined as needed.

NOTES

1. Under Reorganization Plan No. 3 of 1970, responsibility for carrying out the provisions of the Resource Recovery Act was assigned to the administrator of the Environmental Protection Agency.

2

The Regulation of Hazardous Waste Management

Historical Perspective

For the most part, hazardous wastes are a relatively new segment of the total solid waste stream. Whereas solid wastes can be traced back to the middens in the camps of primitive man at the dawn of human emergence, hazardous wastes are largely a product of the industrial age, when man first began to utilize fossil fuels and synthesize chemical entities. Thus it is that D.G. Wilson's delightful history (1977) of solid waste management begins with Moses' instruction to bury wastes as a means of disposal, and it traces developments through Crete, Rome, England, Medieval Europe, and colonial America; but no mention is made of materials that would be defined as hazardous wastes until subsequent sections devoted to the discussion of industrial waste management. The excerpts from publications and official documents focus on the odor and nuisance problems associated with rubbish. In a review of tort and statutory law concerning environmental issues, Krier (1971) provided a glimpse of the awakening to the fact that wastes may pose a specific hazard. Although it was litigated under common nuisance law, the case of Versailles Borough v. McKeesport Coal and Coke Company (83 Pittsburgh L. J. 379 [1935]) dealt with specific hazards associated with gob piles from coal mining. Gob is the solid interlayed coal and shale material wasted during early coal mining because of its low fuel value. It became a nuisance because it often spontaneously ignited and would burn for great periods of time. Because many types of solid, nonhazardous refuse are also spontaneously ignitable, it is doubtful to what extent the gob would meet modern hazard criteria based on flammability. However, gob releases sulfur dioxide when burned and therefore generates toxic off-gases. In the more recent case

11

of Warchak v. Moffat (379 Pa. 441, 109 A.2d 310 [1954]), suit was brought as a result of toxic hydrogen sulfide emissions from culm banks (a form of refuse from a coal breaker) that destroyed the paint on a nearby house. Similarly, Oregon has been the site of two related cases involving negligence and strict liability (Martin v. Reynolds Metals Company, 135 F.Supp. 379 [D. Ore. 1952]) and nuisance and trespass (Renken v. Harvey Aluminum Inc., 226 F.Supp. 169 [D. Ore. 1963]) as a result of waste discharges bearing fluorine, hydrogen and calcium fluoride, and silicon tetrafluoride causing health and property damage. Aside from the precedence of the legal points made during the litigation, these often-cited cases illustrate the shift from the nuisance and odor problems of solid wastes referenced by Wilson (1977) to those of toxic hazards arising from industrial wastes. Contemporary legislation and regulation in the United States reflect the same shift in emphasis as it was developed in response to the evolving recognition of waste-related problems.

Wilson (1977) related the first known U.S. ordinance in refuse as one adopted by the Corporation of Georgetown in 1795 prohibiting the extended storage of refuse on private property and the dumping of refuse in the street. Soon after, carters were brought under contract to periodically remove refuse from streets and alleys. Solid waste management remained largely a local issue throughout the next century. The major solid waste problems arose in the large cities, where dense population combined with lack of available land for disposal to create a high-demand/low-supply situation. Ordinances and systems for management varied widely between metropolitan areas. Federal action first emerged in 1899 with the Rivers and Harbors Act, which prohibited the disposal of solid objects in waterways such that they would create a hazard to navigation. Although this would appear very indirect, it formed the basis for the limited degree of solid waste management regulations over the next 50 yr. It became known as the "Refuse Act" and provided the legal foundation for Corps of Engineers discharge permits, the forerunner of NPDES permits.

The first major piece of legislation directed specifically to solid waste management was the Solid Waste Disposal Act of 1965 (PL 89-272). This act continued the philosophy that solid waste management was basically a local or state problem, but it recognized that a national effort was required to coordinate a program that could transcend the technical and economic capabilities of local communities. To facilitate improvements in local and state programs, the act authorized specific actions in six areas:

1. Grants to demonstrate new and improved waste disposal technology on local and state levels
2. Grants to foster regional solid waste management systems in areas fragmented into small communities

3. Grants for conduct of statewide solid waste management need surveys and subsequent state plan development
4. Funds for direct research and grants to develop new approaches to solid waste management
5. Funds for direct support and grants focused on creating training programs
6. Technical assistance to local and state entities with solid waste management problems

In this way, PL 89-272 established several key tenets for the federal role in solid waste management. The most obvious is maintenance of the local entity as the functional organization in the management system. Although the need for regional management plans is clearly narrated, Congress intentionally left this to the local decision makers rather than create federal boundaries, as had been done for Air Quality Regions. The character of federal involvement was defined as that embodied in economic support and technology transfer. The administration would act as a resource pool to be drawn on at local discretion. No federal standards were set to create uniformity. Similarly, no formal recognition of a hazardous waste management problem was evident.

Continually mounting concern over protection of the environment led to a second legislative measure, the Resource Recovery Act of 1970 (PL 91-512). Although the title reflects a swing in solid waste management philosophy from the disposal language of the 1965 act, the Resource Recovery Act took the form of a series of amendments to the earlier act. The heavy reliance on state and local control of solid waste management was kept intact, but new sections were added to expand the scope of federal involvement. Specific additions included:

1. The statement of purpose for the act was expanded to cover provision for promulgation of guidelines for solid waste management.
2. Research and grant activity was revised to encourage heavy emphasis on waste reduction and resource recovery, with a special study and demonstration projects mandated on recovery of useful energy and materials.
3. The secretary was directed to provide recommended guidelines for all phases of solid waste management, as well as model codes, ordinances and statutes for implementation of those guidelines.
4. All executive agencies and their contractors were directed to ensure compliance with guidelines promulgated under the act.
5. The secretary was mandated to provide a comprehensive report and plan for creation of a system of national disposal sites for the storage and disposal of hazardous wastes.

The final amendment, that calling for a study of a national hazardous waste disposal system, offers two important features: (1) a recognition that some solid waste management problems cannot be handled at the local level and (2) the first direct mention of hazardous waste management in federal legislation. These changes are embodied in Section 212 of the amended Solid Waste Disposal Act, which has been reprinted in the previous chapter. The results of the subsequent report to Congress were to become the forerunner of current hazardous waste law. In authorizing the study and plan, Congress was expressing three discrete concerns:

1. Hazardous materials and wastes, regardless of origin, represented a substantial and growing threat to public health and environmental quality.
2. Management of such wastes was inadequate, and the need for responsible stewardship would increase in the future.
3. The problems posed in the management of hazardous wastes did not necessarily recognize state boundaries, and the materials in question were derived from a variety of sources, both private and public; therefore, consideration had to be given to the feasibility of establishing a national system of disposal sites.

As a consequence, the report to Congress was undertaken in the spirit of defining a new national goal: In order to protect the public health and welfare and the quality of the environment, present practices in disposal of hazardous waste materials must be significantly improved, and a framework for responsible stewardship established.

The report was prepared by the then newly formed Environmental Protection Agency (EPA), drawing on the inputs of six private contractors. It was presented to Congress on June 30, 1973. The report concluded the following: (1) A significant hazardous waste management problem did in fact exist and was growing steadily; (2) technological solutions for the problem were available, for the most part, but represented a significant increase in processing costs; (3) the legislative and economic incentives existing at that time were insufficient to foster environmentally adequate disposal practices in most instances; (4) the cost-effective solution appeared to be a program centered around regulation of hazardous waste treatment/ disposal; (5) a small but viable hazardous waste management industry existed and could expand if a regulatory program created a larger market; however, economic and other uncertainties precluded accurate assessment of the private sector's response; (6) alternatives were available for governmental involvement in providing services, but the necessity of such a course of action was in doubt. Specific conclusions and summary statements from the report are reprinted in Appendix A.

The report to Congress clearly amplified the new directions resulting from Section 212 of the amended Solid Waste Disposal Act: identification of the hazardous waste management subset of overall solid waste management, and the need for a federal role in subsequent programs. Pursuant to the latter point, the EPA appended to the report a draft of the proposed Hazardous Waste Management Act of 1973. The EPA was not without support for such a move. In the previous month, the National Association of

Counties Research Foundation (1973) had concluded that the dimensions and complexities of the hazardous waste management problem set were such that regulation and enforcement of standards should be at a national level. In fact, the foundation suggested that local involvement should be minimized, save for the authority to impose stricter standards on a local-option basis.

This set the stage for major federal legislation on the management of hazardous wastes. In anticipation, the Environmental Protection Agency identified funding and initiated an ambitious program to provide a more quantitative focus on waste generation and the status of management practices. As data were collected, they were passed on to congressional staffers to provide impetus for passage of the desired legislation. Concomitant pressures for toxic substance control legislation and comprehensive solid waste management legislation muddied the issue and helped delay action for several years. In the end, the architects of environmental legislation within the government chose to maintain the close ties between hazardous wastes and solid wastes. The proposed Hazardous Waste Management Act of 1973 became but one element, albeit an important one, of proposed comprehensive legislation for solid waste management. This took its final form as Subtitle C of the Resource Conservation and Recovery Act of 1976 (PL 94-580), which replaced the amended Solid Waste Disposal Act in its entirety.

RCRA

The Resource Conservation and Recovery Act (RCRA) constitutes a clear and unmistakable shift to federal standards and regulations for solid waste management, with enforcement at the state level when implementation plans are deemed acceptable. The heavy federal role is particularly evident in Subtitle C: Hazardous Waste Management. Subtitle C is divided into 11 major sections prescribing a complete regulatory control program. Each section is summarized below.

3001. Identification and Listing of Hazardous Wastes

The administration is required to promulgate criteria that identify the characteristics of hazardous wastes and allow the listing of those wastes. At the same time, regulations are to be issued that identify hazardous waste characteristics and list specific hazardous wastes. Fulfillment of these mandates creates a working definition of hazardous wastes that can be applied to any candidate waste for classification. State governors are afforded the option of petitioning for classification or listing of a specific material as a hazardous waste. Section 3001 defines the breadth of the law's coverage, because it determines which materials will be defined as hazardous. These regulations have been promulgated and are discussed more fully in Chapter 3.

3002. Standards Applicable to Generators of Hazardous Wastes

The administrator is required to promulgate regulations establishing standards for hazardous waste generators respecting:

Recordkeeping practices relating to production, composition, and disposition of hazardous wastes
Labeling of containers used for storage, transport, or disposal
Use of appropriate containers
Providing information on the composition of wastes to parties transporting, treating, storing, or disposing of wastes
Use of a manifest system to provide for tracking of wastes throughout the management cycle, save those wastes disposed on site
Generating periodic reports on hazardous waste activities

Section 3002 delineates the responsibilities of the waste generator and, in so doing, establishes sufficient data on the sources of hazardous wastes to accommodate cradle-to-grave management.

3003. Standards Applicable to Transporters of Hazardous Wastes

The administrator is to promulgate regulations establishing standards for the transporters of hazardous wastes that will include:

Requirements for recordkeeping concerning wastes shipped
Prohibitions on transport of improperly labeled wastes
Operational mode for compliance with the prescribed manifest system
Requirement to deliver wastes only to permitted facilities designated on the manifest form

The administrator is required to coordinate the foregoing regulations with the secretary of transportation in those cases where hazardous wastes are deemed to be subject to the Hazardous Materials Transportation Act. Section 3003 provides the important information link to connect the tracking system between point of origin and final resting place.

3004. Standards Applicable to Owners and Operators of Hazardous Waste Treatment, Storage, and Disposal Facilities

The administrator is required to promulgate standards and regulations for facilities that treat, store, and dispose of hazardous wastes. Regulatory control is prescribed for:

Maintenance of records on the quantities, nature, and disposition of all hazardous wastes received

Reporting, monitoring, and inspection of manifests for wastes received to
   assure compliance
Management of wastes in a manner acceptable to the administrator
Location, design, and construction of facilities
Contingency plans for action to minimize unanticipated damage
Operation of facilities, including such aspects as ownership qualification,
   continuity of operation, training, and financial responsibility
Compliance with permit requirements

Section 3004 constitutes the regulatory framework for treatment and disposal
of hazardous wastes, thus complementing the controls placed on generation
and transport in previous sections. In many respects, it is the heart of the issue
raised by the report to Congress. Other sections merely provide the mechanisms and data that will facilitate implementation and enforcement of Section
3004 regulations.

3005. Permits for Treatment, Storage, and Disposal of Hazardous Wastes

The administrator is required to establish a permit system such that all parties
engaged in treatment, storage, or disposal must obtain permits in order to
continue in those activities. Permits will define the extent and nature of those
activities. The permit application will include a projection of waste loads,
types and frequencies anticipated, and a description of the site. Permits will be
issued when applications meet requirements, and revoked when operations
are found out of compliance. Existing facilities are granted permit status until
such time as the agency has officially processed an application thereto.
Section 3005 established the means by which the agency can enforce the 3004
regulations, namely through certification with a permit process. To date,
most facilities (8000 by December 1983) have been granted interim status
awaiting development of guidelines and procedures to support a full-fledged
permit program. By late 1984, EPA and other authorized states began to call
in applications for Part B permits, but only five land disposal facilities were
fully permitted. It was estimated at that time that it would take 10 years to
process permits for the 2000 sites likely to apply.

3006. Authorized State Hazardous Waste Programs

The administrator is required to promulgate guidelines to assist states in the
development of state hazardous waste programs. States wishing to administer
and enforce the hazardous waste regulatory program can be authorized to do
so if this application is accepted on the grounds that (1) the state program is
equivalent (criteria for equivalency will be defined by the agency) to the
federal program, (2) the state program is not inconsistent with the federal or
other state programs, and (3) the enforcement of the program is adequate to

promote compliance. Interim authorization is granted to states with existing programs that are substantially equivalent to the federal program. Authorized state programs, when enforced, will carry the same force and effect as the federal program. Should enforcement be found inadequate, authorization can be revoked.

Section 3006 provides the mechanism by which the federal program can be brought back to the state level for enforcement, much as has been done with air quality regulation.

3007. Inspections

The administrator or his designated representative is provided the right to gain entry to facilities and records where hazardous wastes are generated, treated, stored, or disposed for the purpose of inspection and sampling. Records, reports, and information resulting from inspections are available to the public, unless the operator can demonstrate to the administrator's satisfaction that release would divulge information privileged to protection. Section 3007 provides one means of monitoring regulatory compliance.

3008. Federal Enforcement

Enforcement procedures are prescribed through use of failure-to-comply notifications and modification orders issued by the federal or authorized state government. Civil penalties are available for assessment when violations continue beyond a specified time limit. In addition, permits may be revoked. Criminal penalties may be assessed to transporters delivering to nonpermitted facilities, persons disposing of hazardous wastes without a permit, or persons found to have falsified information on permit applications. Section 3008 provides the incentive for complying with Subtitle C regulations.

3009. Retention of State Authority

States are prohibited from enacting regulations less restrictive than those promulgated under the act, except for cases where federal regulations have yet to take effect. Section 3009 renders the federal program the minimum acceptable standard for hazardous waste regulation.

3010. Effective Date

A formula is derived for determining the effective date for preliminary notification of general hazardous waste management activities (90 days after promulgation or revision of 3001 regulations) and the effective date for regulations under this act (6 months after promulgation). Section 3010 establishes the time frame under which regulatory requirements must be met.

3011. Authorization of Assistance to States

Funding is authorized to assist states in the development and implementation of authorized state hazardous waste programs. Allocation formulas are left to the administrator. Section 3011 creates resources for states to address their role in hazardous waste management regulation.

From this summary it can be seen that RCRA Subtitle C addresses many of the issues raised in the 1973 report to Congress. In many respects, it constitutes a revised version of the ill-fated Hazardous Waste Management Act of 1973 by providing for cradle-to-grave management of wastes. Enforcement is facilitated by creation of an extensive information flow system. This begins with the generator, who must declare production of the waste, maintain records on wastes, and initiate the manifest form. The manifest, in turn, travels with the waste as a log of the waste's management history. Finally, the disposer must declare intentions in the permit application and maintain supporting records. In its entirety, the paperwork forms a comprehensive data base that can be employed to determine compliance. Generator notifications taken in the aggregate suggest the rate at which a given industry produces a hazardous waste. If competitors in that industry report no wastes, the regulator has reason to believe there are discrepancies and can arrange for an inspection. Comparison of originator's and disposer's copies of manifests allows identification of waste streams that are being diverted from permitted sites. Because transporters are required by law to deliver only to permitted sites, they, as well as the generator, are motivated to foster compliance.

The framework of Subtitle C is such that it places total reliance on private industry to meet the needs for hazardous waste management. No provisions have been made for federal activities beyond the regulatory role. Hence, the loans and loan guarantees alluded to in the 1973 report to Congress did not survive the legislation process. On the other hand, the act does not preclude local and state governments from undertaking some aspects of hazardous waste management such as private industry might. Such action would be subject to the same degree of regulation and would, for all intents and purposes, parallel facets of solid waste management where municipal collection systems operate in direct competition with private services.

As with many environmental laws, the extent and nature of RCRA's impact will be closely tied to the content of regulations promulgated as a result of the act. The effectiveness of the resulting management program and indirect effects on commerce will be determined by such features as (1) the rationale employed for selecting criteria to define hazardous wastes, (2) the criteria themselves, (3) the use of performance versus design standards, (4) the standards selected for acceptance of permit applications, and (5) the scope of financial responsibility requirements.

While Subtitle C is relatively comprehensive, it leaves some unanswered questions. The issue of federal financial assistance or incentives to sponsor private activities has been discussed. The law also fails to address the problem of abandoned or closed disposal sites that may be causing environmental problems. The "imminent hazard" provisions of Section 7003 gives the administrator a means of restraining operators whose activities present an imminent and substantial endangerment to health or the environment, but they do not include funds or mechanisms for remedial action to stop leachate, vapor loss, or other natural mechanisms that will continue to spread contaminants after cessation of operations. Hence, the law is directed to prevention of future Love Canal incidents, but offers no remedy for those sites whose origins predate promulgation of new regulations or for sites operated outside of the law.

RCRA does not provide the EPA or the Justice Department with the power of subpoena. This may severely cripple efforts to enforce subsequent regulations. The development of hazardous waste legislation and regulations has been continually hampered by lack of data on the current situation. As the prospects of litigation and major penalties become increasingly likely, data will become even more difficult to obtain. Implementation will also be weakened by the lack of adequate funds for the development of state programs. Because the EPA is not properly staffed to administer RCRA in all the states, successful implementation will depend on certification of state programs. However, the $50 million allocated for development of the necessary programs appear considerably short of the actual requirements. This has put increased emphasis on the evaluation of fee systems such as those imposed in Maryland and California.

1984 Amendments to RCRA

In 1984 Congress reauthorized RCRA and amended the law to address some of the deficiencies noted above. Key elements of these amendments are summarized below:

Small Quantity Generators. The EPA is required to promulgate standards for the management of hazardous wastes generated by firms producing quantities greater than 100 Kg/month and less than 1000 Kg/month.

Landfill Bans. The EPA is required to develop regulations banning specified wastes from landfills and/or demonstrate which wastes can be safely disposed in landfills. If statutory deadlines are not met, these wastes will automatically be banned.

Other Land Disposal Restrictions. These provisions impose a ban on disposal of bulk or noncontainerized liquids and require the EPA to promulgate regulations for the disposal of containerized liquids. Placement of bulk liquids in salt domes, bedded salt, underground mines, or caves is prohibited until regulations can be devised. Use of contaminated waste oils for dust suppressants is prohibited. Restrictions are also placed on waste injection.

Retrofitting Surface Impoundments. This provision places interim status impoundments under the double liner, leachate collection, and groundwater monitoring requirements for new impoundments. Exemption procedures are delineated.

Storage of Banned Waste. Specific procedures are defined for surface impoundments which store or treat wastes banned from land disposal.

Minimum Technology Standards. This provision establishes minimum standards for hazardous waste management facilities such as double liners or their performance equivalent for new land disposal facilities or interim status facilities receiving wastes six months after enactment; and 99.99% DRE for new incinerators.

Corrective Action. The EPA is required to promulgate regulations that require evidence of financial responsibility for corrective action and corrective action must encompass affected areas beyond the boundary of the facility. All permits must address releases of hazardous waste or constituents regardless of when the source materials were emplaced. Corrective actions can be required of interim status facilities.

Permits. This provision requires review of all permits every ten years and review of land disposal facility permits every five years. Timetables are established for permit application and review, and permits are required for construction of RCRA facilities.

Exposure Assessments. Permit applications must be accompanied by an asssesment of the potential for the public to be exposed to hazardous substance releases from the facility. If a substantial risk is indicated, health assessments must be conducted.

Waste Minimization. Manifests must include a generator certification that waste volume and/or quantity and toxicity have been minimized to the extent economically practicable. Biennial generator reports are required on waste reduction activities.

New Waste Additions. A schedule is established for considering new listings of specified hazardous waste (chlorinated and other halogenated dioxins and dibenzofurans, coal slurry pipeline effluent, coke by-products, chlorinated aliphatics, dioxin, dimethyl hydrazine, TDI, carbamates, bromacil, linuron, organobromines, solvents, refining wastes, chlorinated aromatics, dyes and pigments, inorganic wastes, lithium batteries, and paint production wastes). Waste must be listed if they contain carcinogens above a given health-risk level. The EP test must be considered and other criteria identified for testing wastes.

Delisting. The EPA must consider all factors in delisting wastes, not just those for which the waste was originally listed.

Burning/Blending. This provision requires notification by individuals who burn hazardous waste derived fuels and mandates EPA standards to regulate these practices.

Used Oil. This provision requires the EPA to evaluate the listing of waste oils as hazardous and promulgate regulations accordiingly.

Burning of Municipal Solid Waste. Burning of municipal solid wastes is exempted from Subtitle C as long as provisions are made to ensure hazardous wastes are not burned at the facility. The EPA is required to perform a dioxin emissions study.

Domestic Sewage. Regulations are mandated to control the release of hazardous wastes mixed with domestic sewage or other wastes routed to a POTW.

Hazardous Waste Exports. Requirements are placed on the export of hazardous wastes to tighten management.

Special Wastes. This provision allows modification in double liner, prior release, and land disposal restrictions for mining, utility, and cement kiln wastes. In addition, uranium mill tailings regulations under UMTRCA need not be modified to reflect the amendments to RCRA.

Related Legislation

Although Subtitle C of RCRA is the first major piece of legislation to address regulation of hazardous waste management, several other acts contain segments that deal with specific aspects of the problem set. These include the following.

CERCLA

Whereas RCRA was deemed necessary, in part, because evidence was found of significant damages resulting from improper disposal of chemical wastes, it is ironic that RCRA included no sections addressing the "sins of the past." With the discovery of sites such as Love Canal and the Valley of Drums, the oversight became evident. Public pressure mounted until Congress constructed and passed the Comprehensive Environmental Response, Compensation and Liability Act of 1980 (CERCLA), PL 96-510. To a great extent, CERCLA addresses some of the financial responsibility aspects of hazardous waste management. In particular, CERCLA establishes two funds to aid in compensation for damages arising from hazardous-waste-related incidents: (1) Superfund and (2) the Post-Closure Liability Trust Fund (PCLTF).

Superfund. Superfund is the central feature of CERCLA. For many people it was the major reason for passage of the act. In essence, Superfund is a pool of money generated by special taxes to ensure that funds are available to pay for removal or remedial action in response to release of hazardous materials. It can be likened to the National Contingency Fund for response to hazardous material spills, with a broader charter for the types of situations that are covered. In the case of Superfund, monies are collected from a tax of $0.79 per barrel of crude oil received at U.S. refineries or per barrel of petroleum products refined outside the country and received for use. In addition, a schedule of specified tax rates is established for 42 other chemical substances. These taxes constitute 87.5% of the fund monies. The remainder comes from the U.S. Treasury. Fund outlays may include payment for damages to natural resources (up to $50 million per accident), limited studies of resource damage, resource restoration, enforcement and abatement actions against releases, epidemiology, response cost and safety, as well as the prescribed remedial action. The fund cannot be used to cover personal injuries. This remains a controversial issue, with critics claiming that failure to consider injuries constitutes a judgment "that property is more significant than human beings" (Senator George Mitchell, Maine). The fund itself guarantees the availability of monies to perform the foregoing actions. If parties can be identified that are responsible for the damages, the fund can seek compensation and replenish itself. In the case of unknown parties or an inability to bear the financial burden, the fund is diminished accordingly.

Superfund was passed on the basis of a large number of known or suspected sites where hazardous wastes were improperly disposed. However, provisions were included to cover sites identified or created in the future and to mandate reporting of said sites. Section 103 of the act requires that, unless exempted, any person who owned or operated a site at which hazardous substances were stored, treated, or disposed must notify the EPA. Notification was required by June 9, 1981. Compliance generated a list of candidate sites for remedial action that has subsequently been reviewed for prioritization. In addition, the act requires that release of quantities exceeding 1 lb or an otherwise specified reportable quantity must be reported immediately to the EPA if not allowed under existing federal permits. Hence, spills not covered by Section 311 of the Water Pollution Control Act Amendments of 1972 (WPCA) are now reportable and can be responded to using federal funds. Substances requiring spill notification include those designated as hazardous under the clear water, clean air, and RCRA laws, toxic pollutants as identified under Section 307(a) of FWPCA, and imminently hazardous chemical substances or mixtures subject to action under Section 7 of the Toxic Substances Control Act.

PCLTF. The Post-Closure Liability Trust Fund (PCLTF) is the least publicized of the two CERCLA funds. It was established to provide for remedial action and damages arising from chemical contamination associated with sites operated and closed in compliance with RCRA. As currently structured, the EPA will qualify sites for coverage under the PCLTF. Criteria will include operation and closure with a RCRA permit, and observations for up to five years that there is no substantial likelihood of significant risk. Upon qualification, the federal government takes responsibility for all post-closure cases after a period of up to 30 years, as well as all claims arising from chemical losses. The latter can include damages to natural resources up to $50 million, continued monitoring and maintenance, remedial action, and compensation for injuries or loss. Hence, unlike Superfund, the PCLTF addresses injury and personal claims arising from chemical migration.

The fund itself is created by a tax of $2.13 per dry ton of hazardous waste disposed at a facility. Collection begins on September 30, 1983, and continues until the fund reaches $200 million. The tax is then suspended until such time as the fund drops below the $200 million level. Additional revenues to the fund include interest and other income related to fund management. Hence, if claims are low, the fund could sustain itself without the tax, after initially reaching the $200 million threshold. On the other hand, if claims are high, the threshold may be inadequate and taxes may have to be increased. Recognizing that the tax rate and the threshold were selected arbitrarily, Congress mandated a study of the likelihood of claims so that readjustments could be considered.

The findings of that study indicate the fund will be inadequate within the first 50 yr of its existence. Since all landfills will eventually leak in some manner, migration of contaminants will occur and ultimately put demands on the fund. These demands will mount until funds are exhausted. This finding has led to increased scrutiny of landfill bans for selected wastes and serious reconsideration of the efficacy of the PCLTF. At the same time, the Department of Treasury concluded that no private insurance programs were available to replace the PCLTF, and that such programs were not likely without well-defined limits on liability. Hence, if the PCLTF is not activated, long-term liability concerns may significantly increase the estimate life-cycle costs of landfills for hazardous wastes.

NCP.   Section 105 of CERCLA requires the EPA to issue a revised National Contingency Plan (NCP) to reflect the emergency response and remedial action requirements arising from the act, as well as the traditional needs arising out of Section 311 of the WPCA. This proved to be a major point of controversy. The EPA had great difficulty in deciding upon a tone and level of detail for the NCP. As a consequence, promulgation was delayed until July 16, 1982. In the interim, state and environmental organizations threatened suits of mandamus and other action to force promulgation. The NCP became a linchpin for all of Superfund, because remedial action was to be in compliance with the NCP. With promulgation, Superfund activity enjoyed rapid growth, even though controversy arose over the content of the NCP. A brief synopsis of that content pertaining to hazardous wastes is provided next.

Subpart B: Responsibility.   The duties of the president as specified in CERCLA (as well as the WPCA and subsequent Clean Water Act) are delegated to the EPA and U.S. Coast Guard (USCG) with a mandate to coordinate with state and local authorities. Other federal agencies are required to make resources or facilities available for assistance when appropriate. Responsibilities include those to coordinate and direct all public and private efforts to abate the threat. Some federal agencies have duties established by statute. All agencies must report releases covered by CERCLA. In addition:

The Department of Health Services (DHS) is delegated authorities for health-related work, e.g., hazard assessments and surveys.
The Federal Emergency Management Agency (FEMA) is delegated authorities with respect to permanent relocation of residents.
The Department of Defense (DOD) is delegated all authorities to implement CERCLA for releases on DOD properties/facilities.

States are encouraged to assign offices to participate in the process and to address situations not eligible for federal funding. Participation by non-government groups is encouraged.

Subpart C: Organization. National planning and coordination is conducted by the National Response Team (NRT), a committee of representatives from affected agencies, with the EPA representative serving as chairman. The NRT is also the central point for all communications about response activities. Regional planning and preparedness actions prior to a response action are the responsibility of the Regional Response Team (RRT). The RRT has a similar structure as the NRT, using regionally assigned agency staff. The On-Scene Commander (OSC) is responsible for developing local contingency plans as well as for direction of all emergency response activities. The National Strike Force (NSF) consists of USCG strike teams available to support the OSC upon request. The Emergency Response Team (ERT) consists of EPA personnel or delegates trained to evaluate, monitor, and supervise response as well as provide limited "initial aid" actions. Scientific support for response is provided by the Scientific Support Coordinator (SSC). Public information activities are conducted by the USCG Public Information Assist Team (PIAT) and the EPA Public Affairs Assist Team (PAAT).

Subpart D: Plans. Regional and, where practicable, local contingency plans are required.

Subpart F: Hazardous Substance Response. CERCLA is to be employed for removal or remedial action when it is deemed necessary because other responsible parties cannot or will not take proper action. Fund conservation dictates that attempts be made to encourage state or industrial action where possible. Discovery may result from Section 103 or 104 required action or other activities and must be followed by prompt notification of the cognizant OSC. At that time, the lead agency must make a preliminary assessment of the scope of the problem. If imminent hazard is determined, immediate removal action to a ceiling of $1 million is to be initiated, but should not exceed 6 months time to complete. The NCP provides a general list of actions that can be taken as a part of immediate removal. As soon as practicable, an evaluation is made to determine if further response is required through planned removal and remedial action. Studies can be authorized as required to support the evaluation. States may submit sites for priority ranking and designate the highest priority within the state. These will be collated into a national priority list, which, if practicable, will always have the top priority from each state in the top 100 sites. Planned removal may be conducted under federal direction and contracting, or if a cooperative agreement is signed, the state may take the lead role. In either event, planned removal must be requested by the governor, along with assurances that the state will cover 10% of all costs of the action, or at least 50% or greater, as determined by EPA, depending on the degree of responsibility borne by the state or political subdivision thereof. If continued action is not required immediately to prevent an emergency, planned removals should not exceed $1 million or 6 months. Guidelines are provided for determining when a planned removal is warranted. Remedial actions are those directed to effecting a permanent solution. Once again, guidelines are provided to help determine the need for remedial action. A list of remedial alternatives is to be prepared, with each entry undergoing screening with respect to costs, effects, and feasibility. A detailed analysis is conducted on those alternatives that emerge well from the screening process. Documentation is maintained throughout the process to support any cost-recovery action. A list is provided of methods for remedying releases.

Subpart G: Trustees for Natural Resources. Responsible federal agencies are designated as trustees for specific resources that may be damaged by a release. States may also act as trustees. In either case, trustees are responsible for assessing natural resource losses, seeking recovery for those losses, and devising and implementing plans for restoration, rehabilitation, and replacement.

Appendix A. The NCP is appended with "Uncontrolled Hazardous Waste Site Ranking System: A User's Manual."

Subsequent to promulgation of the revised NCP, controversy has arisen over key points in the plan. For the most part, the ensuing debate has pitted industry against environmental groups. Major issues include the following:

Alternatives evaluation: The NCP appears to place greatest emphasis on cost as a criterion for selecting response alternatives. While industry applauds this position, environmental groups contend that excessive concern for cost-effectiveness will greatly impact the substantive goal of CERCLA—the protection of human health and the environment.

Cleanup goals: The NCP champions maintenance of flexible goals for restoration activities under Superfund and therefore sets no baseline as a minimum restoration requirement. The environmentalists claim that existing air and water criteria could be employed to establish basement contamination levels not to be exceeded after remedial action has been implemented. Numerous examples can be sited where this approach was taken in the past. Industry supports the flexibility of the current NCP in part because it facilitates the use of cost-effectiveness as a criterion.

Planned removals: The environmentalists have voiced dissatisfaction that states must submit sites requiring planned removal for prioritization. They contend that sites requiring some near-term work but no immediate attention with respect to remedial action should not be allowed to clutter the priorities and divert resources.

RCRA compliance: Environmentalists have been unhappy that the NCP is largely silent on issues of RCRA compliance during CERCLA actions. They fear that the omissions of direct statements concerning the need for permits for off-site disposal, on-site treatment, storage, or disposal following excavation, and waste management for materials generated during remedial action imply that these activities may be allowed outside the RCRA regulatory framework.

The environmentalists have also been disappointed in what they term a "vague and weak" plan defining citizen participation.

Other Features of CERCLA.   Other features of importance in CERCLA include the designation of an assistant administration (AA) of EPA for solid wastes. This marks the first time that an individual in the agency has been charged with the responsibility for solid wastes and given authority on a par with AA's for water, air, and toxic substances. In some respects, this signals the final "coming of age" for solid waste in balancing concern for the physical environment.

CERCLA also contains requirements for a number of studies. These include the following:

Post-Closure Insurance: (a) A Treasury Department study of the feasibility of establishing an optimum system of private insurance for post-closure. (b) A presidential study of whether or not adequate private insurance coverage is available for liability claims. (c) An EPA study of the need for adequate tax revenue for the PCLTF.

Hazardous Waste Facilities: An EPA report to Congress on issues, alternatives, and policy related to the selection of optimum locations for hazardous waste facilities.

Industry Taxes: A joint EPA/Treasury study of potential additions to the hazardous waste lists and appropriate tax rates to support the PCLTF.

Employee Protection: An EPA study evaluating the potential loss of employment from administration and enforcement of CERCLA.

Worker Safety: A joint EPA/OSHA/NIOSH study to modify the NCP as necessary.

Natural Resources: A presidential study of losses to natural resources caused by losses of oil and hazardous substances.

Legal Redress: A 12-member panel of lawyers is required to study the adequacy of existing common law and statutory remedies to provide legal redress for harm to man and the environment from releases.

Superfund Implementation: A presidential report to Congress on how successfully Superfund is being implemented.

Use of Superfund Cleanup Authorities: A joint EPA/attorney-general study to establish guidelines for using imminent hazard, enforcement, and emergency response authorities.

While CERCLA itself speaks to issues not addressed in RCRA, there are obvious areas of interface. Clearly, the previously listed studies on post-closure insurance relate to financial responsibility elements of RCRA regulations. The study of a hazardous waste facility's siting is directly responsive to RCRA concerns while that on industry taxes may lead to changes in Section 3001. Similarly, CERCLA specifies use of RCRA definition of hazardous waste when determining which wastes are subject to the PCLTF tax. On the other side, RCRA regulations have been formulated to reflect CERCLA. For instance, before RCRA permits will be approved for existing facilities, existing contaminant plumes must be removed and/or mitigated. Other areas of implied interface have not been developed. As previously noted, the revised NCP is strangely silent on the need for remedial actions to be performed in accordance with RCRA requirements.

Growing dissatisfaction in Congress with the way EPA has chosen to promulgate hazardous waste management programs, and insights gained in retrospect have raised several issues of concern. These have subsequently formed the basis for proposal amendments and reauthorization bills for CERCLA.

As is evident from the 1984 RCRA amendments, Congress has chosen to move much more deeply into the specifics of regulatory programs rather than create a broad legislative net from which EPA can define a program. This clearly illustrates a growing conflict between the legislative and administrative branches with respect to implementation of environmental programs and has been evidenced in debates over other pending CERCLA amendments.

Key issues discussed in the debate over reauthorization of CERCLA have included broader topics such as the need to increase the size of Superfund tenfold and the need for victim compensation mechanisms. Both topics are extremely controversial and have raised real possibilities of a presidential veto.

Federal Water Pollution Control Acts (PL 92-500 and PL 95-217)

These acts contain several sections dealing specifically with toxic effluents (307a) and hazardous substances (311) from the standpoint of acute and chronic discharges. If a hazardous waste includes a designated material, its direct discharge to navigable waters could constitute violation of regulations under either of these sections. Through Sections 301 and 304, effluent limitations and guidelines are established for direct discharge. Sections 307b and 307c complement these with pretreatment standards for wastes destined for public sewage treatment facilities. Sections 403 and 404 address guidelines for ocean discharge and disposal of dredge and fill materials. The latter may include materials qualifying as hazardous wastes. In addition to these direct impacts on hazardous waste management, the federal water pollution control acts influence hazardous waste management by creating treatment requirements that will result in sludges qualifying as hazardous wastes. Hence, the residuals burden is shifted further from water to land.

Marine Protection, Research, and Sanctuaries Act, PL 92-532

The act establishes regulatory control over the dumping of materials into the ocean. This and the federal water pollution control acts and the Rivers and Harbors Act of 1899 combine to form a relatively comprehensive package eliminating the use of ocean disposal for management of hazardous wastes.

Safe Drinking Water Act, PL 93-523

While this act focuses largely on the development of primary and secondary standards for drinking water quality, the need to protect underground sources of water has given rise to a mandate for development of an underground injection control program. As currently interpreted, this constitutes the regulation of deep-well injection facilities. Currently, coverage is not extended to pits, ponds, or lagoons that may threaten potable aquifers through percolation. That aspect of waste management is left to RCRA.

Federal Insecticide, Fungicide, and Rodenticide Act, PL 92-516

Section 19a of this act requires the EPA to establish regulations for the storage and disposal of pesticide containers, excess pesticides, and pesticides for which registration has been canceled. Because many of these materials are likely to be defined as hazardous wastes, resultant regulations relate directly to those required by Section 3004 of RCRA. Similarly, Section 16 regulations on labeling will interface with labeling requirements in Sections 3002 and 3003 of RCRA.

Toxic Substances Control Act, PL 94-469

The Toxic Substances Control Act was enacted as a means of regulating the entry of toxic substances into society and the environment. Control is made possible by empowering the EPA administrator to place restrictions on the production, distribution, use, and disposal of new toxic substances or toxic substances proposed for new applications. Hence, this act provides a mechanism for the creation of specific hazardous waste management regulations for individual substances. Section 6e represents a mandate for promulgation of the first such set of regulations on storage and disposal of a specific hazardous waste, PCB-contaminated materials.

Clean Air Act, PL 88-206, as Amended

In many respects, this act parallels the federal Water Pollution Control Act with regard to atmospheric emissions. Regulations are to include the creation of air quality standards and emission control requirements. In this regard, specific rules, guidelines, and regulations will address substances potentially defined as hazardous wastes. Perhaps the greatest area of hazardous waste management impact lies in the control of emissions from incinerators and other hazardous waste treatment facilities. As in the case with the water laws, the Clean Air Act creates incentives to reduce residual emissions and in so doing creates sludges and other wastes that will be disposed of on land. Many of these wastes are potentially hazardous in nature.

Hazardous Materials Transportation Act, PL 93-633

This act covers the regulation of labeling and transporting hazardous materials. As such, it addresses the transportation aspects of hazardous waste management. Potential conflict with RCRA was anticipated by Congress and dealt with through specific instructions in RCRA for the EPA to coordinate all transportation-related regulations with DOT.

Occupational Safety and Health Act, PL 91-596

This act does not address hazardous waste management per se, but all hazardous waste management facilities will have to be operated in compliance with standards focused on worker safety and health.

With the listed pieces of legislation addressing segments of the hazardous waste management problem set, there are many opportunities for overlap and potential inconsistencies (Figure 2-1). To date, the only safeguards

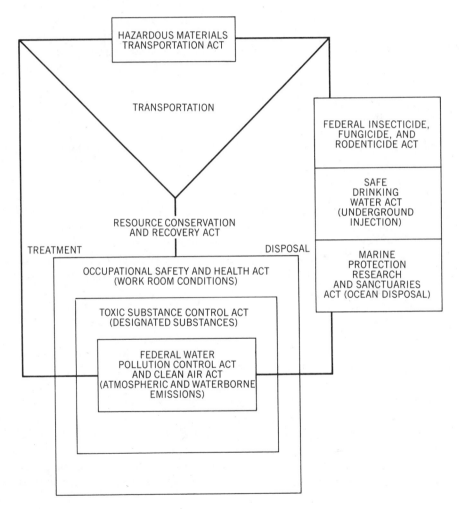

Figure 2-1 / Summary interfaces of legislation addressing hazardous waste management.

against such problems have been interagency task forces and reviews. Historically, this has been insufficient to provide ample coordination of related regulatory thrusts. As a consequence, the summaries of legislation provided here should not be viewed as illustrative of a comprehensive regulatory net with well-defined interfaces. Rather, the authorities and coverage are confused, and there is great need for a hazardous materials management reform act that would provide a single focal point for cradle-to-grave management of all hazardous substances. This would eliminate replicate definitions (e.g., toxic pollutants, hazardous substances, toxic substances, hazardous wastes, etc.), management gaps (e.g., funding for cleaning up abandoned waste sites), conflicting incentives (e.g., effluent guidelines that stimulate production of more concentrated residuals streams while RCRA drives the costs for disposal up), and multiple regulatory requirements that create unnecessary economic impacts (e.g., duplicate labeling and spill reporting mechanisms for transporters of hazardous materials).

State Programs

While RCRA clearly signaled a shift in solid/hazardous waste management regulatory philosophy to one with strong federal elements, the opportunity for more restrictive state programs and state implementation and enforcement created a continuing role for state government in hazardous waste management. As a consequence, there has been a distinct trend in the consideration and passage of state legislation throughout the last decade.

In 1973 the report to Congress concluded that at that time there were no state hazardous waste management programs, except for those that addressed radioactive wastes. This pronouncement was based on a survey of 16 states representative of all regions, sizes, and levels of commercial development. Some legislation was noted in the area of explosives and pesticides, but regulations were generally too narrow to provide a legal basis for management of other hazardous wastes.

By 1976, a *Waste Age* survey indicated that 40 of the 50 states required special processing or disposal of hazardous wastes. Of these, 21 relied on specific state rules and regulations. These figures can be deceiving on the surface, because there was no consistent definition of what constituted a hazardous waste. Perhaps more revealing was the fact that of the 21 states claiming rules and regulations, only 12 listed promulgation dates after the 1973 report to Congress. Hence, nine of the affirmative responses referred to legislation deemed inadequate in the 1973 report. Higher activity levels were evidenced by some 13 states in various stages of drafting legislation and/or regulations.

In late 1978, a survey was conducted of the 50 states, Puerto Rico, and the District of Columbia (Gilroy, 1979). Excluding South Dakota (no response), the information obtained revealed that 19 states had statutes that mentioned hazardous waste specifically in the title, 26 had passages authorizing hazardous waste management as a subset of solid wastes, and one (Michigan) had a general environmental statute covering some classes of hazardous wastes. Of the five jurisdictions with no statutes, four had legislation proposed or under consideration. Additionally, 34 jurisdictions reported regulations in effect that addressed hazardous wastes. Seventeen jurisdictions noted that new or additional regulations were planned, proposed, or out for public comment.

A review of the 45 jurisdictions with regulations in place (Puerto Rico was excluded because of difficulties in translating the response) generated the following characteristics:

Waste Management

1. Forty-five had means of regulating treatment, storage, and disposal facilities.
2. Seven had requirements for or authority to require prior notification for treatment, storage, and disposal facilities.
3. Forty-five had a permit system or authority to develop a permit system for the above facilities.
4. Thirty-four had operating and/or design standards as a part of the permit process.

Waste Transportation

5. Twenty-five regulated or had authority to regulate hazardous waste transportation.
6. Seven required prior notification of hazardous waste transportation activity.
7. Twenty-five utilized or were authorized to utilize permits or other approval systems to regulate hazardous waste transportation.

Waste Generation

8. Thirty regulated or had authority to regulate hazardous waste generators.
9. Thirteen required prior notification of generation activities.
10. Seven required or were authorized to require permits for storage of hazardous wastes.
11. Seven specified or were authorized to specify standards for temporary storage of hazardous wastes.
12. Seventeen had requirements for periodic reporting by hazardous waste generators.
13. Twenty-two of the above 30 required use of manifests or similar information transfer documents.

Definition

14. Forty-five had at least a general definition of hazardous wastes.
15. Thirty-four specified waste characteristics to help define hazardous wastes.
16. Eight specified test protocols for identifying hazardous wastes.
17. Thirteen maintained lists of wastes or constituents that would make a waste hazardous.
18. Nine listed industries or processes whose wastes were hazardous.
19. Five utilized volume or waste characteristics that distinguished between two or more classes of hazardous wastes.

A summary of specific responses to a 1982 questionnaire developed by the National Conference of State Legislatures is provided in Table 2-1.

Much of the diversity stemming from variations in existing state law will be minimized with promulgation of RCRA regulations pursuant to the Section 3006 guidelines for EPA acceptance of state programs. In a 1980 survey by the Chemical Manufacturers' Association, 44 of 48 states contacted indicated that they would seek certification for administering the federal program. Of the 48 states, 47 have acted on pending legislation to implement RCRA, and 13 have or intend to promulgate regulations. By 1982, 49 of the states had enacted some piece of hazardous waste related legislation as categorized in Table 2-1. Hence, current state law may undergo change in the near future. Residual differences can be anticipated in specific areas as a result of more restrictive regulations for defining hazardous wastes and siting facilities and reporting on activities.

The states have also been quick to match federal legislation in the area of emergency response and compensation. By October of 1982, 31 states had created some form of mini-Superfund to address spills and uncontrolled hazardous waste sites. Features of individual state programs vary considerably. Of the 31 plans, 13 are funded by taxes on hazardous waste generators, and 10 are funded by taxes on hazardous waste management firms. The majority of the states supplement the tax base with some form of legislative appropriation. In addition, some states utilize penalties and fines, a portion of permit fees, and grants to help fund responses. Some 22 states use funds for emergency response actions only, and 10 use them for restoration of abandoned sites as well. Functionally, the funds may be used to finance the state's share of federally approved action or to address sites not prioritized by the federal program.

In a related issue, states have taken great interest in worker's right-to-know legislation, which requires employers to inform workers of the hazards to which they may be exposed. By 1982, seven states had enacted such legislation (Table 2-2) and the debate was joined at a national level.

Table 2-1 / State Legislation Enacted from 1976 to 1982.[a]

	Siting Procedure	Funding Mechanisms	Alternatives to Land Disposal	Carrier Requirements
Alabama		x	x	x
Alaska				x
Arizona	x	x	x	x
Arkansas				x
California		x	x	x
Colorado	x			x
Connecticut	x	x	x	x
Delaware				x
Florida	x	x	x	x
Georgia	x	x	x	x
Hawaii				x
Idaho				x
Illinois	x	x	x	x
Indiana	x	x	x	x
Iowa	x		x	x
Kansas	x	x	x	x
Kentucky	x	x	x	x
Louisiana		x		x
Maine	x	x	x	x
Maryland	x	x	x	x
Massachusetts	x	x	x	x
Michigan	x	x	x	x
Minnesota	x		x	x
Mississippi		x	x	x
Missouri		x	x	x
Montana				x
Nebraska	x			x
Nevada		x		x
New Hampshire	x	x	x	x
New Jersey	x	x	x	x
New Mexico		x		x
New York	x	x	x	x
North Carolina	x	x	x	x
North Dakota				x
Ohio	x	x	x	x
Oklahoma			x	x
Oregon	x	x	x	x
Pennsylvania	x	x		x
Rhode Island	x	x	x	x
South Carolina		x	x	x
South Dakota	•			x
Tennessee	x	x	x	x
Texas		x		x
Utah	x			x
Vermont				x
Virginia				x
Washington	x			x
West Virginia			x	x
Wisconsin		x	x	x
Wyoming				

[a]Adapted from National Conference of State Legislatures (1982).

Table 2-2 / States with Worker Right-to Know Legislation[a]

	California	Connecticut	Maine	Nevada	New York	West Virginia	Wisconsin
Information required							
MSDS	X	X	X		X		X
Other				X		X	
Persons covered							
Current employees	X	X	X	X	X	X	X
Former employees			X	X	X		
Employee representative	X		X		X		X
Employee physician	X		X		X		X
Some employees excluded						X	
Substances							
Listed by OSHA	X	X	X	X	X		X
Listed by EPA	X						X
Other listing	X	X	X		X	X	
Defined by criteria	X	X			X		X
Other					X	X	X
Exposure records/registry							
Employee access to employer records				X	X		
Employee representative access to employer records					X		X
State-maintained records			X	X	X		X
Employer's responsibilities							
Employee training and education		X	X		X		X
Non-retaliation		X			X		X
Penalties for violation		X	X		X		X
Manufacturer must supply information	X	X					X

[a] Adapted from National Conference of State Legislatures (1982).

Foreign Regulatory Programs

As one might expect, the extent of hazardous waste management activities outside of the United States directly reflects the degree of industrialization and the level of environmental awareness in a given country. Examples range from highly integrated systems well ahead of American practices to minimal programs that do not recognize the existence of special waste categories.

Canadian efforts closely parallel those in many of the individual states in the United States. The close proximity to the United States and extensive industrial interrelations have made Canadian business a part of the hazardous waste management chain in the United States, both as a source of such wastes and as the operator of treatment and disposal sites. Rising interest in quantifying the management problem and recognition of planning needs have stimulated a flurry of recent activities. There was no comprehensive hazardous waste legislation for all of Canada by 1980. However, through the Environmental Contamination Act, the government does have certain inter-provincial authority. Primacy resides with the provinces. The federal government and several provincial governments are conducting studies to inventory wastes and site facilities. Ontario has taken the most extensive action, with a seven-point program that among other things outlaws the disposal of liquid wastes to land. Provincial-owned disposal facilities are being considered. A high degree of interchange between United States and Canadian officials suggests that much of what results from these efforts will closely parallel work in the United States.

Similar inventory studies have recently been published in Mexico for the area surrounding Mexico City. This work has stirred public interest, but no specific legislation has been forthcoming. In outlying areas, discharge of toxic materials goes on virtually uncontested. Hazardous materials management problems in Central America are largely associated with the use and disposal of pesticides. Incidents of acute poisoning in Central America are reported at a rate 2160 times that observed in the United States. No legislation addresses the waste disposal aspects of the problem. Similarly, most South American countries have no specific toxic waste legislation, even though the problem has been clearly identified as a result of incidents such as mercury contamination of the Bay of Castogena (Colombia) and the coastal areas near Moron, Venezuela. Brazil and Argentina have enacted some legislation focusing on discharges to rivers, but effectiveness is hampered by less than vigorous enforcement. For instance, in Brazil, officials of the federal environmental agency (SEMA) have expressed concern that toxic effluents are severely impacting the environment in the northeastern part of the country but that local officials have withheld action for fear of discouraging industrial development. The economic problems of the early 1980s severely hampered efforts to gain tighter control on these problems.

Hazardous waste management is highly advanced in Europe. A summary of pertinent regulations as of October 1979 is presented in Table 2-3.

There is still much diversity in the specific legislation and extent of implementation between countries, but a thread of continuity is beginning to develop. Members of the European Community (EC) take into account EC directives when designing new acts. The latter have been instituted to (1) harmonize European legislation and thus stop trade barriers between the states and (2) guarantee acceptable levels of risk within the EC. The council of the EC adapted a directive on toxic and dangerous wastes on March 20, 1978. Members of the EC had until March 1980 to indicate how they would implement the directive. Heavily concentrated industries and decreasing availability of land for disposal has forced a strong regulatory posture early on. Specific legislation in Britain dates back to 1974 and the Control of Pollution Act. Implementation and enforcement are left to local government, whose charter it is to survey all wastes and arrange for safe disposal. To date, the more relevant aspects of the law have not been implemented. The Poisonous Wastes Act includes the first segments addressing disposal of hazardous wastes on land. Consideration of reclamation is mandatory. In addition, the government has established a Waste Management Advisory Council to examine problems associated with recovery of wastes and methods of encouraging reclamation.

Several features of Swedish law closely resemble segments contained in RCRA. Industries are required to report the quantities and content of wastes, and chemical waste transporters and disposers must obtain permits. The federal role is particularly strong, because most wastes are handled by SAKAB (Swedish Waste Conversion Company), which is 90% state-owned. Although Norwegian industry is responsible for proper management of chemical wastes, environmental authorities assist by collecting wastes and distributing them to commercial treatment plants. Denmark has been a forerunner in toxic waste management, with pesticide-related regulations dating back to 1953. The major authority for hazardous waste regulation stems from the Environmental Protection Act of 1973. Current laws prohibit direct discharge of rivers or sewers, and above-ground storage is closely controlled. Chemical wastes are collected by municipalities and shipped to a central plant for treatment.

Disposal of waste chemicals and oils in The Netherlands is subject to a series of laws that require generators to deliver wastes only to licensed disposers. The Chemical Wastes Act of 1977 relates to chemical wastes and used oil. The main purpose of this act is to introduce much stricter controls over the transportation, disposal, and treatment of chemical and oily wastes. Nevertheless, the act envisages measures to limit the production of these wastes. These may include regulations requiring (1) a prohibition to manufacture or market certain goods and (2) a prohibition to manufacture or

Table 2-3 / Existing National Regulations on Toxic and Dangerous Waste.

Germany

1972 law on the disposal of waste, amended in *1977*, covers the problems of transportation, notification, creation of sites, etc.

The arrêté on the control of disposal of *2 June 1978* settles the problem of the transfer of toxic and dangerous waste.

Powers of the Länder (e.g., Rhineland-Westphelia).

Belgium

Law of *22 July 1971* on toxic waste.

Arrêté royals of *9 February 1976* presenting a general regulation on toxic waste sets concentrations for arsenic, cadmium, and berylium.

Denmark

Publication in *1975* of a regulation on chemical products.

The arrêté on chemical waste of *1978* covers the transport of toxic and dangerous waste.

France

Law of *15 July 1975* on the disposal of waste and the recovery of materials.

Decree of *19 August 1977* on information to be supplied on the subject of waste that generates nuisances.

Draft arrêté on thresholds determining the alligation for waste to be disposed of in an approved plant.

Ireland

A regulation on the disposal of waste now being studied.

Italy

Draft law (n*144) on waste now being studied.

Luxembourg

Draft law on the disposal of waste.

Netherlands

Law on chemical waste of *11 February 1976*.

Decree of 26 May 1977 drawing up a list of substances classed in four categories (A,B,C,D).

Each category of substances is accompanied by a concentration threshold:

Class A—50 mg/kg

Class B—5,000 mg/kg

Class C—20,000 mg/kg

Class D—50,000 mg/kg

United Kingdom

The Deposits of Poisonous Wastes Act, *1970*.

The Control of Pollution Act, *1974*—section 17 provides for a very strict system of notification for various kinds of waste (e.g., lead, cadium).

A document is being drafted on the fixing of concentrations, pursuant to section 17 of the Control of Pollution Act.

EAC

Directive on waste of *5 July 1975* (0.5.1194 of *25 July 1975*)

Directive on toxic and dangerous waste of *20 March 1973* (0.0.YL34 of *31 March 1978*) market certain goods if they do not comply with the requirements of a General Administrative Order. Other pertinent acts include the Waste Disposal Act and the Soil Protection Act.

Spain has not had active laws in place for hazardous waste management but is in preparation for joining the EC; legislation is being drafted on protection of air, water, and soil quality. This, of necessity, will be designed to meet standards set by the EC. Irish efforts stem from the creation of a task force in April 1971 whose charter would be to investigate problems of waste disposal. A series of reports (the most recent in April 1975) has resulted from subsequent meetings, along with a recommendation that government become more involved in legislation for disposal and reclamation.

Belgium passed a hazardous waste management act in 1979 and has several active governmental working parties studying specific issues. Further legislation is planned for specific aspects of waste disposal and reclamation. Preliminary indications suggest that, in part, this legislation will establish a joint government-industry company to collect and manage hazardous wastes. Although Luxembourg does not currently have legislation on the books, they historically have followed Belgium's lead.

French efforts began in earnest in 1975 with passage of a law placing management responsibility for dangerous wastes on the generators. Ultimately, a national agency is to be established with a charter for disposal and recycle of wastes. Officials currently have authority to demand waste inventories from generators and require disposal in certified facilities. A national system for toxic waste collection is also under consideration. Although West Germany boasts several very sophisticated waste management facilities, specific regulation is still in the discussion stages. Future activity in Italy is expected to authorize organization of regional toxic waste processing and disposal facilities as a taxed service. In the meantime, Italian industry claims to be increasingly less competitive with other European firms because of stricter constraints on waste management. Swiss law is quite general, prohibiting disposal to stagnant waters or areas of public access.

As in other parts of the world, the recession of the 1980s slowed promulgation of new regulations and decreased enforcement of existing ones. While a number of sites requiring remedial action have been identified, no major program has emerged in any of the European countries to hasten restoration.

Relatively low concentrations of industry in the Middle East and Africa have forestalled public recognition of hazardous waste problems and subsequent legislation. South Africa does boast a fledgling service industry reportedly employing secure landfills, but recent reports question the adequacy of some of the sites being utilized.

Asia offers examples of some of the best and worst of foreign management programs. In India, several major incidents and widespread contamination near metropolitan areas have failed to stimulate definitive regulation of dumping or other waste discharge practices. In Bangladesh, programs to provide for secure landfill and resource recovery are under development. The government of Hong Kong has drafted a bill to regulate the disposal of toxic wastes. Chinese efforts to control environmental releases were set back by recent modernization programs. Efforts there, as well as in Japan, Taiwan, and the Philippines, are focused on air and water emissions. Hazardous waste legislation per se has not been developed. Similarly, a recent report by Australian scientists has decried the lack of specific legislation in that country addressing issues of land pollution from waste disposal.

REFERENCES

Gilroy, M. J. 1979. State regulation of hazardous waste disposal. In *Proceedings of the 1979 conference on hazardous material risk assessment, disposal and management.* Miami Beach: Information on Transfer, Inc.

Krier, J. E. 1971. *Environmental law and policy.* New York: Bobbs-Merrill.

National Association of Counties Research Foundation. 1973. *Basic issues on solid waste management affecting county government.* Washington, D.C.: Environmental Protection Agency.

National Conference of State Legislatures, 1982. *Hazardous waste management: A survey of state legislation 1982.* Denver, Colorado.

Waste Age, 1976. U.S. hazardous waste control practices. *Waste Age* Vol. 8, No. 4, April 1976.

Wilson, D. G. 1977. *Handbook of solid waste management.* New York: Van Nostrand Reinhold.

3

Defining Hazardous Wastes

Differentiating Hazardous Wastes

Given the proper legislative backdrop for management of hazardous wastes, implementation of regulatory control may proceed. The initial task involves the selection of a clean, quantitative means of identifying those wastes that are hazardous. The importance of developing a strong rationale for designating hazardous wastes has often been overlooked, and yet the breadth of the regulatory net is determined by that definition. Subsequently, the economic impact of regulation is directly influenced by the quantities and types of wastes that will require special treatment. As is often the case, the magnitude of the economic impact is inversely proportional to the health and environmental impacts being addressed. Thus, implicit with selection of a good definition is identification of the general zone where marginal costs to generators approximate marginal benefits to society.

The primacy of the designation methodology was recognized by Congress when it wrote Section 3001 of RCRA mandating the development of a definitive set of criteria for differentiating hazardous wastes from the rest of the solid waste stream. In route to a discussion of such criteria, it is important to note that numerous other categories of toxic and hazardous materials have been created by legislation in the last decade. A review of the pertinent acts and their intent is therefore instructive.

Two groups of materials are defined in the federal Water Pollution Control Act Amendments of 1972 (PL 92-500) that are often confused with hazardous wastes: (1) toxic pollutants and (2) hazardous substances. Toxic pollutants to be designated as a result of Section 307 of that act are materials considered to threaten sufficient harm to human health or the environment to warrant specific effluent limitations. These are clearly defined as pure compounds and are designated only in the context of discharges to water.

Therefore, these materials themselves do not constitute hazardous wastes; however, the sludges produced in removing them from liquid streams may be hazardous wastes. Originally, the EPA selected 12 substances for designation as toxic pollutants. Subsequent litigation resulted in the Flannery decision, a consent decree that was later reinforced by the Clean Water Act of 1977, wherein some 129 "priority pollutants" were designated for control under Section 307. Water quality criteria and industry group effluent guidelines are being developed for each of these contaminants.

Similarly, the Clean Air Act defines a category of hazardous air pollutants for which the EPA is to promulgate specific emission standards. The capture of these constituents in sludges or some other contained form will subsequently create a hazardous waste.

Hazardous substances are defined in Section 311 of PL 92-500 as

> such elements and compounds which, when discharged in any quantity into or upon the navigable waters of the United States or adjoining shorelines or the waters of the contiguous zone, present an imminent and substantial danger to the public health or welfare, including, but not limited to, fish, shellfish, wildlife, and beaches.

Once again, materials so designated are pure substances and not wastes per se. However, spillage of otherwise designated hazardous wastes may also constitute spillage of hazardous substances based on the pure components included in the waste. In other words, hazardous wastes may include hazardous materials as components. Indeed, off-spec or wasted batches of hazardous substances also qualify as hazardous wastes. Conversely, recovered materials after spillage of hazardous substances are likely to constitute hazardous wastes. To date, over 300 elements and compounds have been proposed for designation as hazardous substances.

A related category of legally defined materials that may be confused with hazardous wastes is that addressed by the Toxic Substances Control Act (TSCA). The enabling legislation provides regulatory options for control of these materials throughout their use and disposal and therefore creates the potential for preemption of hazardous waste regulations. It would appear, however, that Congress intended that hazardous waste regulatory mechanisms would prevail once a substance is wasted. The focus of toxic substance legislation is production and distribution activities. The intent is to minimize the amounts of unnecessary toxic substances in the marketplace. Consequently, while the legislation affords TSCA regulation throughout the materials existence, for the most part toxic substances become hazardous wastes when discarded. Figures 3-1 and 3-2 are provided to summarize the interrelations of the aforementioned groups as defined by existing law.

The interrelation of spilled chemicals and designated hazardous wastes is brought to focus in CERCLA, where hazardous substances are defined as follows:

PURE	WASTE	
TOXIC POLLUTANTS	DISCHARGED TO WATER
HAZARDOUS SUBSTANCES	SPILLED TO LAND OR WATER
HAZARDOUS AIR POLLUTANTS	ALL OTHER MODES OF RELEASE

Figure 3-1 / Function relations of hazardous and toxic materials in industry.

(14) "Hazardous substance" means (A) any substance designated pursuant to section 311(b)(2)(A) of the Federal Water Pollution Control Act; (B) any element, compound, mixture, solution, or substance designated pursuant to Section 102 of this Act, (C) any hazardous waste having the characteristics identified under or listed pursuant to Section 3001 of the Solid Waste Disposal Act (but not including any waste the regulation of which under the Solid Waste Disposal Act has been suspended by Act of Congress), (D) any toxic pollutant listed under Section 307(a) of the Federal Water Pollution Control Act, (E) any hazardous air pollutant listed under Section 112 of the Clean Air Act, and (F) any imminently hazardous chemical substance or mixture with respect to which the administrator has taken action pursuant to Section 7 of the Toxic Sub-

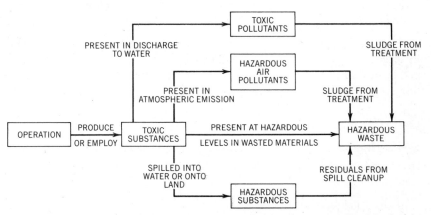

Figure 3-2 / Relation of various hazardous and toxic materials defined by legislation.

stances Control Act. The term does not include petroleum, including crude oil or any fraction thereof which is not otherwise specifically listed or designated as a hazardous substance under subparagraphs (A) through (F) of this paragraph, and the term does not include natural gas, natural gas liquids, liquefied natural gas, or synthetic gas usable for fuel (or mixture of natural gas and such synthetic gas). . . .

Section 102 allows the administrator to add additional entries on the basis of evidence that they may pose a substantial hazard when released in the environment.

Given these distinctions, confusion may still occur as a result of the tendency for people to use the terms "toxic" and "hazardous" interchangeably. The words are not synonymous. The former refers to intrinsic characteristics, whereas the latter also includes extrinsic ones. This distinction has been expressed succinctly by the Food Protection Committee of the NAS (1970):

> *Toxic* defines the capacity of the substance to produce injury including effects such as teratogenicity, mutagenicity, and carcinogenicity.
>
> *Hazardous* denotes the probability that injury will result from use of (or contact with) a substance in a given quantity or manner.

The implications of these definitions may be misleading, because some chemical properties leading to injury (e.g., flammability, explosiveness, and reactivity) are not toxicological in nature. Webster narrows the scope by noting that toxic materials produce injury as a result of poisonous properties. In turn, "poisonous" pertains to injury resulting from chemical action of a material via intake or direct contact. In summary, then,

"Toxic" refers to an intrinsic property.
"Hazardous" results from both intrinsic and extrinsic properties.
"Hazardous" may encompass the potential for injury resulting from toxic *or other* actions.

Having segregated hazardous wastes from other material groups created by legislative action, and having acknowledged the distinction between toxic and hazardous, it is now possible to define hazardous wastes both in generic terms and in a specific quantitative manner.

Defining Hazardous Wastes

Recognizing that "hazardous" implies both intrinsic and extrinsic factors, any complete definition of hazardous wastes should answer five basic questions:

1. Hazardous to what?
2. Hazardous for what reason?

3. Hazardous to what degree?
4. Hazardous at what times?
5. Hazardous under what conditions?

Hazardous to What?

Injury or harm can be sustained by a variety of target receptors. Primary concern has typically been associated with human health and property. More recently, attention has also turned to other living organisms and the environment in general. The final factor is somewhat of a catchall that, if taken literally, subsumes the previous three. Whereas in years past, legislative definitions could easily have been restricted to human health, the present atmosphere calls for consideration of all receptors.

Hazardous for What Reason?

Once the receptors have been defined, it is necessary to delineate the attributes of the hazard that are to be considered. Candidate attributes include the following:

Ability to bioconcentrate
Toxicity (via all routes)
Flammability
Explosiveness
Reactivity
Irritation or sensitization potential
Corrosivity
Genetic-change potential
Etiology
Radioactivity

Radioactivity often is not considered, because a large body of law and supporting regulations exist to deal with these materials. Radioactivity may also be considered as a subset of toxicity, as can genetic-change potential. Genetic-change potential and etiology are often grouped in a more general category such as "otherwise damaging" because of vagueness in defining them. Specification of which of these attributes are of concern in the definition of hazardous wastes clearly identifies the types of criteria to be applied in designating specific wastes.

Hazardous to What Degree?

In dealing with hazardous wastes, it must be recognized that hazard is not an either/or situation, but a matter of degree. That is, hazard for any given material is a continuous function dependent on the conditions of exposure. Therefore, some sense of threshold must be established to identify the point at which hazard becomes significant enough to warrant regulation. This can be dealt with qualitatively for definitions that are destined to be included in legislation through use of terms like "substantial" or "significant." A more quantitative treatment is required for a truly working definition appropriate for regulatory control.

Hazardous at What Times?

A definition of hazardous wastes should also address the question of when the hazard is evidenced. This can be viewed in a functional manner (e.g., during handling, transportation, treatment, storage, and/or disposal), or it can be viewed in a temporal sense (e.g., present hazard or potential future hazard). The two views are not completely independent. Handling, transportation, and treatment activities are concerned almost exclusively with present hazards. Storage and disposal activities may involve both present and potential future hazards. Many of the documented incidents involving damage or injury from improper hazardous waste management surfaced some time after storage or disposal was effected. Because damage is not mitigated by the nature of the activity under way at the time, the temporal view (using present hazard or potential future hazard to delineate what times) appears most appropriate. This assumes that all activities related to the use or management of hazardous wastes fall within the scope of the legislation.

Hazardous under What Conditions?

It is important to note the conditions under which the hazard is evidenced. Pertinent factors include quantity, concentration, form, and the presence of other materials that may add to or detract from the hazard. These factors are very complicated and consequently difficult to deal with in any but a general manner.

With the passage of the Resource Conservation and Recovery Act of 1976, Congress has created a baseline definition of hazardous wastes answering the foregoing questions:

(5) The term "hazardous waste" means a solid waste, or combination of solid wastes, which because of its quantity, concentration, or physical, chemical, or infectious characteristics may—

(A) cause, or significantly contribute to an increase in mortality or an increase in serious irreversible, or incapacitating reversible illness; or

(B) pose a substantial present or potential hazard to human health or the environment when improperly treated, stored, transported, or disposed of, or otherwise managed.

The definition encompasses a broad scope. In addition to addressing hazards related to physical and chemical properties, it includes infectious hazards and hazards due to the quantity of a waste. On the other hand, the major emphasis is placed on toxicological hazard (A) and especially those hazards evidenced through impacts on the human population. Congress did not list the kinds of specific hazards of concern, nor did Congress choose to designate the hazard level of concern beyond "significant increase" and "substantial present or potential." Hence, the task of quantifying these terms is left to the EPA as an implicit part of Section 3001 regulations.

Finally, the definition focuses on the emergence of these hazards when the waste is improperly managed. This implies certain subtleties not originally a part of hazardous waste designation. Earlier definitions were aimed at segregating wastes that posed these hazards when the wastes were managed along with other nonhazardous solid wastes. The implication was that hazardous wastes had sufficiently different properties that they required special treatment. Hence, in the report to Congress, it was stated that

> The term "hazardous waste" means any waste or combination of wastes which pose a substantial present or potential hazard to human health or living organisms because such wastes are lethal, nondegradable, persistent in nature, biologically magnified, or otherwise cause or tend to cause detrimental cumulative effects. General categories of hazardous waste are toxic chemical, flammable, radioactive, explosive, and biological. These wastes can take the form of solids, sludges, liquids, or gases.

With the RCRA terminology, there is greater latitude and consequently more of a burden on the EPA for selecting definitive criteria. All solid wastes can cause environmental damage when improperly managed. The EPA must determine at what point that impact can be defined as a substantial present or potential hazard. Without a listing of the properties of concern, the agency must decide if such impacts as increases in biological oxygen demand (BOD), depletion of the ozone layer, corrosiveness, phytotoxicity, or effects on microlife constitute a substantial hazard. Such decisions are key to identification of a proper set of criteria for hazardous waste designation.

Other organizations, such as the National Solid Waste Management Association, have chosen to maintain the posture of the definition in the report to Congress by enumerating the major hazards of concern (Chemical Waste Committee, 1976):

> any waste or combination of wastes which, because of its quantity, concentration, or chemical characteristics, poses a substantial present or potential hazard to human health, living organisms, or the environment

because such wastes are bioconcentrative, highly flammable, extremely reactive, toxic, irritating, corrosive, or otherwise damaging.

Similarly, a majority of the states have retained segments describing the primary hazards of interest. Thus, by 1980 35 of the 48 states with formal definitions for hazardous wastes included a listing of hazard types to be addressed. Of the eight states that had adopted the definition in RCRA, two chose to supplement it with a listing of hazards. Eight employed a definition that specified hazardous wastes as those wastes that require special handling or cannot be managed along with "normal" wastes. Although these descriptors are not quantitative, they establish a background philosophy for identifying hazards of concern: Wastes are hazardous if they produce substantial injury to public health or the environment when processed in the same manner as nonhazardous solid wastes. This distinction is not present in the broader RCRA definition, and thus states such as California have developed designation criteria that involve hazards resulting from practices that are not allowed for any wastes, such as direct discharge to rivers or abandonment.

In Europe, an umbrella definition has been provided in the European Community council directive of March 20, 1978:

> Toxic and dangerous waste means any waste containing or contaminated by the substances or materials listed in the Annex of this Directive of such a nature, or such quantities or in such concentrations as to constitute a risk to health or the environment.

This is magnified and given a certain amount of individual character by individual European states through statutory definitions such as those summarized below.

France

>   . . . categories of waste may be defined by decree and the enterprises that produce, import, transport or dispose of wastes which belong to these categories and which are in a state such that they cause, or at the time of their disposal may cause, a nuisance such as . . . injurious effects on the soil, plants, or animals, to degrade the scenery or the countryside, to pollute the air or water, to create a noise or odor, or [are] harmful to human health or the environment [Art. 8 and 2; Law No. 75-633; July 16, 1975].

Federal Republic of Germany

Special wastes are such wastes from commercial or trade companies that because of their nature, composition, or quantities are especially hazardous to human health, air, or water or that are explosive or flammable or may cause diseases. Their disposal must be subject to additional requirements according to the Act (Federal Act on the Disposal of Waste, 1972, as amended 1976).

The Netherlands

Chemical wastes are (1) wastes consisting wholly or partly of chemicals indicated by General Administrative Order and (2) wastes produced by chemical processes designated by General Administrative Order (Chemical Waste Act, 1977).

United Kingdom

Waste "of a kind which is poisonous, noxious or polluting and whose presence on the land is liable to give rise to an environmental hazard." (Deposit of Poisonous Waste Act, 1972). Special wastes are those that "may be . . . dangerous or difficult to dispose of" (Control of Pollution Act, 1974).

Differences in emphasis are readily apparent when these definitions are compared. The British place major emphasis on wastes that may not be landfilled while Germany and France address hazards that might arise at any time during the management cycle. The Dutch merely leave the issue to a listing or designation of wastes.

As in the case of determining the meaning of any descriptive term, the final selection of an accepted definition for hazardous wastes rests in arrival at a consensus within the user community. There is no single right definition. There are many acceptable alternatives. The key is to focus on one alternative that conveys similar connotations to the most people. In this regard, several summary observations are in order:

The preemptory power of RCRA gives the definition of hazardous wastes embodied there a legitimacy that renders it the baseline for work from 1976 forward.

A generic description such as that in RCRA leaves a great deal of freedom for the regulator to interpret terms such as "substantial hazard" and to select the types of hazard of concern.

A majority of the user community appears to favor limiting the scope of such a definition by designating the hazards of greatest concern as a part of the definition.

Based on current definitions employed by individual states, regulation should focus on materials that are toxic, flammable, explosive, corrosive, radioactive, or infectious.

There is an apparent but not always expressed intent to designate as hazardous those wastes that display the foregoing properties when they are managed as a part of the nonhazardous solid waste stream.

Based on these considerations, a generic definition of hazardous wastes has been selected for use here. Throughout the remainder of this text (unless otherwise noted), reference to hazardous waste will mean

> those wastes or combination of wastes which, because of their quantity, concentration, or chemical characteristics pose substantial present or potential future hazards above those associated with the management of municipal solid wastes as a result of intrinsic properties such as toxicity, flammability, explosiveness, corrosiveness, and reactivity.

While the terminology is considerably different than that of the RCRA definition, the intent is the same. The wastes deemed hazardous as a result of this definition will be essentially the same. The modified form is offered to provide a clearer focus for the development of regulatory programs.

In general, existing definitions and those that may be generated as a result of consideration of the factors discussed here are aimed at meeting the needs of enabling legislation. Subsequent regulation and ultimate control of hazardous waste management require the use of a quantitative means for designating specific wastes as hazardous.

Designating Hazardous Wastes

The process of designating hazardous wastes is difficult and ultimately involves some of the classic trade-offs encountered in environmental regulation, such as that between ease of implementation and equity between parties. As in the case with selection of a generic definition, there is no single right way to designate hazardous wastes. There are good and bad ways to accomplish designation, based on the primary concerns of the community.

To date, two basic approaches to designation have been developed: (1) listing and (2) use of criteria thresholds for intrinsic waste properties. Each approach offers advantages that must be weighed against inherent weaknesses.

Listing

Designation of hazardous wastes through listing was one of the first options explored by the EPA and its contractors. In this approach, candidate waste streams are reviewed by source and segregated into hazardous and non-hazardous categories. Those deemed to be hazardous are then so designated generically. Hence, if selected drilling muds are assessed to be hazardous, drilling muds in general are listed as hazardous. This process is continued until all hazardous waste source streams have been identified, evaluated, and listed.

The major advantage of this approach is its ease of implementation and enforcement. Once wastes have been designated, all wastes can readily be categorized by source. Both generators and regulators can quickly determine the status of a waste. No testing is required. As a consequence, costs associated with identification of hazardous wastes are low. An example of the listing by process approach is provided in Appendix B.

This ease of implementation is gained at the expense of equitable treatment and technical correctness. Categorization by source ignores the reality that waste streams from two generically similar sources may have greatly different properties. The use of different additives, varying process conditions, and postprocess treatment can significantly change the nature (i.e., the intrinsic properties) of a waste. Generic designation by source fails to recognize these differences and ultimately stands as a disincentive to the generator who may modify his process to reduce the hazardousness of his waste.

The listing approach is also quite limiting in that it does not accommodate new wastes or combinations of wastes. The list of designated wastes is current at a point in time when the evaluation is made. From that point forward, it is outdated as new processes and new products continue to change the nature and number of candidate waste streams. As a consequence, an updating schedule is required, with periodic evaluation of wastes to reaffirm or change designations.

Listing approaches have been relatively common in the European community (Patrick, 1975). The proposed bill for a special chemical waste act in The Netherlands called for a listing of names of substances that would be addressed in an order of council. Subsequently, wastes have been categorized and threshold concentrations specified. A federal-state committee on disposal of special refuse in West Germany has produced a list of 35 types of waste (by source of origin) that cannot be disposed with municipal waste (Patrick, 1975). Similarly, by 1976 Denmark had listed 39 process wastes and classes of chemicals as hazardous. In the United Kingdom, the Greater London Council, which administers the Deposit of Poisonous Waste Act in that geographic area, utilizes a classification system identifying wastes by constituents. Similarly, Japan has defined hazardous wastes as those including one or more of the following constituents: alkyl mercury, mercury and its compounds, cadmium and its compounds, lead and its compounds, organic phosphorus compounds, chromium VI compounds, arsenic and its compounds, cyanides, and polychlorinated biphenyls.

The first implementation of a listing approach in the United States took place in California pursuant to the California Hazardous Waste Control Act of 1973. Prior to the amendments of 1977, this act required that the Department of Health "shall prepare, adopt, and may revise when appropriate, a listing of the wastes which are determined to be hazardous." As a result, the department published a generic list (by waste type) in February 1975, including the wastes designated in Appendix C. Problems immediately arose. The most serious of these was described by Collins (1975):

The list of hazardous wastes and the list of extremely hazardous wastes adopted by the Department have been extensively revised and will be heard publicly and adopted in the next few months. Since the law requires the Department to adopt these lists, proposed revisions must be subjected to public hearings. Due to new materials being developed as well as others being discontinued, it is virtually impossible to maintain an up-to-date list.

To prevent this, Dr. Collins recommended that other states should authorize agencies to develop and maintain lists without adopting the lists themselves as regulations. This would not circumvent the need for continual updating, but it would eliminate costly public hearings and the time delays associated with a formal regulatory change. Because this option was not available in California, the department included with the generic listing several blanket clauses that gave greater flexibility to the testing: (1) the waste contains substances listed in Articles 2 and 3 of the attached regulations, or (2) the waste contains substances known to be hazardous or extremely hazardous as defined in Sections 25117 and 25115, respectively, of the attached hazardous waste law.

The first of these constitutes a second kind of listing approach for designation of hazardous wastes: the pure compound approach. The pure compound approach emerged during the development of the report to Congress. It is predicated on the assumption that the hazardous properties of a waste stream will be those of the most hazardous constituents. Therefore, a list of hazardous chemicals is developed (Table 3-1), and wastes found to contain any of these are subsequently designated hazardous. As noted previously, this approach has been employed in Japan.

Candidate constituents for use in the pure compound approach may be selected in a number of ways. To date, lists have been generated through collection of lists designated hazardous by other agencies, e.g., DOE and USDA. Lists may also be generated by applying hazard criteria on thresholds and determining which chemicals exceed these numeric values. A third way to generate lists as well as a means of putting lists in perspective is to review the constituents which have been implicated in hazardous waste incidents in the past.

The advantage of the pure compound approach is that data on most of the selected hazardous chemicals are much more readily available than data on waste streams. Therefore, literature data can be employed and waste analysis reduced to chemical characterization. This eliminates costly hazards testing of waste streams. Conversely, reliance on published chemical data is also the major weakness of the pure compound approach. Assigning the properties of a single constituent to a complex waste mixture fails to account for interactions between constituents that may markedly alter the hazardous nature of a waste. Such interactions can include the following:

Table 3–1 / Sample List of Nonradioactive Hazardous Compounds Employed for the Pure Compound Listing Approach from the 1973 Report to Congress.

Miscellaneous inorganics
Ammonium chromate
Ammonium dichromate
Antimony pentafluoride
Antimony trifluoride
Arsenic trifluoride
Arsenic trioxide
Cadmium (alloys)
Cadmium chloride
Cadmium cyanide
Cadmium nitrate
Cadmium oxide
Cadmium phosphate
Cadmium potassium cyanide
Cadmium (powdered)
Cadmium sulfate
Calcium arsenate
Calcium arsenite
Calcium cyanides
Chromic acid
Copper arsenate
Copper cyanides
Cyanide (ion)
Decaborane
Diborane
Hexaborane
Hydrazine
Hydrazine azide
Lead arsenate
Lead arsenite
Lead azide
Lead cyanide
Magnesium arsenite
Manganese arsenate
Mercuric chloride
Mercuric cyanide
Mercuric diammonium chloride
Mercuric nitrate
Mercuric sulfate
Mercury
Nickel carbonyl
Nickel cyanide
Pentaborane-9
Pentaborane-11
Perchloric acid (to 72%)
Phosgene (carbonyl chloride)
Potassium arsenite
Potassium chromate
Potassium cyanide

Potassium dichromate
Selenium
Silver azide
Silver cyanide
Sodium arsenate
Sodium arsenite
Sodium bichromate
Sodium chromate
Sodium cyanide
Sodium monofluoroacetate
Tetraborane
Thallium compounds
Zinc arsenate
Zinc arsenite
Zinc cyanide

Halogens & interhalogens
Bromine pentafluoride
Chlorine
Chlorine pentafluoride
Chlorine trifluoride
Fluorine
Perchloryl fluoride

Miscellaneous organics
Acrolein
Alkyl leads
Carcinogens (in general)
Chloropicrin
Copper acetylide
Copper chlorotetrazole
Cyanuric triazide
Diazodinitrophenol (DDNP)
Dimethyl sulfate
Dinitrobenzene
Dinitro cresols
Dinitrophenol
Dinitrotoluene
Dipentaerythritol hexanitrate (DPEHN)
GB (propoxy(2)-methylphosphoryl fluoride)
Gelatinized nitrocellulose (PNC)
Glycol dinitrate
Gold fulminate
Lead 2,4-dinitroresorcinate (LDNR)
Lead styphnate
Lewisite (2-chloroethenyl dichloroarsine)
Mannitol hexanitrate
Nitroaniline

Table 3-1 / (Continued)

Nitrocellulose	Chlorinated aromatics
Nitrogen mustards (2,2′,2″-	Chlordane
trichlorotriethylamine)	Copper acetoarsenite
Nitroglycerin	2,4-D (2,4-dichlorophenoxy-acetic acid)
Organic mercury compounds	DDD
Pentachlorophenol	DDT
Picric acid	Demeton
Potassium dinitrobenzfuroxan (KDNBF)	Dieldrin
Silver acetylide	Endrin
Silver tetrazene	Ethylene bromide
Tear gas (CN) (chloroacetophenone)	Fluorides (organic)
Tear Gas (CS) (2-chlorobenzylidene	Guthion
malononitrile)	Heptachlor
Tetrazene	Lindane
VX (ethoxy-methyl phosphoryl-N,N-	Methyl bromide
dipropoxy-(2-2)-thiocholine)	Methyl chloride
	Methyl parathion
Organic halogen compounds	Parathion
Aldrin	Polychlorinated biphenyls (PCB)

Additive effects, where constituents operate through similar mechanisms and thereby stress the receptors as if they were the same total quantity of just one of the constituents (an example would be a mixture of hydrocarbons of the same general molecular weight)

Synergistic effects, where the total effect of a combination of constituents is greater than the sum of their individual effects (examples would be the mixture of chlorinated hydrocarbon pesticides and solvents, and cadmium and zinc in water)

Antagonistic effects, the functional opposite of synergistic effects, where the total effect is less than the sum of effects of the constituents (an example would be arsenic and selenium)

Chemical interaction effects, where the presence of one constituent modifies the hazard potential of another through direct chemical reaction or modifies the availability of that constituent to the receptor [examples would be acid and alkali in the same solution (neutralization) and sulfate in solution with barium (immobilization)].

Similarly, a mixture of nonhazardous constituents may react to produce a hazardous product in the waste stream. These shortcomings can be resolved only by direct hazard testing of a waste stream.

A second major weakness in both the pure compound approach and the listing by source approach is a subtle one that becomes visible upon comparison to the criteria approach. In the evaluation for selection of generic waste streams or hazardous constituents, it is necessary to establish some means of measurement, some criteria by which candidates are determined to qualify as hazardous. For equity and resolution, these criteria should be quantitative. If regulators rely on existing lists of hazardous chemicals, they are tacitly excepting someone else's criteria developed for different purposes. If a new set of criteria are developed, then, in essence, a criteria approach has been taken. However, the equitable treatment of the latter has been forgone to avoid the costs of waste stream hazard testing.

The state of Washington has developed a scheme to accommodate many of the advantages of the two systems in the context of a criteria approach. This method is described in the following section. Recognizing the inherent difficulties in relying solely on a listing approach to designation, most states have gone to alternative systems. The 1977 amendments to the California Hazardous Waste Disposal Act added the requirement for adoption by regulation of "criteria and guidelines for the identification of hazardous wastes and extremely hazardous wastes." Similarly, RCRA requires the EPA to "promulgate criteria for identifying the characteristics of hazardous wastes and for listing hazardous waste."

As such, the listing approach is evolving as a part of the designation procedure that can be employed in one of two ways:

1. Lists are provided as the designation mechanism, but their constituency is determined through the application of criteria.
2. Lists are employed merely as a convenience to generators and administrators, indicating which wastes should be subjected to direct hazard testing prior to designation.

Criteria

The criteria approach to designation of hazardous wastes is a quantitative one that can be applied directly to wastes or to chemicals for selection of entries in a pure compound approach. The methodology relies on comparison of specific material characteristics to a selected threshold value. When the threshold is exceeded, the material is designated as hazardous. For example, one could define as hazardous all wastes with a flash point of 100°F or less. Once selected, any constituent or waste stream could be evaluated for designation against this threshold. Values for comparison can be measured empirically or calculated from component data.

Use of criteria eliminates many of the weaknesses associated with listing approaches. The criteria can be applied to any future waste as well as those currently being generated. Hence, there is a mechanism for continual updating without the need for regulatory change or public hearings. Once the criteria have been accepted through the public hearing and comment process, individual designation exercises are straightforward. Tests are run, and wastes pass or fail based on intrinsic properties. Legal challenges and adversarial proceedings are minimized. Criteria and associated testing procedures are published so that generators, shippers, public officials, and disposal site operators as well as regulators can evaluate any given waste to determine its proper designation.

As suggested previously, the major weakness of the criteria approach is the cost and time associated with testing waste streams directly. Toxicological and physical testing can require significant expenditures, depending upon the number and type of tests involved. Single-dose screening bioassays may be accomplished for as little as $1000 per sample (oral, dermal, inhalation, and irritant tests). Chronic and sublethal or genetic activity tests could exceed $500,000 for a single material.

Following recognition of the impact of these potential costs, additional mechanisms have been devised to minimize testing and to leave the decision whether or not to test to the generator who will bear the cost. A common means of minimizing testing is to offer a listing approach on an advisory basis. Lists are provided indicating that all waste streams enumerated should be treated as hazardous unless the generator wishes to test them and in so doing show that they do not meet criteria. The generator can weigh the cost of testing against the cost differential for available disposal options and make the decision to test or not test on an economic basis.

A second methodology has been devised as an intermediate measure between application of advisory waste listings and direct waste testing. In this case, formulations are provided to allow the generator to calculate hazardous property parameters utilizing data on the waste's components. The derived values are then compared to modified threshold values to determine if direct testing is required. The modified thresholds are conservatively selected to provide for a safety margin, because calculated values do not take into account any of the possible interactions between constituents. An example of calculated guidelines can be found in an early draft of the proposed California regulations (California Department of Health, 1978). The direct test criteria threshold for oral ingestion has been established at $LD_{50} \leq 5000$ mg/kg. The calculated LD_{50} is defined as

$$LD_{50} = \frac{10}{\displaystyle\sum_{X=1}^{n} (\%AX/LD_{50AX})}$$

where $\%AX$ is the percent by weight of component AX. LD_{50AX} is the literature LD_{50} value for component AX. In this way, the calculated threshold is actually 10 times the direct measurement threshold. The factor of 10 provides a margin of safety should components be synergistic or produce more hazardous by-products. In subsequent drafts, California dropped the safety factor of 10.

Utilizing these intermediate steps, a generator can schedule evaluation activities to minimize expenditures. If his waste is listed in a guideline list, he can calculate thresholds to determine the accuracy of the designation list. If he fails the calculated test, he may proceed to a direct waste stream test. At each point he has the opportunity to make an economic trade-off analysis between the costs of proceeding and the benefits of having waste designated nonhazardous. The cost-conscious generator will also prioritize the parameters he evaluated from least costly tests to more expensive ones. In so doing, if he fails a criteria associated by less expensive tests, he need not proceed with more costly tests for other hazard parameters.

Application of a criteria approach requires two types of selection activity: (1) selection of hazardous parameters of concern and (2) selection of threshold values for those parameters. With regard to the former, there are many possible candidates, as evidenced by the variety of case histories presented in Chapter 1. Some perspective can be gained by reviewing statistical data on reported damages from mismanagement, such as those presented in Table 3-2 from some 421 case studies (Environmental Quality, 1977) (totals exceed 421 because some incidents involved more than one damage mechanism). Although damage classification does not disaggregate data to individual hazard types or receptor categories, some general conclusions can be drawn. For one thing, water is most often affected as the transport medium and receptor. Air contamination plays a minor role. The predominance of groundwater problems suggests further that effects on potable water are major concerns. This is borne out by complementary data revealing that as a result of these 421 cases, 140 wells were affected. Following this conclusion, the EPA draft criteria for defining hazardous wastes have narrowed toxicity criteria to this: the presence of constituents that form a leachate with contaminant concentrations equal to or greater than 10 times drinking water standards. No provisions are made for aquatic toxicity, phytotoxicity or inhalation toxicity or for the numerous chemicals that have not been incorporated in drinking water standards.

This is a considerable deviation from the position taken in the report to Congress. At that time (1973), the EPA offered a preliminary criteria system described as the "hazardous waste decision model" (Figure 3-3), which included 11 hazard categories for evaluation. This more expansive view of the breadth of hazards to be considered has been carried forward by individual states. Of the 35 states that by 1980 had employed a hazard listing of some kind in the generic definition of hazardous wastes, the frequencies of occurrence for individual hazards were as follows:

Toxic 30
Explosive 28
Infectious 24
Radioactive 21

Table 3-2 / Distribution of Damage Mechanisms among Hazardous Waste Incidents.

Damage Type	Total	Surface Impoundments	Landfills, Dumps	Other Land Disposal[a]	Storage	Smelting, Slag, and Mine Tailings
				Disposal Method		
Groundwater pollution	259	57	64	117	10	11
Surface water pollution	170	42	49	71	—	8
Direct-contact poisoning	52	1	6	40	5	—
Air pollution	17	3	5	9	—	—
Fires, explosions	14	—	11	3	—	—
Total		89	99	203	15	15

[a]Includes abandonment and spray irrigation.

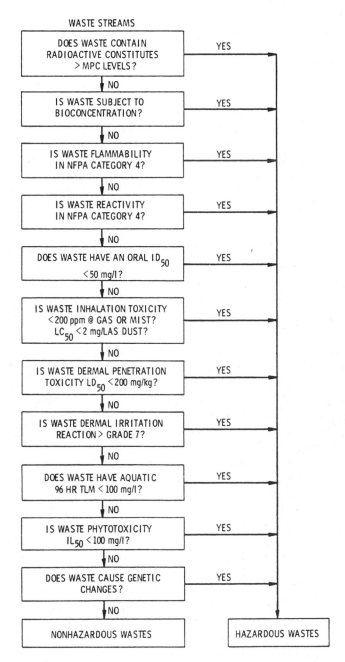

WASTE STREAMS

DOES WASTE CONTAIN RADIOACTIVE CONSTITUTES > MPC LEVELS?	YES
↓ NO	
IS WASTE SUBJECT TO BIOCONCENTRATION?	YES
↓ NO	
IS WASTE FLAMMABILITY IN NFPA CATEGORY 4?	YES
↓ NO	
IS WASTE REACTIVITY IN NFPA CATEGORY 4?	YES
↓ NO	
DOES WASTE HAVE AN ORAL ID_{50} < 50 mg/l?	YES
↓ NO	
IS WASTE INHALATION TOXICITY < 200 ppm @ GAS OR MIST? LC_{50} < 2 mg/LAS DUST?	YES
↓ NO	
IS WASTE DERMAL PENETRATION TOXICITY LD_{50} < 200 mg/kg?	YES
↓ NO	
IS WASTE DERMAL IRRITATION REACTION > GRADE 7?	YES
↓ NO	
DOES WASTE HAVE AQUATIC 96 HR TLM < 100 mg/l?	YES
↓ NO	
IS WASTE PHYTOTOXICITY IL_{50} < 100 mg/l?	YES
↓ NO	
DOES WASTE CAUSE GENETIC CHANGES?	YES
↓ NO	
NONHAZARDOUS WASTES	HAZARDOUS WASTES

Figure 3-3 / Graphic representation of the hazardous waste decision model from the 1973 report to Congress.

Flammable 17
Corrosive 17
Irritant 10
Bioconcentrative 10
Genetically active 3
Reactive 2

These data can be somewhat misleading as a result of the impreciseness of terms employed. Some states assume that "toxic" covers a spectrum of discrete hazards such as "bioaccumulative" and "genetically active," just as "explosive" and "reactive" and "corrosive" and "irritant" may be somewhat synonymous. In developing quantitative criteria, the toxic category is often broken down further to cover oral, dermal, inhalation, and aquatic toxicity.

Before reviewing the threshold values that have been selected for designation criteria, a brief discussion of each hazard type will be instructive.

Toxicity

Toxicity is the ability of a waste to produce injury upon contact with a susceptible site in or on the body of a living organism. Toxicity hazard is the risk that injury will be caused by the manner in which a waste is handled. Wastes may be acutely or chronically hazardous to plants or animals via a number of routes of administration. Phytotoxic wastes can damage plants when present in the soil, atmosphere, or irrigation water. Phytotoxicity is the result of a reduction of chlorophyll production capability, overall growth retardation, or some specific chemical interference mechanism.

Wastes that are acutely toxic to mammals may be active when inhaled, ingested, and/or contacted with the skin. Acute effects are generally evidenced within hours of inhalation or after a single dermal or oral dose. Data pertinent to a single route of administration may not be applicable to alternative routes. Hence, asbestos dust is toxic at very low levels when present in the air, but in water asbestos particles may not pose an ingestive threat at low levels.

Wastes may be chronically toxic to mammals if they contain materials that (1) are bioaccumulated or concentrated in the food chain or (2) cause irreversible damage that builds gradually to a final, unacceptable level. Classic examples of chronic toxicants are the heavy metals and halogenated aromatic compounds. Wastes can also be highly toxic to aquatic organisms. Much data exist on the effects of various materials on fish and fish food organisms. (Toxicity in these cases may result from transfer of toxic materials across the gill membrane surface, and thus toxicity information may not always be deduced from existing information on mammalian toxicity.)

Because of the various routes of exposure that may ultimately lead to hazardous effects, toxicity must be viewed as a function of the transport medium, the physical characteristics of the waste, and the type of disposal practices involved. Although water is perhaps the most pervasive vector, atmospheric emissions may well travel faster and spread farther. Direct contact is the most easily controlled route of exposure.

Toxic wastes can be derived from practically any industry. Toxicity may be the result of pure constituents within the stream, the total effects of several similar waste stream components, or the combined actions of two individually nontoxic materials (binary synergism).

Acute toxicity is typically measured through use of an LD_{50} or an LC_{50}. These terms refer to the median lethal dose and the median lethal concentration, respectively. They are quantitative measures of the dose (concentration) at which 50% of a test population will die from exposure to a chemical under prescribed conditions. For oral and dermal contact, a single dose is administered, and the test subjects are observed for 14 days. For inhalation, the test animal may be exposed continuously for periods of up to eight or more hours. Similarly, phytotoxicity and aquatic toxicity tests are conducted continuously for a prescribed period. A test period of 96 h has become quite standard for the latter test and is often described as the indicator of acute effects as opposed to chronic effects. Chronic toxicity measurements are not nearly as standardized.

Explosiveness and Reactivity

Explosive wastes may be detonated by several mechanisms: thermal shock, mechanical shock, electrostatic charge, or contact with incompatible materials. Like flammable wastes, highly reactive ones may threaten life and property in an acute sense and a latent sense in that the detonation may occur before or after "disposal." In the first case, handling, shipment, or disposal operations can initiate violent reactions, resulting in an explosion. In the second case, reactive materials may be buried in a landfill and, like a time bomb, await the appropriate conditions for detonation.

Typically, the kill radius for explosive wastes will be less than that for comparable volumes of flammable liquids. Prior experience with transportation-related explosions indicates a casualty radius of 100–200 ft (Arthur D. Little, 1972). Detonation of a single waste may be followed by secondary explosions or fire. The magnitude of the hazard existing after completion of disposal activities may exceed the handling hazard if sufficient waste inventory is accumulated.

Reactive wastes include explosive manufacturing wastes, contaminated industrial gases, and old ordnance. There is no universal property of reactivity that allows a single quantitative scale for categorizing these materials. Mechanical-shock-sensitive materials may be described by results of a drop test, such as that associated with the Picatinny Arsenal scale. Other types of reactive materials are simply classified using narrative descriptions. The most common rating is one developed by the NFPA (National Fire Protection Association, 1973). This system divides materials into five categories, 0–4, based on the degree of hazard anticipated in handling them. The extreme hazard group (category 4) includes the following:

Materials that can be detonated by electrostatic charge
Oxidizing materials such as chlorates, perchlorates, bromates, peroxides, nitrates, and permanganates
Self-reactive materials
Materials capable of autopolymerization
Primary explosives that may be detonated by friction, impact, shock, or heat (rated 5 inches or less on the Picatinny Arsenal scale)
Materials that react violently with air or water

The National Academy of Sciences (1973) has also created a classification system for characterizing the reactivities of chemicals with themselves, water, and other chemicals. Grade 4 materials may react with otherwise nonreacting materials, may react vigorously with water, or may undergo self-oxidation.

Infectiousness

Infectious wastes are those materials that contain disease-causing organisms or matter. Pathogenic wastes are those containing organisms, bacteria, or virus that may cause disease. Although infectious and pathogenic wastes have been excluded from the purview of this text, it must be recognized that this hazard type is one of the most frequently listed concerns in generic definitions of hazardous wastes. Wastes that are infectious or contain infectious materials pose a hazard to handlers and the public if they are not isolated and/or disposed of in a manner that destroys the viability of the infectious matter. At the same time, it is recognized that infectious organisms are ubiquitous in the environment and thus are present at varying levels in all manner of materials. This is particularly true of wastes such as sewage sludges and municipal refuse. Therefore, it is necessary to define criteria carefully such that virtually all wastes do not fall subject to hazardous waste regulation. Two approaches have been attempted in the past:

1. Exempt specific wastes such as sewage sludges.
2. Use the source as a part of the criterion.

Options for the latter approach would, for instance, designate as hazardous all infectious wastes resulting from medical experimentation or from diagnosis, care, or treatment of diseased humans or animals. These would include the following:

1. Laboratory wastes such as pathological specimens, infectious cultures, and disposable fomites
2. Surgical and obstetrical wastes such as pathological specimens and disposable fomites
3. Equipment, instruments, utensils, and fomites of a disposable nature from the rooms of patients or subjects with suspected or diagnosed communicable disease.

For the purposes of the foregoing, pathological specimens include tissues and specimens of blood elements, excreta, and secretions obtained from patients or subjects. Infectious cultures include those used for detection, maintenance, or isolation of infectious organisms or suspected infectious organisms, such as microorganisms and helminths capable of producing infection or infectious disease. Fomites include any substance that may harbor or transmit infectious organisms.

Aside from the operational wastes described earlier, this category would also cover such special wastes as carcasses from livestock epidemics and materials from biological warfare agents. As written, it would not include wastes such as dead limbs from trees affected by Dutch elm blight, and yet these wastes may also pose an infectious threat and have required special treatment in the past.

Whereas the use of source-derived criteria for designating infectious wastes has been widely accepted, it should be noted that many states supplement criteria with an exemption for sewage sludges. This is often based on an economic decision because of the large volumes of material involved and the higher costs of hazardous waste disposal. Tacit in this exception is the belief that the infectious hazard can be reduced or eliminated with standard procedures used at the sewage treatment plant. For instance, many pathogenic organisms are destroyed with anaerobic digestion. On the other hand, parasites such as *Ascaris* are not. It is also important to note that sewage sludges may well meet other hazardous waste criteria such as those related to the presence of heavy metals and toxic chemicals. Therefore, when sewage sludges are to be exempted, the exemption should be accompanied by regulations that assure proper treatment and disposal of these materials outside hazardous waste management regulations.

Radioactivity

As noted in Chapter 1, radioactive materials have been intentionally deleted from the scope of the technical discussions of this text. This reflects the fact that radioactive waste management is a technology of its own that could easily fill several volumes. It also recognizes that these materials are, for the most part, addressed by the Atomic Energy Act of 1954 as amended and therefore are regulated by the Nuclear Regulatory Commission. A caution is warranted here, however. The 1954 Act is not inclusive for all radioactive materials. The EPA has the authority to regulate releases of such materials as radium and isotopes produced in accelerators. Therefore, there is a limited gap in regulatory control that some states have chosen to cover through hazardous waste regulation.

Ionizing radiation results from an instability of the nucleus of an atom. The drive toward stability causes a radioactive release that may be manifested in one of many forms. The four major types of radiation are the following:

1. Alpha particles consist of two protons and two neutrons and are the largest and heaviest of the emissions. Interaction with orbital electrons slows alpha particles considerably, and thus they do not travel more than 3 inches when emitted in air. Consequently, they are incapable of penetrating the dead outer layer of human skin. However, ingestion, inhalation, or adsorption of alpha emitters can be extremely hazardous because of their potential ability to damage internal organs unprotected by epidermal layers. Elements with an atomic number of 84 or greater are typical alpha emitters.
2. Beta particles are electrons emitted at high speeds. Their small size and great velocity allow beta particles to travel as far as 10–100 ft in air, and yet can penetrate human skin by as much as half an inch. Excessive external doses can produce skin burns, and internal doses can be highly hazardous even at very low levels. Effects include debilitation of reproductive capability and injury to specific organs.
3. Gamma radiation is electromagnetic energy rather than matter. In contrast to alpha radiation, gamma radiation poses an extreme external as well as internal exposure hazard because of its ability to travel great distances and deeply penetrate human tissue. Gamma rays may not be as hazardous when present internally because of their ability to exit the body without colliding with electrons and causing damage. Many radioisotopes of common elements are gamma emitters.
4. Neutrons separated from the nucleus and traveling at very high speeds constitute a fourth form of radiation. Although human exposure is rare, it is extremely dangerous because of tissue-penetrating capabilities exceeding several feet.

Several major health hazards may result from exposure to radiation: (1) large acute external doses may result in burns or damage to internal organs; (2) large acute internal doses may result in damage to internal organs; (3) low-level chronic internal doses may accumulate in the body until toxic action results; (4) radiation can interfere with the normal functioning of the nuclei of human cells, leading to malignancy; and (5) irradiation of reproductive organs can lead to sterility or possibly harmful mutations.

Wastes containing radioactive materials may cause any or all of the foregoing effects. Acute exposure can result from improper handling by employees or improper disposal to nonsecured locations. Chronic exposure can potentially result from leaching of landfills, volatilization of radioactive materials, or proximity to unmarked repositories.

Radioactive wastes can be categorized by activity level (the rate at which the nucleus of the isotope decays) and therefore are subject to quantitative characterization. Relative safe levels have been defined as maximum permissible concentrations (MPC), which can be used as a point of reference for designating wastes as hazardous.

Flammability

Highly flammable wastes can pose both acute handling hazards and latent disposal hazards. Handling problems involve safety hazards to personnel at the site of origin, during transport, and at the disposal site. An example of a latent disposal hazard is the potential damage caused by unintentional or spontaneous combustion of flammable residues at a disposal site. Fear of such consequences has led to a ban on landfilling of flammable liquids in many areas.

Both acute and latent hazards relate to injury, destruction of property, and/or rapid depletion of resources. A 9,000-gal tank truck and a 30,000-gal tank car of flammable liquid are likely to be associated with kill radii of 115 and 230 ft, respectively, if ignited during handling or transport operations (Arthur D. Little, 1972). Secondary effects beyond the initial disaster area may include ignition of nearby inflammables and detonation of heat-sensitive substances in the vicinity. Hazards related to disposal sites may exceed those of transportation and handling if sufficient waste volumes are involved. Flammable wastes may include contaminated solvents, oils, pesticides, plasticizers, complex organic sludges, and off-specification chemicals.

Flammable wastes may be characterized for relative categorization or compared on the basis of flammability properties. Properties typically measured as an indication of material flammability include flash point and autoignition temperature.

Flash point is defined as the minimum temperature at which a liquid will give off sufficient vapor to form an ignitable mixture in the air above the liquid's surface. An ignitable mixture in this context is a vapor–air mixture within the flammable range. Some solids such as camphor and naphthalene volatilize sufficiently to have flash points, but the measure refers most directly to liquids. Flash point is measured by standard analytical techniques wherein an ignition source is used to determine when the vapor–air mixture is in fact ignitable. Flash point should not be confused with fire point. The latter is the temperature at which a liquid can sustain combustion. It is generally several degrees higher than the flash point, because additional heat is required to sustain the vapor–air mixture in the flammable range over time.

The autoignition temperature is the minimum point at which a material will self-ignite. This measure is highly sensitive to the conditions under which the test is run and therefore does not represent a universal value for a material. As a consequence, autoignition temperature has not received as much use as flash point.

The most widely accepted scheme for categorization of flammable materials is that employed by the National Fire Protection Association (1973). This classification system includes five groupings, 0–4, similar to the reactivity groupings. Placement in a group is based on a combination of properties. The most flammable materials, category 4, are characterized as "very flammable gases, very volatile flammable liquids, and materials that in the form of dusts or mists readily form explosive mixtures when dispersed in air."

Actual categorization of specific materials has been based on judgment of a reviewing committee. In general, category 4 materials are in one of the following categories:

1. Flammable gases
2. Flammable liquids with boiling points below 100°F and vapor densities ≥ 1.1 (density is measured as the ratio of the weight of a volume of vapor to an equal volume of dry air under similar conditions)
3. Flammable liquids with flash points below 100°F and vapor–air densities ≥ 1.1
4. Materials spontaneously combustible in air

The vapor and vapor–air density data are meant to account for the hazard of vapors traveling along the ground to an ignition source and then flashing back. This could be a real hazard in landfill operations, where heavy equipment exhaust or sparks could ignite escaping vapors. The category 4 rating is roughly equivalent to a grade 4 rating on the NAS fire hazard scale (National Academy of Sciences, 1973). The NAS ratings are keyed to materials with flash points and boiling points below 100°F.

Corrosiveness

Corrosiveness is defined as the ability of one agent to eat away or erode another material through chemical action. The reactor material may be biological membranes. However, this action is generally addressed with criteria for irritation. Therefore, corrosiveness in this context refers to action on inanimate materials. The hazard posed by corrosive materials is largely evidenced during handling and storage activities when containment may be breached. This can result both in release of the corrosive material itself and in the opportunity for these materials to release other hazardous wastes such as pyroforic, toxic, or explosive materials. Some concern has also been expressed with respect to the ability of corrosive materials to solubilize otherwise immobile heavy metals.

The desirability of considering corrosive properties in hazardous waste designation has been a point of contention during the regulatory development period. The hazard posed during handling and storage is already addressed by OSHA and DOT rules and guidelines. Indeed, DOT standards are often employed for selecting corrosive criteria. Hence, inclusion for these concerns constitutes an area of overlap. Consideration on the grounds that these materials may be disposed in metal drums is a moot point, because drum corrosion will occur over time regardless of the contents. Corrosive materials will merely accelerate this natural process. If the concern is focused on release of contents in a landfill, that concern is misplaced. Drums are not an effective barrier to release in the landfill environment. Indeed, the use of drums during waste disposal is a practice designated to facilitate loading, transport, unloading, and placement through temporary containment. As a consequence, the decision for including corrosiveness as a hazard of concern rests with the appraisal of the adequacy of existing regulations for storing and transporting these materials and concerns for solubilization of metals.

Corrosiveness is most often measured in terms of the material's ability to corrode a given thickness of steel during a set period of time. Testing is accomplished by standard procedures, as specified by National Association of Corrosion Engineers (NACE) Standard TM-01-69. Corrosiveness is also often associated with acidity or alkalinity. Subsequent criteria can therefore be defined on the basis of pH. Both of these measures are quantitative and accommodate selection of a criteria threshold.

Irritation

Some wastes (namely, those containing allergens capable of sensitizing skin, agents that cause contact dermatitis, or substances that are corrosive to living tissues) can cause severe discomfort if contacted. Examples of these wastes are concentrated acids and alkalis, waste warfare agents, and waste substances with allergenic properties. Such wastes pose a hazard when discharged to waterways or uncontrolled landfills where accidental exposure cannot be prevented.

When the primary irritant response of a material can be traced to acidity or alkalinity, pH can be employed as a quantitative measure for criteria comparison. Similar measurements are more difficult for non-acid-mediated irritation and sensitization. In the latter case, designation is made by definition. Materials are classified as sensitizers if contact with them is found to render the receptor more susceptible to allergic reactions on future exposures or on exposure to other materials. Some materials are also photosensitizers. In this case, the receptor becomes sensitized to exposure to light of wavelengths between 280 and 430 nm (2800–4300 Å).

For materials that irritate eyes or skin but are not adequately characterized with pH measurement, arbitrary scales have been devised to connote relative irritant power. The FDA methodology (21 CFR 191.1) employes a 10-point ordinal scale with values assigned based on the severity of dermal response. Higher, more hazardous ratings are associated with moderate or severe edema (swelling) and erythema (redness) after 24-h exposure. A second system available for quantifying irritant capability has been devised by Smyth and co-workers (1941, 1944, 1948, 1949, 1951, 1954, 1962, 1969). With this approach, specific dermal responses are observed at varying chemical concentrations and are assigned grades accordingly. For instance, grade 8 categorizes materials that produce necrosis (death or decay of tissue) when applied in a 1% solution. Grade 10 materials produce a similar response when applied in a 0.01% solution.

Bioconcentration

The term "bioconcentration" is used to describe the hazard posed by materials that can be concentrated in a single organism or magnified by successive levels in the food chain until they reach toxic levels. The hazard is one of chronic exposure, and it generally occurs when the contaminant is present in the environment at low levels.

Wastes may possess this characteristic as a result of the presence of bioconcentrative constituents such as cadmium, lead, mercury, polychlorinated biphenyls (PCB), or carbon tetrachloride. Improper disposal of these wastes can lead to release of low levels of bioconcentrative materials to the environment. Organisms may then pick up and concentrate these materials until concentration reaches a sufficient level to cause death or debility. Concentration to the threshold toxic level often occurs in higher life forms such as fish, birds, and mammals, including man. This hazard may be evidenced in hazardous waste management as a result of long-term disposal to productive surface waters or through landfarming where crops are ultimately employed for consumption.

Bioconcentration factors are defined as the ratio of the concentration of the contaminant in an organism to the concentration of that contaminant in the surrounding environment or food (e.g., for aquatic communities, the factor equals the concentration in fish tissue divided by the concentration in water). The definition is not relevant to materials with cumulative effects or materials for which substantial nutritional requirements have been established. In general, bioconcentrated materials as defined here are those for which the detoxification-excretion mechanism is either nonexistent or extremely slow.

Bioconcentrative materials can be grouped into two categories, based on retention mechanisms. The first includes the heavy metals, such as mercury and lead. These materials, through a strong affinity characteristic with sulfhydryl groups and disulfide bonds, are capable of inactivating or denaturing enzymes and proteins, thus blocking normal metabolic pathways, interfering with control mechanisms, and crippling cellular integrity. The second category of bioconcentrative substances is represented by persistent organic materials such as DDT and PCBs. These materials concentrate through an affinity for nonpolar solvents and low solubility in water. The contaminants quickly migrate to fatty tissues or lipid cellular fractions, where they typically cause hepatic disorders (disorders in the functions of the liver).

Evaluating the data on bioconcentration can be very difficult, because no standard bioassay or testing procedure has been adopted by which the bioconcentration potential of a material can be consistently assessed. One alternative proposed for selection of bioconcentrative materials is response to octanol–water partition tests. Others rely on analysis of residuals in model environment studies. Until such a standard testing procedure is developed, literature sources documenting environmental buildup of a material or laboratory studies indicating less than complete elimination or detoxification of a material by one of the higher organisms of animal life must be used to select substances under this criterion.

Genetic Activity

Wastes may contain materials with carcinogenic, mutagenic, or teratogenic properties, evidenced as malfunctions of the genetic process either in mitosis or meiosis. When chemically induced, such effects may be the result of chemical modification of DNA nucleotides in the target species. Exposure routes are usually direct and continuous.

Dye plant wastes and petroleum sludges can contain genetically active materials. The California Hazardous Wastes Working Group (Governor's Task Force on Solid Waste Management, 1970) noted that

> Most proofs of carcinogenesis in humans are limited to occupational exposures but there is probably a general population exposure of unknown magnitude. Various reports substantiate this assumption in one way or another and give emphasis to the urgent need for comprehensive chemical, experimental, and epidemiologic studies to determine actual hazards.

More recent estimates of "environmentally" induced cancer range from 60% to 90% of all cancer cases. Hence the concern for releases of materials with even low levels of known or probable carcinogens.

Knowledge of genetic activity is currently such that numeric values or rating systems have not been devised to quantitatively compare potencies or assess risk. In part, this reflects fundamental disagreement among experts as to the nature of the hazard. A primary facet of the controversy rests with the existence or nonexistence of threshold concentrations for chemically induced carcinogenesis. These are the opposing views:

1. Chemical carcinogenesis, like radiologically induced cancer, is a function of probability upon exposure; i.e., with each exposure there is a given statistical chance for a carcinogenic response. Hence, even a single exposure increases the probability of response, and all exposure is to be avoided regardless of concentration. On the basis of this assumption, chemicals can be assigned a cancer risk value which can be compared between chemicals to rank potency. This value is related to the slope of the curve in Figure 3–4.

2. Chemical carcinogenesis is directly related to the degree of exposure only for materials above a set concentration threshold. Hence, low-level exposures are of no consequence, and safety standards, potency ratings, and categorization can be based on the numeric value of the threshold.

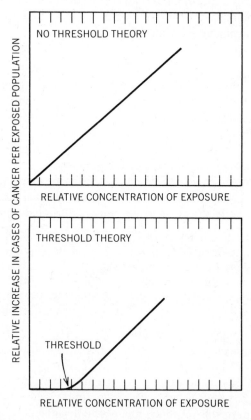

Figure 3–4 / Characteristic curves for current theories of carcinogenic response.

Characteristic dose–response curves for the two theories are presented in Figure 3–4. Belief that there may be no measurable threshold for carcinogens has fostered responses such as the Delaney clause, which forbids the addition of any known carcinogenic material to food products in any concentration. (Recent results of studies conducted for the National Cancer Institute on over 24,000 mice support this view.)

Without accepted numeric ratings for genetic activity, criteria have of necessity followed the lines of the pure constituent approach. That is, wastes are defined as genetically active if they contain known carcinogens, teratogens, or mutagens above a given concentration. Selection of the threshold concentration is highly controversial because of the considerations previously noted. There is also a great deal of controversy over what data are required to declare a constituent a known carcinogen. Standard tests have been and continue to be developed and endorsed by such groups as the National Cancer Institute, but the interpretation of results continues to rely heavily on the judgment of the evaluator.

Extraction Procedures

Extraction procedures have long been employed as a means of dissolving and/or separating complex materials for the purposes of analysis. In solid and hazardous waste management, extraction procedures take on added importance, because they are employed to simulate leachate generation and therefore provide a basis for prediction of mobility after disposal. It is this predictive aspect of extraction procedures that has aroused controversy with respect to use in regulatory definitions for segregating hazardous and non-hazardous wastes. Because of the added costs associated with the management of wastes classified as hazardous, generators of waste are concerned about the degree to which any given procedure represents what will happen in the disposal environment.

Extraction procedures employed for differentiating hazardous and non-hazardous wastes have four major elements: (1) specification of how the waste is prepared for the evaluation, (2) specification of the solution used for leaching, (3) specification of the means and extent of contact, and (4) specification of how resulting extracts are to be analyzed. The first three of these factors are reviewed next. Issues related to analysis are left to other texts.

Waste Preparation

The rate of leaching is directly related to the surface area of the material being leached. As a consequence, the degree to which a waste is reduced in size will greatly affect the results of an extraction procedure. The EPA extraction procedure calls for the use of a sample that has a surface area

≥ 3.1 cm^2 or passes through a 9.5-mm (0.375-inch) standard sieve, while California has proposed a threshold particle size associated with passage through a #10 sieve (2000 microns). The sample processing scheme can be seen in Figure 3–5, taken from a proposed procedure for the state of Washington.

At issue with respect to sample preparation is the degree to which size reduction may discourage the fixation of wastes. Fixation processes may involve materials that encapsulate and physically entrap contaminants rather than chemically binding them. In these cases, size reduction will breach the coating and allow leach solutions to reach otherwise contained contaminants. As such, extraction will be much higher than can be anticipated from typical disposal site conditions, where the fixed mass is not likely to be broken down to smaller sizes as a result of weathering and other natural forces. To circumvent these problems, it has been recommended that size reduction be applied only to those wastes that fail a specified structural integrity test. Wastes that pass can be evaluated in a monolithic form (McKown, 1981).

Leachate Solution Selection

Once the sample is prepared, it is exposed to the leach solution in order to initiate extraction. The chemistry of the solution can greatly affect subsequent leaching. For instance, low-pH solutions are likely to solubilize metal

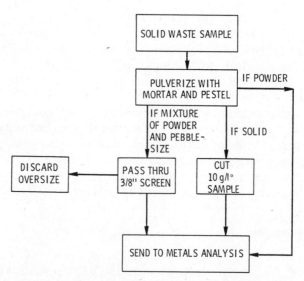

Figure 3–5 / Proposed solids preparation method for metals analysis (WAC 173-302, September 1979). *Liter of leachate (acetic acid solution).

salts, and alkaline waters will produce higher concentrations of certain organic chemicals. Similarly, the presence of chelating agents will sponsor increased mobility of metals and other contaminants through complexation. In the simple case of disposal of a single waste at a given site, the leachate can be made to simulate rainwater through use of distilled water. Somewhat higher acidity (pH 4.5–5.0) may be desirable to simulate acid rainfall conditions encountered in more industrialized areas. Where codisposal with organic-based wastes is anticipated, more sophisticated solutions are employed. These "synthetic garbage juices" are mixed to simulate the organic acids derived during anaerobic decomposition of refuse. A typical formulation calls for use of acetic acid or sodium acetate (U.S. EPA and state of Washington), and some states have gone to more active ingredients. For instance, the state of California employs citrate, a stronger complexant than acetate.

A great deal of controversy surrounds the selection of the leachate solution. Industrial commentators have been highly critical of the EPA for using acetate, which is known to free up metals and other toxic chemicals. They propose distilled water as a compromise to the "worst-case" scenario. Other observers support the use of both distilled water and acetate to provide categorization of wastes. Those failing criteria with an acetate leach but passing with distilled water would be restricted to noncodisposal landfilling options; i.e., these wastes would be segregated from an environment where acetate or other organic acid constituents would be anticipated.

The effect of employing different leach solutions has been investigated in recent work sponsored by the EPA. A number of proposed extract solutions were employed in parallel studies on five industrial wastes: dewatered sludge from a POTW (publicly owned treatment works), organic still bottoms, ink pigment waste, baghouse dust, and pharmaceutical waste. Representative results are provided in Tables 3–3 and 3–4 for comparing measured levels of leachate organics and metals from the POTW sludge. Based on these results and results from the other tests, researchers concluded that distilled water was the best extractant for organic contaminants. The acetate buffer solution simulated the proposed EPA extraction procedure (EP) for leaching metals, whereas citrate was considered much more aggressive. In general, the degree of aggressiveness was both medium-dependent and sample-dependent. Also, the leaching method was deemed unsuitable for oily wastes (McKown, 1981).

Method of Contact

The efficiency of an extraction procedure is greatly affected by the means of contact and mixing employed. Kinetic considerations produce quantitative differences between static and mixed contact, as well as between batch and

Table 3-3 / Organics in POTW Sludge Leachates.[a]

Leaching Medium	Relative Amount[b] of Given Analyte Found				
	ClBz	p-DCBz	TCBz	Naph	Phen
EP	0.9	1.7	1.0	0.9	1.1
0.1-M OAc, 4.0	0.8	1.0	1.0	0.9	1.0
Univ. Wisc.	NA[c]	1.1	1.2	0.7	0.9
0.1-M Cit, 4.0	0.6	1.5	2.4	1.2	3.5
0.05-M Cit, 4.0	0.5	1.4	2.8	1.1	4.4
0.1-M Cit, 4.0, NH4	NA	1.3	2.2	1.0	3.3
0.1-M Cit, 5.0	NA	1.6	2.8	1.2	4.0
0.1-M Cit, 5.0, FeSO4	NA	1.3	1.9	0.9	2.4
0.1-M Cit, 5.0, Na2S2O4	NA	0.7	1.2	0.4	2.1
0.1-M Cit, 5.0, Igepal	NA	2.9	5.6	2.4	10
0.05-M Cit, 5.0	NA	2.2	2.8	1.4	3.7
0.05-M Cit, 5.0, Na2S2O4	NA	2.7	6.3	0.8	2.5
0.05-M Cit, 5.0, Igepal	NA	1.3	2.4	0.4	0.8
0.02-M Cit, 5.0	NA	1.0	1.1	0.4	0.2
0.05-M Na2S2O4	NA	0.9	0.8	0.9	0.7
0.01% Igepal	NA	1.2	1.0	1.0	0.9
	Amount Found (μg/l) Using Distilled Water				
	33	20	180	27	15

[a]Adapted from McKown (1981).
[b]Relative to the amount found using distilled water.
[c]Not analyzed.

Table 3-4 / Metals Leached from POTW Sludge.[a]

Leaching Medium	Relative Amount[b] of Given Metal Found					
	B	Mo	Cu	Pd	Cr	Mn
EP	8.5	0.6	0.9	1.9	2.9	117
0.1-M OAc, 4.0	4.1	0.7	0.5	0.9	3.1	109
Univ. Wisc.	6.5	2.5	ND	16	31	128
0.1-M, Cit, 4.0	15	5.6	6.0	86	131	146
0.05-M, Cit, 4.0	12	4.6	5.0	53	106	131
0.1-M Cit, 4.0, NH4	13	5.3	10	91	124	140
0.1-M Cit, 5.0	13	5.3	13	78	124	143
0.1-M Cit, 5.0, FeSO4	22	4.2	ND	5.3	131	150
0.1-M Cit, 5.0, Na2S2O4	24	6.1	42	91	134	143
0.1-M Cit, 5.0, Igepal	22	5.6	8.9	91	131	143
0.05-M Cit, 5.0	18	5.1	6.8	24	37	117
0.05-M Cit, 5.0, Na2S2O4	28	6.3	25	80	143	157
0.02-M Cit, 5.0	6.8	1.7	0.1	0.9	4.3	26
0.1-M OBu, 4.0	4.6	1.0	0.5	1.6	3.8	114
0.05-M, Na2S2O4	3.0	2.3	0.3	0.7	2.5	29
0.01% Igepal	1.2	1.1	0.2	ND	0.7	0.6
	Amount Found (μg/l) Using Distilled Water					
	46	57	53	88	29	35

[a]Adapted from McKown (1981).
[b]Relative to the amount found using distilled water.

column leaching. The magnitudes of these differences will vary with the waste material being extracted. To date, little work has been performed on the standardization of column tests for chemical wastes. Rather, efforts have been focused on mixed batch extractions employing shakers, mixers, or other means of agitation. One comparative study concluded that while column studies are more realistic, they are more complex and less reproducible because of the potential for channelization of flow in the column. In one series of parallel extractions of chemical wastes, participants in an ASTM survey determined that improper mixing was a major cause of poor reproducibility of results. They concluded that use of a wrist shaker or NBS mixer for 48 h minimized these effects. Similar results were reported by Garrett, et al. (1981) as a result of reviewing a number of studies.

There appears to be general agreement with respect to use of a liquid-to-solid ratio of 10:1. While these proportions do not reflect conditions likely to occur in a landfill, they accommodate good mixing and leave ample solution for analysis. Use of room temperature is also generally accepted as appropriate. Landfills may be subjected to a wide range of temperatures, but these are dampened significantly at the bottom of the cells, where the leachate will emerge. If average ground temperature is known, it would be a desirable alternative to room temperature. This degree of sophistication is optional pending the availability of data.

The timing of contact is also an important parameter. As noted previously, participants in the ASTM study believed that 48 h was optimal. Similarly, the state of California originally proposed an approach requiring 48-h contact periods for the first exposure, followed by periods of 4, 8, and 16 days. California has subsequently selected a single 48-h contact period. The EPA study, however, concluded that 24 h should be adequate. Their logic was that whereas ideally each solution should be brought to equilibrium, the time required for that endpoint differs between constituents and waste types. Therefore, a long but convenient contact period is optimum. A full 24 h was selected to accommodate scheduling in a laboratory (Garrett et al., 1981). Certainly, 48 h would also meet those criteria.

The number of leachings is a different question yet. Many proposals to date employ four sequential extractions. These are plotted to determine trends in concentration levels. If the trend is upward, additional extractions may be required. If the trend is constant or downward, further work may be deemed unnecessary.

Extract Analysis

While it is not the intent of this work to review analytical chemistry or discuss the related technology in depth, it is important to note that differences in round-robin studies of leaching procedures have in part been attributed to differences in analytical results. To provide guidance in this area, the EPA undertook studies to compare EPA, ASTM, and standard methods procedures with respect to variability, (DeWalle et al., 1981). As a result of that work, the EPA has recommended methods for leachate analyses as summarized in Table 3-5.

Criteria Selection Rationales

Once hazards of concern have been selected, specific numeric threshold values or other means of determining when a waste meets a given criteria must be selected. Two approaches to the selection process have been utilized in the past. The first was put forward in the 1973 report to Congress when thresholds were selected to be compatible with values used by other agencies, e.g., U.S. Department of Transportation and U.S. Atomic Energy Commission (Figure 3-3). The rationale for this approach was to minimize variations in defining hazardous and toxic substances and thereby foster movement to a universal designation. The drawback in this approach is that thresholds designated by other agencies are directed to meeting regulatory needs other than those addressed by hazardous waste management authorities. Therefore, they may be more or less restrictive than necessary, depending on the agency's charter. The appropriateness of thresholds selected strictly for compatibility would therefore be fortuitous.

The second and more recent approach to selection of thresholds is the development of a general risk assessment framework. With this alternative, emphasis is placed on identifying the likely means by which a hazard would be evidenced and selecting thresholds that, under those circumstances, would render that hazard probable. For instance, how toxic must a waste be before it will not be diluted to safe levels in leachate while moving to a potable well? The variations available to this approach and the judgment required in applying them are innumerable. Rather than explore them all, approaches taken by various groups are summarized here for illustrative purposes.

U.S. Environmental Protection Agency

On May 19, 1980, the EPA published proposed guidelines for defining hazardous wastes. Although numerous tests were considered for possible future use, criteria were established for only four hazards of concern: ignitability, corrosivity, reactivity, and toxicity. The basic rationale behind selection of criteria thresholds was the development of release scenarios to determine what level of properties would pose a hazard under anticipated management practices. Ignitable wastes were defined as liquids with a flash point below 140°F (60°C), ignitable compressed gases [49 CFR 173.300(b)], oxidizers (49 CFR 173.151), and nonliquids that can ignite through spon-

Table 3-5 / Recommended Methods for Leachate Analysis.[a]

Leachate Parameter	Recommended Method
pH	Electrometric method on fresh sample, using glass electrode and temperature correction
Oxidation-reduction potential	Electrometric method on fresh sample, using platinum electrode with the calomel reference electrode
Turbidity	Nephelometric method on a fresh sample
Conductivity	Electrometric method using platinum electrode and temperature correction
Free volatile fatty acids	Chromatographic method; COD determination should also be conducted
Chemical oxygen demand (COD)	Manual dichromatic reflux method using the ferrous ammonium sulfate titration
Total residue	Drying method at 104°C
Volatile residue	Drying at 550°C without prior filtration of sample
Organic nitrogen	Kjeldahl manual titration
Ammonia nitrogen	Distillation titration method
Sulfate	Gravimetric method
Total phosphorus	Ascorbic acid method using the persulfate digestion step
Chloride	Potentiometric titration
Alkalinity	Potentiometric method with titration to inflection point of about pH 4.5
Nitrate	Cadmium reduction method after separate determination of the nitrate ion
Sodium and potassium	Atomic absorption or manual flame or automated flame emission method
Calcium, magnesium, barium	Atomic absorption spectrophotometric
Heavy metals (Fe, Zn, Pb, Cr, Cd, Cu, Ni)	Direct aspiration atomic absorption spectrophotometric method

[a]Adapted from DeWalle et al. (1981).

taneous chemical changes, absorption of moisture, or friction. The flash point of 140°F was chosen over that of 100°F (38°C) used by DOT to define flammable liquids for regulation of transportation because there is evidence that waste materials are exposed to the higher temperature during handling. Indeed, it has been recommended that the DOT consider modifying its definition to the higher temperature, because ambient temperatures may exceed 100°F in many areas (Kuchta and Burgess, 1970).

The threshold for corrosivity was proposed as materials that corrode SAE 1020 steel at a rate greater than 0.25 inch per year (103°F) or are aqueous solutions having a pH value greater than or equal to 12 or less than or equal to 3. The former limit was taken directly from corresponding DOT criteria. The latter limits are said to reflect data on skin corrosivity, aquatic toxicity, and heavy-metal solubility.

Reactive wastes are defined by narrative description, with emphasis on ability to detonate, react violently with water, or release toxic fumes when contacted with water. Cyanide- and sulfide-bearing wastes are specifically identified for their potential release of toxic gases when mixed with dilute acids. Class A (49 CFR 173.51) and class B (49 CFR 173.53) explosives are also listed as reactive wastes.

Numerous approaches to selection of toxicity criteria were considered. In the end, it was determined that the major threat posed by improper hazardous waste management lay within contamination of potable water supplies through leachate intrusion. Consequently, toxic wastes were defined as those wastes that, when subjected to a standard leaching test, produced a leachate with constituent concentrations in excess of 100 times the EPA National Interim Primary Drinking Water Standards. This yields the quantitative limits listed in Table 3–3. The multiplier of 100 was selected as representative of a conservative level of dilution that may be expected between the point of leaching and potable use from a nearby well. It is based on review of limited data on leachate levels from monitored landfills. The standard leachate test is conducted using distilled water adjusted to pH 5.0 with acetic acid (representative of organic acids produced during degradation of refuse in a landfill). Liquids are filtered or centrifuged from waste after a 24-h agitation period and analyzed for the constituent specified in Table 3–6. Additional information on extraction procedures is provided in a subsequent chapter on analytical procedures.

The criteria employed by EPA are meant to test the hazardous nature of any waste. At the same time, the EPA system includes a list of wastes by source or type that have been generically defined as hazardous. Generators who have wastes covered in the list must treat them as hazardous unless they can successfully demonstrate that the waste is in fact not hazardous. The latter is done through a petition. The petition must clearly identify the waste

Table 3–6 / Toxic Constituent Criteria for Defining Hazardous Wastes (U.S. EPA).

Contaminant	Extract Threshold (mg/l)
Arsenic	5.0
Barium	100
Cadmium	1.0
Chromium	5.0
Lead	5.0
Mercury	0.2
Selenium	1.0
Silver	5.0
Endrin	0.02
Lindane	0.4
Methoxychlor	10
Toxaphene	0.5
2,4-D	10
2,4,5-TP (Silvex)	1.0

and present results of testing to show that it qualifies for delisting. This process is done on a case-by-case basis. Required tests are those that evaluate the properties for which the waste was originally listed; i.e., if the waste was tested as hazardous because of flammability, the ignitability (flash point) test should be conducted. The agency currently recommends that generators obtain copies of successful petitions to use as a model. The process is accelerated if a temporary delisting is requested, rather than a permanent one. Applications for permanent delisting can be submitted after a temporary delisting is granted, so that unnecessary costs are not incurred while awaiting a decision. It was originally thought that the initial surge of delisting activity would be followed by a low-level plateau. Experience has shown that initial registration of wastes was conservative. As costs have increased, more generators are viewing delisting as a means of cutting unnecessary costs. Wastes commonly considered for delisting are plating wastewater treatment sludges and incineration residue.

California

The hazardous waste identification system devised by California blends both pure compound and criteria approaches into a tiered structure. Tiers move from the simpler, inexpensive evaluations to detailed waste stream analysis. Results of the simpler analyses are compared to thresholds with greater safety margins to ensure that the imprecision of the test does not allow hazardous materials to escape the regulatory net. The general structure of the tiers or phases is illustrated in Table 3-7. The logic flow from tier to tier is illustrated for toxic evaluation in Figure 3-6.

Individual thresholds were derived from scenario development. For toxicity, six subcategories were incorporated: oral, dermal, inhalation, aquatic, carcinogenic, persistent, and bioaccumulative. A threshold value of $LD_{50} \leq$ 5000 mg/kg was selected for oral toxicity in a standard acute rat feeding test. This level was selected by applying a safety factor of 100 to levels deemed to pose a high health hazard by the National Institute for Occupational Safety and Health (NIOSH) (1975). In like fashion, California selected a dermal threshold of $LD_{50} \leq$ 4300 mg/kg and inhalation thresholds of $\leq$10,000 ppm (vapor). The threshold for aquatic toxicity was set at $LC_{50} \leq$ 500 mg/1. This level was chosen for its compatibility with the limit set by the EPA to select hazardous substances with respect to spills as mandated by Section 311 of the Federal Water Pollution Control Act Amendments of 1972 (PL 92-500).

Criteria for carcinogens were set by a listing/concentration threshold approach. Wastes are designated as hazardous if they contain a constituent listed in Table 3-8 equal to or in excess of 1000 ppm. Selected constituents were chosen on the basis of designation as known or suspected carcinogens by the Occupational Safety and Health Administration (OSHA) or NIOSH

Table 3-7 / Evaluation Steps for Defining Hazardous Wastes, State of California.[a]

	Toxicity	Flammability	Pressure-generating/ Reactivity	Corrosion/Irritation
Phase I[b] Component identification and evaluation	Establish by analysis or other means the identity and concentrations of any potentially hazardous or extremely hazardous components of the waste.			
	Evaluate from reported data the toxicity of waste components	Evaluate from reported data the flammability of waste components	Evaluate from reported data the pressure-generating properties and reactivity of waste components	Evaluate from reported data the corrosivity and irritability of waste components
Phase II[b] Screening options	Determine if any of the carcinogenic, persistent, or bioaccumulative substances in the system's tables are in excess of TLCs	Measure flash point for liquid wastes and flammability limits for gases; evaluate solids for flammability	Measure or evaluate from reported data the pressure-generating properties and re-activity of waste	Measure pH of waste
Phase III[b] Calculation options	Calculate the approximate oral, dermal, and inhalation toxicities of waste	—	—	—
Phase IV[b] Animal test options	Measure fish toxicities and mammalian oral, dermal, and inhalation toxicities	—	—	—

[a] Adapted from California Department of Health (1978).
[b] Phase I must, by regulation, be carried out. Phases II, III, and IV are optional tests. In lieu of doing the tests, a producer must accept a hazardous designation for the waste if phase I shows that any components are hazardous.

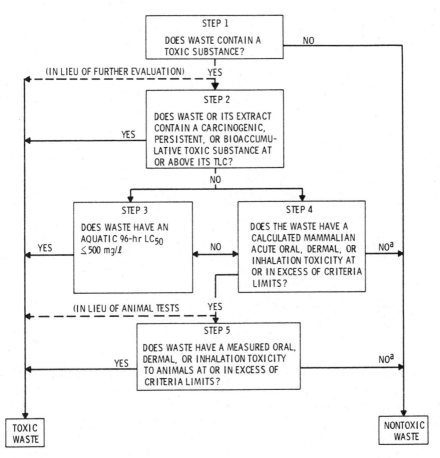

Figure 3-6 / Logic flow for toxicity evaluation, state of California. [a] Providing step 3 also shows a "no" result.

criteria documents. Threshold limits were based on the 1000 ppm (0.1%) minimum level set by OSHA to exempt compositions from workroom regulation as a carcinogen.

Persistent and bioaccumulative materials were similarly dealt with using a listing/concentration threshold approach. Constituents of concern (Table 3-9) were selected as those materials that persist in the environment and for which man has no major nutritional requirement. As is apparent from Table 3-9, two threshold concentrations are assigned. First, a soluble threshold was established for leachable constituents as 100 times the drinking water or chronic fish toxicity limits. The multiplier of 100 accounts for likely dilution of leachate prior to arrival at a potable well or surface water. This

Table 3-8 / California Criteria for Designating Hazardous Wastes Based on Carcinogenic Hazards.[a]

If a waste contains any of the following substances having concentrations equal to or greater than 0.1% by weight, the waste is designated as extremely hazardous:

2-Acetylaminofluorene (2-AAF)
Acrylonitrile
4-Aminodiphenyl (4-ADP)
Arsenic
Benzidine and its salts
Carbon tetrachloride
Chloroform
bis-(Chloromethyl) ether (BCME)
Chloromethyl methyl ether (CMME)
1-2,-Dibromo-3-chloropropane (DBCP)
3,3'-Dichlorobenzidine (DCB) and salts
4-Dimethylaminoazobenzene (DAB)
Ethyleneimine (EI)
α-Naphthylamine (1-NA)
β-Naphthylamine (2-NA)
4-Nitrobiphenyl (4-NBP)
N-Nitrosodimethylamine (DMN)
β-Propiolactone (BPL)
Vinyl chloride (VCM)

[a]Adapted from California Department of Health (1982).

value was based on work performed during development of criteria for the state of Minnesota (Battelle Memorial Institute, 1976). The second threshold is a total value set at 100 times the soluble limit for most contaminants. This level was selected arbitrarily as a safety margin. As noted in Table 3-7, California also devised a means for calculating toxicity that could be used for phase II screening of wastes. The algorithm is of the basic form:

$$\text{CaLD}_{50} = \frac{100}{\displaystyle\sum_{X=1}^{n} \frac{\%\text{AX}}{\text{LD}_{50\text{AX}}}}$$

where CaLD_{50} is calculated toxicity value, $\%\text{AX}$ is present by weight of constituent AX, and $\text{LD}_{50\text{AX}}$ is the published toxicity value for constituent AX. The waste is deemed hazardous by this calculation if the calculated value is found to be less than or equal to the criteria threshold for that toxicity parameter.

The flammability criteria were chosen in a manner similar to the proposed EPA criteria (flammable gases, readily ignitable solids, pyrophorics, and strong oxidizers). California chose not to seek compatibility with DOT flammable liquid criteria and selected a flash point of $< 140°F$ (60°C). Explosive and reactive criteria are an expansion on the EPA work incor-

Table 3-9 / California Designation Criteria for Wastes Bearing Persistent and Bioaccumulative Toxic Substances.[a]

If a waste contains any of the following substances having concentrations equal to or greater than those listed in this table, the waste is designated hazardous.

Substances	Soluble Threshold Limit Concentration [mg/kg(ppm)]	Total Threshold Limit Concentration [mg/kg(ppm)] (wet weight)
Asbestos		10,000
Aldrin	0.14	1.4
Antimony compounds	100	500
Arsenic and compounds	5	500
Barium (excluding barite) and compounds	100	10,000
Beryllium and compounds	7.5	75
Cadmium compounds	1	100
Chlordane	0.25	2.5
Chromium (VI) compounds	5	500
Chromium (III) compounds	25	2500
Cobalt compounds	80	8000
Copper compounds	2.5	250
DDT, DDE, DDD	0.1	1
2,4-Dichlorophenoxyacetic acid	10	100
Dieldrin	0.8	8
Dioxin (TCDD)	0.00	0.0
Endrin	0.02	0.2
Fluoride salts	180	18,000
Heptachlor	0.3	0.3
Kepone	0.47	4.7
Lead compounds, inorganic	5	1000
Lead compounds, organic	—	13
Lindane	0.4	4
Mercury and compounds	0.2	20
Mirex	2.1	21
Methoxychlor	10	100
Molybdenum compounds	350	3500
Nickel compounds	20	2000
Pentachlorophenol	1.7	17
Polychlorinated biphenyls (PCBs)	1.2	12
Selenium compounds	1	100
Silver compounds	5	500
Thallium compounds	7	700
Toxaphene	0.5	5
Trichloroethylene	204	2040
2,4,5-Trichlorophenoxyacetic acid	1	10
Vanadium compounds	24	2400
Zinc compounds	25	2500

[a]Adapted from California Department of Health (1982).

porating class C as well as class A and B explosives (per DOT definition 49 CFR 173), water-reactive materials, and class 2, 3, and 4 reactive materials as defined by the National Fire Protection Association (NFPA). Compatability with existing definitions was also the rationale behind selection of the steel corrosion test. The pH criteria of pH ≤ 2 or ≥ 12.5 differ from that of the EPA by being less restrictive.

Minnesota

Prior to the adoption of EPA criteria, the Minnesota Pollution Control Agency chose to use a scenario approach based on landfill disposal of hazardous wastes. Toxicological criteria were specified for oral, dermal, inhalation, acquatic, bioaccumulative, and carcinogenic hazards. The oral threshold of $LD_{50} \leq 500$ mg/kg was selected as the toxic level required before a 100-kg man drinking 2.5 l of water a day would consume a toxic amount of the contaminant, assuming that leachate is diluted by a factor of 100 in traveling to a potable well. The dilution factor was selected after a review of empirical leachate data (NIOSH), 1975). The dermal threshold of $LD_{50} \leq 1000$ mg/kg was selected as the toxic level required to be lethal to an operator completely doused with a chemical. The inhalation thresholds of $LD_{50} \leq 2000$ mg/m^3 (dust and 1000 ppm (vapor) were based on acceptable levels of landfill operators exposed over a brief period. The aquatic criteria of $LC_{50} \leq 100$ mg/l was selected under the assumption that leachate undergoes the standard 100:1 dilution on traveling through the groundwater and an additional dilution of 100:1 on entering the surface waterbody. Similarly, criteria for specified bioaccumulative agents were selected using the 100:1 and 10,000:1 dilution factors for groundwater and surface water on drinking water and chronic fish limits, respectively. The resultant criteria are enumerated in Table 3–10. Criteria for carcinogens were based on the same premise used by California. OSHA- and NIOSH-designated carcinogens were listed, and the lowest of OSHA concentration limits (1000 ppm) for carcinogens was employed across the board.

Corrosive criteria were set at pH ≤ 3 and ≥ 12 using the same groundwater and surface water dilution factors applied to recommended fish and human exposure limits for acidity and alkalinity, as well as the DOT definition of ability to corrode 0.25 inch or more of SAE 1020 steel in a year. Irritation criteria were selected as materials capable of producing first- or second-degree burns or having a score of 5 or more on the rabbit skin contact evaluation.

Flammable limits differed from DOT standards in that liquids with a flash point at or below 200°F were designated as hazardous wastes, as were spontaneously combustible solids. The higher temperature was selected to reflect the contention that during handling, materials will be exposed to tempera-

Table 3-10 / Minnesota Criteria for Recognized Bioaccumulative Materials.

	Drinking Water Criteria (ppb)	Threshold in Leachate Based on Drinking Water (ppb)	Freshwater Aquatic Life Criteria (ppb)	Threshold in Leachate Based on Freshwater Aquatic Life (ppb)
Aldrin	1	100	0.01	100
Cadmium and compounds	10	1,000	0.4	4,000
Chlordane	3	300	0.04	400
DDT	50	5,000	0.002	20
Endrin	1	100	0.002	20
Heptachlor	0.1	10	0.01	100
Lead and compounds	50	5,000	0.30	300,000
Mercury and compounds	2	200	0.05	500
Mirex			0.001	10
Methoxychlor	100	10,000	0.005	50
PCBs	—	—	0.002	20
Toxaphene	5	500	0.01	100

tures in excess of ambient conditions. Oxidizers and explosives are defined narratively. Minnesota also categorically defined waste oil as hazardous. This last measure reflects the diversity of oils as wastes and the excessive costs that would be associated with bioassay and analysis of oils to determine if they meet criteria.

Washington

The Department of Ecology in Washington took a unique and totally different approach. They built a system from the basic premise that the proposed regulations for Section 311 of the Water Pollution Control Act amendments were directly applicable to hazardous wastes. The Section 311 approach defined a harmful quantity of hazardous materials as described in Table 3-11. Interpolating, Washington noted that a harmful quantity would be reached if one had 100 lb of 1% category A material, 100 lb of 10% category B material, or 100 lb of pure category C material. Using these equalities and an initial assumption that a material of $LD_{50} \leq 50$ mg/kg is an extremely hazardous waste, the state then extrapolated to designate categories A and B as extremely hazardous, category C as potentially hazardous based on quantity, and category D as dangerous. Designation of complex wastes is made by determining the combined percentage of all constituents in each category and converting to the equivalent total concentration as category

Table 3-11 / Proposed Definitions of Harmful Quantities of Hazardous Substances.

Category	Oral LD_{50} (mg/kg)	Aquatic LC_{50} (ppm)	Harmful Quantity (lb)
A	≤5	≤1	1
B	5–50	1–10	10
C	50–500	10–100	100
D		100–500	1000

A (i.e., A + 0.1B + 0.01C). If this value exceeds 1% and the waste quantity is greater than 100 lb, the waste is deemed extremely hazardous.

The enabling legislation in Washington also required consideration of persistence and genetic effects. Rather than identify specific compounds, Washington chose to designate all "heavy metals, halogenated hydrocarbons, and aromatic hydrocarbons as persistent and potentially having genetic effects." Wastes are regulated accordingly if they exceed 1% and 100 lb of these constituents (Melhoff et al., 1979).

Belgium

The definition employed in Belgium is of interest in that it follows the custom of most European countries in taking a listing approach but further modifies this by employing concentration thresholds. This has been a point of controversy when proposed to other members of the commission of the European community. Dissenters have contended as follows:

Concentration is but one factor determining the actual hazard posed by a waste; it should be replaced by a clear definition of the properties of wastes that render them hazardous.

Current knowledge is insufficient to set firm concentration thresholds.

Concentration thresholds do not take into account local conditions that can greatly impact hazard potential.

Concentration thresholds would encourage evasion of rules through partial treatment or dilution.

Be that as it may, Belgium has maintained reliance on concentration levels, as outlined in Article 2 of the law of July 22, 1974, as follows:

Article 2. The products and by-products which are not used or which cannot be used, the residues and the wastes arising from an industrial, commercial, craft-trade, agricultural or scientific activity shall be regarded as toxic waste:

1. if they are composed principally of one or more of the following chemical substances:
   (a) the chemical substances distinguished by the symbol T (toxic)

listed at Annex I and referred to in Article 723 bis, 4, of the *Règlement général pour la protection du travail* (general regulation on health and safety at work);

(f) more than 100 mg of mercury or of soluble compounds of thallium, expressed as Tl; or

(g) more than 500 mg of cadmium or its soluble compounds, expressed as Cd; or

(h) more than 250 mg of soluble compounds of beryllium, expressed as Be; or

(i) more than 1000 mg of organo-halogen compounds, except polymerized materials and the substances referred to in paragraphs 3, 4 and 5; . . .

3. if they contain pesticides or phytopharmaceutical products enumerated in the lists at Annex II to the Royal Decree of 5 June 1975 on the conservation, commerce and use of pesticides and phytopharmaceutical products;

4. if they contain more than 10% of organic solvents;

5. if they contain more than 1 mg per kg of dry matter of one or more of the chemical substances enumerated in the list of carcinogenic substances at Article 148 decies of the general regulation on health and safety at work;

6. if they arise from chemical processes of the pharmaceutical industry, the phytopharmaceutical industry or research laboratories.

Packaging which has contained toxic waste and has been polluted thereby and is not to be re-used shall be classed with toxic waste.

Natural ores and worked metals shall not be regarded as toxic waste.

The Belgian concentration levels are applicable throughout the country. The basic criterion is solubility. The toxic wastes are defined by the quantity of chemical substances per kilogram of dry matter.

Criteria in Perspective

Similar rationales and criteria thresholds or modifications thereof have been employed by states and countries other than those presented here to designate hazardous wastes. A summary of representative criteria is given in Table 3-12. Further perspective can be gained by noting the relation between criteria selected by various states and the characteristics of common chemicals and fuels as compared in Figures 3-7 through 3-10 or by reviewing the constituents that have been reported to have been at the source of hazardous waste damage incidents in the past (Table 3-13).

As noted previously, European nations appear to have taken the course of listing wastes for the purposes of definition. In 1972 the Institution of Chemical Engineers in the United Kingdom published a Provisional Code of Practice for Disposal of Wastes that classified wastes as hazardous and

Table 3–12 / Summary of Key Criteria as Reported by States.

State	Oral LD$_{50}$ (mg/kg)	Dermal LD$_{50}$ (mg/kg)	Inhalation LC$_{50}$ Vapor (ppm)	Inhalation LC$_{50}$ Dust (mg/m^3)	Aquatic LC$_{50}$ (ppm)	pH	Flammability/ Flash Point
California	≤ 5000	≤ 4300	≤ 10,000		≤ 500	≤ 2, ≥ 12.5	≤ 100° F
Minnesota	≤ 500	≤ 1000	≤ 1000	≤ 2000	≤ 100	≤ 3, ≥ 12	≤ 200° F
Washington	≤ 500 hazardous ≤ 5000 dangerous				< 100 hazardous ≤ 1000 dangerous	< 3, > 11, with equal weight of water	≤ 100° F
New York	≤ 50	≤ 200	≤ 200	≤ 2000	≤ 1000		
Oklahoma	≤ 500	≤ 1000	≤ 1000	≤ 100	≤ 100		
Oregon	≤ 500	≤ 200	≤ 200 mg/m^3	≤ 2000	≤ 250		
Tennessee	Prose Definition						
Arizona	Concentration in mg/l > 0.35 LD$_{50}$ in mg/kg				Concentration > 10 times (LC$_{50}$)	< 2, > 12 < 3, > 12, with equal weight of water	≤ 140° F ≤ 140° F
Maryland	Narrative						
Illinois	Employ ratings in sax (1968)					< 3, > 10, requires analysis	
Wisconsin	Narrative						
Texas	Employ Hazard Index						
U.S. EPA	Leachate concentrations exceed 10× drinking water standards					< 2, > 12	≤ 140° F

nonhazardous. The former group was further divided into broad groups: flammable, explosive, oxidizing, poisonous, infectious, corrosive, and ratio-active. An example of a subsequent classification scheme implemented by the Greater London Council is provided in Table 3-14. In The Netherlands, proposals for a chemical waste act called for a listing of wastes to be regulated. A comparison of list coverage for representative European countries and the state of California is provided in Table 3-15.

Prioritizing Concern

As is evident from preceding sections, even quantitative criteria result in the designation of a wide variety of wastes as hazardous. These wastes may have characteristic properties (oral LD_{50}, aquatic LC_{50}, etc.) that span several orders of magnitude. As a consequence, significantly different levels of potential hazard are associated with the overall category of hazardous wastes.

At the same time, there are limited resources available for the management of hazardous wastes. For instance, approved sites for landfill are few in number, and it is becoming increasingly difficult to open new ones. This provides incentive to identify the most hazardous wastes to ensure that they receive the greatest amount of attention and the best treatment. Less hazardous wastes should not consume resources needed to properly manage the more toxic or dangerous wastes. Such states as California and Washington have incorporated this philosophy in their enabling legislation by creating a category of extremely hazardous wastes. This provides the regulators with flexibility to specify different levels of management control for wastes posing significantly different hazards by dividing the total waste stream into three categories: (1) extremely hazardous wastes, (2) hazardous or dangerous wastes, and (3) nonhazardous wastes.

Subgrouping in this manner has gained support from several quarters. On reviewing EPA draft regulations, the Manufacturing Chemists Association (MCA) proposed a system that would control hazardous waste management by adjusting treatment, storage, and disposal standards according to degree of hazard involved. This would provide sufficient flexibility to prevent shortfalls in disposal sites for the extremely hazardous wastes. At the core of the system would be the division of wastes into three categories:

Class I. Highly toxic wastes that are persistent and bioaccumulative and present an extreme hazard to health or to the environment. These wastes require special care in storage, handling, and disposal.

Class II. Wastes that present a moderate hazard and can be handled in a normal manner with standard containers and equipment. These include moderately toxic wastes that are persistent or bioaccumulative and highly toxic wastes that are not persistent or bioaccumulative.

NIOSM[22]
EXTREME
HEALTH
HAZARD

1   TEPP

5   STRYCHNINE,
PARATHION,
THALLIUM

NAS[9]
TOXIC
CHEMICALS   NIOSM[22]
HIGH
HEALTH
HAZARD

10   ENDRIN, POTASSIUM
CYANIDE

25

CALIFORNIA
(EXTREMELY
HAZARDOUS)
NEW YORK
WASHINGTON
(EXTREMELY
HAZARDOUS)

50   MERCURIC CHLORIDE,
NICOTINE, ACROLEIN,
ALDRIN, DIELDRIN

100   TOXAPHENE, LINDANE,
KEPONE, ARSENIC
TRIOXIDE

NAS[9]   NIOSM[22]
MODERATELY   MODERATE
TOXIC   HEALTH
CHEMICALS   HAZARD

OREGON,
WASHINGTON
(HAZARDOUS)

250   COPPER SULFATE, DDT

500   DIETHYLAMINE,
DIMETHYLAMINE

MINNESOTA,
OKLAHOMA

750   ANILINE,
FORMALDEHYDE

1000   ABS (LINEAR), CRESOL

2000   CHLOROFORM,
NAPHTHALENE

3000   PROPANOL,
ACETIC ACID,

NAS[9]   NIOSM[22]
SLIGHTLY   SLIGHT
TOXIC   HEALTH

BUTYL
ALCO-
HOL

CHEMICALS   HAZARD

4000   METHYL ETHYL
KETONE, KYLENE,
CYCLOHEXANE

CALIFORNIA
WASHINGTON
(DANGEROUS)

5000   BORIC ACID

NAS[9]   NIOSM[22]
PRACTICALLY   NO SIGNIFICANT
NONTOXIC   HEALTH HAZARD

> 5000   BENZENE, CARBON
TETRACHLORIDE,
TOLUENE, ACETONE

Figure 3-7 / Relation of state criteria to oral toxicity of common chemicals

DERMAL LD50 *(mg/kg)*

NIOSH[22] EXTREME HEALTH HAZARD		1
		5
NIOSH[22] MODERATE HEALTH HAZARD		10 ENDRIN, ETHYLENEIMINE
		43 ACETONE CYANOHYDRIN
	CALIFORNIA (EXTREMELY)	50 DIELDRIN, PARAQUAT, CAMPHOR
NIOSH[22] MODERATE HEALTH HAZARD		100 ALDRIN, ENDOSULFAN
	OREGON, NEW YORK	200 HEPTACHLOR, NICOTINE
		300 PENTACHLOROPHENOL
		340
		400
NIOSH[22] SLIGHT HEALTH HAZARD		500 DIAZINON, CHLORDANE
		600
		800 DDT, ETHYLENEDIAMINE
	OKLAHOMA, MINNESOTA	1000 LINDANE, TOXOXAPHENE, ANILINE
		1500 CRESOL
		2800
NIOSH[22] NO SIGNIFICANT HAZARD		2000 KEPONE, MALATHION, NAPHTHALENE
	CALIFORNIA →	4300

Figure 3-8 / Relation of state criteria to dermal toxicity of common chemicals

Class III. Wastes that present minimal hazard to health or to the environment when adequately controlled. These include certain corrosive, ignitable, and reactive wastes and those of low or moderate toxicity that
are not persistent or bioaccumulative.

Similarly, the Council on Wage and Price Stability submitted a release to
the EPA expressing the concerns of the Regulatory Analysis Review Group

FLASH POINT (°F)

		−50	COAL TAR LIGHT OIL, ETHYLETHER, GASOLINE, CARBON DISULFIDE, CYCLOHYXANE
		0	HYDROCYANIC ACID, VINYL ACETATE
NAS[9] HIGH TO EXTREME HAZARD		10	BENZENE, METHYL ETHYL KETONE, CRUDE PETROLEUM
		50	METHYL ALCOHOL, ETHYL ALCOHOL
	WASHINGTON, U.S. EPA, DOT	100	BUTYL ALCOHOL, DIESEL FUEL 1-D, DIESEL FUEL 2-D
NAS[9] MODERATE HAZARD	TENNESSEE, ARIZONA, CALIFORNIA	140	
		150	ANILINE, FUEL OIL NO. 6, CRESOL, PINE OIL
	MINNESOTA	200	DODECYL MERCAPTAN, DIMETHYL SULFOXIDE
NAS[9] SLIGHT HAZARD		250	CHLOROPHENOL, ANTHRACENE, ROSIN OIL, MINERAL OIL, ASPHALT
		500	CASTOR OIL, COCONUT OIL, SOYBEAN OIL, PEANUT OIL
		1000	

Figure 3–9 / Relation of state criteria to flash point for common materials

(RARG) on the draft RCRA regulations with respect to the need for considering degree of hazard. The final report prepared by the RARG placed major emphasis on the need for a more refined classification system that would include subgroups of hazardous wastes. The RARG's contention was that such a system would provide more protection at lower cost, because the best (and most expensive) sites would be reserved for the most hazardous wastes. This, in turn, would reduce the complications and costs of handling less hazardous wastes and thus provide less incentive for evasion. It was the

LC50 (ppm)

		LC50 (ppm)	
NAS[9] TOXIC CHEMICALS		0.001	ENDRIN, DDT
		0.01	ALDRIN, SILVER, CHLORINE, TOXAPHENE
		0.1	MERCURY, ACROLEIN, CYANIDE
		1	
		5	NAPHTHALENE, CADMIUM, LEAD, COPPER SULFATE
NAS[9] MODERATELY TOXIC		10	ACRYLONITRILE, CRESOL, PHENOL, ZINC
		25	AMMONIA, FORMALDE-HYDE, BENZENE, NICKEL
		50	ACETALDEHYDE
		75	ACETIC ACID
	WASHINGTON (EXTREMELY HAZARDOUS) MINNESOTA, OKLAHOMA	100	ANILINE, CHLOROFORM, CARBON TETRACHLORIDE
	OREGON →	250	METHANOL, PROPYL ALCOHOL
NAS[9] SLIGHTLY TOXIC	CALIFORNIA →	500	PYRIDINE, CITRIC ACID
		750	ADIPONITRILE
	NEW YORK, WASHINGTON (DANGEROUS)	1000	TOLUENE, CYCHOHEXANOL
NAS[9] PRACTICALLY NONTOXIC		5000	BORIC ACID, ETHANOL, ISOPROPYL ETHER, TRICRESYL PHOSPHATE
NAS[9] NONTOXIC		10000	ETHYL ETHER

Figure 3–10 / Relation of state criteria to aquatic toxicity of common chemicals

Table 3–13 / Contaminants Involved in 421 Damage Incidents.[a]

		Disposal Method				
Contaminant	Total	Surface Impound-ments	Land-fills, Dumps	Other Land Disposal[b]	Smelting, Slag, Mine Tailings	Storage
As	19	5	4	10		
Cd	5	3	1	1		
Cr	33	11	9	12	1	
Cs	1		1			
Cu	20	6	4	7	3	
Fe	40	10	20	6	4	
Hg	11	1	1	9		
Mn	26	3	15	4	4	
Ni	13	5	2	5	1	
Pb	22	5	6	8	3	
Zn	22	9	5	5	3	
Cl^-	27	11	6	9	1	
CN^-	19	6	4	9		
F^-	8	5		3		
NH_3	14	6	2	6		
NO_3^-	16	6	2	7		1
$SO_4^{=2-}$	18	9	2	5	2	
Inorganic acids	27	9	4	10	4	
Misc. inorganics	83	21	25	29	6	2
PCBs	3		1	2		
Petrochemicals	27	10	5	10		2
Phenols	31	9	10	12		
Misc. organics	88	19	25	39		5
Bacteria	11	1	2	8		
Pesticides	71	1	6	57		7
Radioactive	9	2	3	1	1	2
Unspecified leachate	25	5	18	1	1	
Total	689	178	183	275	34	19

[a]Adapted from U.S. EPA (1977).
[b]Disposing on vacant properties, on farmland, spray irrigation, etc.

RARG's conclusion that use of a single category was an expedient to simplify implementation and enforcement. The net social cost, however, would exceed the associated savings to the agency. As a consequence, the RARG recommended that the EPA pursue the concept of differentiating hazardous wastes by degree of hazard.

Such approaches would segment the waste stream into discrete packages with associated levels of priority. Priority can also be established on a continuum. An approach to this was first offered in the 1973 report to

Congress. Wastes were compared on the basis of a ranking factor defined as

$$R = Q/CP$$

where R is ranking factor, Q is annual production quantity for a waste, and CP is the critical product for the waste being considered. The critical product was further defined as the product of the lowest environmental concentration at which a waste would manifest any hazard of concern and an index representing the waste's mobility in the environment. Hence, a waste that is miscible in water or highly volatile would be considered more of a hazard than a relatively immobile waste.

The prioritization formulation is simple, but it requires further development to create a consistent format for the mobility index. Once a proper dimensionless index has been selected, the resulting ranking factor has units of volume per year representative of the volume of environment (air, water, or soil) potentially contaminated to a hazardous level. Klee (1976) pointed

---

Table 3-14 / Sample Designation Scheme from the Greater London Council.[a]

Types of waste
1 Solid toxic
  (a) Cyanides
  (b) Metal-bearing and other inorganic
  (c) Asbestos
  (d) Pharmaceutical and laboratory reagents
2 Acid solutions or sludges
  (a) Metal-bearing
  (b) Without metals
3 Alkaline solutions or sludges
  (a) Metal-bearing
  (b) Without metals
4 Aqueous solutions or sludges—neutral
  (a) Inorganic
  (b) Organic
  (c) Mixed organic and inorganic
  (d) Cyanide solutions
  (e) Metal-bearing
5 Oily wastes
  (a) Mineral
  (b) Fatty (*i.e.,* animal/vegetable)
  (c) Oil–water emulsions
6 Tarry wastes
7 Solvent wastes
  (a) Combustible
  (b) Incombustible
8 Organic materials
9 General factory waste contaminated by various toxic materials

[a]Adapted from Patrick (1975).

	Denmark	EEC[b]	Federal Republic of Germany	France[b]	Netherlands	United Kingdom	California
Aluminum-containing waste	X		X	P[c]	P	X	P
Antimony and compounds		X		X	X	X	X
Arsenic and compounds		X	X	X	X	X	X
Asbestos	X		X	X	X	X	X
Beryllium waste		X	X	X	X	P	X
Cadmium waste		X	X	X	X	X	X
Chlorine					X	X	X
Chromium III waste	P				X	X	X
Chromium VI waste	P	X	X	X	X	X	X
Copper waste	X				X	X	X
Cyanide compounds	P	X[d]	X	X[d]	X	X	X
Dye manufacturing waste	X		X		P	X	
Fluorine					X	X	X
Halogenated solvents	X	P	X	X	X	X	X
Herbicides	X	X	X	P	X	X	P
Isocyanates	X	X		X	X	X	
Laboratory waste		X		X	X		
Lead waste	X	X	X	P	X	X	X
Magnesium waste			X	P	X	X	X
Mercury waste	X	X	X	X	X	X	X
Metal surface treatment waste	X	X	X	X	P	X	X
Nickel waste	X			P	X	X	X
Nonhalogenated solvents	X	P	X	P	X	X	X
Oil refinery waste	X		X	X	X	X	X
Organic peroxides	X			P	X	X	
Paint manufacturing waste	X		X	X	P	X	X
Pesticides	X	X	X	P	X	X	X
Pharmaceutical manufacturing waste		X	X	X	X	X	
Phenol-containing waste	X	X	X	X	X	X	P
Phytopharmaceutical waste		X		X	X	X	
PCBs	X		X	X	X		X
Rubber manufacturing waste	X		X	P		X	
Silver-containing waste	X				X	X	X
Sulfur-containing waste	X		X	P	X	X	X
Thallium waste		X		X	X		X
Vanadium and compounds				X	X	X	P
White phosphorus					X	X	X
Zinc waste	X			X	X	X	X

[a]Adapted from NATO (1977).

[b]Draft list.

[c]P indicates that a country's category falls *partially* under the heading in the left column, or that the waste is considered hazardous under certain *restrictions*.

[d]Excluding ferrocyanide and ferricyanide.

out that one of the flaws of this approach is that by relying on the smallest concentration at which a hazard is evidenced, all remaining data on other hazard types are ignored. This constitutes a tacit assumption that all hazards are of equal concern. When comparing between wastes, the ranking fails to recognize that even though two materials may have similar critical concentrations, one poses greater additional hazards at slightly higher critical concentrations.

This defect is also shared in part by the model developed by Pavoni, Hagerty, and Lee (1972) (the PHL model). In this methodology, criteria were established in five areas:

1. Human toxicity $= 12 \times$ Sax (1968) rating
2. Groundwater toxicity $= 6$ (4 $-$ log of the smallest critical concentration)
3. Disease transmission potential $=$ index assigned on scale of 0 to 105
4. Biological persistence $= 16(1 -$ BOD/TOD), where BOD is biochemical oxygen demand, and TOD is theoretical oxygen demand
5. Mobility $= 7 - C +$ log solubility, where C is net charge of waste determined from molecular formula in reaction with water at pH 7

The hazard ranking itself is calculated as the sum of these criteria. Some of the hazardous data not covered by the smallest critical concentration value are recoupled through the additional factors; so this model is more comprehensive in scope than the previous one. No rationale is given for the selection of weighting factors incorporated in the summing process.

An approach to prioritization developed by Booz Allen Applied Research (1972) (BA model) focuses on waste hazards and the medium affected by disposal technology. The total effects rating (TER) is defined as

$$\text{TER} = (AT_\text{H})W_1 + (AF_\text{H})W_2 + (AT_\text{E})W_3 + (WT_\text{H})W_4 + (WF_\text{H})W_5 + (WT_\text{E})W_6 + (ST_\text{H})W_7 + (SF_\text{H})W_8 + (ST_\text{E})W_9$$

where

$A =$ air
$W =$ water
$S =$ soil
$T_\text{H} =$ toxic effects to humans
$F_\text{H} =$ flame/explosion/reaction hazards
$T_\text{E} =$ toxic ecological effects (nonhuman)
$W =$ weighting factors

Each of the nine criteria is rated subjectively on a scale of 1 to 3 (minimal, moderate, and severe hazard). No guidelines are given on how to select the rating. All weights were initially assigned a value of 1, implying equal utility among all hazard types. The hazard rating (HR) is then derived as

$$\text{HR} = (\text{TER})(\text{HER})$$

where HER is the sum of production and distribution criteria values measured on scales of 1.0 to 1.5 and 0 to 0.5, respectively (i.e., large production quantity -1.5, widely distributed -0.5, HER $= 2.0$).

Some of the intent of the MCA proposal for waste grouping, as well as prioritization of individual wastes, is encompassed in the system employed by the Texas Water Quality Board (Board order No. 75-1125-1) for classifying industrial wastes with respect to possible impacts from improper land disposal. The three categories are defined as follows:

Class III. Essentially inert and essentially insoluble industrial solid waste, usually including materials such as rock, brick, glass, dirt, certain plastics, rubber, etc., that are not readily decomposable

Class II. Organic and inorganic industrial solid waste that is readily decomposable in nature and contains no hazardous waste materials

Class I. All waste materials not classified as class II or III, normally including all industrial solid waste in liquid form and all hazardous wastes

The subgrouping here segments the nonhazardous wastes, however, rather than the hazardous wastes. The latter category is subsequently dealt with through a prioritization ranking. Each class I waste is assigned a hazard index (HI), calculated as

$$HI = \frac{50}{\sum\limits_{1}^{N} (C_i / Tox_i)}$$

where C_i is concentration of a constituent in the liquid fraction of a waste or in the leachate, and Tox_i is toxicity of the constituent expressed in terms of the oral LD_{50}, LD_{Lo}, or TD_{Lo}.

This approach involves the development of a waste stream toxicity value and direct comparison between wastes on that basis. As such, it considers only oral toxicity and is a derivation of Finney's mathematical model (1977) for additive joint toxicity, wherein

$$\frac{1}{LD_{50} \text{ waste}} = \sum_{x=1}^{n} \frac{P_x}{LD_{50A}}$$

where P_x is the fraction of constituent X in the waste. In addition to a rating system, this formulation allows one to calculate waste toxicity from published data or constituents without having to bear the cost of direct bioassay. In this respect, it is analogous to the waste toxicity calculation procedure offered as phase II criteria evaluation by the state of California.

In an analysis of potential RCRA site failures and their effect on fund balance in the PCLTF, Battelle devised a simplified scheme for ranking hazardous waste constituents by hazard (ICF, 1984). The rating R was derived as

$$R = [\log (S/cc) - \log (Kd)]$$

where $S =$ solubility of the constituent, $cc =$ critical concentration of the constituent, either detection limit or toxic threshold depending on the triggering event of interest, and $Kd =$ distribution constant for the constituent between soil and water.

The rating was based on breaches of contaminant and subsequent contamination of drinking water supplies. As a consequence, the cc values refer to human health concerns. In essence, the rating prioritizes constituents by mobility, that is, the earliest that arrives at a toxic concentration receives the highest rating.

Categorization of wastes is also being contemplated as an outcome of landfill ban regulations. Groups such as EPA and the state of California are evaluating criteria to identify those wastes that will be banned from land disposal. This constituents a degree of hazard concept for a subset of waste disposal options.

Numerous other approaches to prioritization can be and have been devised. All have several elements in common. The first is the selection of hazards of concern to be considered. The second is the mechanism by which exposure probability is evaluated. The third is the selection of a weighting system for integrating inputs on the various hazards. In all three areas there is opportunity for employing arbitrary index values or quantitative data. Subjective judgment enters in and thereby creates a bias or prejudice in the ranking based on the orientation of the developer. Hence, even when two systems are devised to use the same input data, they may result in significantly different rankings. Klee (1976) has suggested that this may be avoided by use of an additive utility model that provides an explicit and logically consistent structure for assessment. To date, such an approach has not been applied as a part of the hazardous waste regulation development process. However, the rapidly developing area of chemical risk assessment appears to offer an attractive framework for equitable prioritization.

European approaches to categorization of wastes reflect their heavy commitment to listing approaches. For instance, Article L231-6 of the French labor law established five categories of waste: (1) chemical compound groups, as well as a list of some 200 specific chemicals, (2) radioactive contaminated materials, and (3–5) selected industry-specific wastes.

Similarly, wastes in the Netherlands are grouped as follows:

Class A: Mercury and cadmium compounds, oil compounds of comparable damage potential; includes toxic, bioaccumulative, and persistent materials present at levels greater than 50 mg/kg.

Class BI: Heavy-metal compounds whose toxicity occurs at levels greater than those for class A; chemical wastes when concentration exceeds 5000 mg/kg, as well as organic compounds whose extraction, manufacture, or synthesis is industry-related, other than polymers.

Class BII: Elements or compounds that release acute, volatile toxic materials when contacted with water.

Class C: Inorganic compounds present at levels in excess of 20,000 mg/kg, and some organic compounds similar to those in class BI.

Class D: Reactive elements and compounds (except sulfur) that are not toxicologically harmful, and inorganic acids with corrosive properties at concentrations greater than 50,000 mg/kg.

An issue related to prioritization of hazardous wastes is the determination of minimum exemption quantities. The initial EPA draft regulations and some states have implied that small quantities of hazardous wastes are of sufficiently low priority that they can be exempted from regulation. Exemption quantities have been defined as anywhere from 1 lb of category A waste (Washington) to 100 kg ($\sim$ 220 lb) per month (EPA). (1000 kg/month prior to the 1984 RCRA amendments). In the latter case, a distinction is made for extremely hazardous wastes, where the exemption quantity is reduced to 1 kg. Prior to 1980, the EPA proposed no distinction. As a result, a generator would have been exempted if he produced 990 kg/month of dioxin, one of the most toxic chemicals known to man. The Regulatory Analysis Review Group took EPA to task on this proposal. They concluded that small-quantity exemptions are justified only when tied to consideration of the degree of hazard involved. Exemption quantities should be smaller for the more hazardous wastes, in a fashion similar to the categorization scheme used by Washington (i.e., 1 lb for category A, 10 lb for category B, and 100 lb for category C). Once again, at the core of the question is the trade-off between the economic impact of complying with regulations and the net social benefit of regulating small quantities of waste. The case for tying exemption quantities to degrees of hazard is underscored, because the net social benefit goes up with the intrinsic hazard of the waste.

Similar considerations have also given rise to the declaration of special wastes. Because of large volumes and the massive potential economic impact, regulatory authorities have taken to exempting certain extremely high volume wastes that might otherwise be designated hazardous. Creation of these special categories allows less restrictive and less costly disposal practices to be used. The proposed RCRA regulations have specified three types of wastes for exemptions: (1) sludges from publicly owned treatment plants, (2) agricultural crop residues and manures intended for spreading back on the land, and (3) mining overburden and tailings scheduled for return to the mine site. In addition, the following wastes have been classified as special wastes subject to a different set of standards: (1) cement kiln dust, (2) utility waste (flyash, bottom ash, and scrubber sludge), (3) mining wastes (including uranium tailings), and (4) gas and oil drilling muds and oil production brines.

Epilogue

From the preceding, it should be clear that defining, classifying, and prioritizing hazardous wastes can be a complex and demanding task. At the same time, it is an important undertaking, because it will determine the size of the regulatory net and hence the economic impact of hazardous waste regulations. Various approaches have been proposed and attempted. No single approach is right. All have merits. Some are better than others in that they come closer to meeting the underlying objective: segregating a subset of the solid waste stream that requires special management practices in order to prevent unnecessary impacts on health and environmental quality. With the passage of RCRA, the stage is set for the EPA to create a foundation system for national use, and this would appear to diminish the need for in-depth study of approaches and continued investigations into alternatives. However, the legislation clearly notes that states must adopt equivalent programs *at a minimum*, but they are not precluded from opting to be more restrictive (that would constitute use of more inclusive definitions). Hence, there is a continuing need for evaluation and selection of alternatives. Indeed, even in the context of the federal criteria, the apparent emerging shortfall in disposal sites will provide incentive for continual evaluation of definitions to assure that the public good is served. Beyond the management of wastes under RCRA, there also looms the challenging task of rectifying all regulations and legislation pertaining to hazardous materials control. This ultimately begs the development of compatible definitions to put all categories of toxic/hazardous pollutants/contaminants/materials/substances in a consistent framework.

REFERENCES

Arthur D. Little, Inc. 1972. *Alternatives to the Management of Hazardous Wastes at National Disposal Sites.* EPA contract No. 68-01-0556.

Battelle Memorial Institute, Pacific Northwest Laboratories. 1976. *Toxicological Criteria for Defining Hazardous Wastes.* Minnesota Pollution Control Agency.

Booz Allen Applied Research, Inc. 1972. *Study of Hazardous Waste Materials, Hazardous Effects, and Disposal Methods.* Report to the U.S. Environmental Protection Agency under contract No. 68-02-0032.

California Department of Health. 1978. *California Characterization and Assessment System for Hazardous and Extremely Hazardous Wastes.*

Chemical Waste Committee. 1976. *A legislative Guide for a Statewide Hazardous Waste Management Program.* Washington, D.C.: National Solid Waste Management Association.

Collins, H. F. 1975. Experiences of a state hazardous waste program. In *Proceedings of the International Waste Equipment and Technology Exposition.* Los Angeles: National Solid Waste Management Association, pp. 137–142.

DeWalle, F. B.; Zeisig, T.; Sung, J. F. C.; Norman, D. M.; Hatlen, J. B.; Chian, E. S. K.; Bissel, M. G.; Hayes, K.; and Sanning, D. E. 1981. *Analytical Methods Evaluation for Applicability in Leachate Analysis.* PB 81-172 306. Springfield, Virginia: National Technical Information Service, May 1981.

Environmental Quality—1977. 1977. Eighth annual report to the Council on Environmental Quality, Washington, D.C.; December 1977.

Finney, R. A. 1977. Industrial solid (hazardous) waste guidelines. *Waste Age,* Vol. 8, No. 4, pp. 20–27, April 1977.

Garrett, B. C.; McKown, M. M.; Miller, M. P.; Riggin, R. M.; and Warner, J. S. 1981. Development of a solid waste leaching procedure and manual. In D. W. Schultz (ed.), *EPA Symposium on Hazardous Waste Research.* Springfield, Virginia: National Technical Information Service, pp. 9–17.

Governor's Task Force on Solid Waste Management. 1970. *Selected Problems of Hazardous Waste Management in California.*

ICF, Inc.; and Battelle Pacific Northwest Laboratories, 1984. "Post-Closure Liability Trust Fund Simulation Model," U.S. EPA.

Jackson, D. 1984. Comparison of column and batch methods for predicting composition of hazardous leachate. In D. M. Goertemoeller and N. P. Barkley (eds.), *Land Disposal of Hazardous Waste, Proceedings of the Tenth Annual Research Symposium.* EPA-600/S9-84-007.

Klee, A. J., 1976. Models for evaluation of hazardous waste. *Journal of the Environmental Engineering Division,* ASCE, Vol. 102, No. EE1, pp. 111–126, February 1976.

Kuchta, J. N.; Burgess, D. 1970. *Recommendation of Flash Point Method for Evaluation of Flammability Hazard in Transportation of Flammable Liquids.* Safety Research Center report No. 54131.

McKown, M. M. 1981. Chemical classification and characterization. Paper presented at the Battelle Technical Inputs to Planning Conference, Columbus, Ohio, March 18–19, 1981.

Melhoff, L. C.; Cook, T.; Knudson, J. 1979. A quantitative approach to classification of hazardous wastes. In *1979 Sanitation Industry Yearbook.* New York: Solid Waste Management, Communication Channels, Inc., Vol. 21, No. 13, pp. 70–86, December 15, 1979.

National Academy of Sciences. 1973. *Evaluation of the Hazard of Bulk Water Transportation of Industrial Chemicals (A Tentative Guide).* Washington, D.C.: U.S. Coast Guard.

National Fire Protection Association, 1973. *Fire Protection Guide on Hazardous Materials,* 5th ed. Boston: NFPA.

National Institute for Occupational Safety and Health. 1975. *An Identification System for Occupationally Hazardous Materials.* DHEW publication No. (NIOSH) 75-126.

NATO. 1977. *Disposal of Hazardous Wastes—Manual on Recommended Procedures for Hazardous Waste Management.* NATO, No. 62, prepared by the U.S. and Canada, June, 1977.

Office of Solid Waste. 1980. *Test Methods for Evaluating Solid Wastes, SW-846.* U.S. Environmental Protection Agency.

Patrick, P. K. 1975. Treatment and disposal of hazardous wastes in Western Europe. *Journal of Hazardous Materials,* Vol. 1, No. 1, pp. 45–58, September 1975.

Pavoni, J. L.; Hagerty, J. D.; and Lee, R. E. 1972. Environmental impact evaluation of hazardous waste disposal in land. *Water Resources Bulletin,* Vol. 8, No. 5.

Sax, N. I. 1968. *Dangerous Properties of Industrial Materials,* 3rd ed. New York: Van Nostrand Reinhold.

Smyth, H. F.; and Carpenter, C. P. 1944. The place of the range-finding test in the industrial toxicology laboratory. *Journal of Industrial Hygiene and Toxicology,* Vol. 26, pp. 269–273.

Smyth, H. F.; and Carpenter, C. P. 1948. Further experience with the range-finding test in the industrial toxicology laboratory. *Journal of Industrial Hygiene and Toxicology,* Vol. 30, pp. 63–68.

Smyth, H. F.; Seaton, J.; and Fischer, L. 1941. The single dose toxicity of some glycols and derivatives. *Journal of Industrial Hygiene and Toxicology,* Vol. 23, No. 6, pp. 259–268.

Smyth, H. F.; Carpenter, C. P.; and Weil, C. S. 1949. Range-finding toxicity data: list III." *Journal of Industrial Hygiene and Toxicology,* Vol. 31, pp. 60–62.

Smyth, H. F.; Carpenter, C. P.; and Weil, C. S. 1951. Range-finding toxicity data: List IV. *AMA Archives of Industrial Hygiene and Occupational Medicine,* Vol. 4, pp. 119–122.

Smyth, H. F.; Carpenter, C.P.; Weil, C. S.; and Pozzani, U. C. 1954. Range-finding toxicity data: list V. *AMA Archives of Industrial Hygiene and Occupational Medicine,* Vol. 10, pp. 61–68.

Smyth, H. F.; Carpenter, C. P.; Weil, C. S.; Pozzani, U. C.; and Striegel, J. A. 1962. Range-finding toxicity data: list VI. *American Industrial Hygiene Association Journal,* Vol. 23, pp. 95–107.

Smyth, H. F.; Carpenter, C. P.; Weil, C. S.; Pozzani, U. C. Striegel, J. A.; and Nycum, J. S. 1969. Range-finding toxicity data: list IV. *American Industrial Hygiene Association Journal,* Vol. 30, pp. 470–376.

U.S. Environmental Protection Agency. 1977. *State Decision-Makers Guide for Hazardous Waste Management.* U.S. EPA.

4

Quantifying Hazardous Wastes

Introduction

Given the proper legislative backdrop and a clear definition of hazardous wastes, the federal, state, or local official is ready to design a hazardous waste management program. The size and sophistication of that program will depend in part on the types and quantities of wastes to be managed. It follows that early in the developmental stages of any program it is advisable to initiate a survey effort to identify and quantify the hazardous wastes within that area of interest. Many such surveys have been conducted at all levels of aggregation using a variety of techniques. A review of experiences resulting from this previous work can be valuable for designing subsequent surveys as well as for providing better insight into the data obtained from them. Once one has a clear concept of what is required to obtain a good information base, the data bases themselves have more meaning.

Surveying Hazardous Wastes

The accuracy of data obtained during hazardous waste surveys appears on the surface to be a function both of the level of aggregation of the study and of the survey technique employed. In the case of the former, however, variations in the level of accuracy are directly controlled by the surveyor and the resources available to him. A national survey can be conducted as a series of very detailed local studies and subsequently can result in more definitive data if sufficient resources are available. The larger levels of aggregation (i.e., national and regional) have typically provided less accurate data for two reasons: (1) The objectives of the effort were more generic in scope and did not require great detail or accuracy. (2) The resources available

107

for the survey were not sufficient to support the level of detail obtained in more localized efforts.

The same is not true for variations in accuracy due to the survey technique employed. The latter variations are representative of the technique and can be modified only by altering that technique. Hence, in reviewing data bases, it is far more important to determine the technique employed to obtain the impact than to consider the level of aggregation reviewed.

The individual survey techniques that have been employed in the past were developed to meet different desired levels of accuracy, as well as to function within the discrete resources available (time, funding, staff, and staff skills). As such, they represent the developer's solution to a trade-off question between resource allotment and output accuracy.

Survey techniques can be characterized in such way as to indicate which provide greater accuracy. To accomplish this, each technique is described with respect to four features:

1. Medium employed to collect data
2. Extent to which the population of interest is sampled
3. Extent to which the responder is preconditioned
4. Incentives available to encourage accurate responses

Within each parameter, specific techniques will vary from quite general to quite accurate, as illustrated in Figure 4–1. Discussions of each parameter follow.

Data Collection Medium

Hazardous waste surveys are generally conducted by one of four means: (1) use of industrial experts or published work, (2) dissemination of mailed questionnaires, (3) conduct of telephone inquiries, and (4) personal interviews and site visits. The four approaches are listed here in order of ascending degree of accuracy. Similarly, the order also reflects ascending costs and demands on other resources. Selection between the four must be based on an accurate matching of the resources required and the resources available. To date, most surveyors have focused on the use of mailed questionnaires.

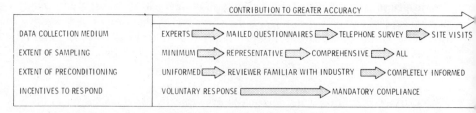

Figure 4-1 / Effect of techniques on survey accuracy.

The use of experts and published data is predicated on the assumption that sufficient knowledge exists to identify hazardous waste types and quantities for a representative facility in each industry. Data can then be extrapolated across the industry. Obviously, this approach is tied directly to a low level of sampling within an industrial category. Specific sources of information can include former industrial officials familiar with plant operations, environmental specialists, trade association staff, texts on industrial waste, reports from surveys conducted in other geographic areas, and calculations based on known stoichiometry for processes. The major weaknesses of such an approach relate to the lack of interaction between reviewer and generator, inconsistencies in the data bases, and the selection of "credible" experts.

The first major hazardous waste survey conducted in the United States employed the expert opinion approach (Battele Memorial Institute, 1973). The objective of this effort was to provide an order-of-magnitude estimate of hazardous waste volumes in the United States as input to the 1973 report to Congress. The broader nature of the scope and the limited time available to meet the congressional mandate necessitated use of this less accurate approach. Data sources included all of those listed earlier, save previous inventories. Waste profile sheets were developed describing each hazardous waste, annual production quantities, and distribution patterns. A summation of these was then made to estimate total national generation figures, waste form summaries, and geographical subtotals.

Mailed questionnaires are the most common data collection technique employed in hazardous waste surveys. Standard forms are devised and sent to all known or suspected generators of wastes, who are requested to provide data on those wastes. Returned information is then tabulated and extrapolated to account for those who did not receive questionnaires or failed to respond. Use of the questionnaire allows information flow from the source and hence can be more accurate than use of experts, but it also relies heavily on a clear understanding of requirements by the generator and hence careful design of the forms employed. The approach suffers from lack of interaction between the respondent and the requestor. The quality of data obtained is also linked closely to the cooperation of the respondent. Voluntary questionnaire surveys of this kind often underestimate waste generation levels because respondents are reluctant to highlight waste problems and bring on subsequent regulatory problems. This problem is exacerbated by the fact that generators often do not have a good understanding of their own wastes.

Mailed surveys have been conducted in a number of states, including Ohio, California, Texas, and Minnesota. The Minnesota survey was conducted in 1976 and illustrates many of the key features of efforts in other states. The survey itself was developed by a joint committee formed by representatives of the sponsoring agency, the Minnesota Pollution Control

Agency (MPCA), and the Minnesota Association of Commerce and Industry (MACI). Questionnaires were sent out under the MACI letterhead, and respondents were assured that data sources would not be disclosed when raw data were passed on to the MPCA. In this way, generators would not feel threatened by direct regulatory action as a result of providing accurate responses. A copy of the questionnaire and other materials supplied in the base survey package is provided in Appendix D. Recipients were not provided any detailed definition of what constituted a hazardous waste. Decisions on the inclusion or exclusion of reported wastes were left to agency staff through case-by-case analysis.

The integrity of data obtained in surveys of this kind depends to a great extent on the size of the sample population. In Minnesota, 15% of the recipients responded to the survey representing 26% of the employment (Battelle Memorial Institute, 1977) (Table 4-1). From the foregoing statistics, it would appear that the larger companies were more prone to respond. This has been attributed to their high public visibility and the greater availability of environmental staff to provide inputs. The survey response in Ohio included 4285 returns, which was 33% of the industries contacted (*Waste Age*, 1976). Hazardous waste generation was declared by 10% of the companies surveyed. In both cases, response was deemed adequate for extrapolation purposes with respect to most industrial classifications. Because of poor response in several specific industrial categories, Minnesota chose to supplement the survey with a limited number of direct interviews and with data from surveys in other states.

Telephone surveys offer the opportunity for direct interplay between the reviewer and the respondent. This provides a means of clearing up uncertainties in how to respond and allows the reviewer to gain greater detail on points of interest. The survey can be conducted in as formal or informal a manner as the reviewer desires. Each call may be entirely different in approach and format as long as key items are discussed. Formal presentations may involve the use of a written questionnaire that is merely recited over the telephone, with responses being recorded. Informal presentations may consist of relatively free-form conversations. Experience suggests that many respondents are reluctant to provide information over the telephone.

Table 4-1 / Response to the Minnesota Industrial Hazardous Waste Survey.[a]

	Number of Firms	Percentage of Total	Employment	Percentage of Total
Survey recipients	5568	100%	242,858	100%
Respondents	857	15%	63,500	26%
Respondents declaring hazardous wastes	139	2%	28,475	12%

[a]Adapted from Battelle Memorial Institute (1977).

As a consequence, telephone surveys often must be followed by written correspondence requesting specific information of company officials. On the other hand, it should be noted that whereas written requests are directed through official channels, telephone interviews often can be conducted with plant personnel, who are less cautious about revealing information about the operation. In these cases, data obtained often are of greater accuracy and value than data that come through formal written responses. Indeed, in some cases, plant personnel provide data that management has declined to offer. The value of telephone surveys over mailing has been demonstrated by some states, with mailing alone producing less than 50% responses, but telephone follow-up increasing the response to as much as 70% (U.S. EPA, 1977).

Telephone surveys have been conducted in Massachusetts. The most extensive hazardous waste telephone survey completed to date was conducted for EPA Region X in the states of Washington, Oregon, Idaho, and Alaska. The survey itself was initiated with the use of indirect procedures to focus the subsequent telephone interviews. The first step in this procedure was the compilation of a list of industries in the region with the *potential* to generate hazardous wastes. Industries were selected on the basis of their Standard Industrial Classification (SIC) (U.S. Office of Management and Budget, 1972). An SIC list corresponding to industries that had the potential to generate hazardous wastes was cross-referenced with the manufacturing directories of the states of Alaska, Idaho, Oregon, and Washington (Alaska Department of Economic Development, 1974; Idaho Department of Commerce and Development, 1973; Oregon Department of Economic Development, 1974; Washington Department of Commerce and Economic Development, 1974). This cross-referencing resulted in a list of industrial operations within the four states that were considered to have the potential to generate hazardous wastes.

Given the list of candidate industrial sources, the effort was directed to the identification of hazardous waste quantities and management practices in the industrial sector. This constituted the bulk of the study effort. Each of the industries identified as a potential generator of hazardous wastes was considered individually, using inputs from the following data sources:

State manufacturing directories. Aside from identifying individual firms in a category, these directories typically gave brief descriptions of the size of the operation, employment level, basic products, and, in some instances, the type of process.

State surveys. The states of Idaho, Oregon, and Washington had previously conducted industrial solid and hazardous waste studies of varying scope and depth (Idaho Department of Environmental and Community Services, 1973; Oregon Department of Environmental Quality, 1974; Washington Department of Ecology, 1974). The data included hazardous waste quantities and current disposal practices.

NPDES permits. The list of potential industrial hazardous waste generators was crossed with the National Pollution Discharge Elimination System (NPDES) files maintained at the EPA regional office in Seattle. These files contained information identifying specific manufacturing and waste treatment procedures used by individual companies. Information from "trip reports" often included discussions of airborne and solid waste management. Data on plant effluent characteristics were also useful indicators of the existence of a hazardous solid waste, as well as the potential for increased sludge volumes with stricter water pollution controls in 1977 and 1983.

National industrial surveys. Data were extracted from completed and ongoing industrial hazardous waste studies sponsored by the EPA Office of Solid Waste Management Programs. These studies provided (1) general background information on hazardous wastes produced by selected industries and (2) specific information on some of the Pacific Northwest hazardous waste generating industries that had been surveyed directly. Collection of the latter data required direct contact with the individual contractors who conducted the various studies, because the final reports submitted to the EPA did not characterize hazardous wastes from individual industrial operations.

Municipalities. Some of the larger municipalities in the Pacific Northwest maintain industrial effluent monitoring programs. Data collected from these municipalities helped identify hazardous effluent constituents and determine which industrial operations within the municipal jurisdiction were practicing effluent pretreatment.

Waste processors. Records maintained by some of the waste processing companies within the region were used to identify and quantify the wastes from a number of industrial sources.

The literature. Various reports and publications were located and used to provide background material and operating characteristics for specific industries.

The data collected from the aforementioned sources allowed for division of the candidate industries into four basic groups.

1. Known hazardous waste generators
2. Suspected hazardous waste generators
3. Industries for which little was known
4. Industries known not to generate hazardous wastes

At this point, the direct telephone survey was initiated. All firms in group 1 were contacted. No firms in group 4 were contacted. Spot contacts with individual firms in groups 2 and 3 were made. If these indicated hazardous waste generation potential for an SIC category, all firms in that industrial category were contacted. Firms were asked to provide information relating to the following:

Type of operation
Number of employees
Products and production capacity
Hazardous wastes generated
Hazardous waste disposal practices
Effluent characteristics, including treatment practices

The interviewer was provided a sheet for each firm giving all of the data collected from the sources listed earlier. This allowed the interviewer to anticipate responses, verify questionable data, and probe areas that might otherwise be glossed over. After the initial telephone contact, a "profile" on the company was prepared. This profile contained a summary of pertinent information derived from the aforementioned sources and the telephone survey. Copies of these profiles were sent to each industry for inspection and comment as to accuracy and completeness.

As noted previously, many firms required written requests before releasing data. However, quantitative information was received from over 90% of the firms contacted. Response to the mailed profiles was considerably poorer. The researchers concluded that direct interaction between parties ensured a higher degree of responsiveness (Stradley et al., 1975; Dawson and Stradley, 1976).

The most comprehensive approach to conducting a hazardous waste survey is the use of direct interviews. The study team can capitalize on direct interaction with the generator, as in the case of the telephone survey, but they can also review records, observe plant operations, and meet with a broader cross section of plant personnel. Once again, the format can range from informal discussions to walk-through inspections and joint completion of detailed questionnaires. Success rates of over 90% have been reported for on-site surveys (U.S. EPA, 1977). The major drawback with this technique is cost. It is estimated that a trained interviewer can inspect 20 plants per month.

Direct interviews have been conducted by Alabama, Tennessee, Florida, Washington, and the federal government. Minnesota used direct interviews to verify data and fill in gaps identified during an earlier mail survey effort (Battelle Memorial Institute, 1977). The state of Washington interviewed some 450 firms as part of an overall industrial waste survey in 1974. (Washington Department of Ecology, 1974). The federal surveys were part of an extensive program to improve on the data base employed in the report to Congress. Surveys were conducted through contractors selected for specific industrial categories, including the following:

Primary metals smelting and refining
Paints and allied products
Organic chemicals, pesticides, and explosives
Electrical and electronics

Electroplating and metal finishing
Inorganic chemicals
Rubber
Batteries
Pharmaceuticals
Textiles, dyeing, and finishing
Petroleum refining
Special machinery
Leather tanning and finishing
Plastics
Waste oil re-refining

Representative firms were identified to yield data covering all economic regions, sizes, and technologies common to the industry. Teams were then sent to interview plant personnel and make direct observations on hazardous waste management practices. Extensive trip reports were filed for use in determining waste production rates, but data were not reported until after aggregation on a regional or state basis. Thus, sensitive data were protected. The number of actual visits varied between industries but never exceeded 7%, except for the waste oil re-refining industry. When generators refused to participate, contractors were required to estimate waste production levels using expert opinion. Experience gained in the conduct of these federal industrial surveys was discussed in EPA publications (U.S. EPA, 1977) and is reflected in the model data collection forms provided in Appendix E.

Sample Size

A second variable in which the survey team can balance trade-offs between cost and accuracy is the size of the sample taken. Data can be collected from review of a single firm or from contact with every firm in an industrial category. Limitation to something less than the entire population may be a matter of choice (e.g., Washington elected to interview only 600 of the industrial establishments in the state) or a reflection of the cooperation received from the generators (e.g., 857 of the 5568 survey recipients responded in Minnesota). In either case, construction of a picture of actual hazardous waste generation rates from the sampling data requires an extrapolation technique.

The survey conducted in Florida (Carter et al., 1979) serves as a good example of how sample size can be selected. For that effort, sample size for each SIC category was selected according to the following:

In the event that no more than 30 firms are in a particular stratum (2-, 3-, or 4-digit), a census must be taken for that stratum.

In the event that more than 30 firms are in a particular SIC stratum (2-, 3-, or 4-digit), a minimum of 30 firms must be surveyed in that stratum.

No stratum is to contain less than three firms.

In the event a 4-digit SIC stratum contains less than three firms, the 4-digit SIC codes are to be collapsed to a 3-digit stratum such that the 3-digit category contains more than the minimum number of firms (continue to 2-digit level if necessary).

Census (or complete inventory) was attempted, when possible, to reduce uncertainty. Use of data from strata with three or more firms only was required to assure confidentiality of data from specific firms. Based on this approach, samples were allocated as illustrated in Table 4-2. In the end, only 352 of the 433 firms initially selected were interviewed directly. Another 20 firms were interviewed by telephone, and 40 could not be located. Some 14 firms refused to participate.

Extrapolation of hazardous waste survey data is accomplished by converting existing generation rate data to a waste generation factor (WGF) defined as

$$WGF = AGR/RP$$

where AGR is annual generation rate and RP is representative parameter indicative of the sizes of individual generators and the industry as a whole. Numerous approaches have been employed in identifying a representative parameter. A common choice is total production level. When production data are utilized, the waste generation factor becomes a measure of waste per unit production (e.g., pounds of hazardous waste per ton of product). The advantages and disadvantages of this and other representative parameters are compared in Table 4-3.

Regardless of the extrapolation procedure employed, it is important to note that most industries cannot be grouped together and accounted for as a single entity. Waste generation factors often differ considerably with the size

Table 4-2 / Sample Size Selected for Florida Hazardous Waste Survey.[a]

SIC	Description	Number of Firms	Number of Firms to be Interviewed[b]
2491	Wood preserving	19	19
2753	Engraving and plate printing	16	16
2813	Industrial gases	21	21
2819	Industrial inorganic chemicals, NEC	27	27
282	Plastics, synthetic resin, rubber and man-made fibers	14	14
2821	Plastic material and synthetic resins		
2822	Synthetic rubber		
2834	Pharmaceutical preparations	16	16
2851	Paints, varnishes, lacquers, enamels, and allied products	74	40

Table 4-2 / (*Continued*)

SIC	Description	Number of Firms	Number of Firms to be Interviewed[b]
2869	Industrial organic chemicals, NEC	4	4
2879	Pesticides and agricultural chemicals, NEC	24	24
2891	Adhesives and sealants	12	12
2892	Explosives	4	4
2893	Printing ink	13	13
2952	Asphalt felts and coatings	9	9
3021	Rubber and plastic footwear	5	5
3041	Rubber and plastic hose and belting	4	4
3069	Fabricated rubber products, NEC	20	20
3111	Leather tanning and finishing	6	6
3292	Asbestos products	5	5
3312	Blast furnaces, steel works, and rolling mills	10	10
333	Primary smelting and refining of nonferrous metals	4	4
3332	Primary smelting and refining of lead		
3333	Primary smelting and refining of zinc		
3339	Primary smelting and refining of nonferrous metals, NEC		
3341	Secondary smelting and refining of nonferrous metals	6	6
335	Rolling, drawing, and extruding of nonferrous metals	5	5
3356	Rolling, drawing, and extruding of nonferrous metals, except copper and aluminum		
3357	Drawing and insulating of nonferrous wire		
3412	Metal shipping barrels, drums, kegs, and pails (drum refinishers)	6	6
3471	Electroplating, plating, polishing, anodizing, and coloring	63	35
348	Ordnance and accessories, except vehicles and guided missiles	5	5
3482	Small arms ammunition		
3483	Ammunition except for small arms, NEC		
3612	Power, distribution, and specialty transformers	12	12
3613	Switchgear and switchboard apparatus	11	11
367	Electronic components and accessories	8	8
3675	Electronic capacitors		
3677	Electronic coils, transformers, and other inductors		
3691	Storage Batteries	10	10
3692	Primary batteries, dry and wet	4	4
3732	Boat building and repairing	268	20
3861	Photographic equipment and supplies	21	21
8221	Universities	17	3
971	National security	15	6
—	Captive electroplating shops	—	10
Total		*758*	*435*

[a]Adapted from Carter et al. (1979).

[b]Three of the captive electroplaters also appear in SIC 971.

Table 4-3 / Relative Desirability of Various Parameters as a Base for Extrapolation of Hazardous Waste Generation Volumes.

Representative Factor	Advantages	Disadvantages
Production level	Direct measure of the activity that is generating the waste	Data are not always available; industry may hold production as proprietary
Value added	Similar to production; helps eliminate nonproductive activities that do not generate wastes	Data generally available only in the aggregate for large geographical areas
Employment	Readily available on a firm-by-firm basis	Does not segregate sales and administrative staff from productive staff
Income	Tied to production level	Not always available; not directly tied to production (different profit structures for different firms)

of an operation and the processes employed. Hence, it may be necessary to subgroup an industry according to factors affecting waste generation and then deal with each group independently. Data for large corporate firms that can economically install recycling processes should not be combined with those for family businesses. Similarly, recognizing these differences in waste production characteristics, the survey team must analyze and carefully select the sampling points to assure that data are collected for each unique subgroup in an industrial category. This necessitates some familiarity with the industrial processes and structure involved if the survey is to be taken to anything less than the total industry.

Respondent Preconditioning

As with many types of activities, the quality of data obtained from a hazardous waste survey can be greatly enhanced if the respondents are fully cognizant of the pertinent technical issues. For example, the more clearly and quantitatively the surveyor can define what constitutes a hazardous waste, the more likely the generator is to correctly identify the portions of his waste streams that are of interest.

Respondent preconditioning is usually accomplished by one of two methods. In mail surveys, a set of instructions is accompanied by general information, definitions, and regulatory background data to assist the respondent. An example is provided in Appendix D. In telephone surveys and direct interviews, background information is provided during the conversation on an as-needed basis. Both approaches are more productive if they are preceded by a general awareness of the hazardous waste management problem and some previous respondent experience in evaluating the impact of regulations on internal operations. When surveys are taken in areas where little or no awareness has been aroused, respondents often are reluctant to participate and often offer widely divergent opinions of what wastes should be included.

Survey data quality can also gain measurably from preconditioning of the survey team. This is particularly true in telephone and direct interviews, where the team must probe and inquire until they develop a concise picture of the waste production activities. To accomplish this, the team must know what questions to ask. Respondents may reply that they have no hazardous wastes, but if the interviewer can ask how a particular by-product is handled, he may identify important information otherwise lost. In the EPA Region X survey, this was accomplished by providing the interviewer with general information on the industry, results of previous surveys, and data on the individual firm collected from other sources. Many times, the interviewer was also provided a list of specific questions that should be addressed.

Incentives to Respond

The cooperation of respondents and the accuracy of responses can be tied directly to the incentives provided the waste generator. Voluntary responses to informational questionnaires often are low-level and casual. Many potential respondents fear that data supplied will only be used to make regulations more restrictive and to facilitate implementation. In this context, the waste generator has nothing to gain from complying.

The alternative approach is to make compliance mandatory. This is the situation when reporting is required by law. For instance, provisions in RCRA mandate that all hazardous waste generators report the quantity and characteristics of their wastes to the EPA. Similarly, the manifest system designed to track transport and disposal of hazardous wastes obligates the generator to initiate forms on each waste shipped off-site. Copies of these documents constitute an extensive record of hazardous wastes other than those disposed on-site. The latter are enumerated in records on the disposal site itself. Although, practically speaking, generators can and will intentionally fail to report all hazardous wastes, the civil and criminal penalties involved provide economic incentives to properly complete and file required reports and records. Data generated in the state of California are presently some of the best available, because mandatory reporting and manifests are an integral part of the Department of Health's ongoing program. Mandatory manifests in Texas are providing a similar data base in that state. The metropolitan counties of Minnesota encompassing the Minneapolis–St. Paul area also have a good data base, because waste generation permits have been required there for several years. The national data base has improved greatly since the EPA has promulgated regulations giving a clear definition of hazardous wastes and requiring reporting and manifests. Elsewhere, the United Kingdom has legislated a mandatory notification system to provide the necessary data base for management in the future.

Incentives may be provided somewhere short of legal requirements to facilitate cooperation in responding to surveys. Both Minnesota and Ohio designed their survey packages in conjunction with appropriate trade associations: the Minnesota Association of Commerce and Industry; the Ohio Manufacturers Association, Ohio Petroleum Council, and American Petroleum Institute. The accompanying information package included introductory statements from the associations demonstrating how accurate responses could benefit the industry (see Appendix D).

Current Estimates of Hazardous Waste Generation Rates

In 1973, the U.S. Environmental Protection Agency reported to Congress that there were some 10 million tons of nonradioactive hazardous wastes produced in this country each year and that these represented some 10% of the total industrial waste stream. Subsequent intensive industry surveys have suggested that these data greatly underestimate the actual production levels. Results of the EPA industrial category surveys suggested that hazardous waste production was actually closer to 40 million wet metric tons per year, 30 million wet metric tons for the 14 industries surveyed, as detailed in Table 4–3. This can be put in perspective by considering the total estimated residuals generation in the United States, as presented in Figure 4–2. In a September 1983 release, the estimate was quadrupled to 150 million metric tons for the year 1981. This figure excluded wastes from small-quantity generators ($<$ 2000 lb/yr), wastes that were incinerated, and wastes illegally disposed. By 1984 the estimate was raised to 250 million metric tons.

On the basis of the earlier industrial category surveys, the bulk of all potentially hazardous waste is produced in four industrial categories (83% of the 14 industries surveyed): primary metals; organic chemicals, pesticides, and explosives; electroplating and metal finishing; and inorganic chemicals. Caution is warranted when comparing values between industries, because each survey was conducted independently, with varying guidelines with respect to what constitutes a hazardous waste. Hence, the data bases are not always comparable. The currently proposed definition of hazardous waste is less expansive than that used by most contractors. Hence, waste generators' data may be overrated. On the other hand, the 1981 reported waste volume was four times as high as estimated, and 71% of the total waste came from the chemical and petroleum industries (SIC 28–29), 22% from the metal-related

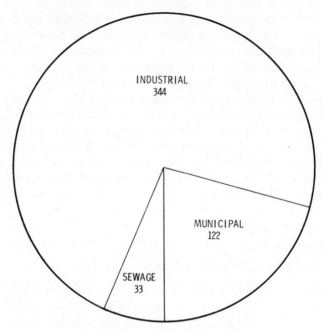

Figure 4-2 / Relative production of residuals in United States (net weight in millions of metric tons per year). (From U.S. EPA, 1976.)

industries (SIC 33-37), and 7% from other industries (Westat, 1984). Within the industries surveyed, potentially hazardous wastes constitute 14% of the total waste stream. This is somewhat higher than the 10% estimated for industries as a whole, and it underscores the rationale for selecting these 14 industries for survey. Between the specific industries, the hazardous waste fraction varies from a low of 5% (inorganic chemicals) to as much as 100% (electroplating, batteries, and petroleum refining).

Military operations in the United States constitute an additional sub-category of hazardous waste generation. Aside from numerous sites that may require exhumation and disposal of contaminated soils, it is estimated that some 160 million tons of military ordnance requires disposal. An additional 10 million tons/yr of waste munitions are produced and 1.1 million tons/yr of naval paint and plating wastes.

The hazardous waste management industry often deals with input materials in five generic groupings: (1) metals/metal finishing, (2) paint/solvents/coatings, (3) organics, (4) petroleum, and (5) inorganics. The makeup of wastes according to this categorization is presented in Table 4-4.

Based on an analysis of the effects of controls required by air and water

Table 4-4 / Sources of Hazardous Wastes by Type.[a]

Waste Disposal Market Category	Estimated Dry Tons Available for Contract Management (millions of metric tons wet)	Industry	Percent of Total for Category	Types of Wastes
Metals/metal finishing	3.5	Batteries	<1%	Acid solutions
		Electroplating	80–90%	Metal-bearing sludges
		Primary metals, smelting, refining	10%	
		Special machinery	3–6%	
		Electronic	2%	
Paints/solvents/coatings	0.1	Paint and allied products	100%	
Organics	2.7	Organic chemicals	80–90%	Organic solvents
		Pharmaceuticals	3–5%	Pesticides
		Rubber and plastics	8–12%	Biological
		Textiles, dyeing, finishing	5%	Plastics, rubber
Petroleum	0.8	Petroleum refining	>87%	Oily wastes
		Waste oil re-refining	<13%	
Inorganics	0.8	Inorganic chemicals	80–90%	Aqueous solutions of salts, metals, etc.
		Leather tanning	10–20%	

[a]Adapted from Foster D. Snell, Inc. 1976.

pollution regulations and projected industrial growth, it was estimated that hazardous waste production for the 14 industries would grow 48% between 1974 and 1983, for an annualized rate of 4.5% per year (Table 4–5). With subsequent changes in effluent guidelines and delays in the implementation of regulations, this growth was not likely realized. The effect of pollution control residuals on total waste quantities can be noted from the projections themseves. In 1974, residuals were estimated at 19.5 million metric tons, or 13% of total industrial wastes for the 14 industries. By 1983, residuals were estimated to reach 35.7 million metric tons, 19% of total industrial wastes. The actual reported value of 250 million metric tons overwhelms the 1983 estimate.

Estimates of future hazardous waste production must be viewed cautiously. They are subject to great change as a result of several factors. Total quantities will depend on the ultimate regulatory definition of hazardous wastes. Surveys to date have been conducted without a clear definition, and thus results may overstate or understate waste production, depending on whether the regulatory definition is broadened or narrowed. Second, it is difficult to anticipate process changes that may be initiated in response to RCRA regulations. Designation of waste as hazardous will raise the associated management costs. This creates new incentives to reduce hazardous waste volumes, which in turn will affect the total quantities of hazardous wastes generated in the future. Present trends suggest that both factors (a narrowing of the regulatory definition and potential high costs for disposal) are reducing the total quantities of hazardous wastes below current projections. Evidence of such changes is available from West Germany, where, despite an increase in hazardous waste management facilities, the amounts of wastes brought to these facilities have been significantly lower than the estimates derived from surveys (Council of Environmental Advisors, 1978).

On a geographic basis, most hazardous wastes from the 14 industries (76%) were produced in EPA Regions III, IV, V, and VI, as depicted in Figure 4–3. As one would expect, this pattern parallels the density of industrial activity within the country. Waste is not distributed evenly on a per capita basis. The national average runs 0.14 ton per year per person, and those for individual states vary considerably, as shown in Table 4–6. Once again, these data may be misleading in that they are based on different definitions of hazardous wastes. Therefore, the most meaningful comparisons are those based on data from a given survey (e.g., the EPA Region X study of Washington, Oregon, Idaho, and Alaska). With respect to individual states, 60% of all hazardous wastes in the United States are produced in New Jersey, Ohio, Illinois, California, Pennsylvania, Texas, New York, Michigan, Tennessee, and Indiana. The self-reported data from 1981 indicated distribution of hazardous waste generators as follows:

Table 4-5 / Projected Growth in Hazardous Waste Production.[a]

Industry	Amount (million metric tons/year)						Percentage Growth 1974-1983[b]
	1974		1977		1983		
	Dry	Wet	Dry	Wet	Dry	Wet	
1. Batteries	0.005	0.010	0.082	0.164	0.105	0.209	2000
2. Inorganic chemicals	2.000	3.400	2.300	3.900	2.800	4.800	40
3. Organic chemicals, pesticides, and explosives	2.150	6.860	3.500	11.666	3.800	12.666	77
4. Electroplating	0.909	5.276	1.316	4.053	1.751	5.260	92
5. Paint and allied products	0.075	0.096	0.084	0.110	0.105	0.145	40
6. Petroleum refining	0.625	1.757	0.715	1.841	0.811	1.888	30
7. Pharmaceuticals	0.062	0.065	0.070	0.074	0.104	0.108	68
8. Primary metals smelting and refining	4.454	8.335	4.732	9.104	5.536	10.418	24
9. Texile dyeing and finishing	0.048	1.770	0.500	1.870	0.179	0.716	373
10. Leather tanning	0.045	0.146	0.050	0.143	0.068	0.214	51
11. Special machinery	0.102	0.163	0.094	0.153	0.157	0.209	54
12. Electronic components	0.026	0.036	0.036	0.078	0.050	0.108	92
13. Rubber and plastics	0.205	0.785	0.242	0.944	0.299	1.204	46
14. Waste oil re-refining	0.057	0.057	0.074	0.074	0.144	0.144	253
Totals (to date)	10.763	28.755	13.795	34.174	15.909	39.089	48

[a]Adapted from U.S. EPA (1977).
[b]Figures based on dry weight quantities.

123

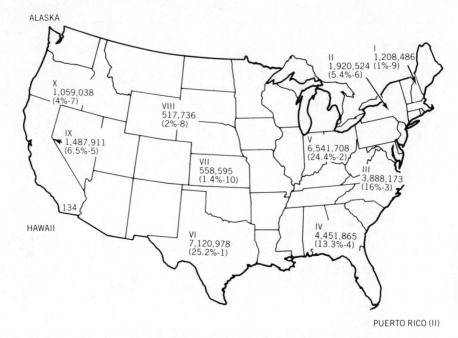

PUERTO RICO (II)

Figure 4-3 / Geographic distribution of hazardous wastes from 14 industrial categories. Production in metric tons for 1974, wet basis.

Region	Number of Hazardous Waste Generators ($\geq$ 2000 lbs/yr)
I	1430
II	1670
III	1400
IV	1830
V	3240
VI	1240
VII	730
VIII	240
IX	1990
X	330

Based on the results of the 14-industry survey, the EPA concluded in 1977 that only 10% of all potentially hazardous wastes was being disposed in an environmentally adequate manner (Table 4–7). The vast majority of all hazardous waste (48%) was placed in unlined impoundments. An additional 30% was placed in nonsecure landfills. The 1981 survey indicated that of the 55

Table 4-6 / Annual Per Capita Hazardous Waste Generation.

Geographic Area	Estimated Hazardous Waste Production (metric tons)	Population	Waste Production Rate (tons per capita)
Unites States total[a]	29,000,000	211,000,000	0.140
Washington[b]	106,000	3,400,000	0.031
Oregon[b]	18,000	2,100,000	0.085
Idaho[c]	6,000	750,000	0.008
Minnesota[b]	136,000	3,100,000	0.035
Alaska[c]	1,000	350,000	0.003
California[d]	1,500,000	21,000,000	0.070
New Jersey[e]	2,455,000	7,500,000	0.305
Massachusetts[f]	151,000	5,800,000	0.026
Florida[g]	574,343	7,200,000	0.080
New York[h]	1,200,000	18,500,000	0.065

[a]Data from U.S. EPA (1977).
[b]Data from Battelle Memorial Institute (1977).
[c]Data from Stradley et al. (1975).
[d]Data from Collins (1974).
[e]Data from Buchanan and Metry (1978).
[f]Data from Fennelly (1977).
[g]Data from Carter et al. (1979).
[h]Data from Booz Allen & Hamilton (1979).

Table 4-7 / Disposition of Hazardous Wastes.[a]

Disposal Practice	Percentage of Total Wet Weight of Potentially Hazardous Wastes[b]
Potentially inadequate	
Unlined surface impoundments	48.3
Nonsecure landfills	30.3
Uncontrolled incineration	9.7
Deep-well injection	1.7
Landspreading	0.3
Use on roads	<0.1
Sewered	<0.1
Total	90.4
Environmentally adequate	
Controlled incineration	5.6
Secure landfills	2.3
Recovery	1.7
Lined surface impoundments	<0.1
Wastewater treatment	<0.1
Autoclaving	<0.1
Total	9.6

[a]Adapted from U.S. EPA (1977).
[b]Based on annual generation in 14 key industries during the period 1973–1975.

million metric tons (MMT) disposed, 32 MMT (58%) went to injection wells, 3 MMT (5%) went to landfills, 19 MMT (35%) went to surface impoundments, 0.4 MMT (1%) went to land farms, and 0.07 MMT (1%) went to other disposal technologies. By far, it is the land that bears the brunt of the burden from hazardous waste management. Once again, variations between geographic areas are considerable. In Minnesota, disposal was divided among landspreading (35%), landfill (28%), recycling (13%), lagooning (10%), incineration (6%), and municipal sewer, hauler, chemical treatment, and other (8%). For EPA Region X, disposition consisted of land applications (landspreading and landfilling) (45%), recycle (37%), lagooning (3%), incineration (<1.0%), and municipal sewer, hauler, chemical treatment, and other (16%). The differences reflect resource availability, cost, climate, and the nature of industrial activities. The vast majority of the total potentially hazardous waste stream represented by 12 of the 14 industries surveyed is disposed of by the generator. As detailed in Table 4-8, only 27% of the national total is currently available for outside contract disposal. This figure can be somewhat deceiving, because only four industries manage less than 50% of their hazardous wastes in-house (organic chemicals, pesticides, and explosives; inorganic chemicals; primary metals smelting and refining; and textiles, dyeing, and

Table 4-8 / Hazardous Wastes Available for Outside Contract.[a]

Industry	Estimated Hazardous Wastes Available for Outside Contract (millions of metric tons per year)		Percentage of Total Hazardous Wastes Available for Outside Contract (wet basis)
	Dry	Wet	
Batteries	0.003	0.006	65%
Inorganic chemicals	0.300	0.510	15%
Organic chemicals, pesticides, and explosives	0.430	1.372	20%
Electroplating	0.636	3.693	70%
Paint and allied products	0.071	0.091	95%
Petroleum refining	0.366	0.780	60%
Pharmaceuticals	0.053	0.055	85%
Primary metals smelting and refining	0.348	0.407	5%[b]
Plastics	0.031	0.147	20%
Rubber	0.046	0.046	95%
Leather tanning and finishing	0.043	0.139	95%
Special machinery	0.097	0.155	95%
Textiles, dyeing, and finishing	0.012	0.098	5.5%
Total	2.445	7.499	27%[b]

[a]Adapted from Foster D. Snell, Inc. (1976).
[b]Figure reflects reduced estimate for primary metals hazardous wastes derived subsequent to publication of the Snell study.

finishing). However, these four industries account for 71% of all potentially hazardous wastes from the 12 industrial categories.

The fraction of wastes managed in-house is likely to change significantly with implementation of RCRA regulations. It is difficult to accurately predict the changes. Growing concern over the liability from incidents such as Love Canal has influenced many firms to maintain custody of wastes by using sites under their own control. On-site disposal will also eliminate the need to use manifest forms for each shipment. On the other hand, siting requirements may be such that most industries will not have suitable sites within their holdings. In that event, transportation to certified contractors will become more attractive.

Based on the variations between geographic areas noted here, it is clear that great care must be taken before data from one survey can be employed for estimation purposes in another geographic area after changes in regulatory programs. Ideally, states and local entities will perform their own surveys when assessing the nature of hazardous waste management within their constituency. If not, they should seek data from areas that have conditions similar to those prevailing at the time of interest.

Although direct comparisons are difficult because of differences in defining hazardous wastes, per capita production in many European countries appears on the order of the 0.14 to 0.56 metric ton/capita/yr reported for the United States. A waste disposal inventory conducted in France by the Environment Ministry estimated annual generation at 14.7 million metric tons or roughly 0.24 metric tons per capita (*Chemical Week*, 1979). Of that, 5.7 million metric tons were derived from the chemical industry. Twenty-seven percent of the toxic or potentially toxic waste is discharged into sewers. On the other hand, Sweden generates some 0.5 million metric tons per year, which is approximately 0.06 metric ton per capita. Production in the United Kingdom is reported to be 21 million tons/yr or 0.38 ton/capita/yr (*Chemical Week*, 1979). Of the total, 75% is disposed in landfills, 15% is incinerated, and 10% is dumped at sea. Some 6524 tonnes/yr of wastes from industry are delivered to country waste disposal authorities for disposition. (Department of the Environment, 1979). These data are at odds with the commission of the European Community working party estimate of 21 million tonnes/yr for all member states (Working Party on Toxic and Dangerous Waste, 1979). But the latter value was based on very incomplete returns to a voluntary survey. These differences may also reflect variations in the wastes that were included for quantification; for instance, the waste production outlined in Table 4–9 (Watson, 1978). It is unclear how a division was made to derive the estimate of 21 million tons/yr.

The difference in per capita production of hazardous wastes in Europe appears to reflect the higher degree of industrialization between nations. It is further interesting to note that the ratio of industrial wastes to domestic

Table 4-9 / Solid Waste Production in the United Kingdom.[a]

Waste Category	Annual Production (tonnes)
Domestic	19
Industrial wastes	45–50[b]
Building wastes	3
Pulverized fuel ash from utilities	12
Mining wastes	66
Quarrying wastes	50
Agricultural wastes	250
Sewage sludge	25

[a]Adapted from Watson (1978).
[b]The Department of the Environment estimates 23 tonnes/yr industrial waste.

wastes is higher (2.82) in the United Kingdom than in the United States (2.63) and appears to reflect a lower usage of packaging and disposables in that society. The ratio of industrial waste to sewage sludge, however, is much lower in the United Kingdom (2) than in the United States (10.4) because of the large proportion of rural, unsewered population in the United States. Waste types also differ significantly between nations. In Poland, for instance, 80% of the wastes (30 million tons per year) are metallic because of the heavy emphasis on plating and metalworking (*Environmental Science and Technology*, 1980).

Table 4-10 / Industrial Waste Production in Three English Counties.[a]

SIC Order Number	Industrial Grouping	Cheshire	Staffordshire	Shropshire
III	Food, drink, and tobacco	11.5[b]	18.0	49.6
IV	Coal and oil processing	11.3	3.6	—
V	Chemicals and allied industries	38.4	8.8	0.7
VI	Metal manufacture	6.4	45.3	15.8
VII	Mechanical engineering	0.57	3.6	2.4
VIII	Instrument engineering	—	0.3	0.2
IX	Electrical engineering	0.51	1.3	3.8
X	Shipbuilding	—	1.3	—
XI	Vehicle manufacture	0.85	2.4	0.87
XII	Other engineering industries	2.9	3.6	5.0
XIII	Textiles	0.57	0.74	2.6
XIV	Leather and leather goods	—	1.7	—
XV	Clothing and footwear	0.99	0.67	0.30
XVI	Bricks, pottery, glass, and cement	14.4	4.3	42.9
XVII	Timber and furniture	6.6	—	—
XVIII	Paper, printing, and publishing	16.4	4.5	1.3
XIX	Other manufacturing industries	1.3	1.5	3.8

[a]Adapted from Department of the Environment (1979).
[b]Results are shown as annual waste production per employee (tonnes) for the manufacturing industries.

Table 4-11 / Relative Production of Hazardous Waste Types in the United Kingdom.[a,b]

Waste Category	Cheshire (%)	Essex[c] (%)	Greater London[c] (%)	West Midlands[d] (%)
Acid wastes	0.4	7.2	2.1	9.9
Alkaline wastes	2.1	2.2	7.6	2.4
Metallic sludges or liquids	3.8	8.2	10.8	30.9
Oily wastes	7.9	13.0	14.1	22.3
Tarry wastes	0.8	1.6	3.6	—
Solvent wastes	3.3	1.2	1.9	0.4
Other sludges or liquids	40.4	46.7	25.9	8.6
Miscellaneous organic materials	2.2	1.5	6.4	—
Cyanide-bearing waste	—	0.1	1.3	0.9
Asbestos	5.3	—	12.9	—
Metal-bearing solid waste	—	—	7.1	21.8
General contaminated factory waste	4.5	—	6.3	1.8
Inert, nonnotifiable, or miscellaneous waste	29.2	18.3	—	—
Total (%)	100.0	100.0	100.0	100.0
Total (thousand tonnes/annum)	630[e]	570[e]	266	270

[a]Adapted from Department of the Environment (1979).
[b]This indicates the rough proportions of different wastes that were notified in four areas of England during some part of 1973 or 1974.
[c]Much of the waste that arises in Greater London is deposited in Essex.
[d]Covers the areas of Birmingham, Aldridge, Meriden, and West Bromwich only.
[e]The annual totals are extrapolated from a 6-month total for Cheshire and a 1-month total for Essex.

Some perspective on the waste production of individual industrial segments in the United Kingdom can be found in Table 4-10, where survey data for three counties are compared. Waste types are enumerated in Table 4-11. With respect to disposition, data indicate that 94% of the wastes in Cheshire were disposed of to the land, whereas 4% went to sea and 2% were incinerated. There is some indication that these relative percentages may not be representative of Britain as a whole, but the heavy emphasis on land disposal is characteristic.

REFERENCES

Alaska Department of Economic Development. 1974. *Directory of Alaska Commercial Establishments.* Juneau: ADED, July 1974.

Battelle Memorial Institute. 1973. *Program for the Management of Hazardous Wastes.* U.S. Environmental Protection Agency, contract No. 68-10-0762. July 1973.

Battelle Memorial Institute. 1977. *The Impact of Hazardous Waste Generation in Minnesota.* Minnesota Pollution Control Agency, October 1977.

Booz Allen & Hamilton. 1979. *Options for Establishing Hazardous Waste Management Facilities.* New York State Environmental Facilities Corporation, September 1, 1979.

Buchanan, R. J.; and Metry, A. A. 1978. Closing the gap in hazardous waste management in New Jersey. In *Proceedings of the 1978 National Conference on Control of Hazardous Material Spills* (Miami, April 11–13, 1978), Information Transfer, Inc., pp. 196–201.

Carter, C. E.; Fink, L. L.; Teaf, C. M.; and Herndon, R. C. 1979. *Hazardous Waste Survey for the State of Florida, Final Report.* State of Florida Department of Environmental Regulation, October 1979.

Chemical Week. 1979. Europe gets a grip on wastes. *Chemical Week.* New York: McGraw-Hill. Vol. 1, No. 3, pp. 44–45, August 8, 1979.

Collins, M. J. 1974. California's legislative and regulatory policy for hazardous waste management. In *Proceedings of the Third National Congress on Waste Management Technology and Resource Recovery* (San Francisco, November 14–15, 1974), U.S. EPA, pp. 145–158.

Council of Environmental Advisors. 1978. *Summary of the Environmental Report 1978.* Federal Republic of Germany, February 1978.

Dawson, G. W.; and Stradley, M. W. 1976. A regional approach to hazardous waste management—how do you derive your data base? In *Proceedings of the Fourth National Congress on Waste Management Technology and Resource Recovery,* U.S. EPA (SW-8), pp. 10–21.

Department of the Environment. 1979. *Waste Management Paper No. 22— Local Authority Waste Disposal Statistics 1974/75 to 1977/78.* London: Her Majesty's Stationery Office.

Environmental Science and Technology. 1980. Hazardous wastes here and there. *Environmental Science and Technology,* Vol. 14, No. 8, pp. 906–909, August 1980.

Fennelly, P. F. 1977. Surveying Massachusetts' hazardous wastes. *Environmental Science and Technology,* Vol. 11, No. 8, pp. 762–766, August 1977.

Foster D. Snell, Inc. 1976. *Potential for Capacity Creation in the Hazardous Waste Management Service Industry.* PB-257 187. Springfield, Virginia: National Technical Information Service, August 1976.

Idaho Department of Commerce and Development. 1973. *Manufacturing Directory of Idaho.* Boise: IDCD.

Idaho Department of Environmental and Community Services. 1973. *Idaho Solid Waste Management Industrial Survey Report.* Boise: IDECS, June, 1973.

Oregon Department of Economic Development. 1974. *Director of Oregon Manufacturers.* Salem: ODED.

Oregon Department of Environmental Quality. 1974. *Hazardous Waste Management Planning 1972–73.* Salem: ODEQ, March 1974.

Stradley, M. W.; Dawson, G. W.; and Cone, B. W. 1975. *An Evaluation of the Status of Hazardous Waste Management in Region X.* U.S. EPA, Region X, December 1975.

U.S. Environmental Protection Agency. 1976. Growth potential in the hazardous waste management service industry. Paper presented at the National Solid Waste Management Association International Waste Equipment and Technology Exposition, Chicago, June 2, 1976.

U.S. Environmental Protection Agency. 1977. *State Decision-Makers Guide for Hazardous Waste Management.*

U.S. Office of Management and Budget. 1972. *Standard Industrial Classification Manual.* Washington, D.C.: U.S. EPA.

Washington Department of Commerce and Economic Development. 1974. *Directory of Washington Manufacturers.* Olympia: WDCED.

Washington Department of Ecology. 1974. *A Report on Industrial and Hazardous Wastes.* Olympia: WDE, December 1974.

Waste Age. 1976. 1975–1976 Ohio industrial waste survey results. *Waste Age*, Vol. 7, No. 12, pp. 10–11, December 1976.

Watson, D.C. 1978. *Recent Developments in the Management of Hazardous Wastes.* AERE-R9269, United Kingdom Atomic Energy Authority, Harwell, September 1978.

Westat, Inc. 1984. National Survey of Hazardous Waste Generators and Treatment, Storage and Disposal Facilities Regulated Under RCRA in 1981, U.S. EPA.

Working Party on Toxic and Dangerous Waste. 1979. *Toxic and Dangerous Wastes in the Community.* Annex 3, Commission of the European Community, ENU/324/79-EN, Br/rac, May 28, 1979.

5

Facility Siting

Introduction

It has been estimated that the hazardous waste management industry currently has capacity for about 75% of the hazardous wastes that will require contracted services under RCRA regulations. At the projected hazardous waste production growth rate of 5% per year (U.S. EPA, 1977), the present capacity shortfall will continue to increase unless existing facilities are expanded and new facilities built. As a consequence, implementation of RCRA will create intensified pressure for the siting of new hazardous waste management facilities. And yet, siting has become the major stumbling block in the growth of the waste management industry. There are numerous examples of well-founded siting efforts failing to result in a working site because of public opposition.

One of the most publicized failures involved a joint federal-state program. In 1974 the EPA provided the state of Minnesota with $3.7 million to build a model chemical waste landfill that would demonstrate the premise that engineered barriers could be constructed to contain wastes in an industrial area with less than optimum geology. The demonstration of public acceptance was also to be a major feature of the study. After four years of planning and review, public opposition to candidate sites became insurmountable. The state and local agencies finally terminated the project and returned residual funds to the EPA. Similar but less publicized failures have been experienced by private industry throughout the United States. Typically, sites are qualified through state and federal permit processes only to be stymied by local reaction. Technical data and expert opinion carry little weight in this arena. Ultimately, zoning regulations and political pressure are used to prevent completion of the project. This was the case in Detroit when test evaluations demonstrated that cement kilns at the Peerless Cement Company could effectively destroy polychlorinated biphenyls (PCBs). Endorsements for full-scale work were obtained from federal, state, and local agencies, but public reaction ultimately led to complications at the local level that have tabled the project indefinitely (Black, 1978). Ongoing facilities and programs are not exempt from these pressures. Growing public apprehension over PCB disposal in the Earthline Corporation landfill at Wilsonville, Illinois, led to a citizens' suit that has closed the facility despite approval by the U.S. EPA and Illinois Geological Survey after a comprehensive evaluation The issue of removal of PCBs already buried at the site is unresolved. Such failures are not restricted to the United States. In Sweden, the government-owned (90%) Swedish waste conversion company SAKAB was authorized by legislation in 1975 to create a central plant for treating toxic wastes. SAKAB has yet to find a single location where residents will accept siting such a facility.

As a result of past siting failures, it is clear that successful siting is based on adequate resolution of both hard-science- or technology-related issues and soft-science considerations related to winning public acceptance. The importance of the latter considerations cannot be overemphasized. Much of the technical community believes that technology is currently available to allow safe and cost-effective siting of facilities but that public apprehension overrides technical decisions. In this context, the solutions to hard-science issues require better packaging to ensure understanding and acceptance by a nontechnical audience.

Siting Technology

Siting hazardous waste facilities can be a very complex and costly activity. Under close investigation, Pojasek (1980) has enumerated some 15 individual steps that can be attributed to the process;

Preoperational, general
   Public information
   Demand assessment
   Site criteria establishment
Preoperational, specific
   Site nomination
   Land acquisition
   Local information
   Local approvals
   Appeal of local approvals
   State approval
   Negotiations
   Extralegal opposition
Operational
   Project operation
   Enforcement of conditions
   Response to accidents

Postoperational
Closure

The focus of the work here is on those activities required to bring a site to an operational state. For these purposes, the siting process has been divided into three major phases: (1) site screening, (2) site evaluation, and (3) site qualification. In the case of the former, work is focused on the screening of candidate sites, whereas the second focuses on site ranking based on specific criteria. With the latter, an attempt is made to assess the impact that likely will occur as a result of a successful siting. Development of an environmental impact statement is an integral part of this process. Successful completion of both phases requires the resolution of a series of questions:

What are the potential impacts of concern associated with the activity?
What are the site characteristics that will mitigate or exacerbate these impacts?
What criteria can be set to define when impacts are acceptable?

The first of these questions was extensively addressed in Chapter 3 while selecting appropriate definitions for hazardous wastes. The impacts are those associated with migration of and subsequent exposure to materials that are toxic, flammable, explosive, reactive, radioactive, infectious, or irritating. As a consequence, the most desirable sites are those that will minimize exposure of these materials to man or his environment.

At this point it is important to make a distinction between the various kinds of facilities that are employed for hazardous waste management. Treatment facilities are little more than chemical processing plants. As such, they do not pose problems unique from those associated with siting industrial facilities. It is the disposal facility that is the crux of the siting issue. By its very nature, the disposal facility involves an intentional loss (either partial or total) of control over materials that are hazardous to man and his environment. Hence, the landfill, landfarm, lagoon, and injection well are the facilities that pose the real challenge to the siting process.

Criteria Selection

The first major attempt to assemble siting criteria for hazardous waste disposal facilities was made during preparation of background data for the 1973 report to Congress (Battelle Memorial Institute, 1973). At that time it was determined that pertinent criteria could be grouped into one of 12 general categories, as outlined in Appendix F.

Each of these, in turn, requires analysis with respect to specific component characteristics.

Waste considerations are important in the selection and/or evaluation of a specific site, as opposed to generic screening for sites as a whole. Key elements of concern have been addressed extensively in Chapters 2 to 4. In general, geographic distribution is of interest in determining optimal sites from an economic and transportation risk standpoint. These latter factors are tied to intrinsic hazard and transportation system characteristics. For treatment facilities receiving raw hazardous wastes, risk considerations encourage the use of a number of sites placed close to high-intensity waste generation activities to reduce the total ton-mileage experienced from high-hazard materials. Disposal facilities receiving stabilized or inerted residuals can be limited to a few key sites with highly favorable natural features. The volume, uniqueness, and frequency of wastes are pertinent from an economic standpoint. Most processing and disposal operations are subject to economies of scale. This provides incentive to site fewer centralized facilities to reduce capital and operational costs. Regulatory aspects of reference here concern restrictions placed directly on the waste materials themselves. Some wastes are subject to very restrictive management possibilities. A case in point is PCB wastes. These materials can be disposed of only in specially designed incinerators or secure landfills that have met specific criteria (Appendix G) and obtained a supplementary permit. Regulatory constraints also exist in the form of transportation law, discharge permits, local zoning ordinances, and land-use-related statutes addressing such issues as shoreline management.

Process considerations are pertinent to site-specific efforts as well as to generic surveys. The approach to be taken may quickly narrow candidate sites if the process is geologically dependent. For instance, if deep-well injection is contemplated, the survey can be narrowed to specific types of geologic areas, as illustrated in Figure 5-1. Process alternatives will also dictate the types of effluents that can be anticipated and therefore the geologic barriers of interest. Similarly, a hazard inventory based on the processes will place constraints on the type and location of acceptable sites. Finally, process considerations will identify utility needs that will become important economic factors in the siting process.

Geologic considerations include criteria in two main categories: stratigraphic and topographic. Geomorphic/topographic criteria, soils, and hydrology are dealt with separately because of their unique implications in the siting process. Each of the latter can be quantified to some extent, but the more general geologic considerations must be handled subjectively until efforts become very site-specific.

Studies in the 1960s began to encompass the potential geologic problems associated with siting a waste disposal facility. Within that time frame, some common geologic criteria associated with solid waste disposal by landfill and land burial emerged. Engineering geology, which is the application of physical geologic factors to promote efficient land use and ensure stability of various structures and facilities, is routinely used in efforts to reduce or prevent problems associated with development and to reduce the impact of develop-

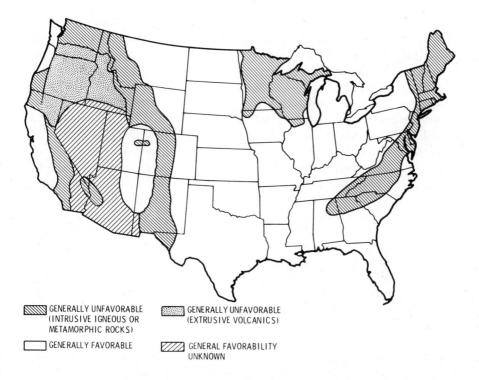

GENERALLY UNFAVORABLE
(INTRUSIVE IGNEOUS OR
METAMORPHIC ROCKS)

GENERALLY UNFAVORABLE
(EXTRUSIVE VOLCANICS)

GENERALLY FAVORABLE

GENERAL FAVORABILITY
UNKNOWN

Figure 5-1 / Deep-well disposal sites. (From Battelle Memorial Institute, 1973.)

ment on adjacent areas. By subjective integration of the criteria developed in these diverse areas, a more general set of geologic considerations was developed for the report to Congress to serve as a guide for evaluating the potential adequacy of sites for hazardous waste disposal systems on a regional basis.

The general rock sequences in each region are considered to be water supply sources and tertiary barriers to unplanned waste release. Centers of sedimentary basins are thought to be the most desirable because of the thicker stratigraphic sections available. Areas with outcrops of igneous and metamorphic rocks are considered unfavorable because of the generally restricted amount of space available for subsurface disposal, even at shallow depths. Basalt, limestone, and glacial outwash plains are considered unfavorable because of associated high permeabilities and potential rapid hydraulic transport of wastes from an unplanned release. Limestone terrain and glacial outwash plains with thick unsorted gravel and floodplain deposits are also considered undesirable from the standpoints of site engineering and cost of construction.

Within any region, the presence of a gross extent of surface or near-surface faulting is undesirable. In areas where major faults are known, there is a higher probability of localized faults and fractures being found during detailed site study. Faults and fractures can provide high-permeability paths to the groundwater system and increase the potential for pollution from a waste processing/disposal system. Earthquake history, both frequency of occurrence and magnitude, is considered to be related to the potential for additional faulting in a given area. Areas of frequent earthquake occurrences also may prove unsuitable for deep disposal of waste products because of the potential triggering effect. Earthquakes per se are a perceived risk, requiring higher design costs and presenting difficulty with public acceptance. Earthquake potential has been quantified roughly by definition of seismic risk zones, as depicted in Figure 5-2.

Physiographic considerations are tied both to geologic considerations, in that they are indicators of recent geologic activities, and to hydrologic considerations, because of interactions with surface water runoff and ground-water recharge.

Areas of high relief (greater than 10% slope) are associated with active geologic processes that generally impose severe restraints on land use: Site engineering and construction costs are relatively high, special roads or improved transport routes may be necessary (increasing system costs and potential spill hazards), and geology tends to be more complex and variable over local areas of site dimensions, therefore requiring more detailed studies. Folding, fractures, faults, and landslides often associated with high-relief areas coupled with generally higher precipitation can also provide major recharge points for groundwater aquifers. Areas of flat slope (less than 1%) are also considered undesirable to some extent. These areas are located predominantly along floodplains of major streams, near swamps, or in sinkhole areas. Natural drainage is poor, soils are generally unconsolidated, and competing land use for agriculture, recreation, and ecologic preserves is more prevalent. If proper precautions are taken, the land can be used for disposal sites, but at a higher cost and higher safety risk. Moderate slopes of 1% to 10% present the fewest land use constraints for hazardous waste processing/disposal sites.

Soil characteristics are considered to be of importance in four categories: (1) as a secondary barrier to unplanned waste releases, (2) as a landfill disposal medium for relatively insoluble waste components, (3) as a surface on which structural foundations are constructed, (4) as a medium for growing agricultural plants rather than as a waste treatment plant site.

Serving as a secondary waste release barrier is considered to be the most important of the four. Although accident prevention can be engineered into the facility, an absolute guarantee that no accidental waste solution release will ever occur cannot be provided; thus, this function of soil is given the

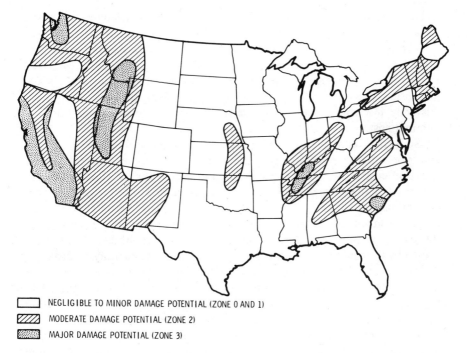

NEGLIGIBLE TO MINOR DAMAGE POTENTIAL (ZONE 0 AND 1)
MODERATE DAMAGE POTENTIAL (ZONE 2)
MAJOR DAMAGE POTENTIAL (ZONE 3)

Figure 5-2 / Seismic probability map. (From Battelle Memorial Institute, 1973.)

highest priority. The movement (percolation) of a released solution through the soil occurs over a finite period of time; therefore, time is available in which to respond to a release. In addition, many waste components are sorbed by soils, which allows even longer response times.

The distance to groundwater and the soil depth criteria affect both solution movement and sorbed component movement. The greater the soil depth, the more time will elapse before groundwaters are contaminated by a release. The soil texture, soil structure, and slope criteria primarily affect the movement rate of liquids. The finer the texture, the lower the soil permeability, and the more slowly the liquids will move through the soil. Similarly, the more structured and layered the soil, the lower the soil permeability. The greater the slope, the faster the liquids will run off the soil surface and the slower will be the infiltration rate. These characteristics tend to spread any release over a much greater surface area. The soil sorption capacity criteria, as measured by the cation-exchange capacity (CEC) and soil pH, affect only sorbable waste components. In general, the greater the CEC, the greater the surface area of a soil, and that promotes the sorption of all sorbable wastes, including cations, anions, and organics. Furthermore, the pH tends to increase the sorption of many sorbable components, particularly the heavy metals.

The remaining three soil functions are considerably less important than the secondary barrier function. In these, engineering techniques can overcome the inadequacies of the soil. For instance, if the soil at the waste processing site is unsuitable for landfill disposal of insoluble wastes, this relatively low-volume waste can be shipped to another site for burial at minimal cost.

Soils that contain large percentages of montmorillonite clay or that consist primarily of silt-size particles will result in higher foundation construction costs. The use of class I agricultural land for the siting of a waste treatment plant permanently removes this land from the growing of agricultural crops. The importance of this criteria will vary with the availability of class I farmland. In the United States, class I farmland is plentiful, and this consideration is not deemed important. (A soil capability classification is used by the Soil Conservation Service to group soils generally according to their suitability for most kinds of agriculture. Capability classes, the broadest grouping, are designated by the roman numerals I through VIII. The numerals indicate progressively greater limitations and narrower choices for practical use. Class I soils have few limitations restricting their use.)

Hydrologic considerations encompass criteria related to quantity, quality, and location of surface waters and groundwaters. Areas judged to be most desirable for processing sites (1) contain water supply sources (groundwater and/or surface water) continually providing at least 200 gpm of high-quality water, (2) are underlain by noncarbonate consolidated rock aquifers, and (3) lack features such as floodplains, large bodies of water, and high groundwater table.

Floodplains are undesirable for two reasons, the first of which is related to safety. A processing site should logically be located above the elevation of the probable maximum flood to protect the plant, equipment, and wastes from being inundated. Second, floodplains are normally located in zones of high water tables, high wave runup, and permeable soils. In addition, structures located in floodplains act as restrictions to water movement.

Zones of high wave runup are undesirable because of the flooding potential and the associated accidental releases of hazardous materials. Zones of high water tables are undesirable because of the possibility of contamination of water supplies in the event of an accidental release. Greater depth of the water table provides more filtration of contaminants by the earth materials. The effectiveness of filtration is related to the type of material, the rate of fluid movement, and the distance traveled prior to entering a water supply system. The rate of fluid movement is dependent on the permeability of the earth materials and the hydraulic gradient. In general, steep gradients are associated with low permeability, and vice versa. The steepness of the terrain also has an effect on the hydraulic gradient; in general, the steeper the terrain, the steeper the hydraulic gradient. A low hydraulic gradient is desirable because such a gradient accompanies flatter terrain. However, a low gradient might indicate a very permeable aquifer that is a valuable source of water. This would increase the risk of transporting a hazardous waste from the site.

It was estimated that each processing site must have an acceptable continuous water supply of 200 gpm. The quality of the water source determines the cost necessary to make it acceptable for use in the various plant processes.

Areas underlain by noncarbonate consolidate rock aquifers are considered superior to those areas underlain by aquifers of carbonate consolidated rock and by unconsolidated aquifers. The latter implies unconfined surficial aquifers, which are undesirable on the depth-to-water-table basis. Carbonate (limestone) aquifers are generally undesirable because they can contain solution cavities and potential high-permeability channels.

Climatological considerations of importance to siting pertain mostly to precipitation, wind rose, air pollution potential, and extreme weather potential (hurricanes and tornadoes). Low precipitation values are desirable. This is true for most plants, based on a variety of factors: simpler plant design; a potential for more effective operation of evaporative systems; less chance of an accidental release or leaching of spillage in concentrated form into the groundwater; better road conditions for transportation of materials; the potential for combining the waste processing plant with a temporary or permanent storage facility at the site. On the other hand, low rainfall portends several disadvantages: If a potentially dangerous respirable material is released into the atmosphere, it may be desirable to have rainfall scavenge the material from the air. For hazardous waste landfills where involatile materials are to be disposed, low precipitation is of great value, because it eliminates the major route of contaminant egress: leachate.

Wind rose criteria are tied to the direction and density of population. A buffer zone is desirable downwind of sites to permit ample dilution of any releases. Similarly, low air pollution potential indicates a greater degree of turbulence, which will minimize the occurrence of concentrated plumes of contaminants. Occurrence of extreme weather conditions (tornadoes, hurricanes, and thunderstorms) is undesirable in that it raises the potential for damage to the facility and dispersion of materials through erosion or flooding.

Transportation considerations can be grouped into two areas: (1) risks associated with movement of hazardous wastes, and (2) economics of movement. Factors related to both favor siting where transportation links are as short as possible and involve the most effective transport modes.

Economic variables considered in criteria selection include relative volumes of wastes generated at given points, relative transportation distance between major waste generation points, the transportation network of the area, and the transportation link between the area and other waste generation points. Costs are generally linear with distance, except when routing requires transfer between modes. Costs may also increase substantially when distances exceed 250 miles, because this defines the average range for a single day's travel. Overnight expenses for drivers come into play for truckers when longer distances are involved.

Risk variables considered in criteria selection include the distance traveled, type of terrain traversed, and number of sizes of metropolitan areas traversed during transit of waste (population density en route).

Distance is an important consideration in determining risk because of the relationship between risk and distance traveled; risk of accident typically increases as distance increases. Transport handling, such as switching, normally increases with distance and correspondingly increases the risk. Also, the time associated with transport increases with distance and thus increases the risk because of the additional chance of accidents. Routes of travel are important factors for two specific considerations: The first is the number and relative size of metropolitan areas along the route. This aspect is important because risk normally increases with the number and sizes of metropolitan areas to be traversed. Metropolitan areas introduce two important risk characteristics: people and congestion. The severity of any accident increases with the number of persons exposed to the accident, and the probability of accident increases with traffic volume. The second major consideration is the type of terrain traversed along the route. The more rugged the terrain, the greater the potential for accidents.

Resource considerations relate to factors stemming from both the natural inventory of a site and the human activity that is associated with the site. Many of these considerations can be generically classified as those of land use. Development and evaluation of criteria in the area of resource utilization are difficult. They must be done carefully to integrate the criteria in this category with those in other categories, particularly earth sciences, ecology, and the human environment.

Included in the definition of resource uses are minerals and their extraction, agriculture (aside from ecological considerations), manufacturing, recreation/tourism, and unique commercial, industrial, residential, and other developments. Compatibility is determined primarily by considering dominant resource utilization patterns and potential and projected demands within an area.

Human environment considerations are particularly difficult to translate into criteria. Evaluation of the effects of man's actions on man has always required a high degree of judgment that has been complicated by historical and geographic variations in societal desires. The establishment of criteria and their relative importance to the human environment, therefore, should be divided between those factors that can be quantified and those that must be dealt with qualitatively. The first group includes such characteristics as population distribution, recreation demand, and distance from areas of high intrinsic value. The second group contains factors in the area of public attitudes and acceptance and hence moves from the technical siting arena to that of the soft-science concerns.

Generally, large urban residential areas such as standard metropolitan statistical areas (SMSA) are considered unsuitable, although it is recognized that many of these have associated well-planned industrial parks that might accommodate waste processing plants. Conversely, much industry is ill-situated in such areas, particularly industries around which residential areas grew early in this century. Although large centers of population generate the bulk of the wastes (implying favorable transportation considerations), there is no identifiable uniform attitude on the part of political entities or the populace concerning hazardous waste processing plants located nearby, even in the industrial parks. Also, even though the real risk for such facilities will be minimized by design and operational planning, the perceived risk will be high and may well dictate the exemption of highly populated areas from consideration.

Other human environment parameters deemed unfavorable include the following: nonindustrial developed shorelands (because of their value for open space, recreation, and other human uses and because of current pro-tective legislation); areas of high scenic and recreational interest; proximal international boundary areas; counties with significant numbers of historical or cultural features; and counties with high densities of schools, hospitals, and similar sensitive facilities.

Biological considerations can be managed in the context of characteristic ecosystems. Ecosystems are highly complex and consist of many physical, chemical, and biological components interconnected by a multitude of functional pathways. In attempting to characterize ecosystems, it is useful to limit the description to the predominant parts of the system. The indicator approach is commonly used when it is neither possible nor desirable to make appropriate measures or estimates of the status of all the components of the system. Instead, selected groups of species, habitats, or other important components are examined in some detail and used as an index to estimate the overall quality or condition of the ecosystem.

Land use was employed as such an indicator for early work in this area (Cleveland et al., 1979). The inference of ecological quality from land use involves a logical, two-step process. The land use patterns followed in the work for the report to Congress mapped the major existing vegetation types in the United States and their uses (e.g., cropland, ungrazed desert shrubland, ungrazed forest, and woodlands). The various vegetative types and their uses were considered excellent indicators of the environmental quality of a specific region. Vegetation, by the nature of the photosynthetic process, is the source of organic energy for an ecosystem. The productivity of the vegetation in terms of energy or biomass is important in regulating the numbers and kinds of animals that can exist in the ecosystem and the use man can make of that system. Vegetation also provides breeding and nesting sites and protective cover for animals, thus enhancing the quality and utility of the system.

An advantage of using land use patterns as an ecological indicator is that it gives some measure of the degree of interference man has imposed on the natural environment. For example, urban development represents an extreme departure from the natural state, and as such has little value as a smoothly functioning balanced ecosystem.

Areas attractive for the siting of industrial facilities such as hazardous waste processing plants are low in ecological quality because of either urban development or physical-chemical limitations imposed by natural conditions. Agricultural lands form an intermediate grouping. Here, disturbance from man has decreased, but the monoculture results in a less diverse system, with reduced stability and resiliency. Forests, which characterize the third grouping, represent an improvement in both diversity and stability over the preceding two groups. Finally, swamplands and marshes, because of their high productivity and extreme sensitivity to disturbance, are considered incompatible with processing plants and disposal/storage facilities. A unique habitat, especially one associated with rare and endangered species, is also unacceptable for siting purposes.

Based on these considerations, some guidelines can be selected defining desirable conditions for siting hazardous waste facilities. One approach to this is presented in Table 5-1. For the most part, these criteria guidelines are directed to land disposal facilities. Treatment facilities without hazardous release are essentially identical to chemical processing plants. As such, they do not require as rigorous a set of siting criteria.

Applying Siting Criteria

Once criteria have been selected, their application can be initiated as part of the active siting effort. That effort is typically divided into three phases, as depicted in Figure 5-3: site screening, site evaluation, and site qualification. The first two of these phases involve direct application of the criteria. For the third phase, the bond with the criteria is less rigorous. In order to facilitate conduct of the criteria application phases, criteria are first segregated into two groups: go–no go criteria and relative value criteria. The former category refers to criteria that must be met before a site can be considered. For quantitative criteria, a watershed value is defined such that all sites with characteristic values to one side of the threshold are unacceptable, whereas those on the other side may be considered. Go–no go criteria and associated watershed values are often derived from the "un-

Table 5-1 / Guidelines for Siting Criteria.

Factor	Desirable Conditions	Unacceptable Conditions
Geologic considerations		
Bedrock depth	>50 ft	<30 ft
Character	Shale; very fine undistrubed sandstone; sedimentary basins	Fissured, fractured carbonate rocks; any joint fractured rock
Seismicity	Seismic risk zones 0–1	Seismic risk zone 3
Tectonic	≥1 mile from active fault	<1 mile active fault
Unique features		Archaeological or paleontological significance
Physiographic considerations		
Location	Upland; clay pit	Wet lowland; floodplain; deep pit or quarry; sand and gravel pit
Relief	Flat to gentle rolling; slope ≤10%	Adjacent to steep slope; deep gullies; slope >25%
Soils		
Depth	>40 inches	<10 inches
Texture	Silt to loam	Very fine clay
Drainage	Moderately well to well	Very poorly
Infiltration rate	0.6–2.0 inches/h	<0.6 or >20 inches/h
Organic matter	~1%	>8%
Slope	2–12%	>25%
Subsoil permeability and character	10^{-3}–10^{-7} cm/s mixtures of sand, silt, clay; glacial till; fine sands; silts	10^{-1} cm/s sands and gravels or 10^{-8} cm/s clays
Hydrologic considerations		
Drainage	Fast-draining materials; dry surface	Heavy clayey or organic mat; area subject to flooding, ponding
Surface water	Valley flat or terraces away from stream	Valley flat near stream; likelihood of flooding (100-yr floodplain)
Distance	>3000 ft from lake, marsh; >2000 ft from a stream	<2000 ft from any surface water, 5 miles to watershed boundary
Groundwater	No indication of high water table	Seepage, springs, marsh, phreatophytic vegetation

Table 5-1 / (Continued)

Factor	Desirable Conditions	Unacceptable Conditions
Depth to water	≥80 ft	ft
Aquifers	Deep bedrock with thick impermeable cover	Use of shallow aquifers; thin permeable cover over deep aquifers
Direction of flow with respect to point of use	Toward site	Away from site
Water supply source	>3000 ft	≤2000 ft
Climatological considerations		
Precipitation	Evaporation ≥4 inches in excess of precipitation	Precipitation exceeds evaporation
Storm event	Low frequency of wind and severe storm events	Within path of major hurricane or tornado events
Wind	Good atmospheric mixing	Population center encompassed in static air zone
	No population centers in downwind direction	Population center ≤0.5 mile downwind
Transportation considerations		
Public facilities	>1000 ft	<1000 ft
Distance federally funded road	>2000 ft	<2000 ft
Availability	Low spill risk on transport routes	
Resource considerations		
Land use	Not adjacent to agricultural land in active use	Bordering parks, recreation areas, wilderness areas, wildlife refuges, or scenic rivers
Human environment considerations		
Demography	Low population density	
	≥5 miles from municipal wells	
Location		Bordering cultural areas, Indian reservations, or international boundaries; <0.5 mile potable wells; <1 mile downstream of intakes in flowing waters
Biological considerations		
Ecology	Low ecological value; low species diversity and uniqueness	Habitat for rare or endangered species

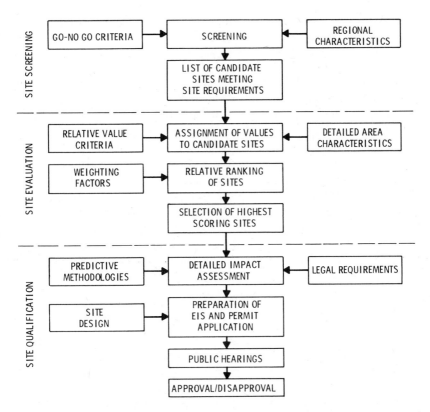

Figure 5-3 / The three phases of the siting process.

acceptable conditions" column of Table 5-1. Relative value criteria are largely derived from the "desirable conditions" column of Table 5-1. These criteria include site characteristics for which the acceptabilities of two sites can be compared on the basis of a numerical comparison of specific characteristics. For instance, the greater the depth of groundwater, the more desirable the site. Relative value criteria include most numeric criteria for which a watershed value also exists. Given this division of criteria, the siting process can begin.

Site Screening. The first phase of the siting process, screening, can be approached in either of two ways: (1) direct identification of promising candidate sites, and (2) identification of candidate sites through elimination of unacceptable alternatives. Neither approach is always optimal; however, experience suggests that the elimination process is often the most efficient for siting hazardous waste facilities. The direct identification method is most appropriate when a single criterion or perhaps two siting criteria are the overriding considerations. For instance, if it is determined that the site must be on federally owned land, direct identification can be used to focus attention on the federal tracts in a given region. More often, however, there will be a number of criteria defining unacceptable sites. In these cases, the elimination approach is optimal.

The elimination approach is an exclusionary process that is activated by removing unacceptable sites from consideration, thus allowing the evaluation to focus on a greatly abbreviated candidate list. The actual application of exclusionary criteria can be performed in a simplistic or sophisticated manner, depending on the complexity of the criteria to be employed. Manual application is best accomplished through use of a system of overlay maps, as illustrated in Figure 5–4. Areas to be eliminated from consideration because of exclusion criteria (go–no go watershed or legal restriction) are plotted on overlays of the region being studied. The individual overlays are then composited to highlight areas that meet all criteria. These then become the candidate sites for evaluation. More sophisticated approaches rely on the use

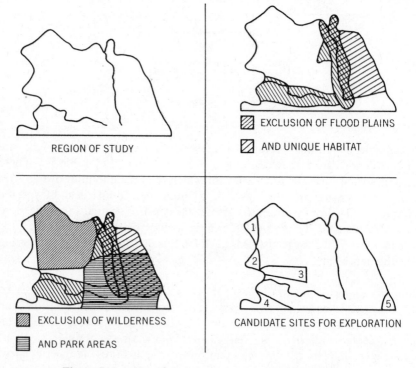

Figure 5–4 / Use of the overlay map exclusion process.

of automated computation devices. Computer techniques were first developed in the 1960s; they fall into two basic categories: (1) approaches using grids or cellular data structures, and (2) approaches using a polygon data structure. The first system facilitates data manipulation for analysis; the second provides greater spatial accuracy (Cleveland et al., 1979). Use of computers in the screening process is most beneficial when large areas are being surveyed at a detailed level or when a large number of exclusionary criteria are being employed and the evaluator wishes to maintain a record of applicable exclusions for each tract.

Site Evaluation. The second phase of the siting process is directed to providing a comparative analysis of candidates identified during the screening process such that individual sites can be ranked according to preference. Siting evaluation in this context is based on use of relative value criteria. As noted earlier, these criteria may be related to watershed criteria employed in the screening phase, but defined a range of acceptable values where one value may be preferred to another yet neither is sufficient to exclude a site (Figure 5–5). A second threshold may also be defined beyond which variations in value have little meaning. In this case, all values in that zone are given the same level of preference. This second threshold is defined for practical reasons. It is not intended to imply that comparison of two sites in the equivalue zone yields no preference. It merely prevents the additional levels of safety provided by the one site characteristic from overriding weakness due to other characteristics when a weighted summation approach is taken. In other words, we would not want a site that is 100 miles from a population center to be rated higher than one that is 50 miles from a population center if the second site has a much greater depth to groundwater than the first.

This raises the issue of how to combine values for each criteria into an overall ranking of sites to allow comparison. Any kind of weighting system will ultimately involve judgments with respect to the relative importance of site characteristics. In the background work for the report to Congress, every county in the contiguous United States was ranked through combinations of scores for individual site characteristics. This is the most extensive hazardous waste siting survey made to date. Experts were used to place a value from

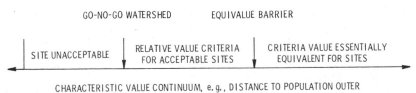

Figure 5-5 / Relation of watershed and relative values for siting evaluation.

0 to 5 for each criterion area. Summation of values was accomplished through use of a weighted value function.

The latter approach requires the development of two factors for each parameter under consideration: the weighting factor and the value function (ranking). The value function is based on a scientific or technological judgment by knowledgeable professionals of the suitability of each site with respect to the criteria. As noted, these were assigned on a scale of 0 to 5. The weighting factor, on the other hand, is subjective and involves the opinion of individuals or groups of individuals as to the relative importance of each criterion area. The individuals supplying the opinions may or may not be knowledgeable in these areas. The value function makes it possible to compare alternative sites in terms of their effects on a particular siting parameter. However, it is not possible to sum all of the effects for an overall evaluation of each alternative site without weighting factors; that is, it is not possible to state technologically that air quality is more important than water quality or that preservation of aquatic biota is more important than avoidance of visual intrusions on the landscape. Because individual opinions may differ as to the importance of each parameter, different groups may produce different sets of weighting factors that in turn may produce different rankings of alternative sites. However, if the groups are large enough, the sets of weighting factors should not be too different.

The weighting factors used for the EPA-sponsored study were statistically derived from the individual responses of 10 professionals. The sum of all weighting factors equaled 100, and the product of the weighting factor (0 to 100) and ranking value (0 to 5) for each siting criterion resulted in a number that could be summed over all of the siting criteria for a given county. This summation for each county represented a quantitative rating that was then numerically compared against a perfect score (500) and against the other counties within a region. Additionally, the weighting assigned to the earth sciences criteria was distributed among the four subcategories using a procedure identical with that used for weighting the major criteria areas. Only earth sciences professionals participated in this determination. These subcategory weightings varied slightly between some regions, depending on the relative importance attached to them by the evaluators (Battelle Memorial Institute, 1973). The weighting factors selected are presented in Table 5-2.

As is evident, the major weakness of this approach is the need for subjective judgment on the relative importance of criteria. Consequently, siting experts have begun to develop means for integrating criteria into a single evaluative framework. These typically take the form of mathematical models designed to predict the movement of chemicals from the site and subsequent exposure of man and the environment. With application of these models, sites can be compared on the basis of relative risk. This type of evaluation is particularly useful in that the output is better understood by the

Table 5-2 / Weighting Factors Employed in Initial U.S. Siting Study.[a]

Criteria		Weight
Earth sciences (physical/chemical)		31
Geology	7	
Hydrology	8	
Climatology	10	
Soils	6	
Transportation		28
Human environment and resource utilization		23
Ecology		18
Total		*100*

[a]Adapted from Battelle Memorial Institute (1973).

public and can be employed directly during phase III site qualification activities. The risk framework provides a meaningful way to combine all site characteristics that are of concern.

Site evaluation models can be relatively simplistic or very complicated. Selection between alternatives involves a trade-off between the cost of the evaluation and the accuracy of the prediction. A task force in the state of Oregon developed a relatively simple model of chemical landfills incorporating (1) hydrodynamic flow velocity derived from the known or estimated porosity and gradient of the porous medium, (2) variable water table level, (3) variable rainfall, (4) reversible adsorption–desorption phenomena, (5) first-order irreversible sorption or chemical reaction, and (6) first-order microbial degradation kinetics (Elzy et al., 1974).

The sanitary landfill model (SLM-1) that resulted is based on a vertical routing of contaminant, similar to the method of Remson et al. (1968), and a horizontal routing corresponding nearly to flow through a series of stirred tanks. A mass balance is performed on the contaminant only, not the water. Horizontal velocity is assumed constant in the landfill and soil and is estimated from input data on permeability, porosity, and hydraulic gradient of each medium. The landfill and the soil are assumed to be homogeneous, with properties uniform throughout each medium. Analysis is conducted in two-dimensional space, with the vertical component and a single horizontal component outward from the landfill. The landfill and soil profile are divided into a grid. Each space is then treated as a mixed reactor. Contaminant is introduced, sorbed onto sediments, and degraded. A year's balance determines the final concentrations. Water is introduced, and after the foregoing calculations the grid is drained to field capacity. The drainage becomes a part of the input to adjacent grid spaces. Output consists of estimates of travel time and concentrations in leachate. Validation studies have shown the model to be useful in predicting leachate movement (Elzy et al., 1974).

A similar simplified failure prediction model was developed to evaluate the adequacy of the Post-Closure Liability Trust Fund. It relies on a simple water balance calculation from infiltration of precipitation followed by contaminant transport in the aquifer (ICF, 1984). A contemporary review is available of simplified methods to analyze subsurface contamination movement (Brown, 1984).

At the other end of the spectrum one can employ the very sophisticated models being developed for siting nuclear waste repositories. That effort, under the sponsorship of the U.S. Department of Energy, deals with four phases of the consequences of release: (1) description of scenarios that would result in a breach of the waste repository, (2) quantification of leach rates associated with contact of groundwaters with the waste form, (3) subsequent transport of contaminants in the groundwater, and (4) estimation of the human dose resulting from waste migration (Burkholder et al., 1979).

Under current concepts for high-level nuclear waste disposal, the major route of contaminant movement has been narrowed to potential dissolution in groundwater and subsequent migration. Therefore, safety assessment efforts have focused on defining the probability and character of natural or human-caused events that would allow water to breach the repository and mobilize contaminants. Plausible scenarios under investigation include those presented in Table 5-3. Analysis of each of these scenarios provides an estimate of waste exposure and flow rates for intruding waters. These are joined with empirical data on waste leaching rates to define the source term for subsequent transport of contaminants outside.

Estimates of hydrologic flow, contaminant transport within that flow, and ultimate human dose are made through application of mathematical models. Several models have been selected to provide for differing levels of complexity depending on the output requirements. A brief summary of the models in use is provided in Table 5-4. These models were selected from a review of over 150 possible approaches. A summary evaluation of the hydrologic and contaminant transport models included in that review has been performed by the Holcomb Research Institute (1978). More recent reviews of models can be found in Brown (1984) and Brown et al. (1983).

The first or simple level of hydrologic contaminant transport model employs an analytic solution of the steady-state flow equation to obtain the groundwater flow velocity for the release scenario of interest. This velocity is then used as input to the GETOUT model (Lester et al., 1975). GETOUT assumes that the contents of the repository are released either as an impulse or at a constant rate to the groundwater, which flows at constant velocity through a one-dimensional column of the homogeneous geologic medium surrounding the respository and discharges to a surface water body. The dissolved radionuclides are assumed to be in sorption equilibrium at all points in the flow path. Radioactive decay (including chain decay of the

Table 5-3 / Potentially Disruptive Phenomena for Nuclear Waste Repositories.[a]

Natural Phenomena	Human-caused Phenomena	Waste-induced Phenomena
Climatic fluctuations	Improper design/operation	Thermal effects
Glaciation	Shaft seal failure	Differential elastic response
Denudation and stream erosion	Improper waste emplacement	Nonelastic response
Magmatic activity	Undetected past intrusion	Fluid pressure changes
Extrusive	Undiscovered boreholes	Fluid migration
Intrusive	Mine shafts	Canister migration
Epeirogenic displacement	Inadvertent future intrusion	Chemical effects
Igneous emplacement	Archaeological exhumation	Geochemical alterations
Isostasy	Weapons testing	Corrosion
Orogenic diastrophism	Nonnuclear waste disposal	Waste package–geology interactions
Near-field faulting	Resource mining (mineral, hydrocarbon, geothermal, salt)	Gas generation
Far-field faulting	Storage of hydrocarbons or compressed air	Mechanical effects
Diapirism	Intentional intrusion	Local fracturing
Diagenesis	War	Canister movement
Static fracturing	Sabotage	Radiation effects
Surficial fissuring	Waste recovery	Material property changes
Impact fracturing	Perturbation of groundwater system	Radiolysis
Hydraulic fracturing	Irrigation	Criticality
Meteorites	Reservoirs	Decay product gas generation
Dissolutioning	Intentional artificial recharge or withdrawal	
Sedimentation	Establishment of population center	
Flooding		
Undetected features		
Faults, shear zones		
Breccia pipes		
Lava tubes		
Gas or brine pockets		

[a] Adapted from Burkholder et al. (1979).

153

Table 5-4 / Characteristics of Release Consequence Analysis Models Employed in Safety Assessment for Nuclear Waste Repositories.[a]

Hydrologic
  PATHS: two-dimensional analytic/numerical method, homogeneous geology, saturated flow
  VTT: two-dimensional finite-difference numerical method, heterogeneous geology, saturated flow
  DAVIS FE: three-dimensional finite-element numerical method, heterogeneous geology, saturated flow

Contaminant transport
  GETOUT: one-dimensional analytical method, chain decay, single speciation, equilibrium sorption, constant leach rate, dispersion
  MMT: three-dimensional numerical method, chain decay, single speciation, equilibrium sorption, time-variant leach rate, dispersion

Water dose
  ARRRG: drinking water, immersion, external shoreline, and aquatic food doses
  FOOD: terrestrial food dose

Air dose
  KRONIC: chronic external dose
  SUBDOSA: acute external dose
  DACRIN: chronic or acute inhalation dose

[a]Adapted from Burkholder et al. (1979).

actinides) is modeled at the repository and during migration through the geologic medium. Trace concentrations of the dissolved nuclides are assumed, and as a result the sorption factors are independent of concentration. A constant axial dispersion coefficient is also assumed. GETOUT is applicable to particulate or fractured media provided that the necessary sorption data are measured properly. The model can be applied to heterogeneous media if the relevant input parameters are averaged properly. The GETOUT code uses the analytical solution of the one-dimensional transport equation and is thus considered the most time- and cost-effective method of determining the release consequences to individuals affected by a contaminated surface water body. However, because of its one-dimensional nature, it cannot be used to accurately calculate the release consequences to individuals such as those using well water.

The second level, or simple intermediate level, uses the PATHS hydrologic code (Nelson, 1976). PATHS assumes a homogeneous geologic medium and saturated steady-state flow and uses a two-dimensional analytic solution for the groundwater potential distribution between the repository and a surface water body. From the potential distribution, the path-line differential equations are solved numerically to give the paths of fluid particles and their position with time between the repository and a surface water body. The GETOUT code is then used to calculate transportation of the radionuclides along any or all paths of interest. The second-level system is less time- and cost-effective than the first-level system, but it provides a more detailed representation of the contaminant transport. The first two levels of modeling complexity idealize the real system and do not require as much input data as the last two levels.

The third level, or combination intermediate-complex level, uses the VTT hydrologic code (Kipp et al., 1972). VTT models two-dimensional saturated groundwater flow through heterogeneous media. The code is based on a Boussinesq approximation model and uses standard finite-difference solution techniques. It has a multiaquifer capability with variable thickness features for the unconfined strata and can solve both transient and steady-state problems. More detailed solution at a finer difference grid spacing is possible by a sequential solution process wherein the results of a large region simulation are used as boundary conditions for a smaller region. The MMT code uses the flow vectors from VTT as input to the nodes of a rectangular grid system and models the transport of contaminants (Ahlstrom and Foote, 1977). The MMT code models dispersion (using a random-walk analogy), radionuclide chain decay, and equilibrium sorption. Because of its numerical nature, all parameters of the model can be varied with position. Thus, data of considerable detail can be accommodated.

The fourth level, or complex level, uses the DAVIS FE code (Gupta et al., 1975). This model considers saturated groundwater flow (transient and steady state) in three dimensions through a heterogeneous geologic medium with both the fluid and geologic medium considered to be compressible. Because the differential equations are solved by finite-element techniques, the code is able to satisfy irregular boundary conditions. The MMT code will again be used to model contaminant transport using flow vectors from the DAVIS FE code as input.

ARRRG calculates annual individual and population doses resulting from radionuclides released to surface water bodies (Soldat et al., 1974). Eight environmental pathways are addressed: eating fish, crustacea, molluscs, and algae, drinking surface water, direct external radiation from shoreline, water immersion (swimming), and direct external radiation from boating. Doses are calculated for skin, total body, gastrointestinal tract, thyroid, bone, liver, lung, and kidney. The output format specifies contributions to the dose by nuclide and by pathway.

FOOD is a companion code to ARRRG, and it calculates annual individual and population doses from eating garden foods and animal products contaminated by air deposition or water irrigation (Baker et al., 1976). Doses are calculated for 14 foods. A key assumption in the code is that the accumulation and dispersion processes in the biosphere are such that the local soils used for farming contain the accumulated radionuclides from a specified number of years or irrigation previous to the time of interest.

KRONIC calculates annual external total body tissue dose from chronic release of radionuclides to the atmosphere (Strenge and Watson, 1973). The dose is calculated as a function of the radionuclides released, the plume height, the plume size, the climatology for the site, and the physical properties. Atmospheric dispersion effects are estimated using data on joint frequency of wind speed, wind direction, and stability for the site. Each sector is considered to be a plume for which the centerline ground-level dose is calculated for specific downwind distances. The output format is a table of annual dose rates as a function of direction and distance from the release point.

SUBDOSA calculates external doses from acute release of radionuclides to the atmosphere (Strenge et al., 1975). Doses may be calculated as a function of depth in tissue and may be summed and reported as skin, eye, and genetic doses. The doses are calculated for six release periods, with separate nuclide inventories and dispersions in each period. Nuclide decay is considered during release and/or transit using a chain decay scheme with branching to account for transitions between isomeric states.

DACRIN calculates the effective radiation dose to the human respiratory tract and other organs from the inhalation of radioactive aerosols (Houston et al., 1974). The program follows the basic precepts of the International Committee on Radiation Protection Task Group on Lung Dynamics and a simple exponential model for retention by the organ of interest. As an adjunct to the organ dose model, mathematical models of atmospheric dispersion are included in order to model transport to humans from either acute or chronic releases. The code calculates the effective radiation dose to any of the 18 organs and tissues from inhalation of any one or combination of the radionuclides.

The hydrologic and contaminant models are directly applicable to non-nuclear hazardous waste disposal facility siting. Inorganic species can be assigned an infinite half-life. Organic species require replacement of decay subroutines with models addressing degradation. The conversion of dose models is more complex. Nuclear models calculate the total cumulative exposure of a receptor to radiation. This concept is not meaningful for most hazardous substances. For instance, for nonaccumulative materials, dose-concentration/time relations are of greatest significance.

Whatever approach is selected for site evaluation, the output should be such that one can compare the relative exposure or risk associated with each site and in this way rank the relative desirability of each site. When this is accomplished, other factors such as cost and public acceptance need to be weighed to select a site for qualification. If a modeling approach has been employed for evaluation, much of the output can be used directly in qualification activities.

Site Qualification.   The final phase of the siting process, site qualification, includes those activities required to fulfill regulatory demands in moving from design to implementation. Such activities include completion of permit applications and development of the environmental impact statement with its incumbent public hearings and regulatory agency review. In this context, site qualification activities parallel closely those of site evaluation, except for some important distinctions:

Qualification is sought for a very limited number of sites, usually one site.
Impact assessment and use of models are taken to a finer level of detail to encompass the entire facility, including engineered barriers and modifications (site screening and evaluation usually are performed on-site on an as-is basis, with no consideration for possible modifications).
The formalized structure of permit processes removes much of the flexibility from the design of the work flow and character of the product.

As a result of the latter distinction, much of the site qualification process is dictated by the number and types of permits involved. Permits may be required at the federal, state, and local levels and, depending on the site and prevailing law, may include those for

Disposal site
Shoreline modification
Surface mining
Water rights
Effluent discharges
Atmospheric emissions
Construction
Zoning variance
Dredge and fill

Proposed federal consolidated permit regulations would have simplified this process somewhat through creation of an approach to cover all pertinent permits at one time. This "one-stop" permitting is under consideration by numerous states. In either event, the single permit form or the combination of permit forms delineates the information that must be generated to complete site qualification activities. Guidelines are typically provided for how data should be collected. An example of such guidelines from the state of Texas is given in Appendix H. For the most part, this information requirement flows directly from site criteria established by the regulatory agency. If the site screening and evaluation have been properly conducted, criteria used in the first process will have been the same as or more stringent than regulatory criteria. This does not ensure site acceptance, however. Not all considerations can be quantified in definitive criteria, and hence the regulator must still exercise judgment.

To some extent, regulatory agents have been successful in minimizing the degree of judgment required in review of permit applications. Hard pressed to perform rapidly expanding duties with relatively fixed resources, administrators often have traded technical correctness for ease of review and implementation.

With respect to permitting activities, this has encouraged the use of design standards in lieu of performance standards. Once design standards have been promulgated, permit application review is greatly simplified, because the regulator need merely check proposed design features against standards and find them in compliance or nonconforming. This simplicity comes at a price. First, design standards neglect site-specific characteristics that are fundamental to the siting process. An illustration of the dilemma posed by design standards can be found in proposed criteria for secured landfills. Many state governments have considered criteria calling for a minimum of X ft of clay with a maximum permeability of Y cm/s. And yet it is openly recognized that some of the finest disposal sites in the United States can be found in the arid West, where high permeabilities are more than compensated for by a lack of precipitation. A single set of design criteria simply is not sufficient to cover the combination of characteristics that may produce a desirable site.

Second, design standards act as a deterrent to innovation and technological advancement. They create a static environment, with negative incentives to identify new means of achieving the desired level of security. Whereas performance standards provide the constant opportunity to devise new means for achieving acceptable levels of performance at a lower cost, new approaches to meet design standards must circumnavigate the tortuous path of regulatory changes or variances. Recognizing the risks associated with that alternative, few industries will commit resources to development when design standards prevail.

Finally, design standards often follow the development of performance standards. In order for design standards to be sufficient to protect against significant impacts without causing an undue economic burden, they must be carefully selected. This is best achieved by knowing what level of performance they will evoke and making a judgment on the acceptability of that level of performance. If this evaluation is implicit in the design standard development process, then performance standards can be readily derived.

Proposed facility siting standards under Section 3004 of RCRA currently rely heavily on a design standard approach. Individual site selection criteria are summarized in Table 5-5. Engineering requirements are added to these site requirements to specify such features as monitoring, runoff control, cover, and liner systems. Whereas some flexibility has been evidenced to provide for exceptions, the criteria offer little clarification of the actual performance characteristics that would be acceptable. Concern has been exposed in many areas. The 500-yr floodplain limitation has received special criticism, because many manufacturing plants and refineries are currently located in these areas and employ on-site disposal. State and local require-

Table 5-5 / Proposed RCRA Siting Criteria.

Geologic
 Outside active fault zone (reasonable probability of damage to site from movement on a fault)
Hydrologic
 Outside of regulatory floodway (100-yr floodplain)
 Outside coastal high-hazard area unless inundation will occur (subject to high-velocity waters,
  e.g., hurricane wave wash or tsunamis)
 Outside 500-yr floodplain unless designed to prevent inundation during a 500-yr flood event
 Outside recharge zone for sole-source aquifer unless there is proof of no endangerment
Environmental
 Outside a wetland
 Outside areas where endangered and threatened species would be adversely affected
Landfill-specific
 Prevent direct contact with navigable waters
 Bottom of fill at least 5 ft above the historical high water table (unless proof can be provided
  that no contact with the water table will occur)
 At least 500 ft from functioning public or private water supply or livestock water supply (unless
  proof is provided that no contact of landfill or leachate will occur and a monitoring system
  is in place)

ments may prove even more restrictive. Guidelines have been codified for permit writers with the intent of allowing some flexibility, but in many cases officials are reluctant to stray from those guidelines. Hence, they become functional standards.

As noted previously, design standards often are preferred by regulators because they are more readily implemented and enforced than performance standards. As regulators move to performance standards, they need to devise means of evaluating permit applications in a comprehensive way. One approach is the use of a decision model to formalize review of each specific concern related to site integrity. A sample decision model for use in evaluating hazardous waste landfills has been described by Perket (1977). The information flow is summarized in Figure 5-6. More detailed information on a single environmental pathway evaluation is illustrated in Figure 5-7. The approach is well suited to performance standards, because the decision process requires a direct comparison of estimated and acceptable releases.

Issuance, review, and acceptance of an environmental impact statement (EIS) is also a part of the site qualification process. In many respects, this document constitutes a formalized reporting of the data considered and generated during the first two phases of the siting process. It is the objective of the EIS to quantify, to the extent possible, the impacts anticipated from the proposed action as well as those of possible alternatives. In this way, decision makers and the public are informed of the social and environmental costs associated with an action for due consideration with respect to potential benefits.

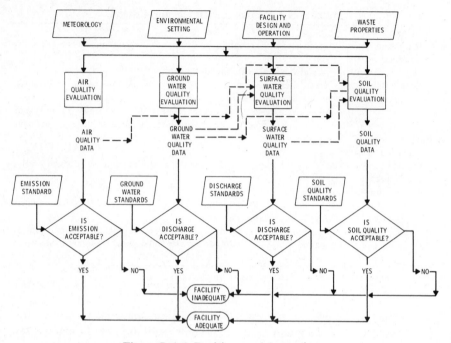

Figure 5-6 / Decision model overview.

The requirement for an EIS stems from the National Environmental Policy Act (NEPA) of 1969, which mandates preparation of "detailed environmental impact statements on proposals for legislation and other major federal actions significantly affecting the quality of the human environment." This has subsequently been interpreted by the courts to include virtually all activities involving federal participation, including funding or issuance of permits. Most states have enacted similar legislation, thus extending the applicability of the requirement.

Upon review of the EIS development and review process, the Council on Environmental Quality (CEQ) took measures to expedite the process while enhancing its responsiveness to the intent of the enabling legislation. Effective July 30, 1979, CEQ issued regulations for implementing NEPA procedures and guidelines on preparation of environmental impact statements. The guidelines call for the issuance of a draft statement for comment, hearings as appropriate, and attachment of substantive comments to the final statement.

Detailed instructions for the content of an EIS are provided in 40 CFR 1500.8. Briefly, the EIS should include the following:

A description of the proposed action
A description of the environment to be impacted

A discussion of the relationship of the proposed action to land use plans, policies, and controls in the affected area

An estimation of the probable impact of the proposed action on the environment (positive and negative, direct and indirect)

A description of alternatives to the proposed action and their associated impacts on the affected environment

A discussion of probable adverse environmental impacts that cannot be avoided

An evaluation of the relationship between local short-term uses of man's environment and the maintenance and enhancement of long-term productivity

A description of any irreversible and irretrievable commitments of resources associated with the proposed actions

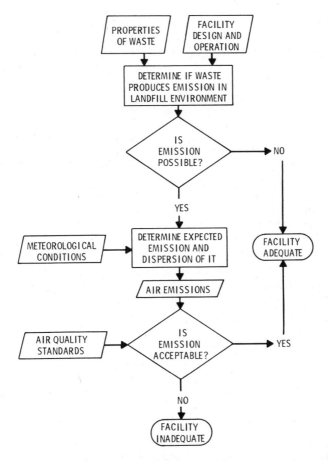

Figure 5-7 / Air quality evaluation.

An indication of other federal interests and considerations thought to offset adverse environmental effects

The guidelines also require that review include all federal agencies with related jurisdiction, affected state and local agencies in the area, and the public. The latter inclusion necessitates publication of notifications to solicit comment from interested parties.

The development of an EIS and methods available for assessment of impacts are topics of sufficient import to warrant volumes in their own right. As such, no attempt will be made here to cover these aspects of the site qualification process.

Existing Siting Systems

Various states and foreign countries have formalized siting criteria systems for hazardous waste facilities. A review of these will help focus attention on criteria commonly viewed as important, as well as the methods currently used for interpreting the basic systems previously described. Three basic approaches have emerged to date:

1. Analytical matrices attempting to quantify numerical criteria such that all play a role delineated by assigned weighting values (Baden-Württemberg).
2. Judgmental evaluation criteria (California, Illinois, and United Kingdom).
3. A decision-tree approach that uses elements of go–no go criteria for candidate site selection followed by judgmental criteria for final selection among acceptable sites (preliminary approach for U.S. EPA). (The modeling approaches have largely been restricted to consideration of nuclear repositories.) Several examples will help illustrate the implementations of such approaches.

California.   Hazardous wastes must be disposed of in class I sites, defined as those sites that can provide complete protection for groundwater and surface water quality as well as public health and environmental resources essentially in perpetuity. Prerequisities for class I sites have been enumerated as follows (NATO, 1977):

Geologic conditions must be naturally capable of preventing hydraulic continuity between liquids and gases emanating from the waste in the site and usable surface waters or groundwaters.

Geologic conditions that are not naturally capable of preventing lateral hydraulic continuity between liquids and gases emanating from wastes in the site and usable surface waters or groundwaters, or the disposal area, must be modified to achieve such capability.

Underlying geologic formations that contain rock fractures or fissures of questionable permeability must be permanently sealed to provide a competent barrier to the movement of liquids or gases from the disposal site.

Inundation of disposal areas shall not occur until the site is closed in accordance with requirements of the regional board.

Disposal areas shall not be subject to washout.

Leachate and subsurface flow into the disposal area shall be contained within the site unless other disposition is made in accordance with requirements of the regional board.

Sites shall not be located over zones of active faulting or where other forms of geologic change would impair the competence of natural features or artificial barriers that prevent continuity with usage waters.

Sites made suitable for use by man-made physical barriers shall not be located where improper operations or maintenance of such structures could permit the waste, leachate, or gases to contact usable ground or surface water.

Sites that comply with the foregoing requirements but would be subject to inundation by a tide or a flood of greater than 100-yr frequency may be considered by the regional board as *limited* class I disposal sites.

In addition, the state Water Resources Control Board has issued guidance (March 1975) on desirable geohydrologic features of class I sites. Acceptable impermeable conditions are defined as

Fine-grain soils classified as CL, CH, or OH by the Unified Soil Classification System

Permeability 10^{-8} cm/s or less is considered to be impermeable

Not less than 30% (by weight) of the soil passing through a number 200 sieve

A liquid limit of not less than 30 as determined by ASTM test D423

Plasticity index of not less than 15 as determined by ASTM test D423

Illinois. The state of Illinois has developed a similar listing of criteria for location of class I and class II landfills that may accept hazardous wastes. Their criteria, however, contain greater emphasis on design standards, rather than performance standards:

Prevention of surface water entering landfill; any runoff from landfill must be properly treated before discharged.

Trench method of disposal.

Waste confinement (liner).

High ion-exchange capacity preferred.

Depth to permeable bedrock—minimum of 10 ft (3 m).

Groundwater monitoring facilities required.

Minimum of 300 ft (91.5 m) between edge of trench and surface water body.

Site must be above 100-yr floodline.

Landslide and active faults (1 mile minimum from active faults; landslide area unacceptable).

Trenches must be composed of in situ natural earth materials. Hazardous liquid wastes must be containerized.

Baden-Württemberg. Siting evaluation methodologies within West Germany differ between the individual states. The scoring system employed in Baden-Württemberg is representative of a quantitative approach. Scores are given to each of the criteria listed below, weighted and summed to provide the comparison between sites. The major subject areas evaluated with regard to both treatment and disposal facilities are as follows (NATO, 1977):

Size of the site
Possibility of expansion
Ability to obtain the site (ownership, state of development of land)
Proximity to subsurface structures
Proximity to similar facilities
Changes in surface and subsurface structures as a result in the landfill
Influences occurring during the life of the landfill
Influences of the landfill after shutdown
Site characteristics with respect to water management
Possibilities of percolation water and process water drainage
Dispersion climatology
Background level of air pollution
Natural capacity of the site to abate emissions from landfill operations
Availability of earthen materials
Degree of difficulty of earthwork engineering
Site conditions that could cause hazards
Possibility of landfill site regarding long-term maintenance

United Kingdom. British officials employ a classification scheme for sites such that a wide variety of wastes can be assigned to a site, depending on quantity, solubility, toxicity, concentration, and other factors. Potential sites are reviewed with respect to potential for lechate attenuation and classified in one of three categories:

1. Provides significant containment of wastes and leachates
2. Allows slow leachate migration, but offers signification attenuation
3. Allows rapid leachate migration, but offers significant attenuation

Further consideration is judgmental with respect to location, size of site, impairment of visual or other amenities, access and transport, land use, and long-term controls.

Acceptable sites are subsequently characterized with respect to the following: depth to water table; water table contours and seasonal variability; location and distance to withdrawal or use points; ratio of evapotranspiration to precipitation and runoff; stratigraphy and structure to base of shallowest confined aquifer, including data on old mineral workings; base flow data on nearby perennial streams; chemistry of acquifer; confining beds; and probable leachate composition. For sites expected to receive large quantities of waste, permitting authorities may also request results of experimental measurements of permeabilities, effective porosities, cation-exchange capacity of each lithography in the saturated and unsaturated zone, moisture content, in situ soil moisture tension in the unsaturated zone, three-dimensional distribution of head to the base of the shallowest confining aquifer, storage and transmissivity, dispersibilities (field measurements seasonally), and definition of recharge and discharge areas. Applications with these data are reviewed by professional engineers and scientists, with weighting applied judgmentally on the suitability of a site. Decisions are subject to appropriate planning and site licensing conditions.

Public Acceptance

Although the essence of the siting function is technical in content, it has always involved a sociopolitical element. With passage of recent legislation granting broad powers to regulators and growing apprehension on the part of the public toward technology, the social-political element has grown. Today, sociopolitical considerations are the single greatest deterrent to siting hazardous waste facilities. Much of the impetus for this development can be traced to a series of developments:

1. The emergence of environmental awareness in the 1960s
2. The implementation of the National Environmental Policy Act of 1969
3. The debate of zero discharge (and zero risk) arising from the federal Water Pollution Control Act amendments of 1972
4. The disclosure of several large environmental "disasters," including Kepone contamination of the James River and Love Canal

As a result, the public has become very active as a party to siting activities through the environmental impact statement process, mandatory public hearings, and local government. Hence, public acceptance has become a key element for successful siting. And yet, advances in man's understanding of the social sciences and quantitative means for managing them have lagged far behind developments in the hard sciences. The ability to obtain public approval for hazardous waste facilities remains an art rather than a science and can by no means be guaranteed for any given project.

Some of the first real experiences with large-scale public opposition to facility siting arose from efforts in the early 1970s to site nuclear power reactors. The newly implemented requirements for an EIS spawned legions of intervenors who challenged the completeness of the impact analysis, the accuracy of the alternatives evaluation, the efficacy of nuclear power, and the significance of impact levels. In reaction, architects of impact statements began to produce more massive documents, with extensive lists inventorying the biota and detailed technical discussions. The average citizen was ill-prepared to critique these documents and began to hire technical experts to represent his interests. The field of environmental law blossomed, as did the number of consumer advocates. The technical community became split, with credible spokesmen emerging on either side of the issues and the public losing confidence in the technologists as a whole. The so-called environmental crises had arrived. With no means of deciding between conflicting statements, the public naturally gravitated to the safest position: zero risk, or minor risk in someone else's neighborhood. Much of this view prevails today.

The first examination of public attitudes toward siting hazardous waste facilities was undertaken as part of the background effort for the 1973 report to Congress (Human Resources Research Organization, 1973). Data were collected by means of questionnaire involving both randomly and purposely selected (influential) subjects in 10 counties of the United States that were potential candidates for national disposal sites (NDS). Here are some of the conclusions drawn at that time:

The social climate for establishment of a national disposal site (NDS) System is positive. Most respondents expressed a favorable attitude toward the NDS concept.
Counties differ in their probable receptivity to an NDS.
National disposal sites are perceived as relatively nondangerous and potentially beneficial to the surrounding area.
Accurate and adequate technical information is desired by most people before they agree to an NDS being located nearby. The preferred sources of such information are the mass communications media.
Approaches to obtaining acceptance of an NDS need to be different for individuals influential in decisions concerning local issues than for the average citizen.
Whereas concerns voiced by most people regarding the area impact of an NDS can generally be countered by factual information, the affect (or feeling) associated with such concerns must be given specific attention.
A proposed public information program would be likely to result in ready agreement to the establishment of an NDS in most locations.
The term "national disposal site" proved to have undesirable connotations for many respondents and should be avoided as a term of reference for such installations. One acceptable alternative designation is "regional processing facility" (RPF).

As a part of the study, the contractor developed a behavior model to predict how a county's population would react to a proposed NDS. Based on results of the questionnaires, it was determined that there were great differences in the degrees of influence related to 11 variable characteristics of

Table 5-6 / Relative Importance of 11 Variables in Determining Individual Receptiveness to NDS.[a]

Characteristic/Variable	Relative Importance
Environmental concern	0.0854
Perceived necessity for an NDS in the county	0.0349
Perceived NDS impact on property values	0.0185
Educational level	0.0169
Population change	0.0088
Median income	0.0082
Age	0.0082
Feelings concerning waste-carrying traffic	0.0079
Reported incidence of large plants located nearby	0.0077
Principal county income source	0.0068
Membership in organized group(s)	0.0068

[a]Adapted from Human Resources Research Organization (1973).

individual subjects (Table 5-6). As a result, a profile was constructed to describe the collective characteristics of citizens associated with declining levels of willingness to accept an NDS (Table 5-7). Data such as these may be of great value in developing socioeconomic criteria for use in site screening and evaluation activities.

The public attitudes study also developed a generic strategy for gaining public acceptance for a given site. The core of the strategy was communication with the public. The underlying theme of the communications was this:

We can and must improve our present methods of hazardous waste disposal. A national hazardous waste disposal system, such as proposed through establishment of RPF's (Regional Processing Facilities), represents a technological leap beyond present disposal systems. Its benefits far outweigh inconveniences it may impose.

If a system of sites was to be established, the communications or marketing strategy was to be focused at two levels: national and local. The national campaign was designed to foster awareness on a national level that there was a need for NDS. The local campaign was directed to gaining acceptance of a specific facility. Based on the results of the questionnaires, the Human Resources Research Organization (HRRO) concluded that specific features of the local campaign should be selected in direct response to information gathered from the areas surrounding the proposed site, as was done for the following guidelines:

Attitudes: (1) If attitudes toward an RPF are moderately negative or neutral, the information program should stress the central theme of the campaign: There is a real need for efficient means of disposing of hazardous waste, and an RPF meets the need, without danger of negative impact on the local environment. (2) If attitudes toward an RPF are very positive or

Table 5-7 / Relative Receptiveness of Characteristic Groups in County to NDS[a]

Receptivity Rank	Relative Score	Median Income	Perceived Effects of NDS on Property Value	Concern for Environment	Existing Industry in County	Perceived Need for NDS in County	Age	Number of Organized Group	Population Changes in Last 10 Years	Education
1	31.89	<6000	No effect	Very concerned						
2	28.33	>6000	No effect	Very concerned						
3	28.05			Concerned	Low salience					
4	27.79			Very concerned		Necessary	18–39			
5	26.69		Unknown or negative	Very concerned			≥40	Yes		
6	25.73		Unknown or negative	Moderate to no concern	Mix of industrial and non-industrial	Not necessary				Some college
7	25.53		Extremely willing to allow hazardous wastes moved near community	Moderate to no concern		Necessary				
8	25.19			Concerned	High salience	Necessary				
9	22.54			Low to no concern		Not necessary			Other than moderately increasing	High school or less
10	22.27		Extremely to somewhat willing to have hazardous wastes moved near community	Moderate to no concern		Necessary				
11	22.17			Moderate to no concern	Small to substantial industrial base	Not necessary				Some college
12	21.87		Unknown or no effect	Very concerned		Not necessary	≥40	No		
13	19.20		Unknown or no effect	Moderate to no concern		Not necessary			Moderately increasing	High school or less

[a] Adapted from Human Resources Research Organization (1973).

moderately positive, the central theme should be included, but not stressed.

Information: An adequate amount of clearly presented and understandable information corresponding to the desires of the citizens must be included in the information program.

Proximity: The information program should stress that the RPF would not be located too near the outskirts of any town.

In addition, architects of the campaign should be intimately familiar with such items as these:

Impact on local economy
Impact on land values of property adjacent to the RPF
Impact on land use patterns, perceived appropriateness in use of the land, whether or not competition for available land will be involved, whether or not establishment of the RPF could result in cessation of ongoing or proposed activities (e.g., residential subdevelopments)
Impact on the general environment
The prospect of aesthetic degradation to the environment caused by the RPF
Safety risks imposed by transportation of hazardous waste to the RPF

The accuracy of information released on these topics is of utmost importance, because these are issues that the opposition will focus on. Should the public discover intentional or inadvertent errors in the information released, the credibility of the campaign will be jeopardized and prospects for success damaged greatly. If hard data are not available, credible, recognized authorities may be cited to produce a foundation of respected opinion. Because of the ramifications of errors in the information process, public relations responsibilities should be assigned to a single individual.

Positive features to be emphasized should include new employment opportunities, the prospect of new and cleaner industries, and a safer and cleaner environment. In general, the lower the economic level of a county, the more strongly economic issues should be stressed; the higher the economic level of a county, the more strongly safety and environmental issues should be stressed. These constitute methods for influencing social judgment for acceptance. On review of the variables affecting public acceptance and the behavioral model, it was further determined that three other key variables can be affected by the information campaign. HRRO's recommendations in these areas follow:

Level of environmental concern. The detrimental impact on the environment caused by the actions of various agencies and industries should be stressed. It should also be emphasized that degradation of the environment does not remain a localized problem, but spreads rapidly, with many long-term effects. Many of these environmental problems can be solved by an RPF.

Perceived RPF impact on property values. The aesthetic aspect should not be overemphasized. In other words, the RPF will not look like a country club. Instead, a reasonable picture of what it will be should be presented, with emphasis that there will be no negative impact on surrounding areas in the form of any kind of pollutant. There will be no odor from any kind of emission. It should be pointed out that usually a tree line is left around the perimeter of the property if it is located in such terrain. Otherwise, low screens may sometimes be erected around some of the equipment and processing facilities.

Feeling concerning traffic carrying waste to the RPF. As quickly as feasible, select (if possible) a route for the traffic that by-passes the nearby towns. Determine the load-bearing capacity of the route to see if it will be adequate; initiate negotiations to participate financially in upgrading it. Make certain that people learn of this, and stress that the reason for the financial outlay is to avoid inconvenience (not danger) to residents of nearby towns.

In addition to these guidelines for the content of the campaign, HRRO developed important recommendations for the dissemination strategy. At the outset of the campaign, the siting team should identify key influential people in the local area, those in government, business, industry, education, environmental groups, service organizations, and the news media. This provides the groundwork for future public interactions by building credibility, and it can help identify potential problem areas in advance. Typically, influential leaders desire information about the proposal different from that desired by the general public. Although they are interested in technical aspects, they often are also concerned about things such as financing and regulations. Pertinent information should be on hand when contact is first made.

The second phase of information dissemination involves communications with the public at large. Based on HRRO's research, it was concluded that in lower-income areas, emphasis should be placed on the use of presentations to local groups, service organizations, and associations. In higher-income areas, more reliance should be placed on use of the news media. To the extent possible, key influential citizens should be brought into the effort through endorsement or mere presence at the presentations. Hence, the initial part of the campaign must win these individuals over by convincing them that the proposal is worthwhile and that their support is part of their public responsibility. In addition, influential citizens must be fully informed in the project. It is often helpful to include influential citizens in the decision-making process to ensure a certain sense of commitment.

It should be noted that events occurring subsequent to the HRRO work have changed the atmosphere within which siting must be conducted. Whereas HRRO noted a basically positive attitude toward the concept of a system of NDS, hazardous waste disposal sites (an unfortunate choice of terminology) have gained a very negative connotation. Publicity surrounding the Kepone contamination of the James River and radiation releases from the Three Mile Island nuclear reactor has created an image of technology out of control. Public apprehension has risen to the point that citizen groups have challenged the integrity of industry and successfully closed permitted sites such as the landfill at Wilsonville, Illionis, even though the state and federal regulatory agencies sided with the operator in concluding that the facility was safe. In the more activist segments of the citizenry (typically the middle-income and upper-middle-income brackets), the contention that hazardous waste disposal is one of the costs of the current standard of living is losing viability. The back-to-nature philosophy and stoicism arising from the stark realities of the energy dilemma have created a mood conducive to denial of the fruits of industry as a means of eliminating undesirable environmental impacts. Hence the emergence of such proposals as strict enforcement of energy conservation measures and abandonment of nuclear power. Although these changes in attitude may reside largely in a very vocal minority, they must be reckoned with. They represent a polarization of opinion since the conclusion of the HRRO survey.

In the 1973 report to Congress, the EPA concluded that the best strategy for management of hazardous waste was to promulgate comprehensive regulations and allow private industry to respond to market forces. As such, the concept of a system of national disposal sites was shelved. Subsequently, follow-on studies and additional development of the work by the Human Resources Research Organization were not pursued with respect to the management of chemical wastes. Social science work has continued, however, in the area of nuclear repository siting. The results of this work can be of value. Indeed, the results of a 1978 survey indicated that at that time people were more concerned over the potential problems of toxic wastes than the problems of radiological waste materials (Table 5–8). These data also illustrate the key role that community leaders can play in opposition to industrial waste disposal facility siting. In related work, it was found that

Table 5-8 / **Results of a Survey Reflecting Concern That a Given Waste Type Could Cause Major Problems.**[a]

	Affirmative Responses			
	General Public		Key Leadership	
	1977	1978	1977	1978
Toxic industrial wastes	71%	68%	91%	91%
Radiological wastes	52%	45%	68%	68%
Slag and tailings	53%	43%	41%	35%
Industrial packaging materials	30%	29%	12%	6%

[a]Adapted from Rankin and Melber (1980).

although the public perceived a strong need for long-term site control, monitoring, and information transfer as to where sites were to be located, there was low, medium, and medium-high confidence, respectively, in technology's ability to fulfill these needs (Rankin and Melber, 1980).

In studies of public acceptance of nuclear power, Battelle determined several important trends with respect to information strategies that can be applied to siting hazardous waste facilities (Rankin and Melber, 1980). In terms of where people obtain their information, these relative rankings emerged:

1. Newspapers
2. Television
3. News magazines
4. Radio
5. Environmental groups
6. Utility companies
7. Government agencies
8. Friends

At the same time, the public puts most faith in what they receive from news media and government agencies and least faith in what comes from environmental groups and the utility companies.

The focus of work performed as a part of the Department of Energy's nuclear waste isolation studies has been the management of impacts resulting from construction and operation of a site. Because these facilities will be larger and more heavily staffed than a chemical waste facility, the impacts related to demographic issues (education, housing availability, property value, etc.) are generally overstated. In similarity to the findings of HRRO, however, Greene and Hunter (1978) noted that these are often the concerns of local officials, in contrast to predominant concerns for safety by the general public. For federally funded efforts such as nuclear waste repositories, these public service and demographic impacts can be managed through compensation. Several mechanisms are available to provide funds in lieu of taxes that help cover the costs of remedial actions to mitigate impacts. For private siting of hazardous waste facilities, such compensation is less likely, because taxes will not be waived. Some financial arrangements may be involved in the way of fees and services to the local facility. Though not required by law, these measures have been suggested as effective means for increasing public acceptance. Should sufficient funds be generated to have a significant impact on the local tax base, a disposal site can be viewed as an economic boon to the area. If subsequent tax reductions are potentially large, communities may be enticed to compete for facilities in their jurisdiction.

Current problems in siting new facilities have raised considerable interest at EPA in incentive programs and compensation to communities and affected individuals. Various forms of payment are possible:

Direct payments to compensate loss in adjoining property values
Community improvements and utility modifications that will better prepare the area for emergencies and provide additional general benefits
Direct payment of fees or special taxes to relieve local financial requirements on citizens
One-time lump-sum payments
Local improvements unrelated to the site, such as recreational facilities, a community center, etc.
Free services for local residents
Property value guarantees against price depression in future years.

Experience in the public hearing process has been gained in recent years through land use planning and landfill siting activities. Hearings have been held to meet the public participation aspects of the planning process required in Section 208 of the Federal Water Pollution Control Act amendments of 1972. The intent of the requirement is to provide a mechanism for the public to define the environmental future it seeks. Subsequent planning is then to be supportive of that future. As one might expect, the process is frustrating, because it is difficult if not impossible to bring the public to a consensus. There often are as many opinions as there are participants. To accommodate this heterogeneity, planners in Pennsylvania sought representatives of five major groups (Kampschroer, 1977):

1. The representative public—elected and appointed officials at every level of government
2. The economically concerned public—business and industry
3. The organized public—civic, social, service, conservation, and environmental groups
4. The academic public—teachers and students
5. The general public

These representatives were then used in steering committees and working groups to help develop a sense of commitment to the project and to identify the issues most likely to arise in subsequent public hearings. It was in this setting that engineers began to identify some of the points of friction. They often found that the public (1) lacked faith in the accuracy of data presented, (2) exhibited general antagonism, (3) failed to understand technical jargon, and (4) opted for no change when technical alternatives were complex or decisions difficult.

No single solution has been found to surmount these difficulties. The 208 public hearings were stormy affairs, often breaking down into emotional duels. It was not unusual for the contractor's staff to be threatened by hearing attendees who became so frustrated over the process that they could not deal with it rationally. As a result, the engineer with interpersonal skills has become a valuable asset. This individual is best used in two capacities: (1) to design and implement a public information program in advance of hearings, thus preparing the potential audience ahead of time and disarming some of the disruptive elements, and (2) to manage the meeting so that full participation ensues, without things getting out of hand or speakers going off on a tangent.

Recognizing the impact of public opposition to siting and the growing number of siting failures, in 1979 the EPA funded an analysis of case histories to provide insights into successful and unsuccessful efforts (Centaur Associates, Inc., 1979). Twenty-one case histories were evaluated by Centaur Associates, representing five discrete sets of circumstances:

Successful siting of a new facility where there was no public opposition (four cases)
Successful siting of a new facility in spite of public opposition (five cases)
Unsuccessful siting of a new facility faced with public opposition (seven cases)
Continuance of an operating facility faced with public opposition (three cases)
Closure of operating facility faced with public opposition (two cases)

Foremost of Centaur's findings was that the single most important factor in addressing public opposition to siting is coordination and communication with the public and local officials. Efforts to that effect must be directed to convince the community that (1) complete information is available on the operation of the site and proposed waste streams; (2) the public and local officials will be substantively involved in the siting process; (3) the operator is a person or organization of lasting integrity; (4) the risk of catastrophic accidents or insidious dangers are slight; (5) there are significant benefits to the local area to offset the risks; (6) the site and its operation are not in conflict with other enterprises or existing activities in the area, nor are there any better and more feasible higher land uses for the site; (7) the government has sufficient resources and expertise to judge independently the merits of site design and operation; (8) there are sufficient government regulations and resources to guarantee safe operations; (9) there are sufficient resources and government regulations to ensure that the facility will be properly maintained after closure; and (10) the technical merits of the selected site and facility are unquestioned.

Centaur determined that it is essential that the foregoing information be conveyed to the public in a timely manner. When the site selection and design are presented largely as a fait accompli, public opposition can become quite intense. The community must feel that they can influence the process and that concurrence has not been taken for granted. Low-profile approaches appear to have been successful only in heavily industrialized areas, where the benefits of the generator industry were readily apparent to potential opposition.

Other points were noted in the Centaur analysis:

Technological studies used in support of siting are of greatest value when produced by a neutral, objective third party.

Certain "political" wastes, such as Kepone, PCBs, and dioxin, stimulate excessive emotionalism and should not be highlighted in disclosures.

Populations in some geographic areas, such as Idaho, are more receptive to siting as a result of a history of trust in government and strong beliefs in individual property rights free of outside interference.

Conversely, areas with a history of more complex political activities are quick to generate strong, well-organized opposition.

Local elected officials are almost always present in the opposition and therefore provide a means of delay and frustration through ordinances and other measures aimed directly at proposed facilities.

No matter how well planned and orchestrated these siting activities are, the outcome cannot be guaranteed. Gaining public acceptance involves more art than science at this time, and the history thus far tells us that many good, technically superior sites will be rejected because of public fears. In the eyes of many, this has occurred too often in the past. The opening of new sites has not kept pace with needs. The regulations mandated by RCRA cannot be met, and as a consequence, the law may provide to elicit more of the truly impactive practices than it resolves. In response, waste management experts have begun to investigate legal options available to circumvent public opposition. Several states have contemplated use of eminent-domain proceedings. The majority of states, however, do not want an active role in procuring and operating sites, even though some 25 have or intend to have authority for siting hazardous waste facilities. The National Solid Waste Management Association (NSWMA) has recommended a more restricted approach that will allow states and the federal government to issue permits that will override local actions aimed at stalling siting indefinitely. The proposed rider might read something like a related passage from Florida law:

No political subdivision of the state shall adopt or enforce any action, rule, ordinance or standard which will operate to prevent the location or operation of a hazardous waste transporter, processor, storer or disposer who is issued a permit.

The powers involved would include the ability to bypass local zoning restrictions. Recently, both Oregon and Michigan have passed laws giving state agencies the authority to approve siting and override local objections. The Oregon statute (SB 925) is actually intended as an aid to local government. Although it provides preemptory powers to the state Department of Environmental Quality, it stresses work in close conjunction with local government to establish sites. The override authority is available only as a last resort. The Michigan law (PA 64) is much more direct in its reliance on state agencies. It deals with hazardous waste sites only. Specific permit applications are reviewed by a board of nine members, with five permanent members representing state interests and four company members from the local area involved. Hence, local interests never have a majority representation. The board must consider local regulations, but the latter may not prohibit construction of a facility. Additional laws may not always be required to remedy siting problems. In New Jersey, the state Superior Court has ruled that municipalities lack the authority to use regulations such as zoning ordinances to prohibit landfill operations within their borders. Maryland has established a siting board with the express authority to override local zoning restrictions. Other states and organizations have stopped short of the override authority, but emphasize the use of mediation to bring the public around. Such an approach has worked in both New Jersey and Maryland, prompting optimism that if the effort is properly conducted, a consensus can be reached. By July 1980, 24 states had passed some form of siting legislation, and acts were pending in 5 other states. A 1982 survey revealed that state siting laws had been adopted in 28 states as summarized in Table 5.9 (National Conference of State Legislatures, 1982.)

The NSWMA has also proposed that use of state or federal lands for sites may be the most feasible approach. Although this is counter to the private enterprise orientation of the trade association, it recognizes the difficulties that have arisen in relations between industry and the public and capitalizes on the ability of the larger governmental entities to circumvent restrictive local policies. The implication is that the larger governmental bodies have a greater resource base from which to make technical decisions and are more capable of selecting between tradeoffs involving a broad segment of society. At issue is the right of a small subelement of society to accept or reject perceived risks. As is apparent, these proposed approaches to siting hazardous waste disposal facilities complicate the issue further by introducing political elements (e.g., states' rights, home rule, etc.).

Even with successful completion of the work described, it must be recognized that the public may never be at ease with hazardous waste disposal sites. There may be no way to convince the average citizen that the associated risks are minimal. Currently, sites with excellent operating records are undergoing public challenges and may be removed from service. As a consequence, it is prudent to redouble efforts in a complementary area: source reduction. Faced with a limited number of existing sites and the potential for few, if any, new sites, burial should be employed only for those materials that cannot be destroyed (elemental materials). Similarly, all means should be explored to minimize waste production and to define hazardous wastes such that only materials posing a real threat to humans

Table 5-9 / States with Hazardous Waste Facility Siting Laws (National Conference of State Legislatures, 1982).

	AZ	CO	CT	FL	GA	IL	IN	IA	KS	KY	ME	MD	MA	MI	MN	NE	NH	NJ	NY	NC	OH	OR	PA	RI	TN	UT	WA	WI
Siting procedures																												
Siting process initiated by																												
State	x											x	x												x	x		
Developer		x	x		x	x	x	x	x	x	x	x	x		x	x	x	x	x	x	x	x	x	x	x	x	x	x
Siting decision by																												
Existing agency	x				x				x			x		x		x			x			x			x			x
Siting board		x	x			x	x	x	x	x	x	x	x		x		x	x	x	x	x	x	x	x	x	x		
Local group					x				x			x				x			x				x					
Permit approval by																												
Existing agency	x	x	x		x	x	x	x	x	x	x	x	x		x	x	x	x	x	x	x	x	x	x	x	x	x	x
Siting board		x	x			x			x																x			
Local group	x																											
Siting impasse resolved through																												
State preemption	x	x	x		x	x	x	x	x			x		x			x	x	x	x	x	x	x		x	x		
Mediation/arbitration													x											x				x
Local veto			x						x						x			x	x									
Public participation																												
Local representatives sit on siting board		x	x					x	x			x	x		x			x		x						x		
Local review board			x		x				x				x		x									x				x

177

Table 5-9 / (Continued)

	AZ	CO	CT	FL	GA	IL	IN	IA	KS	KY	ME	MD	MA	MI	MN	NE	NH	NJ	NY	NC	OH	OR	PA	RI	TN	UT	WA	WI
Notice of permit application given affected parties	x	x	x		x	x	x	x	x	x	x	x	x	x	x	x	x	x	x	x	x	x	x	x	x	x	x	x
Hearings	x	x	x		x	x	x	x	x	x	x	x	x	x	x	x	x	x	x	x	x	x	x	x	x	x	x	x
Citizen suits									x	x												x				x		
Other	x		x	x				x		x			x		x			x	x				x		x		x	
Financial assurances																												
Trust funds	x	x						x	x	x			x								x	x	x		x			x
Financial responsibility mechanisms	x	x	x		x	x		x	x	x	x		x	x	x	x		x			x	x	x		x	x		x
Other	x		x			x	x			x																		
Nonfinancial assurances																												
Inspections	x	x	x		x			x	x	x		x	x	x				x			x	x	x		x		x	x
State ownership	x									x		x										x					x	x
Contingency plans	x		x		x			x	x	x		x	x					x			x	x	x				x	x
Restrictions on future users	x	x			x			x	x	x			x					x	x	x	x		x		x		x	x
Incentives and compensation																												
Local taxes or receipt of fees	x	x			x				x	x	x		x			x	x	x		x			x				x	
Tax prepayments													x															
Other		x			x								x					x			x		x					

178

and the environment are included. It would be foolish to use scarce site capacities for materials of low intrinsic hazard. This also raises the issue of concentration/dilution of wastes. Concentration of hazardous wastes at the source would further reduce siting demands. At the same time, the more concentrated the waste, the greater the potential consequences of a release. Site requirements may be reduced and public acceptance may be enhanced if wastes are diluted to levels below which their hazards have been evidenced. The economic and political implications of these alternatives need to be continually explored such that viable options are available to reduce overall demands for more hazardous waste disposal facilities.

Currently, the U.S. EPA espouses a policy to prioritize the consideration of options for hazardous waste management. This is presented in the format of a waste disposition hierarchy establishing the following alternatives in order of priority (U.S. EPA, 1977):

1. Waste reduction
2. Waste separation and concentration
3. Waste exchange
4. Energy/material recovery
5. Waste incineration/treatment
6. Secure ultimate disposal

In-depth consideration of concentrated disposal versus dilution-dispersion is required in the context of recent experiences and the availability of resources.

Public acceptance problems are not restricted to the United States. European sites have experienced a growing level of public scrutiny and activism. Shipments of Kepone-contaminated wastes to the Herfa-Neurode salt mines were delayed when protests by environmentalists virtually halted activities. The rise of the so-called Green Party and the election of officials who support its environmentally oriented platform have increased the prestige and power of these lobbyists. As a consequence, parties siting new facilities in Europe must pay particular attention to public involvement.

In a French study of how best to involve citizens in the process, three important groups were identified (*Ordeurs Manageres*, 1976):

1. Local elected officials
2. Resident committees
3. Environmental interest groups

These researchers determined that all activities should be designed around a posture that first takes into account the attitudes and opinions of these three groups. When options have been developed, they should be submitted for approval by these groups. Finally, efforts should be made to reduce anticipated resistance to siting before, during, and after initiation of the project. This is done with instructions, information, consultation, and planning during the preconstruction activities. During later stages of development, information dissemination should be accompanied by appeasement. Public educational activities should include the creation of a study and decision-making committee that is allowed to influence the direction of activities. The overall public involvement program can be divided into six activities with the following purposes:

1. Make the public aware of the need.
2. Define the difference between present and future approaches and options.
3. Do not initially release studies of sites, because this will only arouse conflicts prematurely.
4. Reduce candidate sites to best alternatives, but do not explain to the public.
5. Perform detailed studies of variations and analysis, and release these to the public to prevent good solutions from being dropped by local officials.
6. Compare alternatives and select one; release this information to the public.

Financial Considerations

Successful siting of hazardous waste facilities, like the initiation of any enterprise, requires the accumulation of a sufficient quality and mix of resources, including a geologically acceptable location, proper engineering design, social and political power to gain acceptance, and capital. The long-term liability aspects of hazardous waste management can complicate the latter requirement. In addition to the obvious need for financing site construction and securing operating capital, prudence and government regulations require additional financial strength to cover four areas:

Continuity of operation through closure: Operating sites require special modifications when put in a layaway or closed mode. The costs of this action will be site-specific and should be determined at the time of opening. Before a site can be permitted, proof must be provided that sufficient funds will be available to conduct these closure activities. This is to ensure that an operator will not dissolve the organization or declare bankruptcy and leave the site in an unsecure mode, threatening long-term release of wastes.

Post-closure monitoring and maintenance: The potential for loss of hazardous wastes from a site continues beyond the cessation of operations. As a consequence, there is the need for provision of funds to pay for maintenance and monitoring beyond that date. Regulations will specify the exact period of post-closure activities. Proposals to date have typically addressed a 30-yr time frame.

Operational liability: Because of the intrinsic nature of the materials handled, hazardous waste management facilities pose what are perceived as higher than normal risks to operators and the public. In the case of the latter, regulations will require proof of insurance or other financial responsibility to cover third-party damages resulting from sudden and nonsudden releases. Present proposals specify sudden occurrence coverage for a minimum of $1 million per occurrence per site. Nonsudden coverage must meet or exceed $1 million per occurrence and $2 million annually.

Post-closure liability: Finally, funds must be available to cover damages resulting from releases after closure of a site. Because operators may no longer constitute a viable organization at this point, financial responsibility will require creation of a continuing fund to cover any such contingencies.

Of these, the issues related to closure and long-term care have proved the most vexing.

The 1973 report to Congress on hazardous wastes addressed the issue of long-term care in the analysis of private sector involvement in hazardous waste management. Major concern was expressed for the private sector's ability to assure long-term control because of changes in ownership, better investment opportunities, bankruptcy, and other factors. Similar concern is reflected in the program promulgated by the state of Oregon, wherein sites are deeded to the state after closure, and operators must post a closure bond. This resolves the funding of closure activities, but begs the question of how to pay for any necessary remedial actions or liability claims above the value of the land. Indeed, several possible mechanisms have been identified for funding closure and long-term care activities (U.S. EPA, 1977):

Bonding: One means of ensuring the availability of monies for long-term care is the posting of a cash bond or the maintenance of a surety bond by the site operator. The surety bond, if available, probably will be less expensive than a cash bond, because premiums are likely to be less than payments for a loan to cover the cash bond. In either case, the size of the bond should be sufficient to cover the costs of closure, post-closure monitoring and surveillance, and maintenance for the 30-yr observation period. The actual dollar amounts will vary with the site and the nature of the wastes to be managed.

Perpetual-care fee: as an alternative to the burden of providing cash funds on the front end of an operation, a site may collect a perpetual-care fee. This surcharge is based on the quantity and type of waste disposal. It is sized such that by some predetermined time, such as when the site has reached capacity, sufficient monies will be available to pay for closure and long-term care. In this respect, fees are designed as a sinking fund. The weakness of this system lies in its assumption that all sites will be closed after reaching the predetermined cutoff time. Should a site be closed prior to that time, insufficient funds will have accrued to pay for closure and long-term care. In this regard, the decision between bonds and fees involves a trade-off between front-end financial burdens and potential shortfalls in fund requirements.

Bonding/fee combination: Hybrid systems have been devised in an attempt to marry the strengths of the two approaches discussed earlier. Specific configurations could include use of a cash bond, with principal withdrawn in an amount equivalent to fees collected to date. Hence, the bond would be transferred to a sinking fund at a rate commensurate with site usage. Alternatively, a surety bond could be maintained at all times to cover the difference between the sinking fund generated to date and the amount ultimately required for closure and long-term maintenance. This alternative is tantamount to purchase of a declining insurance policy to protect the state against early closeout of operations.

Mutual trust fund: In a modification of any of the foregoing mechanisms, fees or bonds can be pooled within a given state or region to expand the total financial resources that can be brought to bear on a particular site at any given time. This is particularly useful in the context of unexpected repair activities and/or damages.

With the possible exception of mutual funds, the foregoing mechanisms are not well suited to long-term liability questions. Each mechanism is designed to provide for predictable costs associated with activities scheduled in a detailed closure plan, e.g., decontamination of equipment and facilities, removal and disposal of hazardous residuals, modification of site to a secure status, and implementation of long-term care. Financial resources to provide for liability claims are typically embodied in insurance policies. However, such policies may not be available. In his report to Congress in December 1978, the comptroller general noted that policies to cover the proposed $5 million liability level were not available (Comptroller General, 1978). At that time only one of the large disposal firms had a policy for a period of up to 40 years after closure, and the maximum liability level was less than $5 million. At that, the premium was high ($57,000 per year), and the policy could be canceled by the insurer after a 30-day notice. Factors contributing to the unavailability of insurance have been identified: (1) lack of historical data on which actuarial tables and reasonable premium rates can be based, and (2) inability of any private institute to guarantee against open-ended liabilities in perpetuity.

Recognizing these fundamental problems, the National Solid Waste Management Association proposed creation of an industry-wide, federally administered trust fund financed by user fees. This proposal was favorably received by the EPA and has been considered as a potential add-on to future revisions in RCRA. Subsequently, the insurance industry has emerged with a viable alternative. In July of 1980, representatives of the American Insurance Association surprised federal agents responsible for designing closure regulations and permit requirements by announcing that liability insurance would likely be available for the amounts needed. A contributing factor to the future availability of policies may have been the announcement in the same time frame that a risk-assessment methodology had been devised to score sites and facilities on potential risks. This system is being offered by Clement Associates, Inc., of Washington, D.C., to help insurers with little knowledge of the industry. And yet, the March 1982 Department of Treasury Report to Congress mandated in CERCLA found no reason to believe that private could or would replace the Post-Closure Liability Trust Fund.

Whether either or both of these alternatives (the trust fund and insurance will remain viable with issuance of permits under RCRA remains to be seen. At this time, however, the potential for either has considerably brightened prospects for siting new facilities. Financial responsibility requirements can now apparently be met. In turn, this may reduce public concern, because it will provide some assurances that sites will be properly closed, monitored, and if necessary decontaminated.

Neither of the foregoing mechanisms addresses the liability aspects of facilities that have already been closed. Funding for remedial action and damages from these facilities is the impetus for the "Superfund" created by CERCLA.

REFERENCES

Ahlstrom, S.W.; and Foote, H. P. 1977. *Multi-component Mass Transport–Discrete Parcel Random Walk Model Theory and Numerical Implementation.* BNWL-2127. Richland, Washington: Pacific Northwest Laboratory.

Baker, D. A.; Hoenes, G. R.; and Soldat, J. K. 1976. *FOOD: An interactive Code to Calculate Internal Radiation Doses from Contaminated Food Products.* BNWL-SA-SA-5523. Richland, Washington: Pacific Northwest Laboratory.

Battelle Memorial Institute. 1973. *Program for the Management of Hazardous Wastes.* U.S. Environmental Protection Agency, contract No. 68-01-0762, July 1973.

Black, M. W. 1978. Problems in siting of hazardous waste disposal facilities—the peerless experience. In *Proceedings of the 1978 National Conference on Hazardous Material Spills* (Miami, April 11–13, 1978), Information Transfer, Inc., pp. 232–235.

Brown, S. M. 1984 *Remedial Action, Volume 2, Simplified Methods for Subsurface and Waste Control Actions.* Palo Alto: Anderson–Nichols & Co. Inc.

Brown, S. M.; Dawson, G. W.; Drake, R. L.; and Parkhurst, M. A. 1983. *A Review of Environmental Exposure Assessment Methods for the Electric Power Industry.* Richland, Washington: Pacific Northwest Laboratory.

Burkholder, H. C.; Greenborg, J.; Stottlemyre, J. A.; Bradley, D. J.;

Raymond, J. R.; and Serne, R. J. 1979. *Technical Progress Report for FY-77.* PNL-2642, UC-70, Office of Nuclear Waste Isolation, U.S. Department of Energy, April 1979.

Centaur Associates, Inc. 1979. *Siting of Hazardous Waste Management Facilities and Public Opposition.* SW809, U.S. EPA, November 1979.

Cleveland, J. A.; Grover, R. B.; Petrillo, J. L.; and Ladd, E. 1979. Using computers for site selection. *Environmental Science and Technology,* Vol. 13, No. 7, pp. 792–797, July 1979.

Comptroller General. 1978. *Report to the Congress of the United States— How to Dispose of Hazardous Waste—A Serious Question That Needs to Be Resolved.* CED-79-13, USGAO, December 19, 1978.

Elzy, E.; Lindstrom, F. T.; Boersma, L.; Swat, R.; and Wicks, P. 1974. Model of the movement of hazardous waste chemicals in and from a landfill site. In *Disposal of Environmental Hazardous Wastes,* Environmental Health Services Center, Oregon State University, pp. 111–134, December 1974.

Greene, M. R.; and Hunter, T. 1978. *The Management of Social and Economic Impacts Anticipated with a Nuclear Waste Repository: A Preliminary Discussion.* U.S. Department of Energy, May 1978.

Gupta, S. K.; Tanji, K. K.; and Luthin, J. N. 1975. *A Three-Dimensional Finite Element Groundwater Model.* University of California, Davis.

Holcomb Research Institute, 1978. *International Assessment of Groundwater Modeling as an Aid to Water Resource Management.* Butler University.

Houston, J. R.; Strenge, D. L.; and Watson, E. C. 1974. *DACRIN: A Computer Program for Calculating Urban Dose from Acute or Chronic Radionuclide Inhalation.* BNWL-B-389. Richland, Washington: Pacific Northwest Laboratory.

Human Resources Research Organization. 1973. *Public Attitudes Toward Hazardous Waste Disposal Facilities.* NTIS PB-223 638, U.S. Environmental Protection Agency, September 1973.

ICF, Inc.; and Battelle Pacific Northwest Laboratories. 1984. "Post-Closure Liability Trust Fund Simulation Model." U.S. EPA.

Kampschroer, B. J. 1977. 208 planning: The public gets involved. *Water and Wastes Engineering,* Vol. 14, No. 3, pp. 30–42, March 1977.

Kipp, K. L.; Reisenauer, A. E.; Cole, C. R.; and Bryon, C. A. 1972. *Variable Thickness Transient Groundwater Flow Model Theory and Numerical Implementation.* BNWL-1703. Richland, Washington: Pacific Northwest Laboratory (updated 1976).

Lester, D. H.; Jansen, G.; and Burkholder, H. C. 1975. The migration of radionuclide chains through an adsorbing medium. *Adsorption and Ion Exchange,* Vol. 71, p. 202.

National Conference of State Legislatures. 1982. *Hazardous Waste Management: A Survey of State Legislation 1982.* Denver.

NATO, 1977. *Disposal of Hazardous Wastes—Manual on Recommended Procedures for Hazardous Waste Management.* NATO, No. 62, prepared by the U.S. and Canada, June 1977.

Nelson, R. W. 1976. *Evaluating the Environmental Consequences of Ground-*

water Contamination. BCSR-61UC-11. Richland, Washington: BCS Richland, Inc.

Ordeurs Manageres; l'Implantation d'un Centre de Tractment. 1976. Paris: French Ministry of Environment, March 1976.

Perket, C. L. 1977. *Decision Model for Determining the Suitability of Landfilling Hazardous Waste.* Minnesota Pollution Control Agency, June 17, 1977.

Pojasek, R. B. 1980. Developing solutions to hazardous-waste problems. *Environmental Science and Technology* Vol. 14, No. 8, pp. 924–929, August 1980.

Rankin, W. L.; and Melber, B. D. 1980. *Public Perceptions of Nuclear Waste Management Issues.* BHARC/411-80-004, Battelle Human Affairs Research Center, February 1980.

Remson, I.; Fungaroli, A. A.; and Lawrence, A. W. 1968. Water movement in an unsaturated sanitary landfill. *Journal of the Sanitary Engineering Division, ASCE,* Vol. 94, No. SA2, pp. 307–318, April 1968.

Soldat, J. K.; Robinson, N. M.; and Baker, D. A. 1974. *Models and Computer Codes for Evaluating Environmental Radiation Doses.* BNWL-1754. Richland, Washington, Pacific Northwest Laboratory.

Strenge, D. L.; and Watson, E. C. 1973. *KRONIC: A Computer Program for Calculating Annual Average External Doses from Chronic Atmospheric Releases of Radionuclides.* BNWL-B-264. Richland, Washington: Pacific Northwest Laboratory.

Strenge, D. L.; Watson, E. C.; and Houston, J. R. 1975. *SUBDOSA: A Computer Program for Calculating External Doses from Accidental Atmospheric Releases of Radionuclides.* BNWL-B-351. Richland, Washington: Pacific Northwest Laboratory.

U.S. Environmental Protection Agency. 1977. *State Decision-Makers Guide for Hazardous Waste Management.*

6

The Hazardous Waste Management Industry

Introduction

The 1973 report to Congress concluded that a regulatory approach was best to ensure adequate management of the nations' hazardous wastes. This conclusion was based on the finding that a viable waste management industry existed at that time, but was underutilized. Full utilization and subsequent capacity expansion of that industry would occur once enforcement of regulations provided adequate incentives for generators. This philosophy has underlain congressional and agency action since that time. As such, the hazardous waste management industry is expected to play a key role in the future disposition of hazardous wastes.

An identifiable hazardous waste management industry had emerged in the United States by 1969. Entrants to the business came from diverse backgrounds, including the generator industries, the solid waste management field, the transportation segment, materials recovery, and operations and organizations providing tank cleaning services. The mode of entry varied. Some facilities began as solvent recovery operations capitalizing on the economics of secondary markets. Other firms sought to fill the need for transportation between the generator and ultimate disposal sites. As business grew and became more complicated, competitors expanded and integrated vertically. The technology employed ranged from abandonment in isolated locations to sophisticated processing and secure disposal. Practitioners of the former methods conducted their activities in such secrecy that knowledge of the scope and size of their operations is only beginning to emerge with the discovery of abandoned disposal sites. Champions of the more complex acceptable practices have evolved into dominant forces in the current industry.

The following discussion is directed to a description of the recent history and current status of the hazardous waste management industry. It provides a brief view of a dynamic business as it strives to comply with evolving regulations and serve a market undergoing continual change. As such, some of the data are more important for the trends they illustrate than for their intrinsic value.

A Fledgling Industry

The first real survey of the hazardous waste management industry was conducted as part of the background effort for the 1973 report to Congress. At that time it was determined that approximately 10 regional plants existed for treatment/disposal of hazardous wastes. Typically, these facilities served an economic radius of 500 miles. For special wastes, however, shipments over distances as far as across the continent had occurred. The plants themselves were largely situated in the North Central, Mid-Atlantic, and Gulf Coast regions, which not surprisingly have also been shown to generate the bulk of all hazardous wastes in the United States. These figures do not include numerous solvent recovery operations and waste oil refining operations that can be considered a part of the hazardous waste management system, nor do they reflect the many waste haulers and clandestine brokers who were operating outside the law.

The capacity of the industry was estimated at 2,272,000 metric tons per year. This was nearly 25% of the estimated total hazardous waste production at that time (10 million metric tons). Subsequent surveys have revealed that potentially hazardous waste production was actually on the order of 150 million metric tons per year. In this context, the service industry's capacity was roughly sufficient to handle 1.5% of all hazardous wastes or 23% of the approximately 10 million metric tons per year available for outside contract. At the time, only 25% of the industry's capacity was being utilized. Hence, only 4–5% of the wastes available for outside contract disposal were reaching the legitimate operators. In the 1981 survey, Westat (1984) determined excess treatment and disposal capacities as outlined in Table 6-1.

Technology applied by the industry was diverse. Some of the larger and newer facilities, such as Chem Trol in Model City, New York, and Rollins-Purle in Bridgeport, New Jersey, maintained a full range of capabilities, with units including fractional distillation towers, stabilization lagoons, incinerators, metal concentrators, neutralization tanks, and secured landfills. Most wastes were accepted. Resource recovery was practiced on solvents and some metallic salts. Waste oil was burned to help meet fuel requirements when incinerating halogenated hydrocarbons.

The total capital investment in the industry was reported to be $25 million in 1973. Employment was roughly 100 persons. Income data are not readily

Table 6-1 / Distribution of Unused Capacity by Region in 1981 (Westat, 1984).

EPA Region	Unused Treatment Capacity (billions of gallons)	Unused Disposal Capacity (billions of gallons)
I	10	<0.5
II	10	1.0
III	12	6.0
IV	24	15.6
V	71	1.0
VI	24	7.1
VII	3.0	0.1
VIII	<0.5	0.6
IX	3.0	0.3
X	<0.5	0.0
Total	158	26.3

Note: See Figure 6-1 for a delineation of the EPA regions.

available. Several of the major competitors have noted, however, that the bulk of the profits came from sale of recovered solvents. Because of high intrinsic value, feedstocks for this phase of the operation were often purchased, as were wastes with high Btu content that could be employed as fuels. A major expense in the process was the operation of a well-equipped analytical chemistry laboratory to determine the compositions of wastes and the best means of treatment.

Coming of Age

With the selection of the regulatory control approach to management of the nation's hazardous wastes, the U.S. EPA would be placing great demands on the fledgling service industry. Data generated for the 1973 report to Congress were insufficient to determine industries' capabilities to respond to projected needs. Hence, the first comprehensive study of the hazardous waste management industry was commissioned in 1975 (Foster D. Snell, 1976). This was followed by a second in-depth study in 1980 (U.S. EPA, 1980) and an update in 1981 (Booz Allen & Hamilton, 1982).

In 1975 there were approximately 95 firms in the hazardous waste management business operating some 110 facilities. By 1980, 93 firms were operating 127 facilities. This understates the total new siting activity somewhat, because several facilities were shut down. It also obscures some of the consolidation that has occurred. For example, between 1980 and 1981, the nine major firms added 10 new facilities—three new sitings and seven by acquisition. The heavy reliance on acquisition reflects a continuance of the major obstacle to capacity expansion: high public opposition to siting new facilities. As a result of consolidation, the 1980 market share of roughly 50%

for these nine firms is believed to have risen to 60% in 1981. Geographically, these facilities are heavily concentrated (57 of all sites in 1980) in EPA Regions V, VI, and IX. This is not directly correlated with related waste generation volumes, because those three regions account for only 36% of total annual hazardous waste production. Shortfalls were predicted for Regions I, V, VII, VIII, and X. The current distribution of sites (Keller, 1980) and relative density can be seen in Figure 6-1. As of 1980, 19 states did not have incineration, treatment, or disposal facilities. This is down six from a survey made in January, 1977.

In 1975, most firms (94%) operated a single site each. The three largest firms had three to six sites each. By 1980, the industry had coalesced into three tiers, as evidenced in Table 6-2. The four largest firms account for aproximately 45% of industry revenues and waste volumes. They operate nationally and offer a broad range of services, including transportation, storage, treatment, disposal, and spill cleanup.

The next largest firms (8 or 9 privately held companies) are regionally oriented. They account for 12% of industry revenues and volumes. Some operate full-service facilities. Several of the second-tier operations have been the subjects of recent acquisitions by the first tier.

The third and final tier is composed largely of single-facility firms focusing on one or two treatment techniques. These 80 or 81 firms and organizations receive 43% of the revenues and waste volumes.

Figure 6-1 / Geographic locations of all identified commercial hazardous waste management facilities, June 1980. For purposes of this study, the industry includes all facilities engaged in the treatment and disposal of hazardous waste for a fee, but does not include solvent buying, selling, or recovery operations or storage and transfer stations that may be handling wastes classified as hazardous.

Table 6-2 / Industry Profile of the Hazardous Waste Management Industry as of June 1980.

Category and Type of Firm	Number of Firms or Organizations	Number of Facilities	Estimated Hazardous Waste Management Revenues ($ Millions 1980)	Waste Management Services Offered
Majors	4	26	135–151[a]	All services offered
Waste Management Inc. (Chemical Waste Management Inc.)	—	10[b]	44–50[c]	7 secure landfills, 5 chemical treatment, 3 deep-well injection, 4 land treatment, 2 incinerators, resource recovery
Browning Ferris Industries, Inc.	—	9	34–37[c]	3 secure landfills, 4 chemical treatment, 2 deep-well injection, resource recovery
RLC Corporation (Rollins Environmental Services Inc.)	—	4	29–31[d]	2 secure landfills, 3 chemical treatment, 3 incinerators, deep-well injection, resource recovery
SCA Services, Inc.	—	3	28–33[d]	3 secure landfills, 2 chemical treatment, resource recovery
Second tier	8–9	10–13	30–40[e]	Most services offered
Third tier	80–81	88–91	100–124[e]	Typically one or two types of services per facility
Total	93	127	265–315	

[a] Hazardous waste management revenues for the four major firms represent between 7% and 10% of total corporate revenues.
[b] Waste management also has plans to operate a new facility in Denver, Colorado.
[c] Based on annual report data forecasted to 1980 by Booz Allen.
[d] Company estimates.
[e] Estimated from volume and average price data.

The differences between the tiers of companies go beyond size. For instance, the nine largest firms accounted for 81% of all landfilled wastes in 1980, displaying a major concentration of available landfill capacity. They also controlled 93% of all permitted capacity. With acquisitions since that time, landfill concentration has increased further, even though 1981 volume was 10% lower than that in 1980. At the same time, the nine firms accounted for only 56% of all land treatment/solar evaporation and 60% of deep-well injections in 1980. Market fractions for incineration, chemical treatment, and resource recovery were 21%, 23%, and 20% respectively.

The diversification of the hazardous waste management industry with respect to management options can be seen in Figure 6–2. Although it is clear that land disposal alternatives still dominate the practices (58%), incineration, recovery, and treatment have gained a major foothold (42%). The distribution of options among the 10 EPA regions is presented in Table 6–3 along with an estimate of current usage as percentage of capacity. For the most part, facilities continue to operate well under capacity. It is interesting to note

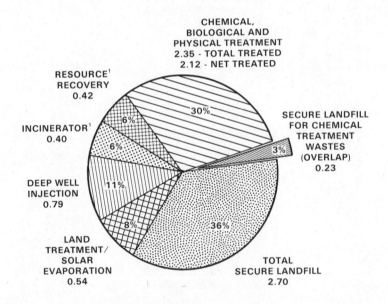

TOTAL WASTE VOLUME = 7.19
(INCLUDES LANDFILL/CHEMICAL
TREATMENT OVERLAP)

Figure 6–2 / Estimated hazardous waste volumes treated disposed by commercial off-site facilities by waste management options, 1980 (millions of wet metric tons). Note: Details may not add to total because of rounding. [1]Also may have some residual that must be treated by other options. This residual is believed to be small and has not been subtracted out of other waste management options.

Table 6–3 / 1980 Estimated Volume of Hazardous Waste Material Processed by Commercial Off-Site Facilities by Waste Management Option in the U.S. EPA Regions (Thousand Wet Metric Tons).

EPA Region	Landfill	Land Treatment/ Solar Evaporation	Incineration	Chemical Treatment[a]	Resource Recovery	Deep-Well Injection	Total Region
I	6	—	23	81	35	—	145
II	375	—	26	619	135	—	1,155
III	170	—[b]	48	467	51	—	736[b]
IV	226		65	157	22	—	470[b]
V	330	—	97	486	170	152	1,235
VI	650	117[b]	98	146	—	635	1,646[b]
VII	62	—	—	36	3	—	101
VIII	[c]	—	—	—	—	—	[c]
IX	822	345	40	294	—	—	1,501
X	59	75	—	62	8	—	204
Total	2,699	537	398	2,346	424	788	7,192
Capacity	27,604	2,437	670	3,921	1,069	4,657	
(% utilization)	(10.2 yr)	(22%)	(59%)	(54%)	(40%)	(17%)	

[a]These are gross volumes and include 10% that will require further treatment.

[b]Volume data from Region IV are included in Region VI to prevent disclosure of confidential data.

[c]Although some landfills in the regions may handle hazardous waste, it could not be determined if these facilities plan to meet RCRA requirements.

Note: Detail may not add to total because of rounding.

that whereas utilization remained relatively constant between 1975 and 1980 for most of the options, that for resource recovery and deep-well injection dropped. In the case of deep-well injection, the utilization changed from an estimated 75% in 1975 to 43% in 1980. Some of this appears to reflect the means of estimation. The 1980 survey team noted that whereas reservoirs were many times larger, other factors such as unloading, filtration, and pumping, made it difficult to exceed 50% of the pumping rate. Therefore, the 1980 estimate is based on the larger value of 50% of pumping capacity or current utilization. The current capacity to store hazardous wastes is estimated at 5–10% of the 1980 volumes processed, or 350,000 to 700,000 wet metric tons.

Utilization in 1981 was lower on a percentage basis than in 1980. Although the number of facilities increased, the volume of waste treated remained relatively constant. In fact, for the nine major firms resurveyed in 1981, volume managed dipped from 3.7 million wet metric tons in 1980 to 3.6 million in 1981. At the same time, revenues grew. Many of the nine firms reported total revenue growth of 20% to 30%. This, in part, can be attributed to the price growth experienced by hazardous waste management activities. A comparison of quoted prices in 1980 and 1981 is presented in Table 6-4. The average price increase was 10–15%. Increases in the prior year had been substantially higher.

The price increases reflect the strengthening of demand caused by implementation of state regulations and public awareness. The reduction in overall volume managed is a signal of more complex influences. Booz Allen & Hamilton (1982) discussed four key factors.

Recession. With the downturn in the national economy in 1981, overall goods production slowed. There was a commensurate drop in hazardous waste generation in that time frame. The effect was particularly evident in the auto, steel, and metal fabricating industries, which are heavy users of commercial waste management services.

Waste reduction. Price increases and growing industrial concern for long-term liability associated with hazardous wastes created the proper environment for implementation of new or previously uneconomical waste reduction technologies.

Shifts to on-site management. Similarly, price and liability concerns have provided increased incentives for large chemical companies to look at complete internal management of wastes. One example of the extent of this trend is the attempt of PP & G Inc. to use an abandoned shale mine in Ohio as a secure repository for production wastes. This trend may be reversed now that Part B disposal permits are being called. Final permits will be more difficult to obtain then interim ones, and many generators are expected to abandon facilities when their final permit applications are called.

Delisting. Some large-volume wastes such as certain paint sludges and pickle liquor sludges were successfully delisted, thus allowing disposal in other solid waste management facilities.

Table 6-4 / Comparison of Hazardous Waste Management Quoted Prices for All Firms in 1980 and for Nine Major Firms in 1981.[a]

Type of Waste Management	Type or Form of Waste	Price		$/Metric Ton	
		1980	1981	1980	1981
Landfill	Drum	$25–$35/55-gal drum	$35–$50/55-gal drum	$120–$168	$168–$240
	Bulk	$40–$50/ton	$50–$75/ton	$44–$55	$55–$83
Land treatment	All	$0.02–$0.09/gal	$0.02–$0.09/gal	$5–$24	$5–$24
Incineration	Relatively clean liquids, high Btu value	$0.20–$0.90/gal	$(0.05)[b]–$0.20/gal	$53–$237	$(13)[b]–$53
	Liquids	$0.20–$0.90/gal	$0.20–$0.90/gal	$53–$237	$53–$237
	Solids, highly toxic liquids	$1.25–$2.50/gal	$1.50–$3.00/gal	$300–$660	$395–$791
Chemical treatment	Acids/alkalines	$0.06–$0.30/gal	$0.08–$0.35/gal	$16–$79	$21–$92
	Cyanides, heavy metals, highly toxic wastes	$0.20–$2.00/gal	$0.25–$3.00/gal	$53–$528	$66–$791
Resource recovery	All	$0.19–$0.80/gal	$0.25–$1.00/gal	$50–$211	$66–$264
Deep-well injection	Oily wastewaters	$0.06–$0.15/gal	$0.06–$0.15/gal	$16–$40	$16–$40
	Toxic rinse waters	$0.50–$1.00/gal	$0.50–$1.00/gal	$132–$264	$132–$264
Transportation			$0.15/ton-mile		

[a]Interviews were conducted in May of 1980 and February of 1982.
[b]Some cement kilns and light aggregate manufacturers are now paying for wastes.

It was estimated that in 1975, employment in the hazardous waste management industry had reached 2000. Approximately 11% of those employees were classified as professionals, most of whom were chemists of chemical engineers. Other professionals typically had business background. Few of the nonprofessional employees were associated with unions. Those who were (one of every five firms had organized labor) were divided among the teamsters, chemical workers, and steelworkers. Unionization was most common in facilities operated by the larger corporations.

The hazardous waste management industry is relatively labor-intensive. Revenues in 1975 exceeded $50,000 per employee. The mean hourly wage was $5.80, with an average benefit package amounting to 25–30% of wages. Some 10% of the firms, all nonunion, included a corporate profit-sharing plan. As of 1975, however, distributions were negligible, because most firms were operating on a loss or break-even basis.

A lack of skilled nonprofessionals in the employment pool had given rise to internal on-the-job training programs for new personnel in 80% to 90% of the firms. Geographic separation appears to have prevented job transfer for the nonprofessional. On the other hand, the professional staff was highly mobile. Turnover was approximately 20%, over twice the average for the chemical industry.

In 1975, the industry boasted a total investment of $90 million in tangible assets, or $45,000 per employee. Approximately 60% of those assets, $54 million, were balanced with debt. A summary of financial data for the industry over the period 1971–1975 is presented in Table 6–5. The data may be somewhat misleading. Industry officials are quick to observe that of the $49 million in fixed assets, roughly $15 million was in transportation equipment. The $34 million remaining in treatment and disposal facilities is the estimated cost for a local plant of less than 400,000 tons per year capacity. Hence, replacement value for fixed assets is much higher than that reported.

It is interesting to note that initial industry expansion occurred primarily between 1971 and 1973, when over 30% annual growth was experienced. Throughout much of the 1970s, revenue growth averaged 29% per year; growth after 1973 can be attributed to price increases as well as to an increase in volume received. Sales and profits remained healthy throughout the decade, with net income varying between 5.5% and 9% of revenues, and profits varying between 18% and 22% of net worth (tangible equity) in 1973. Meanwhile, a low cash position with growing liabilities allowed working capital to decline after 1973, as well as the current ratio (current assets divided by current liabilities). The resulting poor cash flow has limited the availability of low interest rates for long-term debt. And yet, industry expansion was financed through debt and investment of current-period porfits rather than additional capital. The stability of total long-term debt since 1973 reflects a growing reluctance on the part of banks to provide long-term debt to an

Table 6-5 / Financial History of Hazardous Waste Management Industry, 1971-1975.[a]

Item	1971	1972	1973	1974	1975
Number of companies active in industry	76	82	87	91	95
Industry total revenues (millions of dollars)	46	65	88	103	107
Total net income (millions of dollars)	4.1	3.7	5.6	6.6	7.5
Total industry working capital (millions of dollars)	5.2	9.0	13.0	5.0	3.8
Industry current ratio	1.48	1.72	1.55	1.57	1.41
Total tangible assets (millions of dollars)	41	69	77	91	90
Total tangible equity (millions of dollars)	19	30	32	34	33
Total fixed assets (millions of dollars)	26	43	48	50	49
Total land (millions of dollars)	4.7	7.9	8.7	8.4	8.5
Total debt (millions of dollars)	23	39	51	59	54
Total long-term debt (millions of dollars)	14	20	25	32	26
Total short-term debt (millions of dollars)	3.2	5.2	5.8	5.8	7.2

[a]Adapted from Foster D. Snell (1976).

industry with less than a strong financial position and questionable growth potential. As a result, greater emphasis was placed on short-term debt instruments, whose total value rose from $3.2 million in 1971 to $7.2 million in 1975. Although the 1980 and 1981 surveys did not cover these aspects, it is clear from revenue and volume data that the increased prices for service have offered significant opportunities for profit growth. Between 1977 and 1982, the largest of the big four firms, Waste Management, saw revenue growth of 247% and earnings growth of 350%. Whereas solid waste collection and disposal contributed 98% of Waste Management's revenues in 1977, the 1982 figure was 63% and this is expected to drop to 50% by 1987.

Some differences have been noted between firms whose main objective is hazardous waste management and those that are parts of much larger, more diversified organizations. A comparison of key financial ratios between the two groups is given in Table 6-6. In general, the first group (dedicated hazardous waste management firms) produced a higher return on assets in 1975 with greater leverage. Debt was much more heavily short-term-oriented for this group and hence obtained at higher interest reates. The second group, divisions or subsidiaries of large corporations, had a more stable 5-yr financial history.

A number of factors have been found to be of significance in determining the success of firms in the hazardous waste management industry:

Location. Transportation costs are often a significant part of the total costs. The most successful firms are often those near the sources of waste

Table 6-6 / Comparison of Key Financial Ratios for Two Major Groups of Hazardous Waste Management Firms, 1975.[a]

Ratio	Firms with Main Objective of Hazardous Waste Management	Divisions or Subsidiaries of Large, Diversified Firms
Net income/sales	0.080	0.05[c]
Current ratio	1.28	1.83
Working capital ($1,000s per company)	51.7	14,000[c]
Tangible assets/sales	0.76	1.03
Profits/tangible assets	0.106	0.049
Tangible net worth/ tangible assets	0.35	0.41
Fixed assets/total assets	0.60	0.62
Land/fixed assets	0.17[b]	0.15
Total debt/equity	1.64	1.59[c]
Long-term debt/equity	0.62[b]	1.03
Short-term debt/equity	0.30	0.08
Working capital/long-term debt	0.21	0.31
Profits/net worth	0.30	0.12

[a]Adapted from Foster D. Snell (1976).
[b]Estimates based on 10% or less of total market.
[c]Estimates based on <20% of total market.

generation. This may lead to conflict, because producing industries are often in metropolitan areas with relatively high population densities. Siting considerations, however, place priority on areas with low population density, where overall risk is reduced because of a lower exposure potential.

Financing. With a rapidly expanding market and economies of scale, firms have been motivated to expand. However, capital for expansion has been difficult to obtain; the newness of the industry and lack of an established reputation have limited the availability of low-cost long-term debt instruments. Short-term debt mechanisms have raised costs. As a consequence, divisions and subsidiaries of large corporations with greater access to financing have been able to corner a larger share of the market.

Regulatory equity. Differences between state and local jurisdictions in the design, implementation, and enforcement of regulations directed to the waste management industry have created significant cost differentials between competitors. This often creates sizable incentive for firms to locate in areas with minimal control, i.e., large allowable discharge limits, limited design criteria for sites, poor enforcement records, liberal acceptance of low-cost alternatives, and few requirements for accountability. In the past, such discrepancies in policy between neighboring entities have stimulated the shipment of large volumes of wastes to "open" states and encouraged the rise of operators offering inadequate management alternatives. The most successful firms are those that have access to inexpensive means of ultimate disposal, such as landfill, injection wells, and ocean disposal sites.

Siting. As noted in the previous chapter, siting new facilities has become a major hurdle for the industry. Public reaction to the prospect of nearby sites has made it difficult to obtain required permits and has added significant costs to the siting process. As a result, expansion and/or replacement of facilities are tied closely to a firm's ability to gain public confidence. In the future, successful firms will be those that can navigate political waters and maintain a good-neighbor image. This will require sensitivity to local issues, extreme care in maintenance of safety and environmental safeguards, and an active public information effort.

Long-Term Contracting. The vast majority of wastes managed by the service industry are handled on a contract basis. Hence, the more successful firms in the management industry are those that can obtain agreements with large generators. Receipt of such agreements is dependent on many factors that have been evolving rapidly in recent years. With the notoriety surrounding Love Canal and subsequent incidents involving improper management of contracted wastes by third-party contractors (e.g., Silresim, Chemical Control Corp., and the Valley of the Drums), generating industries have become increasingly concerned about loss of control over wastes. This has prompted many to undertake their own internal disposal operations on-site. In cases where this is difficult or costly, generators have next turned to large substantial service firms with recognized expertise and sufficient assets to cover liabilities associated with improper disposal. Hence, there are growing pressures that will favor the larger service firms. Some of these pressures may be relieved with the advent of state and federal permits for operators under RCRA. However, past litigation indicates that use of permitted contractors does not protect the generator from subsequent legal actions. Hence, many will opt to maintain control of hazardous wastes to the extent possible, thus diminishing the number of large, long-term waste management contracts available for the service industry.

Vertical integration. Firms with full-service capabilities (transportation, storage, treatment, and disposal) have been found to be more profitable than specialized operations. This reflects both the greater opportunity to obtain business and the larger overall revenues within which profits can be taken. In many respects, this factor is closely tied to previously listed factors. Full service is often associated with size. Both factors in tandem increase the ability of a firm to obtain long-term contracts. At the same time, capital availability severely limits the ability of small or specialized firms to diversify.

A Glimpse into the Future

As noted previously, the total capacity of the hazardous waste management industry that is deemed environmentally adequate is less than the volume of wastes requiring contract disposal. With the anticipated annual growth in hazardous waste production of 4.5% and the implementation of regulations under RCRA, there will be a growing demand for expansion of contract disposal capabilities. In 1975, Foster D. Snell, Inc. (1976) estimated that by 1983 demand for contract disposal would reach 9 million metric wet tons, over twice the 3.7 million metric wet tons managed in 1975. This would require a total work force of 4000 employees, including 460 professionals. Capital requirements to convert 1975 capacity to a level matching 1983 demand were estimated to be $500 million. Of that total, $100 million would be employed to upgrade existing facilities to meet new environmental standards. Cost projections were derived in part employing the capital cost data presented in Figures 6–3 and 6–4. Industry officials took issue with these estimates. They speculated that $400 million would be needed for upgrading and $600 million for new facilities. That amounts to an investment of $1 billion by 1983

However, from the 1981 data it is clear that the projections were optimistic. The same factors that held down volume growth between 1980 and 1981 have kept volumes from growing beyond twice those managed in 1975. This raises interesting questions concerning the implementation of RCRA and the future of the industry. Because revenues have continued to climb, even when volume has remained steady, it is possible that the industry is tackling a higher fraction of the more hazardous and difficult wastes. For instance, Booz Allen & Hamilton (1982) found that wastes for incineration in 1981 had lower Btu content and were harder to burn. However, they reported that the wastes did not really change much between 1980 and 1981. It is true that generators are reducing overall waste volume and managing more wastes in-house. Thus, the projected shortfalls in commercial capacity may be covered by significant increases in on-site capacity and lower demand for growth. It is too early to tell at this time. As Snell noted in 1975, a major factor in determining demand and subsequent capacity expansion will be the degree to which regulations are enforced (Foster D. Snell, 1976). This remains a question. RCRA has yet to be fully implemented. While proposed land disposal regulations were not issued until 1982 only a handful of final permits have been issued. An apparent trend to less stringent design criteria has industry officials concerned that the regulations may encourage more on-site activities. The commersurate drop in projected demand for off-site service would reduce new siting activity. This uncertainty, along with high entry costs, is likely to continue to discourage small firms from entering the industry. Indeed, in 1980 the EPA concluded that new entrants to the industry would not add any significant amount to existing capacity (U.S. EPA, 1980).

Be that as it may, by 1980 those already in the industry were projecting significant capacity increases for the early 1980's. For instance, estimates of increases between 1980 and 1982 were as follows:

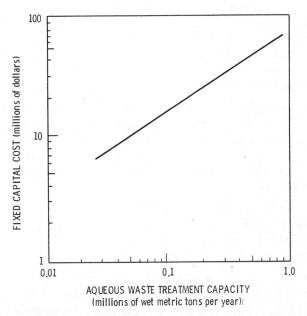

Figure 6-3 / Capital cost of liquid waste treatment facility in 1973. (From Batelle Memorial Institute, 1973.)

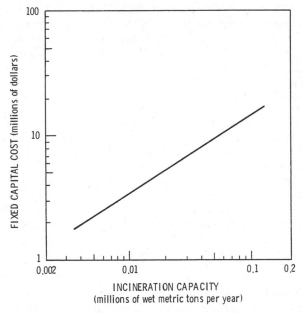

Figure 6-4 / Capital cost of incineration facility in 1973. (From Batelle Memorial Institute, 1973.)

Land treatment: 25%
Incineration: 122%
Chemical treatment: 51%
Resource recovery: 20%
Deep-well injection: no growth

Landfills were projected to have a 9% drop in lifetime capacity. This reflects rapid utilization of existing facilities, difficulties in siting new ones, and uncertainly over the anticipated regulations for land disposal. The lack of growth for deep-well injection continues the trend in reduced capacity utilization during the 1970s. Major factors influencing overall expansion decisions were the following:

Continued public opposition to new siting efforts
Perceptions of how rigorous RCRA enforcement is likely to be
Impacts of financial responsibility requirements
Capital formation for financing new facililites

The final point does not appear to be as restrictive as had been anticipated in 1975. Although hazardous waste management facilities are still ineligible for tax-exempt bonds, firms reported that obtaining financing was not a constraint in 1980. However, it does appear to be an additional factor favoring consolidation to the larger firms.

In a review for the state of New York, Booz Allen & Hamilton (1979) concluded that entry into the hazardous waste management industry will be difficult and likely restricted to a few large firms. Three major factors were noted:

High risks. The inherent risks associated with hazardous waste management with respect to third-party liabilities and potential early closures put increased financial demands on aspiring entrants.

High entry costs. Secure landfills and high-temperature incinerators are capital-intensive facilities. The additional costs associated with siting and the potential for unsuccessful siting efforts make these costs a barrier to firms without large financial resources.

Strict regulations. Federal regulations, and, in many cases, state regulations, place specific financial responsibility requirements on applicants that may preclude small firms. For instance, the original RCRA regulations required proof that a firm had $10 million in assets available to cover liabilities and that these assets not exceed 10% of corporate assets. Although that level has been reduced to $1 million, it may still be a barrier to entry.

In addition to the foregoing, market conditions may further discourage small operators. Major generators concerned over future liabilities have expressed the need to deal with larger disposal contractors whose finanical assets will successfully shield generators from damage suits. All of these factors suggest a culling process in the future that will favor the larger disposal operations.

Waste Exchange

In a time of increasing energy costs and decreasing availablity of raw materials, there is growing economic incentive for reuse and recycle. The simplest means of reuse is identification of a firm that can take a waste material as the input for its own process. This has long been a productive mode of operation with manufactured goods for such groups as the Salvation Army. Adoption by the chemical industry has been more recent, although individual examples, such as lime from the carbide process for acetylene, have been known for some time. A major stumbling block to wholesale adoption of waste exchange has been the lack of an intermediary or information bank to help match supply and demand. Secrecy and competitiveness in industry have hampered the free flow of information.

In a move to surmount these difficulties, government and industry officials have recently turned to a concept first implemented in Europe: the waste exhange or clearinghouse. Simply stated, the waste exchange is an information funnel that strives to collect data on available wastes and on raw material needs by various groups. Whenever a supply and demand appear matched, the two parties involved are notified and an attempt is made to allow them to explore formal exchange arrangements without compromising sensitive information. In some cases the exchange plays an active role in this process, whereas other exchanges simply forward inquires to the firm generating the waste. Once the parties reach a stage where they are talking directly, the exchange withdraws from the line of communication. The exchange may be operated by a governmental or trade agency, or it may be a separate business of its own. In many cases, operating funds are obtained through fees for listing wastes and needs, through finder's fees when a match is made, or through a combination of both. Brokers work on the basis of fees for each waste placed or a front-end retainer.

As outlined by the EPA in its "State Decision-Makers Guide for Hazardous Waste Management," (U.S. EPA, 1977), successful transfer of wastes requires consideration of advantages in four major areas:

Technical feasibility. The matching of the chemical and physical properties of available waste streams with the specifications of raw materials they might replace.

Economic feasibility. Balancing disposal cost forgone and raw materials costs saved against the administrative and transport costs of implementing a waste transfer.

Institutional and marketing feasibility. Guarantees of supply and anonymity; mutual confidence among generator, user, and transfer agent.

Legal and regulatory feasibility. Protection of confidentiality, legality, and unlikelihood of liability suits.

The first waste exchange in the United States was opened in November 1975 by the St. Louis Regional Commerce and Growth Association. In subsequent years, EPA encouragement fostered the creation of as many as 40 different exchanges. However, a large number of these exchanges soon failed. Analysis revealed that failure was often associated with the fact that the exchanges had local and state involvement in their operations. The potential for information on waste generation to be passed on to regulatory agencies became a barrier to those who might otherwise have listed materials with the exchanges. Some exchanges were also improperly managed, such that the necessary follow-through actions were not taken to ensure a successful exchange (Science, 1979). Many may have been understaffed.

In 1981 there were some 20 exchanges operating in the United States and one in Canada. Many were managed by trade associations, chambers of commerce, and universities. Some were operated as profit-making ventures in their own right. In those areas where government–industry trust is high, state and local governments have maintained participation. A list of exchanges operating in 1981 is presented in Table 6–7

The wastes most commonly listed with exchanges include solvents, oils, paper, wood, metal scrap, and surplus chemicals. Confidentiality limits the current data base on how often a successful exchange is accomplished. The St. Louis exchange estimates that 10% to 15% of the materials listed will eventually change hands. Were this representative of activities across the nation, 3% of all industrial wastes would be recycled in this manner (Science, 1979). This figure may well increase with time, because three to four new exchanges are created each year, and impending regulations under RCRA will create even greater economic incentives for exchange. An exchange official in Illinois estimates that 0.2% to 0.5% of the hazardous wastes in that state are being exchanged (Comptroller General, 1978).

The U.S. EPA assigned a full-time employee in January 1978 to encourage the formation and operation of exchanges. Activities have included information gathering, data dissemination, and technical assistance, Some funds have also been provided to sponsor seminars and exchanges around the country and to conduct feasibliity studies in selected areas. This constituted the total federal commitment at that time.

Waste exchanges are not to be confused with agents or brokers. Whereas exchanges act as intermediaries to enhance recovery and reuse of wastes by other industries, agents and brokers work to facilitate the movement of wastes to the commercial management industry. Agents do not actually take possession of wastes, but serve to aid in bringing generators and disposers together. Brokers take possession of the wastes. As such, they may provide

Table 6–7 / Waste Exchanges Operating in the United States in 1981.[a]

Name	Address	Contact
Northeast		
World Association for Safe Transfer and Exchange (WASTE)	130 Freight St., Waterbury, CT 06702	Marcel Veroneau (203) 574-2463
The Exchange	104 Charles St., Box 394, Boston, MA 02114	Howell Hurst (617) 367-2334 or 367-0810
Mid-Atlantic		
The American Alliance of Resources Recovery Interests Inc.	111 Washington Ave., Albany, NY 12210	John Flandreau (518) 436-1557
Enkarn Research Corp.	P.O. Box 590, Albany, NY 12201	J. T. Engster (518) 436-9684
Industrial Waste Information Exchange	New Jersey State Chamber of Commerce, 5 Commerce St., Newark, NJ 07101	Mr. Ludlum (201) 623-7070
National Waste Exchange	P.O. Box 190, Silver Springs, PA 17575	Ron Schaible (717) 780-6189
Northeast Industrial Waste Exchange	700 East Water St., Room 711, Syracuse, NY 13210	Walker Banning (315) 422-8276 or Stephen Hoefer (315) 474-4201
Pennsylvania Waste Information Exchange	222 North Third St., Harrisburg, PA 17101	Janice Berlin (717) 255-3252
Southeast		
Georgia Waste Exchange	Georgia Business & Industry Association, 181 Washington St. SW, Atlanta, GA 30303	Bert Fridlin (404) 659-4444
Iso-Chem Marketing, Inc.	P.O. Box 1268, Suite E, 449 Kingsley Ave., Orange Park, FL 32073	Anthony L. Tripi (904) 264-0070
Mecklenburg County Waste Exchange	Resource Recovery Analyst, Mecklenburg County Engineering Department, 1501 I-85 North, Charlotte, NC 28202	Roy Davis (704) 374-2770
South Central		
Tennessee Waste Swap	Tennessee Manufacturers Association, 708 Fidelity Building, Nashville, TN 37219	Nancy Niemeier (615) 256-5141
Midwest		
American Chemical Exchange (ACE)	4849 Gold Rd., Skokie, IL 60076	Tom Hurvis (312) 677-2800
American Materials Exchange Network	19489 Lahser Rd., Detroit, MI 48219	Vewiser Dixon (313) 532-7900

Table 6-7 / (Continued)

Name	Address	Contact
Chamber of Commerce of Greater Kansas City	920 Main St., Kansas City, MO 64105	Max Norman (816) 221-2424
Environmental Clearinghouse Organization (ECHO)	3426 Maple Lane, Hazel Crest, IL 60429	William Petrich (312) 335-0754
Industrial Material Exchange Service	Division of Land/Noise, 2200 Churchill Rd., Springfield, IL 62706	Larry Moore (217) 782-6760
Industrial Waste Information Exchange	Columbus Industrial Association, 1646 West Lane Ave., Columbus, OH 43221	Newton A. Brokaw (614) 486-6741
Iowa Industrial Waste Information Exchange	Center for Industrial Research Service, 201 Building E, Ames, IA 50011	E. O. Sealine, W. A. Kluckman (515) 294-3420
Midwest Industrial Waste Exchange	10 Broadway, St. Louis, MO 63102	Oscar S. Richards (312) 231-5555
Minnesota Association of Commerce and Industry (MACI)	200 Hanover Building, 480 Cedar St., St. Paul, MN 55101	James T. Shields (612) 227-9591
Ore Corp. ("The Ohio Resource Exchange")	2415 Woodmere Dr., Cleveland, OH 44106	R. L. Immerman (216) 371-4869
Union Carbide Corp. (in-house operation only)	P.O. Box 8361, Building 3005, South Charleston, WV 25303	R. L. Floyd (304) 747-5362
Waste Materials Clearinghouse, Environmental Quality Control Inc.	1220 Waterway Blvd., Indianapolis, IN 46202	Noble L. Beck (317) 634-2142
Southwest		
Chemical Recycle Information Program	Houston Chamber of Commerce, 1100 Milam Building, 25th floor, Houston, TX 77002	Jack Westney (713) 651-1313
West		
California Waste Exchange	Department of Health Services, Hazardous Materials Section, 2151 Berkeley Way, Berkeley CA 94704	Paul H. Williams (415) 540-2043
Information Center for Waste Exchange	Suite 303, 2112 Third Ave., Seattle, WA 98121	Judy Henry (206) 623-5235
Oregon Industrial Waste Information Exchange	Western Environmental Trade Association, Suite 618, 333 SW 5th, Portland, OR 97204	David Clark (503) 221-0357
Resources Conservation Consultants	Suite One, 1615 NW 23rd, Portland, OR 97204	Delyn Kies (503) 227-1319
Zero Waste Systems Inc.	2928 Poplar St., Oakland, CA 94608	Paul Palmer (415) 893-8257

[a]Adapted from *Chemical Engineering* (1982).

206

economies by accumulating batches from small generators until a full load is obtained. Some brokers actually operate management facilities but act as an agent for all wastes requiring technology that they cannot provide in-house.

Industry outside the United States

Although specific legislation in European nations has not been as comprehensive as that in the United States, in some respects European technology has led the field. As a consequence, the hazardous waste management industry in Western Europe is well established. France has 14 waste disposal centers to serve industry (*Chemical Week*, 1979). A program is now in place to certify these facilities. The largest plant opened at LeMay in 1975 and has a capacity of 150,000 tons per year. The government also incinerates wastes in special ships in the North Sea. In Germany, some of the large chemical companies operate their own on-site facilities. However, most chemical companies contract services to outside firms. In a survey for the Commission of the European Community, 104 treatment and disposal plants, tips and storage sites were identified. Many of these, however, do not specialize in hazardous wastes. Some 25 of the contract firms are privately owned. The majority are joint government–industry ventures. Most noted of the German facilities is that at Herfa-Neurode, where wastes such as residual Kepone from Allied Chemical are drummed and stored in a salt mine.

There are 300 government-approved hazardous waste sites in England. One hundred of the latter are on company grounds. Most are landfills, but incineration and chemical processing are increasing. The approximately 28 waste treatment centers have a combined capacity of 840,560 tonnes/yr for incineration, fixation, recovery, and chemical and biological oxidation. Recycling plants form a special subset of the industry, as represented by the Chemical Recovery Association. Membership includes 7 solvent reclaimers and 16 members specializing in the recovery of waste oils. Quantities handled in 1977 were 51 million liters of solvents and 172 million liters of waste oils.

In Sweden, there are some 20 private companies that have special permits to dispose of hazardous wastes. These firms manage 43% of the hazardous wastes, and an additional 11% is handled by SAKAB (Swedish Waste Conversion Company), which is 90% government—owned. Of the rest, 2% of the hazardous wastes are handled by communities and 40% by generators themselves. Some 4% of all hazardous wastes are exported to other European countries for disposal. Haulers must also have special licenses to transport hazardous wastes. Norwegian environmental authorities provide an intermediary function, collecting hazardous wastes and distributing them to commercial treatment plants. Austria currently has one plant operating in Vienna and a second being constructed near Ling.

Denmark is served by four concerns with a current capacity of 96,000 tonnes/yr. One of the four, Foul Bergoe and Son A/S Company, specializes in the treatment of silver-bearing liquids and solids containing silver and lead. The largest firm, Kommunekemi A/S, is a joint undertaking by Danish local authorities. It currently handles up to 60,000 tonnes/yr and will have its capacity expanded to 110,000 tonnes/yr. Hazardous wastes are collected by the municipalities for transport to the Kommunekemi facility at Nyborg.

Dutch efforts have been focused almost entirely on treatment. There are currently some 26 fixed or mobile treatment facilities with a combined capacity of 2.5 million tonnes/yr for toxic and dangerous wastes. Of that amount, 2.15 million tonnes are discharged to the sea. There is no disposal site in The Netherlands. Residuals are sent to West Germany. Joint government–industry efforts are currently directed through a company known as Induval to identify a suitable site within The Netherlands for disposal of future wastes.

Ireland currently has no major centers for hazardous waste treatment or disposal. Many firms process wastes on-site. The remainder use small specialized concerns that treat, dispose, recover, or export (largely to the United Kingdom) particular hazardous wastes.

In addition to both publicly owned and privately owned treatment and disposal facilities, the European hazardous waste management industry includes waste exchanges. These organizations attempt to match the input needs of one industry to the effluents of another. The exchange operated by the Association of the German Chemical Industry (VCI) claims to have placed 30 of the 400 wastes listed since 1972. VCI's Abfallboerse was the first such exchange in Europe, but it has been followed by others in Switzerland, the low countries, France, Austria, the United Kingdom, and Scandinavia. Subsequently, the exchanges have begun to work together. Data from France indicate that 12.4% of toxic wastes are resold.

The Prototypical Processing Plant

As a result of the variety of materials that may be designated as hazardous wastes and the number of alternatives available for managing those wastes, there is no single plant design that could be defined as the typical or standard processing facility design. Indeed, each of the existing processing facilities has its own unique features that distinguish it from the others. (Disposal facilities are much more likely to follow a basic design, because secure landfill and deep-well injection are the only alternatives at this time.) As a consequence, the description of a prototypical plant involves creation of a representative flow scheme that combines the features of various existing facilities but does not portray any given individual plant. Such a prototype was designed as part of the background study for the 1973 report to Congress (Battelle Memorial Institute, 1973).

The prototypical hazardous waste processing facility is designed to accept virtually all hazardous wastes other than nuclear materials. It incorporates the various waste treatment functions from which effluents can be discharged in the vicinity of the processing facility. The facility is basically a chemical processing plant designed to operate safely in a normal industrial area. Effluents from the facility meet applicable local water and air standards. Local solid waste disposal is limited to nonhazardous wastes. Nonhazardous waste brines can be disposed of by ocean disposal or deep-well injection to avoid potential quality impairment of fresh water sources. Land disposal of these brines is a less desirable alternative method that can be used only in arid regions, and infrequently there. Transporting the brines to an arid land disposal site would in most cases be more costly than transporting these materials to deep-well or oecan disposal sites. Because disposal of brine to land is detrimental to soil quality, deep-well and ocean disposal are more desirable alternatives.

The prototypical facility contains equipment and structures necessary for transporting, receiving, and storing wastes and raw material. An important feature is the laboratory to provide analytical services for process control and monitoring of effluent and environmental samples and pilot-scale testing services to assure satisfactory operation of the processing plant. The latter normally is not required in a conventional chemical processing plant, but because of the highly variable nature of the waste feed in this case, pilot-scale testing is considered essential.

Because many inorganic hazardous wastes cannot be rendered nonhazardous by available processing technology, the prototypical treatment facility must be complemented with a disposal facility. The latter facility consists of a secured landfill and the appropriate equipment and structures for burial and surveillance of hazardous wastes.

The two units combined, the processing and disposal facilities, may rely on any of a number of treatment and disposal techniques, as is evident from the review of the alternative candidates listed in Table 6–8. After consideration of a combination of these processess to provide the greatest capability without redundancy, Battelle Memorial Institute (1973) selected nine treatment technologies and three disposal options for incorporation in the prototypical plant (Table 6–9).

A conceptual flow diagram that integrates the various treatment steps in modular form, as illustrated in Figure 6–5, was developed for the prototypical hazardous waste processing facility. The flow pattern represents that normally expected. However, additional piping was included to provide flexibility. In the normal flow pattern, wastes with high concentrations of ammonia or ammonium salts would be diverted from storage to the ammonia stripper for ammonia removal. The ammonia is steam-distilled and appears as a concentrated stream from the condenser, whereas the stripper bottoms, which are essentially free of ammonia, are routed to the chemical treatment

Table 6-8 / Candidate Waste Disposal/Recovery Processes.[a]

I. Physical treatment processes
   A. Gas cleaning
        Mechanical collection
        Electrostatic precipitation
        Fabric filter
        Wet scrubbing
        Activated carbon adsorption
        Adsorption
   B. Liquids–solids separation
        Centrifugation
        Clarification
        Coagulation
        Filtration
        Flocculation
        Flotation
        Foaming
        Sedimentation
        Thickening
   C. Removal of specific components
        Adsorption
        Crystallization
        Dialysis
        Distillation
        Electrodialysis
        Evaporation
        Leaching
        Reverse osmosis
        Solvent extraction
        Stripping
II. Chemical treatment processes
        Absorption
        Chemical oxidation
        Chemical precipitation
        Chemical reduction
        Combination and addition
        Ion exchange
        Neutralization
        Pyrolysis
III. Biological treatment processes
        Activated sludge
        Aerobic lagoons
        Anaerobic lagoons
        Spray irrigation
        Trickling filters
        Waste stabilization ponds
IV. Ultimate disposal processes
        Deep-well disposal
        Dilute and disperse
        Incineration
        Ocean dumping
        Sanitary landfill
        Land burial

[a]Adapted from U.S. EPA (1978).

Table 6-9 / Processes Selected for Inclusion in Prototype Hazardous Wastes Processing/Disposal Facility.[a]

Treatment Processes	Disposal Processes
Neutralization	Ocean disposal
Precipitation	Deep-well injection
Oxidation–reduction	Landfill
Flocculation–sedimentation	
Filtration	
Ammonia stripping	
Carbon sorption	
Incineration	
Evaporation	

[a]Adapted from U.S. EPA (1978).

module. The chemical treatment module serves several functions, including receiving, storage, and distribution of chemical reagents, oxidation of cyanides and other reductants, reduction of chromium 6 and other oxidants, neutralization of acids and bases, and precipitation of heavy metals. The liquids–solids separation module involves sedimentation as the first step, followed by vacuum filtration for dewatering of sludges and multimedia filtration to remove residual particulate matter from the supernatant liquid. Filter effluent is then routed to the carbon sorption module for organic removal and finally to the evaporation module if concentration of the brine is desired. The brine may be concentrated either to a pumpable slurry or to a salt cake, depending on the mode of transportation that will be used to ship this material to a disposal site.

Combustible wastes are processed in an incinerator that includes off-gas cleaning equipment to remove acids and particulates. Ash from the incinerator is transferred as a slurry to the liquid–solids separation module together with scrubber and quench wastewater. Sludges from the liquids–solids separation module, if relatively free of hazardous metals, are buried in a local landfill. Sludges that contain substantial quantities of hazardous heavy metals (except Cr^{6+}) are shipped to a secured landfill. Waste brines and salt cakes are disposed as previously discussed.

For a processing facility with a design capacity of 623 tons per day of waste, wastes would include 122,000 gpd of aqueous wastes, including inorganic sludges, and 74 tons per day of wastes to be incinerated. The processing points where individual hazardous constitutents are removed or inerted are illustrated in Figure 6-6.

Details of the unit processes employed and their design and cost relations are left to subsequent chapters addressing specific treatment and disposal technologies. Similarly, extensive data and discussions on applications of individual processes can be found in texts directed solely to that topic.

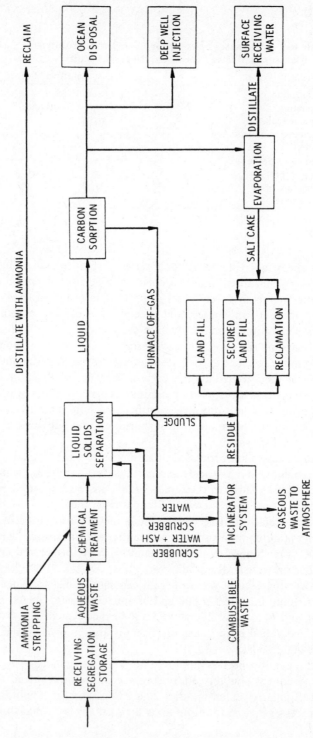

Figure 6-5 / Modular flow diagram for prototypical plant.

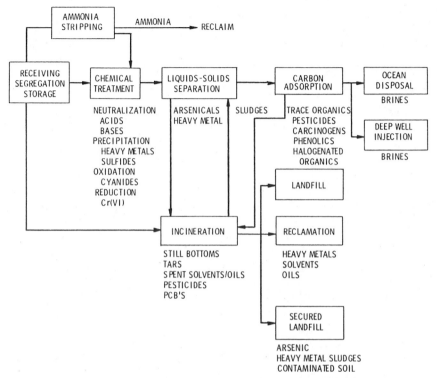

Figure 6-6 / Locus of treatment for specific hazardous constituents.

Preliminary cost estimates have been prepared for three different sizes of prototypical processing facilities. Capital and operating costs were initially estimated for a medium-size facility that would have processing capability for 122,000 gpd of aqueous-inorganic wastes and 74 tons per day of combustible wastes. Capital and operating costs were then estimated for a small facility and a large facility using the 0.6 scale-up factor for all equipment except the multiple submerged combustion units, for which a scale-up factor of 0.9 was used. The small processing facility is sized to process 25,000 gpd of aqueous-inorganic wastes and 15 tons per day of combustible wastes, and the large plant is sized to process 1 million gpd of aqueous-inorganic wastes and 607 tons per day of combustible wastes. All costs are based on the 1973 estimations for the prototype facility (Battelle, 1973) inflated to 1979 dollars.

Capital and operating cost summaries of medium, small, and large facilities are presented in Tables 6–10, 6–11, and 6–12, respectively. The summaries include a breakdown of capital and operating costs for each module. Included in the data for each module are the costs of land, buildings,

Table 6-10 / Cost Estimate Summary for Medium-Size Prototypical Processing Facility (1979 Dollars, ENR Chemical Plant Index 235).[a]

Module	Fixed Capital Cost ($)
Modular capital costs: aqueous waste treatment	
Receiving & storage	5,330,000
Ammonia stripping	1,261,000
Chemical treatment	7,716,000
Liquids–solids separation	14,610,000
Carbon sorption	1,534,000
Evaporation	838,000
Rounded totals	31,300,000
Modular capital costs: incineration	
Incinerator with	7,940,000
scrubber wastewater treatment	

[a]Capacity: 122,000 gpd aqueous waste treatment, 74 tons/day incineration, 260 day/year operation. Total fixed capital cost: $39,240,000.

laboratory, offices, and auxiliary equipment. These cost data are based on preliminary estimates that include many approximations and simplifying assumptions. More accurate estimates would involve detailed material and energy balances around the system and its various internal components so that the individual items of equipment, reaction vessels, and tanks could be sized and estimated with greater accuracy.

The most significant costs are incurred in the receiving-storage modules, the chemical treatment module, and the liquids–solids separation module. In the medium-size receiving-storage module (Table 6-9), labor (including supervision and laboratory personnel) accounts for about 50% of the operating cost, and facilities (capital amortization) account for 28% of the cost of

Table 6-11 / Cost Estimate Summary for Small-Size Prototypical Processing Facility (1979 Dollars, ENR Chemical Plant Index 235).[a]

Module	Fixed Capital Cost ($)
Modular capital cost: aqueous waste treatment	
Receiving & storage	2,057,000
Ammonia stripping	487,000
Chemical treatment	2,978,000
Liquids–solids separation	5,640,000
Carbon sorption	592,000
Evaporation	323,000
Rounded totals	12,078,000
Modular capital cost: incineration	
Incinerator with	3,064,000
scrubber wastewater treatment	

[a]Capacity: 25,000 gpd aqueous waste treatment, 15 tons/day incineration, 260 day/year operation. Total fixed capital cost: $15,142,000.

Table 6-12 / Cost Estimate Summary for Large-Size Prototypical Processing Facility (1979 Dollars, ENR Chemical Plant Index 235).[a]

Module	Fixed Capital Cost ($)
Modular capital costs: aqueous waste treatment	
Receiving & storage	18,815,000
Ammonia stripping	4,452,000
Chemical treatment	27,237,000
Liquids–solids separation	50,391,000
Carbon sorption	5,415,000
Evaporation	5,563,000
Rounded totals	111,870,000
Modular capital costs: incineration	
Incinerator with	28,038,000
scrubber wastewater treatment	

[a] Capacity: 1,000,000 gpd aqueous waste treatment, 607 tons/day incineration, 260 day/year operation. Total fixed capital cost: $140,000,000.

this module. The major cost items in the chemical treatment module are treatment chemicals and processing facilities. The lime handling system alone accounts for almost 70% of the total capital cost of this module. Chemical consumption represents about 45% and facilities about 23% of the operating cost. The major cost factor in the liquids–solids separation module is the facility cost, wherein capital amortization accounts for about half the total operating cost.

On-Site and Off-Site Trade-Offs

Whereas there exists a viable service industry to manage hazardous wastes, there are many factors that must be considered before a specific course of action is taken. Most prominent among these is the issue of economics. There exists a waste generation rate that is sufficiently large to justify on-site processing, as opposed to transportation off-site and use of a contractor. Initial investigations into the economic watershed between on-site and off-site processing were explored as a part of the background work for the 1973 report to Congress (Arthur D. Little, 1972). These researchers determined that the total cost C_i for a particular processing strategy i can be defined as

$$C_i = CHW_i + CTW_i(MW_i) + Cp_i + CHR_i + CTR_i(MR_i) + CF_i$$

where

CHW = handling cost (loading and unloading)
CHR = handling cost of residue
CTW = transport cost of waste permits
CTR = transport cost of residue per mile
CF = final disposal cost

MW = miles waste is transported
MR = miles residue is transported
Cp = processing cost

In comparing two alternatives ($C_1 - C_2$), this relation is modified to

$$C_1 - C_2 = AC = ACHW + CTW(AMW) + ACp + ACHR + CTR(AMR) + ACF$$

Of the costs described, only Cp is typically capacity-dependent. Chemical process data indicate that this dependence can be described as

$$Cp + T^{-0.4}$$

where T is representative of throughput capacity. In turn, $ACp = Cp_1 - Cp_2 = Cp_1[1 - (T_1/T_2)^{0.4}]$. Substituting into the total cost differential, we see that when residuals disposal costs are similar, the cost differential between on-site and off-site processing varies with distance to the processing facility (AM) and the ratio of throughput capacities (T_1/T_2). When a grid of evenly spaced sources is postulated, the variables can be converted to mean separation distance between sources and source size. Setting the cost differential between on-site ($M_1 = 0$) and off-site processing equal to zero, one can solve the relations for source separation as a function of source size and determine the locus of the watershed where on-site processing becomes economical. Such decision maps are provided for major hazardous waste stream types in Figures 6–7 through 6–15. these figures differ from those produced in the 1973 report (Arthur D. Little, 1973) in that they reflect a transportation cost index change of 296.9 to 470.1 for the period of 1971 to 1978 and a chemical processing index change of 319.1 to 560.4 for the same period. Hence, the controlling equation employed by A. D. Little, $g = (1/1.66)[Cp^*/CTW](T^*/T_0)^{0.4}$ (where T^* and Cp^* refer to the standard base waste load used to estimate cost) has been adjusted to $g = (1/1.66)(1.76/1.58[Cp^*/CTW](T^*/T_0)^{0.4}$.

We can also determine the optimum number of sources to be collected (No) for off-site processing from the relation

$$No = \frac{1.23}{(gCT/Cp_0)^{1.09}}$$

as well as the area (A) served by a facility collecting from Ncg plants located g miles apart:

$$A = Nog^2 = (1.23)\frac{(g)(0.91)}{(CT/Cp_0)^{1.09}}$$

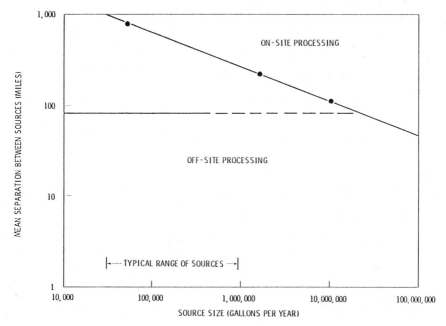

Figure 6–7 / Decision map for concentrated heavy metals. (Adapted from Arthur D. Little, 1972.)

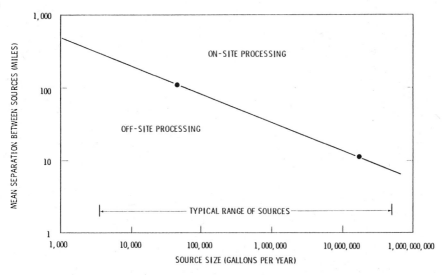

Figure 6–8 / Decision map for dilute heavy metals. (Adapted from Arthur D. Little, 1973.)

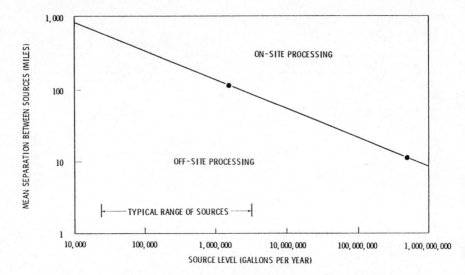

Figure 6-9 / Decision map for dilute metals with organic contamination. (Adapted from Arthur D. Little, 1973.)

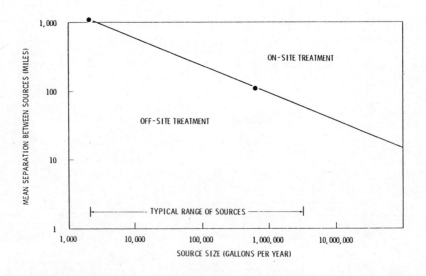

Figure 6-10 / Decision map for asphalt encapsulation of hydroxide sludes (Adapted from Arthur D. Little, 1973.)

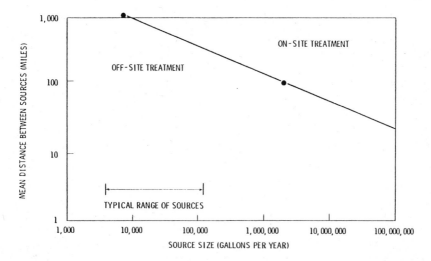

Figure 6-11 / Decision map for cement encapsulation. (Adapted from Arthur D. Little, 1973.)

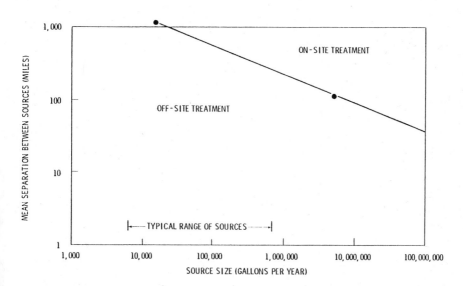

Figure 6-12 / Decision map for concentrated cyanides. (Adapted from Arthur D. Little, 1973.)

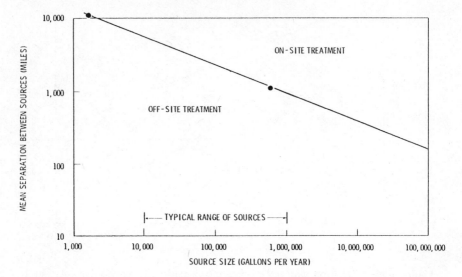

Figure 6-13 / Decision map for liquid chlorinated hydrocarbons. (Adapted from Arthur D. Little, 1973.)

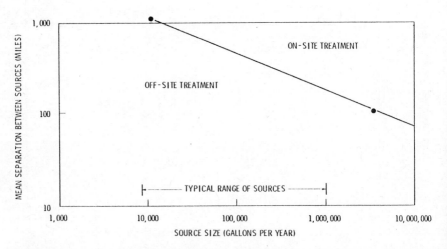

Figure 6-14 / Decision map for dilute cyanides. (Adapted from Arthur D. Little, 1973.)

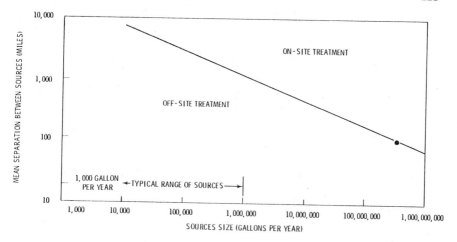

Figure 6–15 / Chlorinated hydrocarbon and heavy-metal slurries.

By similar analysis it can be shown that the maximum economic distance (M) for transportation of wastes to a central treatment/disposal facility is defined as

$$M = \frac{Cp}{CTW} [(\frac{T}{\text{ton-site}})^{0.4} - 1]$$

substituting mean source sizes for the ton-site capacity and average costs for Cp and CTW, we can define the maximum service range for the average source as a function of treatment/disposal facilities capacity (T). Such a relation is derived for "environmentally adequate" treatment utilizing cost relations provided by Powers (1976) and a transportation cost of CT = $0.15/ton-mile in Figure 6–16. Similar curves for incineration of wastes are given in Figures 6–17 through 6–19 using cost data from Berkowitz (1979). As is apparent from the data, for sources generating 500 to 1000 gpd, most off-site facilities can service a radius of 500 to 1000 miles. For infrequent batches (e.g., 1000 gal/yr) the radius expands to a point where a plant can service sources across the continent.

It is clear that economic considerations strongly favor the continued viability of an independent hazardous waste management industry. This is particularly true for the more expensive unit operations such as incineration, as long as constraints on disposal are implemented and enforced. At the same time, other facets of the regulatory framework create barriers to the expansion of the industry. As noted previously, growing local opposition is greatly affecting the industry's ability to site new facilities. Similarly, requirements for closure bonds and financial responsibility loom as deterrents to small

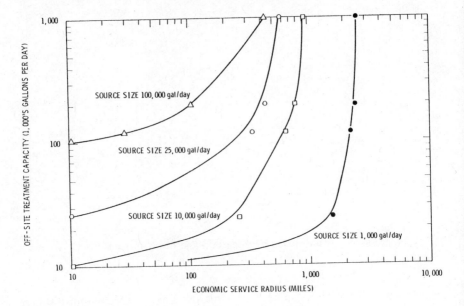

Figure 6–16 / Environmentally adequate treatment.

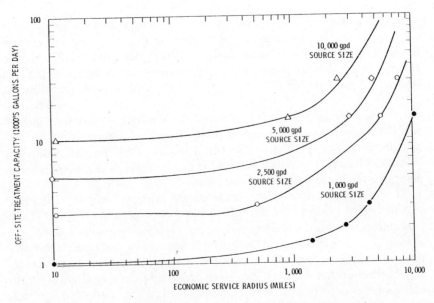

Figure 6–17 / Incineration of chlorinated hydrocarbon sludges.

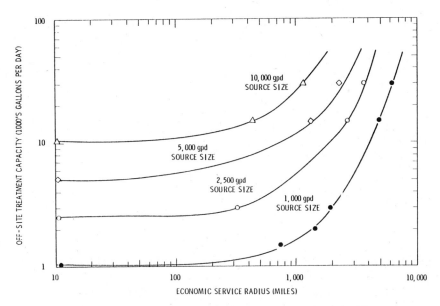

Figure 6-18 / Incineration of chlorinated hydrocarbon liquids.

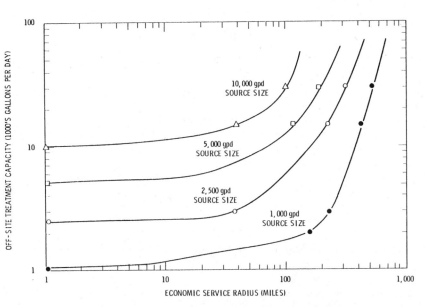

Figure 6-19 / Incineration of hydrocarbon solvents.

**Table 6-13 / Sites Approved for Disposal of PCB Wastes
(December, 1979).[a]**

Firms	Location
NEWCO Chemical Waste Systems, Inc.	Niagara Falls, NY
NEWCO Chemical Waste Systems of Ohio, Inc.[b]	Williamsburg, OH
SCA Chemical Services, Inc.	Model City, NY
Waste Management of Alabama, Inc.[b]	Livingston, AL
Casmalia Disposal	Santa Barbara, CA
Nuclear Engineering Co.	Shelbyville, KY
Chem-Nuclear Systems, Inc.	Arlington, OR
Wes-Con, Inc.	Grand View, ID

[a]Includes capacitors; properly drained transformers, drums, and other containers; dredge spoils; municipal sludges; and contaminated soil or other debris.

[b]These two sites may also accept liquids containing 50-500 ppm PCBs.

private operations. The ultimate configuration of the industry with respect to size, ownership (public or private), and scope of activities will greatly depend on the nature of the regulations mandated by RCRA and its subsequent amendments. In addition, related legislation may be impactive. For example, that promulgated under TSCA on disposal of PCBs has resulted in a limited number of approved facilities to handle that specific waste material (Table 6–13). Similar initiatives on special wastes or waste components may create additional special categories of sites.

REFERENCES

Arthur D. Little, Inc., 1972. *Alternatives to the management of hazardous wastes at national disposal sites.* Review copy. U.S. Environmental Protection Agency, January 19, 1972.

Battelle Memorial Institute. 1973. *Program for the management of hazardous wastes.* Vol. 1, U.S. Environmental Protection Agency.

Berkowitz, J. 1979. Technologies for handling hazardous wastes. In *Scientific and Technical Aspects of Hazardous Waste Management.* American Association for the Advancement of Science, Washington, D.C., Intergovernmental Science, Engineering, and Technology Advisory Panel, pp. 155–188, July 11–13, 1979.

Booz, Allen & Hamilton, Inc. 1982. *Review of activities of major firms in the commercial hazardous waste management industry: 1981 update.* U.S. Environmental Protection Agency, May 7, 1982.

Booz, Allen & Hamilton, Inc. 1979. *Options for establishing hazardous waste management facilities.* New York State Environmental Facilities Corporation, September, 1979.

Chemical Engineering. 1981. Waste exchanges grow in number and scope. *Chemical Engineering,* Vol. 88, No. 14, pp. 68–71, July 13, 1981.

Chemical Week. 1979. Europe gets a grip on wastes. *Chemical Week*, Vol. 1, No. 3, pp. 44–45, August 8, 1979.

Comptroller General. 1978. *How to dispose of hazardous waste—a serious question that needs to be resolved.* Report to Congress, December 19, 1978.

DeRenzo, D. J. ed. 1978. *Unit operations for treatment of hazardous industrial wastes.* Park Ridge, N.J.: Noyes Data Corporation.

Foster D. Snell, Inc. 1976. *Potential for capacity creation in the hazardous waste management service industry.* PB-257 187, National Technical Information Service, Springfield, Va.

Hackman, E. E. 1978. *Toxic organic chemicals destruction and waste treatment.* Park Ridge, N.J.: Noyes Data Corporation.

Keller, J. J., and associates. 1980. *Hazardous wastes service directory.* Neenah, Wi.

NATO. 1976. *Disposal of hazardous wastes—manual on hazardous substances in special wastes.* No. 55, prepared for Federal Ministry of Interior, Federal Republic of Germany, October, 1976.

Powers, P. W. 1976. *How to dispose of toxic substances and industrial wastes.* Park Ridge, N.J.: Noyes Data Corporation.

Science. 1979. Hazardous wastes technology is available. *Science*, Vol. 204, No. 4396 pp. 930–932, June 1, 1979.

U.S. EPA. 1980. *Hazardous Waste Generation and Commercial Hazardous Waste Management Capacity—An Assessment.* SW-894, U.S. Environmental Protection Agency, December 1980.

U.S. Environmental Protection Agency. 1978. *Industrial waste exchanges— fact sheet.* SW-688.

U.S. Environmental Protection Agency. 1977. *State decision-makers guide for hazardous waste management.*

U.S. Environmental Protection Agency. 1976. *Growth potential in the hazardous waste management service industry.* Paper presented at National Solid Waste Management Association International Waste Equipment and Technology Exposition, Chicago, June 2, 1976.

Westat, Inc. 1984. *National Survey of Hazardous Waste Generators and Treatment, Storage and Disposal Facilities Regulated Under RCRA in 1981.* U.S. EPA.

7

Abandoned Disposal
Sites

Introduction

When the Environmental Protection Agency submitted the 1973 report to
Congress on hazardous wastes, the document was accompanied by a proposed hazardous waste management act. At that time, many congressional
leaders expressed concern that air and water legislation was putting increased
pressure on the land as the final repository for wastes, but passage of a
version of the proposed act was 3 years away. Throughout the debate, the
proposed legislation on hazardous wastes ran the gamut from an independent
act to a rider on a bill addressing other subjects. In the end, the hazardous
waste management act became a subtitle of the Resource Conservation and
Recovery Act of 1976. By 1979, prominent members of Congress were
declaring that hazardous wastes were the most urgent environmental issue
facing the nation. The abandoned disposal site problem was the single most
important factor in bringing about this change in attitude.

Widespread public awareness and subsequent congressional rhetoric first
emerged with the events surrounding the Kepone disaster on the James
River. This provided some of the impetus for passage of RCRA as well as the
Toxic Substance Control Act of 1976. However, it was Love Canal that
brought hazardous waste management concerns squarely before the public
and unleashed a flurry of activity on the part of Congress, the EPA, industry,
and public interest groups. The reality of a large-scale evacuation, abandonment of a public school, and reported health effects within the community
raised numerous questions for which answers were not available.

It is ironic that Love Canal and the discovery of other abandoned
chemical disposal sites throughout the nation should be at the source of the
renewed drive to promulgate regulations under RCRA, because RCRA does
not itself address the abandoned site issue. RCRA is basically a prospective

act designed to prevent improper management of hazardous wastes in the future. The authority to address disposal site problems that pose an imminent hazard to health or the environment (Section 7003) is severely limited. First, the site in question must pose an existing hazard. Action cannot be taken in a preventive mode. Second, this authority can be exercised only when the owner or responsible party is identifiable and capable (financially or otherwise) of remedying the situation. Hence, Section 7003 is of little use when the owner is deceased (Valley of the Drums, Kentucky) or bankrupt (Selrisem, Inc., Lowell, MA). Further, the test for "imminent and substantial hazard" places a heavy burden of proof on the agency, and implementation of remedial action requires potentially lengthy judicial action. The combined effects of these factors severely restrict the timeliness and quality of response currently available. For instance, federal action was taken to mitigate leaking barrels in the Valley of the Drums. However, once these were addressed, action was halted, leaving thousands of other barrels that had yet to corrode to a point of loss of integrity.

What, then, are the dimensions of the current abandoned site problem? The following is directed to provide some of the answers to this and related questions.

The Problem Set

Although the 1973 report to Congress included a review of past incidents illustrating the breadth and depth of impacts associated with improper management of hazardous wastes, it only touched on the problems associated with abandoned disposal sites. No attempt was made to quantify or enumerate abandoned disposal sites. Efforts directed to that objective were first initiated in October 1978, when the Environmental Protection Agency requested each of its 10 regional offices to supply a list of sites that were believed to contain hazardous wastes. Within a month, the EPA released the resulting list of 103 potentially dangerous sites and a projection that these represent an estimated 838 sites across the country that may contain significant quantities of hazardous wastes. Some 32,254 sites were reported as having potentially hazardous waste constituents (Table 7–1).

Upon scrutiny, these data were challenged by the General Accounting Office (GAO) as having been derived from an inadequate base. During a GAO evaluation, it was noted that the regional responses were not made with a common methodology. For example, one region merely estimated that there were hazardous wastes in one-third of all the active municipal sites, and another listed only sites known to have large quantities of hazardous wastes and subsequent negative impacts on public health and the environment (*Hazardous Waste Disposal*, 1979). As evidence of the likelihood that the estimate was low, the congressional Committee on Interstate and Foreign Commerce noted that a detailed survey of two counties in upstate New York

Table 7-1 / Summary of EPA Regional Estimates of Hazardous Waste Sites.[a]

Region	Number of Sites That May Contain Hazardous Wastes	Number of Sites That May Contain Significant Hazardous Waste Quantities	Number of Sites with Information Supplied
I	1,200	275	5
II	509	25	4
III	5,000	12	5
IV	14,000	210	16
V	1,800	Unknown	22
VI	320	19	3
VII	8,000	Unknown	7
VIII	25	10	9
IX	400	37	1
X	1,000	250	31
Total	32,254	838	103

[a]From Weight et al. (1979).

yielded a list of 215 disposal sites (78 active, 126 inactive, and 11 unknown). Of these, 35 of the inactive sites had received large quantities of hazardous wastes (*Hazardous Waste Disposal*, 1979).

In a follow-on effort, the EPA selected a contractor (Fred C. Hart Associates, Inc.) to provide a means of extrapolating this data base to a more accurate accounting of the abandoned landfill problem. Data on the 103 sites and an additional 129 sites for which information was available from other sources were collected and reviewed. An alternate means of estimating total hazardous waste sites was also devised. The latter yielded a maximum potential value of 50,664 sites. Hence, the problem was summarized as one of 32,257 to 50,664 total hazardous waste sites, of which 1204 to 2027 may be significant problems (based on an EPA statistic of 4% of all sites posing a significant problem). Analysis of the 232 sites for which information was available on file provided the characterization given in Figure 7-1. With respect to the status of the sites, 22% were classified as active, 30% inactive, and 48% uncertain. Information on financial status was less accessible: 17% were defined as abandoned/abandonable, 17% viable, and 66% uncertain. Twenty-six percent of the sites had received some measure of remedial action (Weight et al., 1979).

After a more detailed examination of costs, the EPA contractor concluded that the average cost of actions to prevent existing problems from becoming worse (level I, e.g., covering or diking) would be $3.6 million per site, and the average cost of an ultimate solution (level II, e.g., exhumation) would be $25.9 million per site. When these data were extrapolated over the 1204 to 2027 sites posing significant hazards (less 8% for radioactive-contaminant-bearing sites and 8% for sites where no action would be required), they yielded a total estimated cost of $3.6 to $6.1 billion for level I activities and

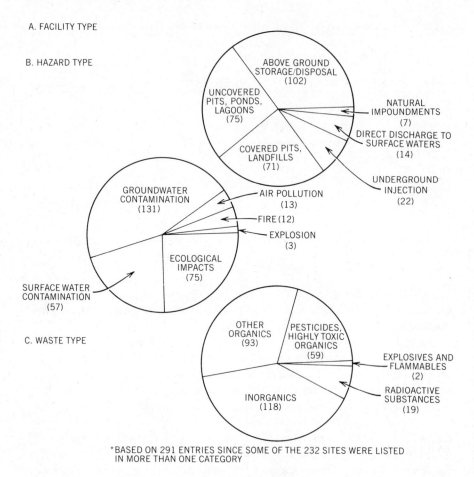

A. FACILITY TYPE

B. HAZARD TYPE

C. WASTE TYPE

*BASED ON 291 ENTRIES SINCE SOME OF THE 232 SITES WERE LISTED IN MORE THAN ONE CATEGORY

Figure 7-1 / Characterization of identified hazardous waste sites. (Based on data from Weight et al., 1979).

$26.2 to $44.1 billion for level II activities (Weight et al., 1979). In a review of 431 potentially dangerous sites in 33 states, the Chemical Manufacturers Association has estimated that site mitigation costs are more likely to average $1 million per site.

By mid-1979, the EPA had expanded the so-called hit list of 103 sites thought to pose a significant hazard to 151 sites. Of that total, 5 were being litigated by the federal government, and 34 were being sued by state governments. An additional 45 sites previously placed on the list had been remedied or exonerated as not posing major problems. As recently as November of 1981, the EPA issued a listing of 114 priority sites that will be the major emphasis for near-term restoration activities. These sites are identified in Appendix I. By late 1984, the National Priority List (the candidate list for all Superfund actions) had grown to 786 sites.

Throughout the evaluation of the estimation process, Congress remained skeptical on the accuracy of the data collection techniques employed. The Oversight and Investigations Subcommittee of the Committee on Interstate and Foreign Commerce under Congressman Bob Eckhardt contended that the only accurate source of information was the generators themselves. Hence, in the spring of 1979, the subcommittee undertook a direct mail survey of the 53 largest domestic chemical manufacturers managing 1605 individual production facilities, or 14% of the nation's chemical plants. Results revealed that since 1950, these 53 companies have disposed of some 762 million tons of chemical process wastes at 3383 sites. Approximately 34% of those sites were company-owned. However, 94% of all wastes were disposed of on-site. Of the other 6%, the generators could not identify where haulers had taken 4.8 million tons. A synopsis of the types of wastes taken to unknown locations is provided in Table 7-2. Without detailed guidelines or an independent evaluation, the amounts of these wastes that are hazardous and the degree of hazard posed by individual sites cannot be determined. The subcommittee concluded that the EPA should investigate 2200 of the sites and 960 of the haulers identified, as well as conduct similiar surveys of other waste generators. Additional quantitative conclusions from the survey must be approached with caution, because the sampling was not necessarily representative, i.e., some subgroups such as plastics and synthetics were overrepresented, and others such as paint and allied products were underrepresented.

Industry has been quick to point out some very positive aspects resulting from the Eckhardt survey. Specifically, producers could identify the disposal sites where 99% of the wastes were deposited. Most of those wastes, 86%, are located at sites still in use (Bolder, 1980). Once again, however, the nature of the data base prevents extrapolation to the whole industry. The survey covered the large chemical manufacturers. Many of the problems associated with improper disposal and off-site processing can be attributed to the smaller firms without the land or resources required to dispose on-site.

Additional surveys have been conducted or are under way at several levels of aggregation: regional, state, and county. However, recent EPA actions to create a full-scale investigative unit working closely with the Justice Department have increased the reluctance of generators to respond to direct surveys. Hence, without the power of subpoena, many agencies are finding subsequent studies less productive. This, in turn, has underscored the need to identify new sources of this information and new methods of investigation.

Several lines of inquiry have been suggested beyond simple identification of specific disposal sites. For instance, investigators can seek:

Table 7-2 / Nature of Wastes from 53 Largest Chemical Manufacturers Taken by Haulers to Unknown Locations.[a]

Waste Component	Percentage of Haulers Taking Wastes to Unknown Locations
General organics	60
Insecticides and intermediates	3
Herbicides and intermediates	3
Fungicides and intermediates	7
Rodenticides and intermediates	1
Halogenated aliphatics	11
Halogenated aromatics	9
Acrylates and latex emulsions	15
PCB/PBBs	1
Dioxins	1
Amides, amines, imides	20
Plasticizers	15
Resins	31
Elastomers	12
Solvents, polar	20
Carbon tetrachloride	1
Trichloroethylene	4
Solvents, other nonpolar	23
Solvents, halogenated aliphatic	11
Solvents, halogenated aromatic	6
Oils and sludges	40
Alcohols	27
Ketones and aldehydes	21

[a]Adapted from *Chemical Week* (1979).

Geographic areas that have received a large amount of fill, i.e., old quarries and former wetlands

Property formerly owned by an industry known to have generated hazardous wastes and likely to have practiced on-site disposal

Areas exhibiting vegetative damage or impaired water quality

Data assisting in the location of the foregoing as well as individual disposal sites may come from a variety of sources:

Public records
Zoning and ordinance changes: Point out areas that formerly had industrial activity or areas where disposal may have been allowed.
Deed files: Identify former and current ownership of lands.
Historical land use maps: Identify industrial areas and quarries.
Utility records and maps: Identify industrially owned areas.
Employment records: Identify former staff.
Tax assessment records: Identify ownership patterns.
Business licenses: Identify ownership.

Department of Commerce records and reports: Identify industrial areas.
Department of Transportation: Identify quarries and waste haulers.
Industrial records
Excavation and construction records: Identify landfill activity or areas of
former industrial use.
Professional society records: Identify former staff and reports on waste
management practices.
Conference proceedings: Identify former staff and waste management
practices.
Consulting firms: Records of former disposal practices.
Labor Unions: Identify former staff.
Other sources
Remote sensing: Identify damaged vegetation.
Aerial photography: Identify areas of fill and disposal pits.
Well logs: Identify former industrial property.
Local historical societies: Identify past industrial patterns.
Interview local residents: Determine prior industrial activities.
Interview local officials: Determine prior industrial activities.
Interview former employees: Determine waste management practices.
Interview personnel assigned to field work (e.g., telephone linemen and
weed control agents): Identify waste storage or disposal areas observed
during work.

There is currently insufficient experience with these techniques to determine their cost-effectiveness. Even with extensive efforts to utilize these
sources, there is little doubt that many sites will remain undiscovered until
accidentally exposed or traced back from environmental problems.

When a discovery is made, either inadvertently or as a result of a survey, a
two-step response process ensues: (1) site assessment and (2) design of remedial action.

The abandoned site problem does not appear to have emerged in Europe
in the same dimensions as in the United States. This may reflect less reliance
on land disposal and earlier controls on disposal activities in Europe. There
also appears to have been some activity in the area of site decommissioning.
The German government was aware of some 50,000 disposal sites during the
1950s. By 1975, that number had been reduced to 4000. Current plans call for
a further reduction to 500. (Only 100 of the sites operating at the present time
are considered effectively sealed.) Those sites that have been discovered
recently tend to be associated with plants that were damaged or closed at the
end of World War II. The United Kingdom has had about 50 sites that have
presented threats to surface waters and groundwaters (*Environmental
Science*, 1980*a*). In an event mindful of Love Canal, the Dutch government
recently purchased 264 homes found to be built over a chemical waste dump
that received up to 5000 drums of industrial waste in 1970. The cost of
purchase is $40 million. Hence, although the overall breadth of the aban-

doned site problem appears less in Europe than in the United States, the implications of individual occurrences are quite the same. Sweden has noted an increase in "midnight dumping" by unlicensed transporters that appears to parallel the tightening of regulations.

Currently, sites are identified by a variety of discovery methods including CERCLA-mandated reports whenever reportable thresholds of hazardous substances are spilled. If reports suggest sufficient risk, field investigative teams are directed to conduct a preliminary assessment. Results of that assessment are input to the Hazard Ranking System (HRS) to obtain a score indicative of the risk posed by a site. Scores are then employed to order sites within the National Priority List (NPL).

The HRS employs a structured value analysis approach using both additive and multiplicative factors. Three categories of hazard are considered: (1) migration of contaminants to off-site receptors, (2) fire and explosion, and (3) direct contact leading to health impairment. Each hazard category is further subdivided into parameters of concern and given a numerical scale range and a weighting factor as illustrated in Table 7–3.

The ranking procedure itself involves assigning a ranking value from the available scale for each parameter. As noted in Table 7–3, most scales allow values of 0 to 3, 0 to 5, or 0 to 8. A user's manual is available to guide the value allocation process. (Example: If separation of waste and aquifer is $\geq$150 feet assign 0, 76 to 150 ft assign 1, 21 to 75 feet assign 2, and 0 to 20 feet assign 3.) The individual values are subsequently multiplied by the weighting factors and summed to form the composite score for each hazard category. The composite scores are converted to percentage scores by dividing them by the maximum possible score for the category, that is, 57,330 for groundwater transport; 64,350 for surface water transport; 35,100 for atmospheric transport; 1400 for fire and explosion; and 21,600 for direct contact. The scores for the three migration routes are combined by taking the square root of the sum of the squares of the composite scores. As of 1983, a migration mode score of 28.5% was sufficient to put a site on the NPL.

New state legislation may provide additional mechanisms for discovery of sites as well as new programs to pursue restoration. By 1982, 32 states had enacted funding mechanisms for cleanup and emergency response as summarized in Table 7–4, and ten had enacted Good Samaritan legislation (Table 7–5).

Site Assessment

Discovery of an abandoned chemical waste disposal site does not in itself necessitate initiation of remedial actions. Abandoned wastes are not synonymous with imminent hazards. Many times, waste hazards are fully mitigated

Table 7-3 / Scoring Values and Weighting Factors for the EPA Hazard Ranking System.

Hazard Category	Factor Category	Factor	Value Scale	Weighting
Groundwater route	Observed release route	Release	0 or 45	1
	Characteristics	Depth to aquifer of concern	0 to 3	2
		Net precipitation	0 to 3	1
		Permeability of unsaturated zone	0 to 3	1
		Physical state	0 to 3	1
	Containment	Containment	0 to 3	1
	Waste Characteristics	Toxicity/persistence	0 to 18	1
		Quantities	0 to 8	1
	Targets	Use within 3 miles	0 to 3	1
		Distance to nearest well/population served	0 to 40	3
Surface water route	Observed release	Release	0 to 45	1
	Route characteristics	Slope and intervening terrain	0 to 3	1
		One-year, 24-hour rainfall	0 to 3	1
		Distance to nearest surface water	0 to 3	1
		Physical State	0 to 3	2
	Containment	Containment	0 to 3	1
	Waste characteristics	Toxicity/persistence	0 to 18	1
		Quantities	0 to 8	1
	Targets	Use within 3 miles downstream	0 to 3	3
		Distance to sensitive environment	0 to 3	2
		Population served/distance to water intake within 3 miles downstream	0 to 40	1
Air route	Observed release	No release	Ignore air route rating	
		Release	45	1

Table 7-3 / (Continued)

Hazard Category	Factor Category	Factor	Value Scale	Weighting
	Waste characteristics	Reactivity/incompatibility	0 to 3	1
		Toxicity	0 to 3	3
		Quantities	0 to 8	1
	Targets	Population within 4-mile radius	0 to 30	1
		Distance to sensitive environment	0 to 3	2
		Land use	0 to 3	1
Fire and explosion mode	Containment	Containment	0 or 3	3
	Waste characteristics	Direct evidence	0 or 3	1
		Ignitability	0 to 3	1
		Reactivity	0 to 3	1
		Incompatibility	0 to 3	1
		Quantities	0 to 8	1
	Targets	Distance to nearest population	0 to 5	1
		Distance to nearest building	0 to 3	1
		Distance to sensitive environment	0 to 3	1
		Land use	0 to 3	1
		Population within 2-mile radius	0 to 5	1
		Buildings within 2-mile radius	0 to 5	1
Direct contact mode	Observed incident	Incident	0 or 45	1
	Accessibility	Accessible to hazardous substances	0 to 3	1
	Containment	Containment	0 or 15	1
	Waste characteristics	Toxicity	0 to 3	5
	Targets	Population within 1-mile radius	0 to 5	4
		Distance to critical habitat	0 to 3	4

Table 7-4 / Funding Mechanisms for Cleanup and Emergency Response (National Conference of State Legislatures, 1982).

	AL	AZ	CA	CT	FL	GA	IL	IN	KS	KY	LA	ME	MD	MA	MI	MS	MO	NV	NH	NJ	NM	NY	NC	OH	OK	OR	PA	RI	SC	TN	TX	WI
Type of fund																																
Perpetual care	x																															
Emergency response		x	x	x	x	x	x	x	x	x			x	x	x	x	x	x	x	x	x	x	x	x	x	x	x	x	x	x	x	x
Other	x	x			x	x			x	x	x	x				x	x						x							x	x	x
Uses for fund																																
Administrative costs	x	x	x														x		x	x				x								
Emergency response		x	x	x	x	x			x	x	x			x	x	x	x		x	x		x		x	x		x			x	x	x
Facility operation/ perpetual care	x			x					x	x	x				x								x									
Cleanup, monitoring of inact./abandoned sites	x				x	x			x	x	x		x		x			x		x		x			x					x	x	
Site identification			x																													
Victim compensation		x	x	x							x		x		x					x		x				x				x		
Personnel training, equipment, supplies		x						x			x	x			x					x		x		x			x					x
Sources of funds																																
Assess facility owner/ operator	x	x		x			x	x	x	x					x	x								x					x		x	
Assess generator	x		x	x			x	x	x		x	x					x		x	x		x	x		x		x		x			
Legislative approp.		x	x	x													x	x	x	x	x	x	x		x			x	x		x	
Spiller reimbursement			x		x				x		x	x					x	x	x	x	x		x		x			x	x		x	x
Penalties/fines	x		x	x				x			x	x		x				x	x	x	x	x	x	x							x	x
Bonds																		x	x	x		x	x				x					
Other																				x			x					x				
Ceiling on fund balance	x	x	x		x	x		x			x	x	x				x		x	x		x			x		x	x		x		
Owner/operator liability standards	x	x	x					x			x	x		x	x			x	x	x	x	x		x		x		x	x	x	x	x

237

Table 7-5 / Good Samaritan Legislation by State (National Conference of State Legislatures, 1982).

	AR	GA	NY	NC	RI	SD	TN	TX	VA	WA
Parties										
Individuals	x	x	x	x	x	x	x	x	x	x
Partnerships	x					x	x	x		x
Corporations	x					x	x	x		x
Actions										
Omission or commission	x	x		x	x	x	x	x	x	x
Must be requested by authorities		x		x	x	x	x		x	x
Materials										
Compressed gases			x	x	x				x	
Liquefied petroleum gas		x				x	x	x		x
Hazardous materials										
Limits on liability exemption										
Act constitutes negligence or intentional misconduct	x	x	x	x	x	x	x	x		x
Act done in ordinary course of business	x	x	x			x	x	x		
Act is in response to accident by actors		x	x	x	x					
Own facilities or equipment		x						x	x	
Other										x

238

by the environment in which they are discarded. Similarly, there may be no receptors threatened by residual hazards. When hazards do exist and potential damages can be identified, complete and quantitative characterization of those damages is necessary to identify the optimum alternative for amelioration. Hence, any remedial action should be preceded by an assessment of the site. The present EPA program aggregates site characterization and assessment activities into two types of output: a remedial investigation (RI) and an endangerment assessment.

Site assessments are directed to evaluating the risk posed by a site. To accomplish that, an assessment must address three tasks:

1. Describe the source term.
2. Quantitatively characterize the transport mechanisms by which contaminants are moving.
3. Analyze the consequences likely to be associated with contaminant migration.

Each of these tasks involves technologies and methods utilized in other fields of engineering.

Source Term Description

The first step in the assessment process is to determine the size, location, and nature of the sources of possible contaminants: What chemicals? How much of each chemical? Where are the chemicals located? Ideally, the responses to these inquires can be found in company and state records. Results of recent surveys, however, have revealed that prior to 1970 these data were rarely recorded. Additionally, those that do exist have often been found to be inaccurate. Although review of historical information is a necessary preliminary step, it is rarely a sufficient one for source term characterization.

The most direct means of supplementing existing records is by drilling sampling holes in the disposal area. A pattern of drilling and subsequent core analysis will reveal the dimensions of the disposal pit and the nature of the deposited waste. However, drilling is both expensive and hazardous. Perforation of existing overburden may release toxic or reactive vapors injurious to the drill operators. As a consequence, less direct means of investigation are desirable.

Current efforts in this regard rely heavily on techniques developed for geophysical surveys. These include use of one or more specific approaches to remotely determine the characteristics of underlying strata. One of the most comprehensive evaluations of potential techniques to date was conducted at the Hanford atomic energy reservation for the U.S. Department of Energy (Sandness, 1979). Field tests were conducted with six individual approaches at that time: (1) metal detector, (2) magnetometer, (3) acoustic reflection,

(4) acoustic holography, (5) ground-penetrating radar, and (6) thermal infrared imaging. Evaluations were made on the basis of screening surveys conducted in a test trench excavated in alluvial deposits consisting of dry sediments ranging in size from fine-grain sand to large cobbles (3–4 inches in diameter). Common objects such as steel drums, gas bottles, steel scrap, and wooden boxes were employed as targets. The results of screening are summarized next.

Metal Detection. Metal detection is the most widely known of the six survey techniques employed. It has long been used as a means of locating mines in wartime. More recently, commercial units have been marketed for use by treasure hunters and coin collectors. Most devices of this type operate on an induction balance principle wherein transmitter and receiver loops are geometrically positioned to create a null condition in the absence of metal objects. The presence of metal disrupts the electromagnetic field, causing deviations in the balance condition that are subsequently detected.

Evaluation of commercially available, hand-held metal detectors at the Hanford site (Figure 7–2) proved them to be useful in locating various metal objects, including large masses of scrap, barrels, and pipes positioned vertically in the soil. The detectors were quite insensitive to elongated objects of small cross section (metal rods), even at short range. Although detection range varied in a complex manner with several factors, it was concluded that objects could be detected at a depth of two to four times the object diameter, width, or other representative dimension. For the most part, metal detection is not well developed for our purposes here. Commercial devices have not been optimized and are largely directed to small objects (e.g., coins) or pipelines. However, some service contractors have devised specialized, large-coil systems to provide greater depth capabilities.

Magnetometry. Magnetic field measurements can be used to locate buried ferromagnetic objects such as tools, steel containers, and steel scrap. This method is based on the fact that an induced magnetization is produced in any magnetic material within the earth's magnetic field. The induced field is superimposed on the earth's magnetic field and, if sufficiently large, can be detected as an anomaly or an aberration in the ambient field.

For a buried object, the induced magnetization per unit volume is given by

$$I = \Delta k B_0$$

where Δk is the difference in volume magnetic susceptibilities of the object and the surrounding material, and B_0 is the intensity of the earth's magnetic field. The value and direction of B_0 are functions of location on the earth; most important, they are functions of latitude. Strongly ferromagnetic minerals, such as magnetite and ilmenite, have susceptibilities in the range 0.1 to

Figure 7-2 / Several varieties of metal detectors are available, including those used by treasure hunters to find coins.

0.8, but these minerals account for only a few percent of the volume of most sediments. Their effects on magnetic field measurements are usually small in surveys for buried steel objects, but must be considered.

The induced magnetic field of a buried object depends on the size, shape, depth of burial, orientation, and susceptibility of the object, as well as on the direction and intensity of the earth's field. Analytical solutions for induced magnetic fields can be obtained for special object shapes, such as spheres and cylinders, and it is generally possible to compute fields due to complex shapes by digital computer methods. However, in the inverse calculation, which is the case of primary interest in geomagnetic surveys, no unique set of parameters can be determined from a measurement of the magnetic field patterns. In other words, an infinitely large number of combinations of parameter values can produce the observed magnetic anomaly. This indeterminacy is characteristic of most geophysical methods. It is one of the factors that motivate a multifaceted approach to geophysical exploration programs. For instance, if radar is used to determine target depth, the magnetic reading can be interpreted to estimate the mass of ferromagnetic material.

The principle of operation of a proton precession magnetometer is that the nuclei of hydrogen atoms contained in a hydrocarbon fluid will align their spin axes in the direction of an applied magnetic field. If that aligning field, produced by a coil around the sample fluid, is removed, the protons will precess like wobbly tops about the direction of the ambient or earth's magnetic field. The frequency of the precession is measured electronically and is precisely proportional to the intensity of the ambient magnetic field. The precession frequently is internally converted to magnetic intensity in units of gammas and is digitally displayed on the face of the control unit of the magnetometer (Figure 7-3).

Use of a magnetometer requires survey on a uniform grid, with spacing determined by the size of the objects sought. Results are mapped and isoquants delineated to determine the presence of buried steel objects (Figure 7-4). Locations are approximate and require interpretation based on the magnetic inclination of the burial site.

Ground-penetrating Radar.   Many aspects of radar technology have become highly advanced and sophisticated in recent years. Military applications and civil applications, such as air traffic control, marine navigation, weather observation, and remote sensing, have stimulated major advances in radar technology. Recent developments in electronics and data processing equipment have enhanced the reliability, power, and flexibility of radar systems. However, these remarks apply primarily to radar systems designed for aboveground applications. Although downward-looking radar devices have been used for many years, there has not been the same intensity of effort to develop these systems as to develop above-ground systems.

In applications that involve the determination of depths of interfaces or the detection and characterization of discrete objects or underground structures, resolution requirements demand the use of short radar wavelengths, generally less than 10 m. Unfortunately, many ground materials are strong absorbers of electromagnetic energy at short wavelengths. Strong absorption by the ground usually requires the use of radar wavelengths greater than about 0.5 m.

The water content of the ground is the most important factor affecting electromagnetic absorption. An increase in moisture content of the soil or other ground material greatly increases both the electrical conductivity and

Figure 7-3 / Magnetometers measure the strength of the earth's magnetic field, thereby detecting ferromagnetic objects that create anomalies.

the dielectric constant of the ground. The magnitude of the effect of ground moisture depends on the composition and porosity of the ground material. The dielectric constant at normally used ground-penetrating radar frequencies is generally in the range 3–30, as illustrated in Figure 7–5. The dielectric constant of pure water is 81.

Measured values of absorption loss as functions of frequency and soil moisture are shown in Figure 7–5. These are laboratory measurements made on a sandy soil from the Hanford area. The expected absorption loss for a saturated volume of this material would be about 8 db/m at 100 MHz. Equivalent measurements were not made for saturated clay soils; however, field observations of radar performance indicated that the absorption loss was at least this high.

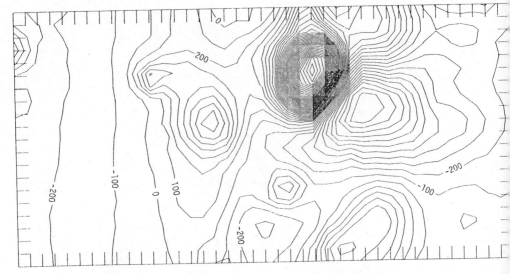

Figure 7-4 / Outputs from magnetometer surveys are displayed as contours to locate ferromagnetic objects.

Additional loss factors that affect the performance or effectiveness of a given ground-penetrating radar system include reflective losses at the air-ground interface, geometric spreading of the transmitted radar beam, the effective backscattering cross section of the reflective target, and the spreading of the reflected signal. A positive factor is a refractive gain due to the focusing effect of the dielectric medium. Field experience suggests that these factors limit vertical penetration to less than 40 ft in dry sandy soils and perhaps 10–12 ft in saturated clays. Range data must be calibrated on a site-specific basis and can be extended with modification of antenna configuration and frequency. The radar approach differs significantly from metal detection and magnetometry in that it detects changes in dielectric constant, rather than specific magnetic properties. As such, it can be used to detect a variety of materials, not just metals. An illustration of graphic output from detection of a buried object is shown in Figure 7-6.

The present state of the art in ground-penetrating radar measurement and data interpretation is such that under the conditions commonly experienced in field surveys, it is often difficult or impossible to determine the size, shape, or composition of a buried reflective object. Research and development efforts will undoubtedly result in improved ground-penetrating radar system capabilities and data analysis procedures in the future. At the present time, measurement devices such as metal detectors and magnetometers can provide useful supplementary information about objects or materials detected and mapped by available radar systems.

Electrical Resistivity.   Electrical resistivity (ER) is an indirect survey technique applicable to exploration of subsurface areas. Depth is limited by the voltage available and safety considerations. These practical constraints typically restrict use to relatively shallow investigations. The ER approach is based on the detection of changes in electrical conductivity properties of soils and buried objects. As such, the technique is most useful in areas with a relatively homogeneous native soil structure.

One of the simplest means of applying ER for subsurface profiling is the Wenner method (Darrler, 1980). In this approach, for copper-coated steel

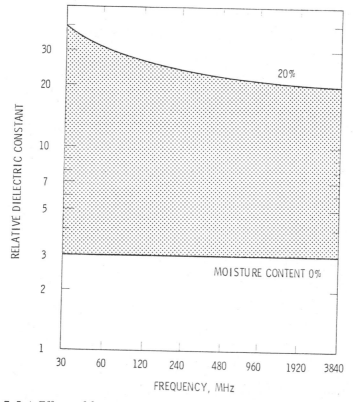

Figure 7–5 / Effects of frequency and moisture content on soil dielectric constant. (From Sandness, 1979, with permission.)

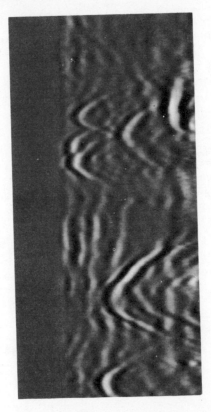

Figure 7-6 / Typical output of radar scan of buried objects. The vertical lines help calibrate depth. The parabolas result from reflections from targets.

electrodes are inserted 4 to 6 inches into the ground along a straight line. Spacing between the electrodes is determined by setting it equivalent to twice the depth of interest (i.e., if a maximum depth of 10 ft is desired, spacing between each electrode should be 20 ft). Typical operating parameters under these conditions might involve a 90-V power source and currents of roughly 20 mA. Power is applied to the outer electrodes. A resistivity meter is connected across the inner electrodes to determine the voltage drop throughout the volume of intervening earth.

The apparent resistivity a is determined as

$$\rho a = 2n\ell V/I$$

where ℓ is spacing between electrodes, V is voltage applied, and I is amperage. The resulting apparent resistivity is in fact an average for all materials through which the current travels. Hence, the presence of buried objects or changes in substrata are deduced by comparing resistivity values in a grid pattern. Experience suggests that sharp changes in data patterns can be anticipated from resistivity changes in the top 10 to 15 ft of soil.

Resistivity can also be applied to the location of contaminant plumes when the materials either are soluble electrolytes of sufficient strength to significantly change groundwater conductivity or are saturated pools of hydrocarbons that will appear as an insulator.

Infrared Imaging.   Changes in soil properties and moisture content can, under favorable conditions, cause distinctive temperature, thermal emissivity, or thermal inertia patterns. When these changes are associated with buried materials and backfill in a disposal trench, infrared radiation can be applied to many burial grounds. Exploratory surveys with this approach were conducted at Hanford during various times of the day from aircraft. Output photos were intensity-modulated to provide clear images of objects (Figure 7–7.)

The observed thermal patterns were found to largely reflect differences in soil patterns and vegetation density, although a series of buried vertical pipes was also located. It was postulated that surface effects could be minimized by repeated use of infrared surveys over a period of time that would allow one to characterize the thermal inertia of the ground and remove it as background. However, at the present time, the surface interferences minimize the value of this approach for finding buried objects.

Acoustic Reflection Profiling.   Acoustic geophysical survey techniques operate through the interpretation of how sound waves travel through the ground. Systems can be designed to operate in a number of modes: reflection (or pulse echo) (Gal'perin, 1974), refraction profiling (Musgrave, 1967), travel-time measurements, acoustic holography (Fitzpatrick, 1974; Fitzpatrick et al., 1972; Metherell and Green, 1970), and mechanical impedance mapping (Keller, 1967; Fieldhouse, 1970; Ballard et al., 1969). The first of these was selected for evaluation at Hanford. Although all five approaches present problems in the dry, sandy sediments of the Columbia basin, this was deemed the most promising, because it is least affected by lateral inhomogeneities in the soil. It is also most easily implemented. The principle involves interpretation of sound waves reflected back from various depths in the soil. Changes in intensity and frequency are associated with variations in the conducting media such as the soil–waste interface.

The field experiments involved a pattern of detection devices (geophones, accelerometers, and hydrophones were all evaluated) and an acoustic source (a steel plate struck with a sledgehammer) depicted in Figure 7–8. All three detectors performed well. Some filtering is required to distinguish between nonvertical reflections. One alternative is to place geophones in a coincidence

Figure 7-7 / Infrared imaging photos.

Figure 7-8 / Use of sledgehammer for acoustic survey source.

mode (equidistant on either side of the source) and select only reflections reaching the detectors simultaneously. An example of normalized output is provided in Figure 7–9.

The coincidence mode proved valuable in providing useful data. The overall prognosis for acoustic reflection at Hanford was not promising, however. The technique requires good acoustic coupling between the source and the ground and the detectors and the ground, as well as an effective medium for transfer in the soil. The dry, homogeneous sediments encountered at Hanford reduced effectiveness. On the other hand, a similar approach at Love Canal in Niagara Falls, New York, proved very successful. The tight, saturated clay in that area, which can hamper radar transmission, provided an optimum means of acoustic conductance. Officials from the New York Department of Transportation were able to develop very accurate data on trench boundaries using explosive sources and detectors.

Acoustic Holography. Acoustic holography uses a principle similar to that with acoustic reflection, but geophones are emplaced at depth in observation holes that allow construction of a three-dimensional image of the subsurface

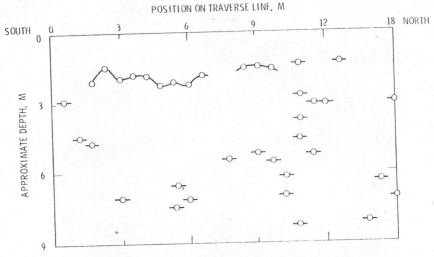

Figure 7-9 / Example of acoustic reflection output.

column being investigated. A schematic of the layout employed at Hanford is presented in Figure 7-10.

Once again, a sledgehammer and steel anvil were used as the acoustic source. Both were wired so that contact closed a circuit and started operation of the computer collecting data from the geophones. This also provided an accurate means of determining travel times to each detector. Analysis of this output allowed computer-assisted construction of an image of the subsurface area illustrated in Figure 7-11. Evidence of the presence of buried materials or packed backfill is emphasized by subtracting the theoretical time of flight for undisturbed ground from the actual observations. The better conducting medium produced by the more densely packed backfill creates a clearly defined trench boundary in the image.

Comparative Analysis. As a result of the comparative studies conducted at Hanford and related efforts, one can characterize the relative advantages and disadvantages of the seven means of source assessment as presented in Table 7-6. A simple decision process for selecting the most cost-effective survey alternative is provided in Figure 7-12.

Based on the findings of the Hanford evaluation, researchers at Battelle's Pacific Northwest Laboratories selected metal detection, magnetometry, and ground-penetrating radar for inclusion in an integrated survey unit for burial-ground investigations. The radar and magnetometer were mounted on a self-propelled vehicle constructed of nonmagnetic materials to minimize interfer-

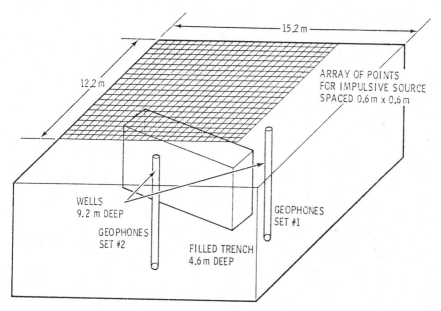

Figure 7-10 / Layout for acoustic holography survey.

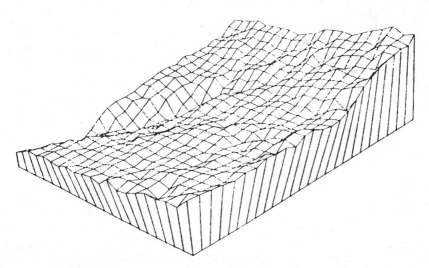

Figure 7-11 / Output from acoustic holography survey.

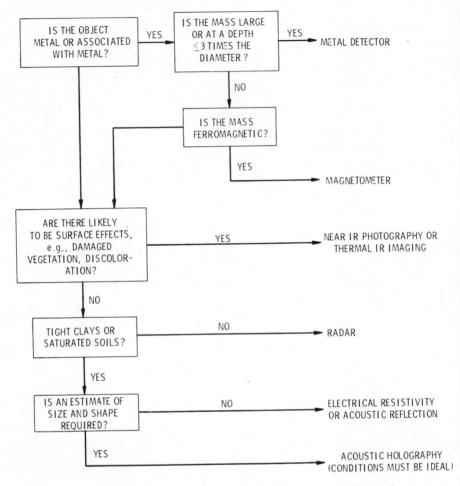

Figure 7-12 / Decision aid for selecting between alternative methods for remote detection.

ence (Figure 7-13). On-board instrumentation included both the electronic equipment to operate the detection devices and a microprocessor unit to receive digital output and prepare it for input to a centralized computer. The processed data are transmitted via telemetry to a truck-mounted computer with appropriate graphics display devices.

Conduct of a burial-ground survey is accomplished in four steps:

1. Physical survey and marking to provide a sampling grid from which all output can be coded and mapped. (This step can be eliminated when

Table 7-6 / Relative Advantages and Disadvantages of Geophysical Survey Techniques for Source Assessment.

Technique	Advantages-Applications	Disadvantages-Limitations
Metal detection	Easy to apply; inexpensive; good for larger metal objects at shallow to moderate depth	Provides no real quantitative data on size or depth of target; misses small-diameter and deep objects; requires presence of metals in buried waste
Magnetometry	Excellent for iron and steel objects of all sizes; can detect larger objects at some depth; may be used for large nonferrous materials in areas of high natural ferromagnetic mineral content	Cannot determine depth of object; for most purposes, limited to iron and steel; requires extensive grid measurements and specialized equipment
Ground-penetrating radar	Provides estimates of location, depth and composition; shows real promise to field interpretive data with more development	Penetration and resolution affected by composition and moisture content of soil
Thermal IR imaging	Used in overflight mode; can cover large areas rapidly; may provide means of locating damaged vegetation from chemical migration	Ground-cover and soil-type changes confuse output analysis; may not identify many types of buried material
Acoustic reflection	Excellent in tight or moist soils; not restricted to metallic targets	Performance deteriorates in dry, course soils
Acoustic holography	Can provide three-dimensional glimpse of buried objects; possible to interpret composition from travel-time data	Requires bore holes; suffers in dry, loose soils
Electrical resistivity	Works well in moist soils, clays; can operate deep if sufficient voltage can be safely applied; can find contaminant plumes for concentrated electrolytes or insulators	Loses resolution in dry, porous soils; output may be difficult to interpret

Figure 7–13 / Battelle geophysical survey unit and support vehicle.

positioning equipment is employed and the computer is used to label data accordingly.)

2. Manual survey of the grid using hand-held metal detectors marking all identified targets and transferring locations to an overlay map.

3. Operation of the survey vehicle over the same grid with data fed directly onto the computer.

4. Data superposition and analysis. This is done largely by a unique software package created to filter and enhance output. Graphic display devices are used first to provide vertical profiles of each radar survey run. These are then combined and sliced horizontally to generate an aerial view of the section surveyed. When specialized enhancement programs are applied, the output appears as color photographs with specific hues depicting targets (Figure 7–14).

This system has been applied on waste sites throughout the United States and West Germany, including Love Canal. It has proved to be quite effective in mapping burial trenches and individual buried items. Electromagnetic induction (EMI) has been added to locate conductive contaminant plumes. Further development on the radar antenna is under way.

With the site mapped and the source characterized, one can proceed with an assessment of contaminant migration to determine the extent to which chemicals are being transported beyond the site.

Contaminant Migration Assessment

Site assessment work can be conducted on a number of levels, depending on the time and resources available. In general, it is directed to determining the current disposition of contamination and the likely migration of contaminants over time. As such, it is important first to understand the various means by which contaminants move from a disposal site. Major transport mechanisms are illustrated in Figure 7-15. The two major media for chemical migration are water and air. Precipitation will carry contaminants in runoff to surface waters or transport them in percolate to the underlying aquifer. Atmospheric travel may result from wind erosion of particulates or vapor migration. The relative importance of each factor will depend on the location of the site with respect to geologic and meteorologic characteristics. In a study of an arid landfill site on the U.S. Department of Energy's Hanford reservation, wind erosion was found to be the single most important mechanism by which contamination spread (Phillips et al., 1977). Wetter areas experience circumstances such as those at Love Canal, where precipitation played the major role in moving contaminants outward from the site. Volatilization may be a factor at any site, depending on the nature of the buried wastes and the integrity of the overburden materials. In a review of major problem sites, the Senate subcommittee on oversight and investigations concluded that the most pervasive damage at these locations was the contamina-

Figure 7-14 / Photo of geophysical output from survey. Targets found include sewer lines, backfilled cistern, clay fill area, and random debris.

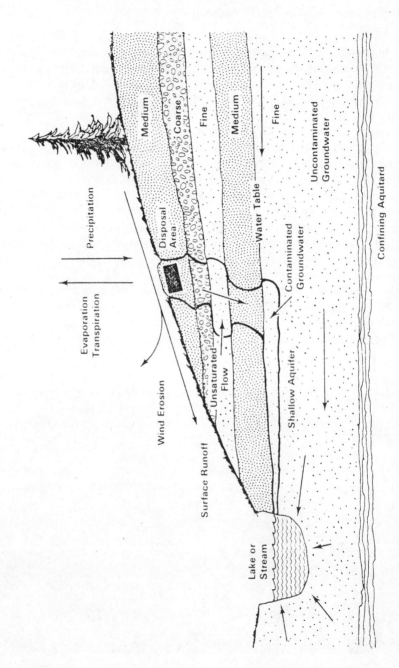

Figure 7-15 / Mechanisms governing contaminant transportation.

tion of groundwater. This may be particularly impactive, because nearly one-half of the U.S. population relies on groundwater for potable supplies.

Regardless of the mechanism involved, the task of measuring contaminant disposition must begin with sampling and analysis. The methods of sampling and analysis are the subject of another section and in many cases have been standardized for use. A critical question remaining, however, addresses the selection of sampling locations and the number of samples. In the case of the former, improper selection may lead to development of a totally erroneous picture with respect to the level and distribution of contamination. In the case of the latter, the trade-off is one between cost and development of an accurate portrayal of the site. There are no simple algorithms to resolve these questions for all sites. Relatively homogeneous areas with simple geohydrologies can be sampled in a statistical manner and an adequate picture developed within the accuracy of the grid spacing selected. As heterogeneity is introduced, the picture becomes much more complex. For instance, fracture zones may create channels for groundwater movement that can overwhelm other routes of flow. If sampling is too sparse to characterize the fracture systems, conclusions will be ill-founded.

In recent years, researchers have turned to the use of mathematical models to assist in the design of sampling programs. In this approach, all known data for a site are collected and entered into the hydrologic model. This may include information on potentiometric surface, porosity, permeability, and levels of contamination. The model is then run and queried to determine (1) parameters to which results are most sensitive and (2) areas where additional data are required to provide solutions with the necessary level of accuracy. In this way, the model is employed to identify the minimum amount of additional sampling and analysis required and the optimum locations for that sampling.

Additional methods other than mathematical modeling are also available to better define sampling needs. One such approach is a statistical algorithm known as kriging. This methodology has been constructed for analyzing field data to produce surfaces and confidence limits as well as loci for sampling that will have the greatest impact on results. Whereas many of the sampling locations identified are obvious from review of the data by an experienced hydrologist, others are hidden in subtle patterns and require quantitative treatment for discovery.

The foregoing techniques help determine where to sample. As noted earlier, there are numerous standardized techniques defining how to sample and analyze. Although that is not the subject of this work, it is important to note that these techniques have an acknowledged weakness stemming from debate over the applicability of laboratory-determined characteristics in the field. Most experts contend that current sampling techniques do not allow accurate measurement of properties such as permeability in the laboratory.

And yet, cost-effective means of measurement in situ are not available. As a consequence, there is a real need for the development of better techniques for the collection of geohydrologic data. Similarly, the introduction of accurate techniques for measuring trace contaminants in the field could greatly enhance characterization work.

Once a site has been adequately characterized, the assessment of health and environmental risks can proceed. The process employed may be as simple as the exercise of professional judgment or as complicated as the application of sophisticated mathematical models. The level of technology employed is determined by several key factors:

Quantity and quality of input data available
Resources available for problem resolution
Level of resolution desired
Time constraints

As noted previously in the discussion of facility siting, there are numerous mathematical models that have been developed to assist in site assessment. These range from relatively simple, one-dimensional constructs to very complex three-dimensional systems. The selection of a given model will once again be based on the factors listed earlier. Regardless of the specific method employed, output should address transport paths, travel times, and arrival distributions of contaminants. Consequence analysis must then be run to determine the potential impact of those projected results. To date, the latter process has been an ad hoc one wherein concentration–duration data are evaluated on the basis of known time–dose relations for test animals and a judgment made as to likely effects. Growing need for a more formal chemical risk assessment methodology has given rise to significant development efforts in this area recently. One of the outcomes of this work has been the emergence of the chemical migration risk assessment (CMRA) methodology designed by Battelle Memorial Institute for the U.S. Enivronmental Protection Agency (Onishi et al., 1979). This methodology was development for the evaluation of impacts from pesticides and pesticide application techniques to aquatic life. Its potential use, however, goes well beyond that limited sphere to hazardous materials in general and may be modified to consider health and terrestrial consequences as well.

The CMRA methodology predicts the occurrence and duration of toxic contaminants in stream systems. At the same time, it predicts the probability of acute and chronic damages to aquatic biota. The methodology consists of four components: (1) overland contaminant transport modeling, (2) instream contaminant transport modeling, (3) statistical analysis of in-stream contaminant concentrations, and (4) probabilistic risk assessment. Figure 7–16 illustrates how these components are connected.

REQUIRED INPUT DATA   ANALYSIS

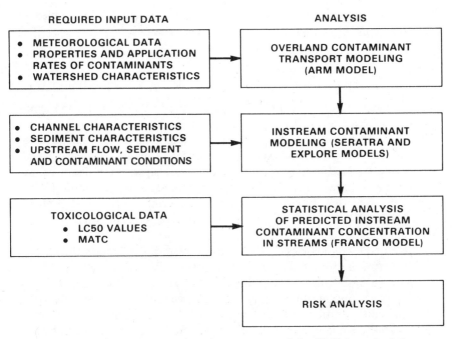

Figure 7-16 / Interrelationships of components of the CMRA methodology.

Overland Contaminant Transport Modeling.   The agricultural runoff management (ARM) model predicts runoff and sediment and contaminant loadings at the edge of a stream. Model inputs include meteorological data,
physical and chemical properties of a contaminant, contaminant application
rates and practices, and watershed characteristics. ARM provides continuous
simulation of contaminant loading by modeling

Hydrologic response of watersheds
Soil erosion
Contaminant adsorption and removal
Contaminant degradation

In-Stream Contaminant Transport Modeling.   In-stream chemical migration
and fate are simulated by the hydrodynamic water quality model EXPLORE
(Baca et al., 1973) and the sediment-contaminant transport model SERA-
TRA (Onishi et al., 1979; Onishi and Wise, 1979a, 1979b). EXPLORE is a
one-dimensional, general water quality model that provides discharge and
depth data to SERATRA. SERATRA is a finite-element model that predicts
time-varying longitudinal and vertical distributions of sediments and contam-

inants. The model consists of the following three submodels coupled to describe sediment-contaminant interaction and migration:

Sediment transport
Dissolved contaminant transport
Particulate contaminant (contaminants adsorbed by sediment) transport

The sediment transport submodel simulates transport, deposition, and scouring for each size fraction (or sediment type) of both cohesive and noncohesive sediments. The transport of particulate contaminants is also simulated for those associated with each sediment size fraction. The contaminant submodels include the mechanisms of (1) advection and dispersion of dissolved and particulate contaminants, (2) chemical and biological degradation, (3) adsorption/desorption, and (4) deposition and scouring of particulate contaminants. SERATRA also computes changes in river bed conditions for sediment and contaminants.

Required input includes channel and sediment characteristics and adsorption/desorption properties of the contaminants. Because the current toxicological data base is not sufficiently advanced to fully utilize the particulate contaminant concentrations, only the cross-sectionally averaged, dissolved pesticide concentrations are further analyzed for risk assessment. However, the capability is there to include particulate contaminants absorbed by suspended and bed sediments that may be important under actual field circumstances, especially for assessment of a long-term aquatic impact.

Statistical Analysis.   The computer program FRANCO (Onishi et al., 1979; Olsen and Wise, 1979) provides statistical summarization of time-varying contaminant concentrations and is the link between simulated in-stream contaminant concentrations and risk assessment. It provides the frequency of occurrence and duration of specified contaminant concentrations in receiving waters. Outputs include the number of times and the percentage of time a given concentration-duration level is exceeded and the concentration levels involved.

Risk Assessment.   Risk is typically determined by multiplying the probability of an event by the consequential effects. FRANCO provides a measure of the probability. The consequential effects are expressed in terms of lethality and sublethality by using a median lethal concentration (LC_{50}) and MATC (maximum acceptable toxicant concentration). The LC_{50} with its associated duration is defined as a concentration at which 50% of an aquatic species will be killed. This represents an acute impact. The MATC range is located between the highest concentration showing no detectable harmful effects and the lowest values displaying some observable effect. Hence, MATC describes the effect–no effect boundary for chronic toxicity. By selecting specific concentration-duration levels to match LC_{50} and MATC values, FRANCO provides a probabilistic risk assessment. As shown in Figure 7–17, FRANCO displays consequence zones of lethality, possible lethality, sublethality, and no effect, with the associated probabilities.

FRANCO typically uses six concentration-duration pairs to define a piecewise concentration-duration curve to provide the number of times, duration, and frequency a given concentration-duration curve is exceeded. Because of the lack of available toxicologic data, the risk assessment is presently limited to the direct effects of the dissolved form of the chemical. Ingestion as a secondary route is not addressed, nor are indirect effects such as bioconcentration and biomagnification.

As noted, an approach related to this may be employed to assess risks to public health and the environment from contaminated sites. To be effective, the current methodology must be made to incorporate groundwater models such as those described in Chapter 5 and should consider effects to humans and terrestrial life. Should the predicted consequences be deemed unacceptable, remedial action will be necessitated.

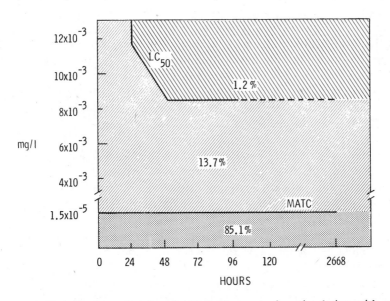

Figure 7–17 / Concentration-duration lethality curve for stimulation with toxaphene. Percentages relate to the proportion of time a given segment of stream will be subjected to noted effects. ($LC_{50} \geq 50\%$ loss of exposed population; MATC above line, sublethal and chronic effects; below MATC line, no effects).

Site Mitigation

Once the source term has been identified and mapped, remedial action can commence. Relevant activities for selection of a preferred option are encompassed in what the EPA has defined as a feasibility study (FS). The evaluation is conducted in three phases:

1. Enumeration of all feasible approaches
2. Assessment of associated costs, effectiveness, and impacts for each feasible approach
3. Selection, detailed design, and costing of the optimal approach

Prior to undertaking these steps, however, it is important to establish the cleanup criteria or restoration goals against which alternatives are to be evaluated. The central question is "How clean is clean enough?"

No soil criteria presently exist for most toxic chemicals. For nearly all contaminants, there are no soil threshold limits defining when hazardous effects will begin to be evidenced. In part, this reflects the fact that there are no simple standard tests to which a contaminated soil can be subjected for designation as hazardous or nonhazardous. Fortunately, there is still a relatively straightforward approach available for establishing criteria. For most contaminants, soil residuals are of concern because of their ultimate ability to contaminate the atmosphere (through volatilization or resuspension) and the hydrosphere (through leaching and runoff). Hence, hazardous chemical levels in soil can be defined as those that will sponsor hazardous levels in air or water. Criteria and guidelines for ambient air and water have been suggested for a number of chemicals. By working backwards with minimal data on dilution potential and attenuation coefficients, criteria can be established. Given these, one can determine the subset of alternatives that can meet objectives and the extent of restoration required.

Alternately, it has been proposed that soils from contaminated sites be subjected to hazardous waste identification tests such as those described in Chapter 3. Cleanup would be required in all areas where the soil failed the test. This approach is limited in applicability by the specific tests considered. For instance, the federal tests relate only to metals for which there are drinking water standards. A more comprehensive battery of tests such as those required by California would be needed to address sites with organic contaminants.

Once criteria have been established, the feasibility study can proceed through analysis of specific mitigation alternatives.

Available options for mitigation can be categorized in one of three groups: (1) extraction and ultimate disposal, (2) in-place destruction, and (3) in-place immobilization. Each group offers advantages with respect to cost effectiveness and feasibility. Individual options must be evaluated on a case-

by-case basis considering the availability of supplies, the environment surrounding the site, the nature of the chemicals involved, pertinent regulations,
and costs. Guidance is currently emerging on this complex subject. The U.S.
EPA has recently let contracts to develop guidance manuals and to evaluate
specific alternatives at known sites. Results of this work will increase the
information available for making initial evaluations.

Extraction and Disposal

Physical removal and disposal is the oldest and most common method of
responding to land-based spills. For most materials, the zone of concern
consists of a volume of contaminated soil. Removal is accomplished by
methods of standard excavation (backhoe, scoop, manual shoveling, blade,
etc.). For instance, simple scraping was employed by the Russian government
to remove radioactive contamination from surface soils (Trabalka, 1979).
Differences from normal excavation procedures include dust suppression, the
need to clean equipment after use, worker protection, and use of containers
or covered trucks to prevent inadvertent losses during transport. When
contamination penetrates the water table, well points or grout may be required to allow successful excavation.

In soil decontamination work at the Los Alamos Scientific Laboratory,
radioactive contamination has been removed through several mechanisms:
(1) manual shoveling into plastic bags, (2) backhoe, and (3) front-end loader.
In the latter case, the ground was first soaked to prevent dust suspension. Soil
was then plowed and pushed into piles for subsequent loading on dump
trucks. Loads were covered and surveyed and tires cleared prior to departure
(Ahlquist, 1979). Similar work by Reynolds Electric and Engineering Co.
utilized vacuum cleaners for shallow contamination. In the more rugged
areas, however, workers had to revert back to shovels and use water flushes
to cleanse rock surfaces (Bicker, 1979). In work on the Hanford reservation
(Toomey, 1979), comparison of these surface removal techniques has resulted
in the profiles presented in Table 7-7

Allen (1979) has reported a unique approach to minimize the potential for
contamination of transport equipment that is such an operation and provides
for rapid cleaning. Cheesecloth fabric is rolled out to cover the bed of the
truck or other areas to be protected. Spray-on fixatives are then applied.
When operations are complete, the sheeting is simply rolled back up and
disposed with the contaminated soil. Available fixatives include polyvinyl
alcohol (can be removed with a basic wash), polybutyl dispersion (removal by
stripping), and waterborne vinyl resin (can be removed by stripping).

When spilled materials consist of light liquids that pass through the soil
and form a slick on the water table, removal can be accomplished with a
system of extraction wells. The first is completed in the underlying aquifer

Table 7-7 / Characteristics of Surface Soil Removal Alternatives.

	Shovel and Box	Dump Truck	Scraper
Equipment		Dump truck, water truck, front-end loader	Scraper, water truck, push caterpillar
Staffing	2 laborers, 1 supervisor	7 operators, 2 supervisors	3 operators, 2 supervisors
Rate	1 yd³/day/shovel	55 yd³/day/truck (maximum)	600 yd³/day/scraper (average)
Limits	Slow and labor-intensive	Rate depends on haul distance, depth, accuracy ±6 inches, heavy equipment	Limited transport packaging; rate depends on haul distance, accuracy ±6 inches
Uses	Known spots of contamination; small areas where protection is important	Areal contamination	Large areas most suitable for clean backfill

and is pumped at a rate that creates a cone of depression large enough to encompass the estimated radius of the spill plume. A second well is then completed in the interior of the cone, where the lighter fluid collects as it flows down gradient. Nearly pure contaminant is pumped from this well until removal is complete. Liquids heavier than water can be removed with a single well if the contaminated zone is underlain by an appropriate aquitard. Use of modeling can be most effective here as a means of evaluating the most effective layout for the required well field.

A related means of removal utilizing wells is currently under investigation. EPA-sponsored research has been directed to design and construction of a mobile facility to complete and operate a series of injection and retrieval wells. One possible configuration for use of this facility is a soil washing operation wherein fluids are injected into the contaminated zone and subsequently retrieved. Fluids are selected on the basis of the contaminants involved and may include solvents, sequestrants, or elutriates capable of removing sorbed or exchanged materials from soil. For example, Thornton (1980) reported that the use of surfactants in an injection mode can increase the amount of spilled gasoline recovered from ground waters. The detergents employed enhanced recovery via two mechanisms: (1) displacement of gasoline from soil and (2) reduction of the zone of capillary action to a narrower band, thus freeing more of the gasoline for collection in a pumpable volume. The EPA mobile facility described has not been field tested at this time. If successful, it could reduce the costs of removal and disposal, because residuals requiring disposal will not include the contaminated soil. Laboratory tests with surfactants were very successful for a number of organic contaminants.

A second contractual effort is under way to investigate soil washing ex situ. In this approach the soil will be removed and contaminants extracted prior to disposal. This will reduce residual volumes, but will still require excavation and handling of large volumes of soil. The technology employed is closely allied to ore benefication and hydrometallurgy for inorganic contaminants and solvent extraction for organic contaminants.

Some experience with soil washing has been gained in decontamination work related to radioactive contamination at Rocky Flats. After determining that stabilization raised too many long-term concerns and that removal/ burial only transferred the problem, a soil working technique was tested. Wet screening at pH 12 reduced 60% to 70% of soils below criteria. An 84% reduction in weight was obtained with scrubbing while decanting fines using such additives as Calgon and HNO_3. Typical results indicated 88.1% of the contaminants in the first wash, 7.1% in the second, 3.5% in the third, and 1.3% in the fourth, with 0.1% remaining in the residuals. Cationic flotation also proved effective for removal of clays, which are often the more heavily contaminated faction of a soil (Thompson, 1979).

Invert wells may also be employed for extraction. In this configuration, contaminant plumes may be provided, flow tubes through underlying aquitards, and allowed to flow under gravity to deep aquifers for disposal. Alternatively, contaminated water can be withdrawn and reinjected in a deep aquifer (*Environmental Science*, 1980*b*).

Some attention has also been focused on the potential for removal by biological systems followed by harvest and disposal. This approach relies on selection of species (usually plants) that accumulate contaminants at levels significantly higher than those present in the soil. Effectiveness is probably limited to the top 12 inches of soil, and the process may take as much as 60 years to achieve 95% removal (Berkowitz, 1979). Consequently, this option is not considered practical.

The foregoing means of induced removal are contrasted with the use of natural extraction processes, i.e., the collection of leachate and contaminated groundwater for treatment and release. This approach relies on the natural elution of contaminants from soil by native groundwater. Implementation requires collection wells and a treatment facility. Costs vary with the contaminant. A major drawback is the time required. In heavily contaminated areas where chemicals are strongly sorbed, natural leaching may take years to cleanse the soil and hence involves the treatment of tremendous volumes of water. This somewhat passive approach also often suffers poor public acceptance.

Once removed, contaminant and contaminated soil should be disposed of in accordance with appropriate RCRA regulations. However, controversy rages over the extent to which CERCLA actions must meet RCRA. The trend appears to be toward compliance. This will often mean use of permitted facilities that are not necessarily the closest, most convenient, or least expensive. Because residuals typically include a large volume of inert geologic medium (e.g., soil), landfill is often the disposal alternative of choice. Incineration and other means of disposal can also be employed, but the cost may be prohibitive. In a recent review, Berkowitz (1979) found that incineration of a half square mile of soil contaminated to 25 ft would take 15 yr to accomplish in a rotary kiln at a capital cost of $10 million and annual costs of $4.5 million. Regardless of the alternative selected, cost is the major drawback for the remove and dispose option because of the large volume of materials that must be handled. Transportation of exhumed waste will also add a new element of risk. The restoration of uncontrolled sites will heavily tax the already limited capacity of hazardous waste facilities.

Extraction and disposal has been identified as the optimal approach to mitigation of Kepone-contaminated soils in the Hopewell, Virginia, area (Dawson et al., 1978), and it was employed in one particularly contaminated zone. It has also been selected as the most acceptable means of amelioration for the PCB-contaminated soils bordering highways in North Carolina.

In-Place Destruction

The most desirable approaches to mitigation of abandoned sites are those involving in-place degradation. These approaches are also some of the most difficult to devise. They rely heavily on the specific chemistry of the material spilled and the environment of the spill. The mechanism may involve physical, chemical, or biochemical action.

On the surface, this approach appears to be new and untried, but in actuality it is one of the common responses made to chemical spills, because organic materials often are left to be degraded by naturally occurring soil bacteria. Although this sounds haphazard on the surface, it is in fact a very effective means of response. Soils contain multitudes of bacterial, fungal, and mold cultures capable of breaking down a wide spectrum of materials. Just as these life forms serve to degrade waste materials in landfarming configurations, they can be left to naturally diminish residuals from spills, thus removing the need for excavation, handling, and ultimate disposal. This is particularly true for aliphatic organic materials and some simple aromatics. For more complex organics, or in nutrient-depleted areas, special cultures and supplements may be required. Emulsifiers are useful for some hydrocarbon nutrients to increase contact with aqueous cultures. Successful field applications have been made on sites containing fuel oil, chlorophenol, and acrylonitrile (Thibault and Elliot, 1980). For many hazardous constituents, however, even acclimated cultures are of little value. In a recent study of in situ techniques, bacteria were found ineffective on monochlorobenzene and yielded only limited action on formaldehyde (Wenstel et al., 1980).

A modified approach to biodegradation has been proposed by Williams et al. (1984). In this system, genetically altered bacteria produce and excrete tailored enzymes to attack specific contaminants. The enzyme is then introduced to the soil as a chemical additive. The economics of the approach become viable with the development of a means to mass produce extracellular enzymes.

The recently designed U.S. EPA landspill treatment unit can accommodate induced biodegradation schemes. Injection wells are used to introduce special cultures of specific-activity bacteria, nutrients, or complete cocktails. Success is very limited, however, because specially acclimated cultures often cannot compete with endogenous species, and action is restricted to surface soils. The EPA unit can also be employed for inducing chemical degradation when feasible. For instance, neutralization agents can be added for acid or caustic spills. Similarly, oxidizing agents such as ozone or chlorine can be introduced to attack organic materials. This is a complex undertaking, however. It is difficult to select a proper dose, obtain good contact, and avoid problems with use of a second hazardous material to mitigate the effects of the first.

A newer set of more sophisticated in-place destruction approaches appears to be emerging. Investigations into means to mitigating the Kepone contamination in Hopewell, Virginia, revealed that when amines were sprayed on Kepone-contaminated soils exposed to sunlight, photodegradation was greatly enhanced, leading to dehalogenation of the Kepone. (Dawson et al., 1978). In subsequent trials with PBB-contaminated soils, similar results were obtained (Christensen and Weimer, 1979). While the approach is limited to contamination near the surface, it could prove to be extremely useful for incidents involving large acreage and materials that are quickly sorbed into soils, such as PBBs and PCBs.

In-Place Immobilization

The third and final class of alternatives available for landspill mitigation includes options that prevent contaminants from being mobilized and ultimately reaching receptor organisms. The latter is accomplished in one of two ways: (1) Contaminants are converted to immobile forms or encapsulated in a highly insoluble matrix. (2) Contaminants are isolated from waters that could serve to transport the spilled material.

Chemical conversion of contaminants to immobile forms is largely restricted to inorganic species with insoluble salts. For instance, heavy metals can be immobilized as sulfide salts. This process incurs many of the difficulties associated with chemical degradation (dosing and hazardous side effects) while lacking permanence. Changes in soil acidity or other parameters could ultimately remobilize the contaminant.

Physical or chemical stabilization of contaminated soils can be achieved with a variety of commercially available materials. Some of the most common are silicate-based cementaceous compounds. Newer materials are sulfur or organic-polymer-based. These can chemically bond contaminants and physically encapsulate them. Most require intimate mixing to effect good stabilization and are therefore difficult to apply in situ. One source, Takenaka Komuten (1973) of Japan, has successfully developed a system for application to saturated sediments in water bodies. Limited work on soil materials was performed with grouting materials and polymeric agents to stablilize tank-farm wastes (Mercer et al., 1970). The process was found to be expensive and was frustrated by the logistics of obtaining widespread coverage without an excessive number of injection wells.

Grout materials and other organic-based sealants are also employed to isolate contaminated soil from the geologic waters that would transport hazardous substances. One approach is the use of a grout curtain. In this case, the sealant is applied in a vertical sheet to limit horizontal movement of waters into or out of the contaminated zone. This technology is often used in conjunction with excavation and construction in a saturated zone where it is necessary to control seepage. When conventional excavation followed by addition of soil (or cement) and bentonite is employed, curtains can be installed for $3.50 to $5 per vertical square foot at rates of 2000 to 3000 vertical square feet per day (*Environmental Science*, 1980b). Such curtains are most effective when taken down to a confining aquitard that prevents vertical escape of contaminant waters. Injection depths are typically limited to 50 to 60 ft or less; sheet piling can be taken to 100 ft, and well points are limited to about 25 ft (*Environmental Science*, 1980b).

The mobile EPA unit design for injection well installation is also equipped to accommodate installation of grout curtains using materials such as cement or bentonite clays. The injected material forms a spherical mass in the soil. The injection pipe is then withdrawn to above the sphere, and a second sphere is injected. This procedure continues until a column is formed like a stack of beads on a string. Finally, multiple columns are merged side by side to form a curtain.

Sealants may also be applied to the surface of a contaminated zone to prevent percolation and subsequent leaching. Clay barriers can be employed for this purpose, as was done at Love Canal to isolate residual wastes from further leaching. A similar approach has been applied to uranium mine trailings to reduce radon gas emissions (Koehmstedt et al., 1977). In that application, emulsified asphalt was employed. This material is a dispersed mixture of water, emulsifier, and asphalt treated to bond with soils and form an impermeable seal. The emulsion can be cationic or anionic, depending on the needs of the material to be stabilized. Difficulties were encountered wherever settling occurred or the surface was breached by heavy equipment. Alternative application means using a soil cover need to be investigated. The EPA is currently researching low-cost portable means for temporary seals after spill incidents. The most promising to date is a simple flexible plastic sheet that can be spread over the affected area.

Ideally, if isolation techniques are to be employed, it would be disirable to create an impermeable envelope around the entire effected area. This will require sophisticated control of the sealant during injection to allow formation of a contiguous wall. The asphalt emulsion materials are some of the most promising for this approach at this time, because they are pumpable and can be excluded from areas by the presence of specially treated waters. If more conventional grout slurries are to be employed, the key to success will lie in placing a layer beneath the contaminated zone to interconnect the surrounding grout curtains. Two novel uses of geotechnical advances appear promising (Thibault and Elliot, 1980). The first is referred to as pancake slurry jetting. High-pressure jets are employed as perpendicular kerfing knives to hollow out a series of overlapping, disc-shaped cavities (Figure 7–18); these, in turn, are backfilled with grout. In the second approach, hydrofracing, two parallel boreholes are similarly kerfed and then pressurized with the grout slurry, the pressurization causing fracturing from the kerfed cavities outward. When the fractures meet and the grout sets, the sealed basement has been constructed (Figure 7–19).

Clay caps are typically employed in conjunction with other isolation barriers to reduce infiltration. Traditional design calls for several feet of compacted clay with an in-place permeability of 10^{-7} cm/sec or less. This layer is covered with topsoil and planted to prevent physical damage to the cap and to promote evapotranspiration. For areas with no natural clay deposits, Opitz et al., (1982) have devised a gelatinous material with alum. The hygroscopic gel acts as both a moisture and a vapor barrier at thicknesses much less than those for clay with comparable performance levels.

Barriers for isolation must be evaluated carefully prior to selection as a remedial action. Recent work has shown that some contaminants interact with grout materials to prevent proper setting or to reduce permeability. In either case, the resulting barrier displays less integrity than the design value and results in contaminant escape. Laboratory compatibility tests are advised to identify such problems in advance.

One approach to stabilization that avoids compatibility problems is the use of in-situ vitrification. This technology is based on implacement of electrodes in the soil. High voltage is then applied across the graphite-treated soil surface to initiate current flow. Subsequent resistance heating causes the soil to melt. The molten mass then becomes the electrical courier and fosters advancement of the melt downward through the soil column. Depth of melt is

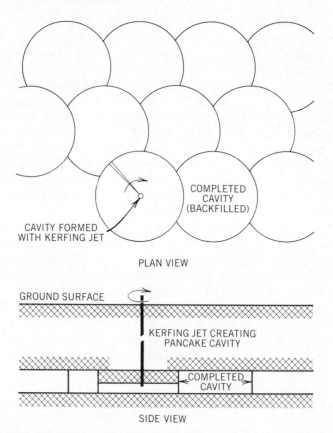

PLAN VIEW

Figure 7-18 / Pancake slurry jetting approach for emplacing impermeable sublayer. (Adapted from Huck et al., 1980.)

controlled by the timing of the applied current. The melt is allowed to solidify forming a vitreous mass of synthetic obsidian. Contaminants trapped in the glass are immobilized. The glass itself is highly resistant to weathering. Pilot tests in the field on radiocontaminants have indicated costs in the range of $160 to $330/m^3 (Buelt et al., 1984). Currently, there is interest in applying this approach to construct vitrified curtains and possibly a bottom liner. In essence, one would be creating a glass envelope around the contamination.

A special application of immobilization techniques involves sites where leaking drums of unknown liquid content are found. Because of the difficulty of identifying the drums' contents and uncertainty over safety measures, it may be desirable merely to seal the drums up for disposal at a secure site. Several approaches have been developed for this. The first employs a pre-formed fiberglass-reinforced polyethylene jacket that when applied is impervious to chemical leaching and holds up well in compression, impact, and puncture tests. In the case of badly damaged drums, the container can be wrapped in a fiberglass casing and an overcoat of plastic painted or brushed on. Fullsize polyethylene containers that can be fuse-welded to seal drums also show great promise.

In addition to technical issues, in-place immobilization raises questions of legality. By its nature, this approach converts a spill site into a permanent disposal site. Hence, permits may be required, and questions must be resolved concerning long-term stability and monitoring. When a similar approach was suggested for use on PCB contamination in North Carolina, it was rejected on the basis that the activated carbon to be employed could not assure permanent fixation and hence might serve only as a postponement to ultimate migration and hazardous effects.

Some of the firsthand experience in evaluating methods for ameliorating surface soil contamination comes via review of actions taken in the Soviet Union after a release of radiological materials. In one attempt, a calcium chloride leach was employed at a rate of 10 tons/acre. After several years, the

STEP 1. KERF CAVITIES AT BOTTOM OF PARALLEL BOREHOLES

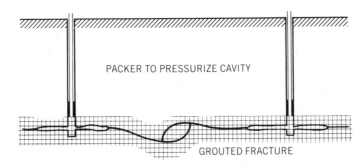

STEP 2. PRESSURIZE AND GROUT FRACTURE

Figure 7-19 / Hydrofrac approach for emplacing impermeable sublayer. (Adapted from Huck et al., 1980.)

contamination had descended to below the root zone. The flora was destroyed in the process, however, and the overall approach was considered ineffective. Better short-term results were obtained through addition of carbonates to tie up the nuclides. The most effective approach was deep plowing to fold the contaminated soil layers under the surface soil and thereby prevent uptake by the flora (Trabalka, 1979). No consideration was given to the long-term implications of percolation and subsequent contamination of groundwaters.

No-Action Alternative

For some sites, no direct remedial action is cost effective or feasible. For instance, in cases of massive aquifer contamination, total restoration is prohibitively expensive. In these cases, no restoration is adviseable per se. As an alternative, attention is turned to preventing exposure. This can be accomplished in a number of ways. At Love Canal and Times Beach, Missouri, the residents were evacuated. In other areas, domestic wells have been replaced with alternate water supplies. In a limited number of cases, treatment units have been installed at the well head to cleanse water prior to use.

A major element of any no-action or exposure reduction alternative is a clear understanding of where contaminants are migrating with time. The use of models to predict transport is often central to that understanding. Advance knowledge of potential exposures allows for action prior to exposure without undo disruption of inhabitants

Although costs and feasibility are highly situation-dependent, Berkowitz (1979) has developed some cost and implementation guidelines to compare the more highly developed options, as presented in Table 7–8. Based on these and other considerations, it is concluded that at present the remove and dispose options are the most viable. The high costs associated with their use and emerging regulations will continue to provide incentives for further advancement of techniques for in-place degradation and immobilization. Research and development may be particularly rewarding in the areas of (1) in-place treatment of refractory or persistent organics to render them susceptible to subsequent biochemical or photochemical attack and (2) flexible grouts or sealants that can be injected in the ground to form a contiguous envelope around contaminated areas. Should efforts be successful in developing these alternatives, the decision for selection of a specific approach to mitigation of an uncontrolled landfill can be analyzed as depicted in Table 7–9.

More accurate estimates of remedial action costs may be obtained by examining the individual unit operations that make up an overall approach. These then can be summed to provide a total cost and modified to reflect site-specific features. Currently, the best available data bases for unit operations

Table 7-8 / Summary of Potential Treatment Methods.[a,b]

Decontamination Method	Project Time (yr)	Capital Investment ($)	Annual Operating Costs ($/yr)	R&D Time Prior to Implementation (yr)	Chance of Success	Notes
Incineration and revegetation	15	10,000,000	4,500,000	1–2	Very high for organics	Heavy metals may require separate treatment
Wet chemical processing	15	10,000,000 to 25,000,000	>3,500,000	3–5	Very high in principle	Specific chemical steps are yet to be worked out
Soil activation	5–20	—	1,400,000(a)	2–4	Moderate(b)	(a) Costs of a single treatment, which might suffice; (b) no really hard data are available
Vegetational uptake	60	—	60,000	3–5	Low	Exclusive of disposal of harvest
Inoculation	Unknown	—	Greater than soil activation	>10	Very low	
Upgradient ground and surface water diversion	Infinite	1,200,000 to 2,200,000	130,000 to 230,000	<2	High for containment; zero for land decontamination	
Downgradient ground and surface water collection & treatment	>50	5,900,000 to 9,700,000	~1,000,000	2–3	High, but very slow	

[a]From Berkowitz (1979).
[b]One-half square mile of contaminated land area, 25 ft deep.

273

Table 7-9 / Decision Matrix for Response to Landspills Assuming Proof of Principles for Current Developments.

Contaminant	Deep Contamination		Shallow Contamination	
	Potable Aquifer Endangered	Arid Zone Or No Mobility	High Vegetation Cover	Low Vegetation Cover
Simple organic	Remove and dispose	Rely on natural degradation	Use natural degradation	Use natural degradation
Refractory organic	Remove and dispose	In-place immobilization	Remove and dispose	In-place destruction
Acid/base	Remove and dispose	No response	Neutralize	Neutralize
Toxic inorganic	Remove and dispose	In-place immobilization	Remove and dispose	Remove and dispose

cost estimation are those generated by SCS Engineers (Rishel et al., 1981) for the U.S. EPA. Detailed tabular output from that effort is available in Appendix J. With time, implementation of the acts on comprehensive environmental response and compensation and liability and related activities will provide data for overall costs as applied in an integrated fashion.

REFERENCES

Ahlquist, A. J. 1979. LASL experience in decontamination of the environment. In *Proceedings of the environmental decontamination workshop.* G. A. Cristy and H. C. Jernigan (eds.), Oak Ridge National Laboratory, CONF-791234. pp. 46–60, December 4–5, 1979.

Allen, R. P.; Arrowsmith, H. W.; and McCoy, M. W. 1979. New decontamination technologies for environmental applications. In *Proceedings of the environmental decontamination workshop.* G. A. Cristy and H. C. Jernigan (eds.), Oak Ridge National Laboratory, CONF-791234, pp. 110–122, December 4–5, 1979.

Baca, R. G.; Waddel, W. W.; Cole, C. R.; Brandstetter, A.; and Cearlock, D. B. 1973. *EXPLORE-1: A river basin water quality model.* Prepared for the U.S. Environmental Protection Agency by Battelle, Pacific Northwest Laboratories, Richland, WA.

Ballard, W. C.; Casey, S. L.; and Calusen, J. D. 1969. Vibration testing with mechanical impedance methods. *Sound and Vibration* 10–21. Reprint obtained from Topp Engineering Sales Corp., P.O. Box 250, Bellevue, WA 98009.

Berkowtiz, J. D. 1979. Technologies for handling hazardous wastes. In *Scientific and Technical Aspects of Hazardous Waste Management.* Washington, D.C.: American Association for the Advancement of Science. Intergovernmental Science, Engineering, and Technology Advisor Panel, pp. 155–188.

Bicker, A. E. 1979. Site decontamination. In *Proceedings of the environmental decontamination workshop.* G. A. Cristy and H. C. Jernigan (eds.), Oak Ridge National Laboratory, CONF-791234, pp. 91–99, December 4–5, 1979.

Bolder, D. 1980. *Chemical waste, fact versus perception.* Presented to the Commonwealth Club of California, San Francisco, April 11, 1980.

Buelt, J. L., V. F. Fitzpatrick, and C. L. Timmerman, "An Innovative Electrical Technique for In-Place Stabilization of Contaminated Soils" AIChE Summer National Meeting, August 19, 1984.

Chemical Week. 1979. Eckhardt unveils waste survey of top 53 firms. *Chemical Week*, Vol. 1, No. 4, p. 22, November 7, 1979.

Christensen, D. C.; and Weimer, W. C. 1979. *Enhanced degradation of persistent halogenated organic materials.* Presented at the 34th Annual Purdue Industrial Conference.

Donigian, A.S.; and Crawford, N. H. 1976. *Modeling pesticides and nutrients on agricultural lands.* EPA-600/2-76-043. U.S. Environmental Protection Agency.

Darrler, R. C. 1980. Indirect techniques for investigating hazardous waste burial sites. In *Proceedings of the 1980 national conference on envrionmental engineering*. New York: ASCE.

Dawson, G. W.; et al. 1978. *The feasibility of migration of Kepone contamination in the James River basin—Appendix A to mitigation feasibility for the Kepone contaminated Hopewell/James River area*. U.S. EPA, June 1978.

Environmental Science. 1980a. Hazardous wastes here and there. *Environmental Science and Technology*, Vol. 14, No. 9, pp. 906–909, August 1980.

Environmental Science. 1980b. Groundwater strategies. *Environmental Science and Technology*, Vol. 14, No. 9, pp. 1030–1035, September 1980.

Fieldhouse, K. 1970. Mechanical impedance concepts. *Instruments and Control Systems*. Reprint obtained from Topp Engineering Corp., P.O. Box 250, Bellevue, WA 98009.

Fitzpatrick, G. L.; Nicholls, H. R.; and Munson, R. D. 1972. *An experiment in seismic holography*. U.S. Department of the Interior, Bureau of Mines, Report of Investigations 7607.

Fitzpatrick, G. L. 1974. First-arrival seismic holograms. *Acoustical Holography*, Vol. 5, pp. 381–400.

Gal'perin, E. I. 1974. *Vertical seismic profiling*. Tulsa: Society of Exploration Geophysicists, 1974.

Hazardous waste disposal. 1979. Committee print 96-1FC 31, Committee on Interstate and Foreign Commerce, House of Representatives, 96th Congress, CPO, September 1979.

Huck, P. J.; Waller, M. J.; and Shimondle, S. L. 1980. Innovative geotechnical approaches to the remedial in-situ treatment of hazardous material disposal sites. In *Proceedings of the 1980 conference on control of hazardous material spills*. U.S. EPA, USCG, Vanderbilt University. Louisville, KY, pp. 421–426, May 13–15, 1980.

Keller, A. C. 1967. *Fundamentals for mechanical impedance analysis*. San Diego: Spectral Dynamics Corporation Technical Publication No. M-2.

Koehmstedt, P. L.; Hartley, J. N.; and Davis, D. K. 1977. *Use of asphalt emulsion sealants to contain radon and radium in uranium tailings*. BNWL-2190/US-20, ERDA/Battelle Pacific Northwest Laboratories, Januray 1977.

Mercer, B. W.; Shuckrow, A. J.; and Ames, L. L. 1970. *Fixation of radioactive wastes in soil and salt cores with organic polymers*. BNWL-1220, April 1970.

Metherell, A. F.; and Green J. 1970. Acoustical holography imaging of geological structures. In *Laser applications in the geosciences*, ed. J. Gauger and F. F. Hall, Jr. North Hollywood: Western Periodicals, pp. 267–279.

Musgrave, A. W., ed. 1967. *Seismic refraction prospecting*. Tulsa: Society of Exploration Geophysicists.

National Conference of State Legislatures, 1982. *Hazardous Waste Management: A Survey of State Legislation 1982*.

Olsen, A. R.; and Wise, S. E. 1979. *Frequency analysis of pesticide concen-*

trations for risk assessment. Prepared for the U.S. Environmental Protection Agency by Battelle, Pacific Northwest Laboratories, Richland, WA.

Onishi, Y.; et al. 1979. *Methodology for overland and instream migration and risk assessment* (draft). U.S. Environmental Protection Agency, March 21, 1979.

Onishi, Y.; and Wise, S. E. 1979a. Finite element model for sediment and toxic contaminant transport in streams. In *Proceedings of conservation and utilization of water and energy resources.* San Francisco: Hydraulics and Energy Divisions, ASCE. pp. 144–150.

Onishi, Y.; and Wise, S. E. 1979b). *Mathematical model, SERATRA, for sediment and contaminant transport in rivers and its application to pesticide transport in four mile and Wolf creeks in Iowa.* Prepared for the U.S. Environmental Protection Agency by Battelle, Pacific Northwest Laboratories, Richland, WA.

Phillips, S. J.; et al. 1977. *Characterization of the Hanford 300 area burial grounds* (draft). PNL-2557, UC-70, U.S. Department of Energy, March 1977.

Rishel, H. L.; Boston, T. M.; and Schmidt, C. J. 1981. *Costs of remedial response actions at uncontrolled hazardous waste sites.* U.S. EPA.

Sandness, G. A. 1979. *Characterization of the Hanford 300 area burial grounds. Task I. Geophysical evaluation.* U.S. Department of Energy, PNL-2957, UC-70, October 1979.

Takenaka Komuten Co., Ltd. 1973. *Report on sludge treatment experiments of Takasago west port,* July 1973.

Thibault, G. T.; and Elliot, N. W. 1980. Biological detoxification of hazardous organic chemical spills. In *Proceedings of the 1980 conference on control of hazardous material spills,* U.S. EPA, USCG, Vanderbilt University, Louisville, KY, pp. 369–374, May 13–15, 1980.

Thompson, G. H. 1979. Soil decontamination at Rocky Flats Plant. In *Proceedings of the environmental decontamination workshop.* G. A. Cristy and H. C. Jernigan (eds.), Oak Ridge National Laboratory, CONF-791234, pp. 104–109, December 4–5, 1979.

Thornton, J. S. 1980. Underground movement of gasoline on groundwater, and enhanced recovery on surfactants. In *Proceedings of the 1980 conference on control of hazardous material spills.* U.S. EPA, USCG, Vanderbilt University, Louisville, KY, pp. 223–235, May 13–15, 1980.

Toomey, J. E. 1979. Project planning approach. In *Proceedings of the environmental decontamination workshop.* G. A. Cristy and H. C. Jernigan (eds.), Oak Ridge National Laboratory, CONF-791234, pp. 167–170, December 4–5, 1979.

Trabalka, J. R. 1979. Russian experience. In *Proceedings of the environmental decontamination workshop.* G. A. Cristy and H. C. Jernigan (eds.), Oak Ridge National Laboratory, CONF-791234, pp. 3–8, December 4–5, 1979.

Weight, A. P.; Humber, D.; and Welshans, G. K. 1979. National survey of hazardous waste problems and associated cleanup costs. In *Proceedings of national conference on hazardous material risk assessment, disposal and management.* Silver Spring, MD: Information Transfer, Inc., April 25–27, 1979.

Wenstel, R. S.; Jones, W. E.; Foutch, R.; Wilkinson, M.; and Kitchens, J. F. 1980. Accelerated restoration of spill-damaged lands. In *Proceedings of the 1980 conference on control of hazardous material spills.* U.S. EPA, USCG, Vanderbilt University, Louisville, KY, pp. 99–102, May 1980.

Williams, G. L., J. E. Rogers, and C. J. English, *Enzymatic detoxification of hazardous organic chemicals in the environment* AIChE National Summer Meeting, August 19, 1984.

8

Hazardous Waste Transportation

Introduction

Throughout the development of the federal hazardous waste management program, the architects of what was to become RCRA have touted the need for a cradle-to-grave management system. This concept embraces the regulation of all phases of the management chain: generation, storage, transportation, treatment, and disposal. Although the major impacts from proper management are largely associated with the final link in that chain (disposal), transportation is a key element in assuring the effectiveness of any regulatory program. The relative importance of transportation in the hazardous waste management cycle is readily illustrated in Figure 8-1. One-half of the wastes generated in New York are transported off-site. An equivalent amount is brought into New York from out of state. An unknown amount of waste is transported through the state en route to other destinations. History has shown that the transportation link is often where the system has broken down. Many of the transgressions of the past decades have been tracked back to unscrupulous operators who maximized financial gains by discharging their cargoes at alternative sites. With no means of enforcing the delivery of wastes to the prescribed destination, generators knowingly and unknowingly allowed the transporters to become in practice the disposers as well. Such was the case in Louisville, Kentucky, where a local transporter dumped several tons of pesticide waste in the sewer system. Clean-up activities cost $3 million.

It is natural that many of the abuses attributed to waste management can be traced to transportation activities. The industry is a large conglomerate of small operators who enter and exit the market with impunity. Local haulers experience little regulatory control and have but small assets at risk from liability claims. Many operators virtually live in their trucks and, as a

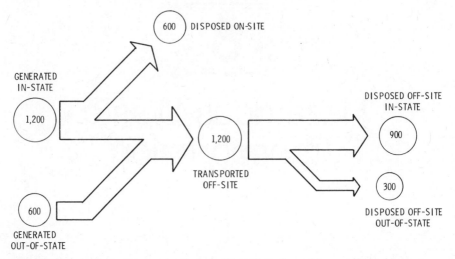

Figure 8-1 / Hazardous waste flows in New York State (thousands tons per year. (From Booz Allen & Hamilton, 1979).

consequence, are difficult to locate from one time period to the next. Thus, their mobility and low profile discourage efforts to monitor activities and assure compliance.

Evidence of both these difficulties in controlling transporters and their role in the improper disposition of wastes was particularly apparent in the spring and summer of 1980 just prior to implementation of RCRA regulations. During that period, large numbers of waste generators and disposal firms with inventories of stored wastes realized that once the regulations were promulgated, enforced compliance would put significant financial burdens on them. Most had neither the facilities to meet required standards nor the funds to contract for services from permitted operators. As a consequence, a massive dispersal program began, with a fleet of "midnight dumpers." Old, illegal truck trailers were filled with drummed wastes and left standing in parking lots and abandoned fields. Tankers were employed to spray wastes on infrequently used roads at night. Other shipments were merely taken into remote areas and dumped. Authorities were virtually helpless in trying to prevent these actions. This tragic course of events illustrates both the key role that transporters play in the management chain and the difficulty in regulating that role with pre-RCRA regulations.

Recognizing the implications of these observations, it is important to consider the nature of the transportation industry and the proposed controls as a part of overall hazardous waste management programs.

The Industry

Little has been written about the hazardous waste transportation industry. Although most transportation in the United States is federally regulated, there is a paucity of data on waste haulers. This reflects several facts:

Many waste haulers are local in scope and therefore do not come under federal jurisdiction.

Wastes have not been regulated as specific commodities, but have been categorized as not otherwise specified (NOS) and are, therefore, impossible to track in waybill statistics.

Liquid waste hauling has often been performed by septage pumpers, scrap dealers, and other operators whose major lines of business on the surface would appear to be different than that of industrial waste hauling.

Entry and exit from the business have been relatively easy in the past, allowing for rapid changeover.

The lack of data on transporters is rapidly being ameliorated as state hazardous waste legislation is enacted to require transporters to be licensed as waste haulers. Numerous states have already instituted such programs, and many others are moving in that direction. A summary of carrier requirements as of 1982 is provided in Table 8-1.

Waste Haulers

There is no single comprehensive source of information on the hazardous waste transportation industry in the United States. However, J. J. Keller and Associates, Inc., of Neenah, Wisconsin, publish a *Hazardous Waste Services Directory* that inventories transporters by state. An analysis of the information contained therein for 1980 provides some insight into the structure of the industry. The directory identifies some 412 companies that operated as waste haulers in the United States through 475 individual terminals. Most companies were associated with a single terminal, but 11 firms had at least two terminals in two or more states, and 10 firms had multiple terminals in a single state.

Waste haulers represent three distinct subgroups: (1) generators who deliver their own wastes, (2) disposers who pick up wastes, and (3) contract transporters who provide a service to either generators or disposers. Of the transporters listed in the directory, some 127, or 31%, offered transportation only. These varied from the one-vehicle local hauler to large incorporated firms with terminals in several states. Another 22 firms (5%) offered full services, that is, treatment/processing, laboratory analysis, consulting, disposal, and reclamation/recovery. The breakdown of additional services provided by a segment of all transporters was as follows:

Table 8-1 / Summary of Carrier Requirements in Individual

	AL	AK	AZ	AR	CA	CO	CT	DE	FL	GA
Permit/license required	x	x	x	x	x	x	x	x	x	x
Fees levied					x				x	
Insurance, bonding, financial responsibility requirements	x					x	x		x	x
Hazard containment and communication requirements										
Manifest/shipping papers	x	x	x	x	x	x	x	x	x	x
Labeling/containerization	x				x	x	x	x	x	x
Vehicle identification	x			x	x	x	x	x	x	
Penalties	x	x	x	x	x	x	x	x	x	
Other	x	x	x	x	x	x	x	x	x	
Emergency preparedness/ response provisions										
Prenotification					x	x		x		
Accident notification	x			x	x	x	x		x	
Contingency plans					x	x				
Emergency orders		x			x	x		x	x	x
Other	x	x	x	x	x	x	x		x	x

	MT	NE	NV	NH	NJ	NM	NY	NC	ND	OH
Permit/license required	x	x	x	x	x	x	x	x	x	x
Fees levied			x	x		x	x		x	x
Insurance, bonding, financial responsibility requirements		x		x	x	x	x	x	x	
Hazard containment and communication requirements										
Manifest/shipping papers	x	x	x	x	x	x	x	x	x	x
Labeling/containerization	x	x	x			x	x	x		x
Vehicle identification	x	x				x	x	x		x
Penalties	x		x	x		x	x		x	x
Other		x	x	x		x	x		x	x
Emergency preparedness/ response provisions										
Prenotification					x					
Accident notification			x			x				x
Contingency plans	x		x						x	x
Emergency orders	x		x	x		x			x	x
Other	x		x	x	x		x			

States (National Conference of State Legislatures, 1982).

HA	ID	IL	IN	IA	KS	KY	LA	ME	MD	MA	MI	MN	MS	MO
x	x	x	x	x	x	x	x	x	x	x	x	x	x	x
		x			x	x				x	x		x	x
		x		x	x	x	x	x	x		x	x	x	x
		x	x	x	x	x	x	x	x	x	x		x	x
	x			x	x	x	x	x	x	x	x		x	x
x	x	x		x	x	x	x	x	x	x	x		x	x
	x	x	x	x	x	x			x	x	x		x	x
		x	x			x	x		x	x	x	x		x
			x										x	
	x					x	x	x	x	x	x		x	
	x				x	x	x	x					x	
x	x			x	x				x		x		x	
					x	x	x		x				x	

OK	OR	PN	RI	SC	SD	TN	TX	UT	VT	VA	WA	WV	WI	WY
	x		x	x	x	x	x		x	x	x		x	
	x		x		x		x						x	
	x			x			x				x			
x	x	x	x	x	x	x	x			x	x	x	x	
	x	x	x	x	x	x				x		x	x	
	x	x	x	x	x	x				x		x	x	
x	x	x			x	x	x	x		x			x	
	x	x	x		x	x	x		x		x		x	
				x	x					x				
	x	x		x	x								x	
x	x	x			x								x	
			x		x	x	x	x	x			x		
x													x	

Treatment/processing   18%
Storage   23%
Laboratory analysis   35%
Consulting   39%
Disposal   26%
Reclamation/recovery   39%

In addition, many firms are associated with spill cleanup work. Most transporters without the disposal capability claim access to an approved disposal site operated by others.

It should be noted that the picture of the industry provided by these data may be somewhat misleading. Lacking regulatory authority and confirmed information sources, the input to the directory was less than complete. For instance, 24 individual terminals are listed for the state of New York. However, in a comprehensive study for that state, Booz Allen & Hamilton (1979) noted that there were 264 registered industrial waste and/or waste oil haulers in New York. Additional unlicensed haulers and "midnight dumpers" were known to exist, but their size and extent were unknown. Hence, the directory appears to have captured only 10% of the industry. Similarly, the 412 firms identified nationwide represented < 10% of the 5000 haulers the EPA believe were operating at the time. This may reflect a number of circumstances. From a review of the firms listed in the directory, it would appear that many generator-transporters are not identified. It also appears that waste oil haulers are only partially covered. The directory identifies only 38 firms that hauled waste oil exclusively in the United States; that is exceedingly low for a well-established industry like waste oil recycling. It is reasonable to assume that these typically small operations would not be identified. Hence, it is concluded that the data presented here on the waste transportation industry are skewed to the larger operators and to the disposer-transporters and contract transporters. Therefore, most of the data are provided as fractions of the total data base rather than absolute numbers.

Most hazardous waste transportation is accomplished via truck. An array of equipment is used, ranging from stake and dump trucks to vans, tankers, and vacuum trucks. Of the companies listed in the Keller directory, roughly 60% are equipped to haul bulk liquids in straight tankers or tractor-trailer tankers (Figure 8–2). Approximately 60% are equipped to haul bulk solids or containerized liquids, and 30% can accommodate both. Most of the firms (66%) service a local or in-state market. The remainder service multiple states or the nation as a whole. Regional operations are particularly prevalent in New England and the Mid-Atlantic states.

A summary of waste types hauled is presented in Table 8–2. As expected, oils are the most commonly carried waste, and PCBs are carried by only 17 of the firms listed. An additional 16 firms (4%) carry specialized wastes only, such as asbestos, flyash, or a particular industrial sludge. These firms are

Figure 8-2 / Common tanker for transport of liquid hazardous wastes. in bulk.

typically small and local in scope. Similarly, firms specializing in oil and solvent shipments are generally locally oriented. The multistate firms and the full-service firms are more likely to serve broader geographic areas.

With the promulgation of regulations stemming from RCRA, the transportation industry will be subject to significant change. Implementation of licensing systems may lead many small haulers to exclude this market segment. In the meantime, the same pressures that will encourage concentration of the disposal industry into larger firms will stimulate vertical integration, so that disposer-transporters will gain a larger share of the market. The scarcity of disposal sites is also likely to increase the number of regional haulers at the

Table 8-2 / Distribution of Waste Types Carried by Hazardous Waste Transporters.

Waste Type	Percentage of Respondents
Acids	51%
Caustics	56%
Oils	85%
Pharmaceuticals	38%
Solvents	72%
Cyanides	35%
Pesticides	33%
Chromic acids	38%
PCBs	4%
All wastes except PCBs	21%

expense of local transporters. Transporters with a limited range will simply not have destinations available within their areas of operations.

Transportation in Europe is accomplished much the same as in the United States, with the exception of the large number of small specialty haulers and unlicensed firms. For instance, in Switzerland, large quantities of liquid wastes are transported in a 15-ton suction-tank semitrailer unit operated by the central disposal facility. Danish wastes are collected by the municipalities and then shipped to disposal sites. In countries such as Germany, private haulers carry much of the hazardous wastes under license agreements with the regulatory agency. As noted in the regulatory section of this chapter, licenses may be for single or multiple shipments.

Barrel and Drum Reconditioners

Many small-lot hazardous wastes and still bottoms and tars are transported in 30- and 55-gal drums. After discharge for treatment or disposal, these drums may be cleaned for reuse. As such, the barrel and drum reconditioning industry constitutes a discrete link in the waste management chain. In the past, this link has occasionally been less than satisfactory. Reconditioners become disposers out of convenience to clear drums for resale. Such was the case with Calumet Container–Steel Container Corporation in Hammond, Indiana, where discharges have contaminated soil and groundwater. At other sites, the business was found not to be lucrative, and large quantities of barreled wastes were simply abandoned. This was the case with such sites as H & M Drum Company in Massachusetts, where over 2000 drums were found, and the Wade site in Chester, Pennsylvania, where the inventory included several thousand drums. Many more members of the industry are legitimate operations providing a needed service for the chemicals and waste management industries.

In a recent study of the barrel and drum reconditioning industry for the EPA, Touhill, Shuckrow, and Associates (TSA) (Touhill, 1980) determined that there are about 250 active reconditioners in the United States. Of that total, 119 (47.6%) are members of the National Barrel and Drum Associates (NABADA) who recondition as a service. Some 92 (36.8%) reconditioners are service organizations not aligned with NABADA, and 39 (15.6%) are user-reconditioners. The relative distribution of reconditioners in the United States is presented in Table 8–3. Ten NABADA member firms were also identified in Canada. It is cautioned that these data are approximate only, based on results of a survey of 25 non-NABADA member firms. Existing data at the U.S. Department of Transportation list 429 firms, of which 326 are not registered with NABADA. Of the nonmember firms, 44% are inactive, defunct, or improperly listed, and 16% only air-test, leak-test, or refill

Table 8-3 / Reconditioner Distribution and Sales by Region (1972 Dollars).

Region	Number of Reconditioners	Sales (in Thousands)
New England	15	5,085
Boston SMSA	6	1,384
Middle Atlantic	86	32,863
New York SMSA	18	6,969
Newark SMSA	13	9,023
Philadelphia SMSA	16	7,248
East North Central	66	28,535
Cleveland SMSA	12	3,035
Detroit SMSA	5	1,708
Chicago SMSA	18	8,521
West North Central	11	2,821
South Atlantic	22	14,940
East South Central	3	280
West South Central	21	7,453
Houston SMSA	4	1,401
Dallas–Ft. Worth SMSA	6	2,019
Mountain	7	1,503
Pacific	32	16,457
Los Angeles SMSA	11	4,729
San Francisco SMSA	6	3,895
Total	263	109,937

drums cleaned by others. The 250-firm estimate derived by TSA is in close agreement with the 263 reconditioners reported by the Department of Commerce in the *1972 Census of Wholesale Trade.*

In a related questionnaire survey of 49 NABADA member firms, TSA (Touhill, 1980) determined that 38.8% were washing facilities (tight-head drums), 18.4% were burning facilities (open-head drums), and 42.9% offered both washing and burning. This represents a decrease in drum-washing-only facilities from a similar survey conducted by NABADA in 1973. The trend is believed to parallel a trend of consolidation from smaller, single-service organizations to the larger, financially stronger integrated operations. Total reconditioning volume for all firms in the United States in 1979 was estimated at 41 million drums. Distribution between reconditioners and treatment type is depicted in Table 8-4. Currently, NABADA member firms are operating at about 70% of capacity.

Drum reconditioning is provided either as a service (laundry) or as a preparatory step to resale. Industry averages indicate that when drums are washed, 44.7% are resold, 51.7% are laundered as a service, and 3.5% are discarded. Burning is employed more often for resale (61.9%), but it also has a higher discard level (5.2%). The high level of service washing reflects the

Table 8-4 / Estimated Annual Drum Reconditioning Volumes.[a]

Sources	Plant Design Capacity		Actual Volume (1979)		
	Wash	Burn	Wash	Burn	Total
NABADA responders	12,567,100	8,559,200	9,220,640	5,854,940	15,075,580
NABADA total	30,520,100	20,786,600	22,393,000	14,219,000	36,612,000
Non-NABADA service	—	—	3,588,000	507,000	4,095,000
Non-NABADA own use	—	—	497,000	—	497,000
Total			26,478,000	14,726,000	41,204,000

[a]From Touhill (1980).

extensive use of heavy-gauge tight-head drums by the petroleum industry. Table 8-5 presents a breakdown of the sources of drums for reconditioning in the United States. The preeminence of the petroleum industry is evident. Drums from the pesticide and industrial chemicals industries are not recycled or reused nearly as frequently. This is in concert with the discovery of these types of drums at abandoned sites throughout the country.

Drum reconditioning is also common in the other industrialized nations of the world. A comparison of new drum production to reconditioning volumes for selected countries and years is provided in Table 8-6. End-use distributions of reconditioned drums in Japan and several European nations are provided in Table 8-7.

Table 8-5 / Drum Use and Fate Distributions.[a]

Drum Content	New-Drum End Use (1979) (% of Total New Drums)	Drums Shipped to NABADA Reconditioners (% of Total Drums Shipped)
Food	5.1	6.8
Oil & petroleum	15.2	36.2
Paint	—	10.0
Ink	6.6	4.8
Adhesive	—	6.8
Resins	—	8.8
Industrial chemicals	40.2	15.6
Pesticides	—	0.5
Cleaning solvents	—	8.8
Janitorial supplies	3.1	—
Other	9.8	1.8
Unspecified	20.1	—

[a]Adapted from Touhill (1980).

Table 8-6 / Volume of Reconditioned Drums and New Drum Production in Selected Nations.[a]

Country	Year	New Drum Production (in Millions)	Reconditioned Drums (in Millions)
Belgium	1978	1.3	1.5
France	1972	—	2.3
	1978	4.8	2.5
Japan	1969	3.3	30.0
	1973	10.2	—
	1974	—	23.0
	1975	7.3	—
	1978	8.8	18.0
Netherlands	1978	3.3	0.8
South Africa	1978	—	1.2
Sweden	1978	—	0.3
United Kingdom	1972	5.5	6.0
	1978	—	6.0
West Germany	1973	4.5	3.0
	1976	7.0	5.2
	1978	7.0	7.0

[a]From Touhill (1980).

Regulation via Manifests

The need to regulate hazardous waste transportation was clearly identified in the 1973 report to Congress. At that time it was concluded that "the Department of Transportation administers a number of Federal Statutes designed to control the transportation of hazardous materials in interstate commerce. These statutes should be amended by DOT where necessary to ensure that hazardous wastes are properly marked, containerized and transported [to an authorized disposal site]." By way of response, Congress included Section 3003 in RCRA, which mandates the creation of requirements for record-keeping, labeling, and compliance standards for transporters of hazardous wastes. These were to be developed after consultation with the Secretary of Transportation and state agencies. Further, all regulations were to be consistent with pertinent regulations promulgated under the Hazardous Materials Transportation Act when the hazardous waste in question was subject to those regulations. The latter caveat led to some conflict, because there are several broad hazardous materials classifications used by DOT that conceivably could cover all hazardous wastes even though they had not been so used up to that time. Subsequently, an interagency agreement has been reached establishing areas of responsibility. Whereas the EPA will manage incidents involving spills, DOT will continue to enforce regulations related to safety. The EPA will also take the lead role for enforcement against transporters

Table 8-7 / End Uses of Reconditioned Drums in Selected Countries.[a]

Country	Year	Chemicals	Oil & Petroleum	Paints	Emulsions	Petro-chemicals	Food	Adhesives & Resins	Other
Japan	1969	16.0	68.6	5.6	1.7	—	—	—	8.1
	1973	37.1	44.3	—	—	12.1	—	—	6.5
	1974	33.5	46.1	—	—	13.1	—	—	7.3
	1975	39.5	38.5	—	—	13.8	—	—	8.3
	1976	30	50	10	—	—	—	—	5
	1979	60	23	—	—	—	5	—	17
West Germany	1976	15	80	—	—	—	—	—	5
France	1976	—	40	—	—	—	—	50	10
United Kingdom	1976	(5)[b]	(1)	(2)	(4)	(1)	—	(3)	—

[a]From Touhill (1980).

[b]Number in parenthesis denotes most frequently processed drums from highest (1) to lowest (5).

when the transportation is ancillary to improper treatment, storage, or disposal, as would be the case with a "midnight dumper."

The regulations were written to ensure that wastes safely reach acceptable treatment and disposal sites. As such, they cover four major areas of concern:

1. Generation of records to show where wastes originated, where they went, and how they got there
2. Use of labels and placards to clearly identify container contents
3. Maintenance of a manifest system to provide a means of tracking all wastes on a real-time basis
4. Assurance that wastes are going to properly permitted sites

Of these items, the manifest is the cornerstone of the entire regulatory effort. It provides the means for tracking waste movement. It acts as a mechanism to ensure proper labeling and routing to permitted sites, and it forms the basis for the required records.

The basic structure of a manifest system is illustrated in Figure 8–3. The manifest itself is little more than an expanded bill of lading. (Indeed, EPA regulations call for manifests on truck transport only. Rail and water trans-

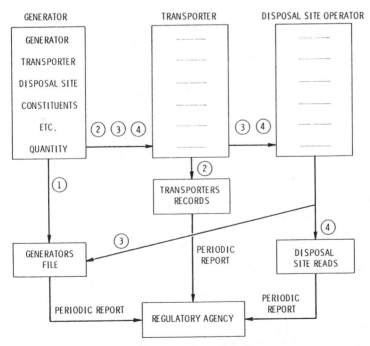

Figure 8-3 / Basic structure of manifest systems.

port will use waybills on standard shipping documents, with the manifest itself mailed from the generator to the point of destination.) In 1984, EPA and DOT came to agreement on a combined manifest form. The document contains information on the waste generator, the transporter, and the destination, as well as a characterization of the materials being transported. The multiple-copy form is completed and signed by the waste generator. On transfer of the load, the transporter signs the manifest and retains all but one copy, which is left with the generator. On arrival at the designated destination, the transporter obtains the treatment/disposal operator's signature, retains one copy of the manifest, and leaves two copies with the load. The treatment/disposal operator retains a copy of the manifest and returns a copy to the generator, thus informing him that the waste has reached its destination. Retained copies of the manifest form the basis for periodic reporting to regulatory agencies. In this way, a materials balance can be conducted to ensure that all wastes are accounted for and have been directed to permitted sites. Throughout the management cycle, the waste is considered to be in the custody of the organization or individual represented by the last signature on the manifest.

The system proposed by the EPA is derived directly from the basic structure in Figure 8–3. More than four copies of the manifest are required if more than one transporter is employed (an additional copy is required for each additional transporter). Records are to be kept for a period of 3 yr. Reports are required of the generator and the treatment/disposal operator annually, using the form depicted in Figure 8–4. In addition, if a generator does not receive a signed copy of the manifest back from the treatment/disposal operator within 35 days of shipment, he must fill an exception report with the EPA regional office. Generators, transporters, and treatment/disposal operators all are given specific identification numbers for the purpose of identification on manifests and reports. These numbers are issued as a result of notification requirements promulgated under RCRA.

The EPA was by no means the first agency to undertake a manifest system for hazardous waste management. Indeed, the EPA system was patterned after a system promulgated in California some years before. The California system requires monthly reporting by generators and disposal operators. Reports consist of copies of all manifests completed in the preceding period. Similarly, transporters must send copies of all manifests within 30 days of delivery of a shipment. In addition, all transporters must be licensed to haul hazardous waste. California has designed the manifest for input to automated data processing to facilitate the generation of summary data and to conduct materials balance checks aimed at discovering illegal activities such as failure to deliver wastes to the designated destination.

The state of Texas instituted its manifest system in 1976. While looking quite similar to the EPA program, the Texas plan includes a provision for the

Please print or type with ELITE type *(12 characters per inch).*

GSA No. *12345-XX*
Form Approved OMB No. *158-R00XX*

&EPA	U.S. ENVIRONMENTAL PROTECTION AGENCY **HAZARDOUS WASTE REPORT**	I. TYPE OF HAZARDOUS WASTE REPORT

PART A: GENERATOR ANNUAL REPORT

THIS REPORT IS FOR THE YEAR ENDING DEC.31, | 1 | 9 | | |

PART B: FACILITY ANNUAL REPORT

PLEASE PLACE LABEL IN THIS SPACE

THIS REPORT FOR YEAR ENDING DEC. 31, | 1 | 9 | | |

PART C: UNMANIFESTED WASTE REPORT

THIS REPORT IS FOR A WASTE RECEIVED *(day, mo., & yr.)* | — | | — | 1 | 9 | | |

INSTRUCTIONS: You may have received a preprinted label attached to the front of this pamphlet; affix it in the designated space above—left. If any of the information on the label is incorrect, draw a line through it and supply the correct information in the appropriate section below. If the label is complete and correct, leave Sections II, III, and IV below blank. If you did not receive a preprinted label, complete all sections. "Installation" means a single site where hazardous waste is generated, treated, stored, or disposed of. Please refer to the specific instructions for generators or facilities before completing this form. The information requested herein is required by law *(Section 3002/3004 of the Resource Conservation and Recovery Act).*

II. INSTALLATION'S EPA I.D. NUMBER

F | | | | | | | | | | | | 1 | | | |

III. NAME OF INSTALLATION

IV. INSTALLATION MAILING ADDRESS

STREET OR P.O. BOX

3 | | | | | | | | | | | | | | | | | | | | | | | | | | |

CITY OR TOWN | ST. | ZIP CODE

4 | | | | | | | | | | | | | | | | | | | |

V. LOCATION OF INSTALLATION

STREET OR ROUTE NUMBER

5 | | | | | | | | | | | | | | | | | | | | | | | | | | |

CITY OR TOWN | ST. | ZIP CODE

6 | | | | | | | | | | | | | | | | | | | |

VI. INSTALLATION CONTACT

NAME *(last and first)* | PHONE NO. *(area code & no.)*

2 | | | | | | | | | | | | | | | | | | | | — | | | — | | | | |

VII. TRANSPORTATION SERVICES USED *(for Part A reports only)*

List the EPA Identification Numbers for those transporters whose services were used during the reporting year represented by this report.

VIII. COST ESTIMATES FOR FACILITIES *(for Part B reports only)*

A. COST ESTIMATE FOR FACILITY CLOSURE	B. COST ESTIMATE FOR POST CLOSURE MONITORING AND MAINTENANCE *(disposal facilities only)*
$			,			.				$			,			.		

IX. CERTIFICATION

I certify under penalty of law that I have personally examined and am familiar with the information submitted in this and all attached documents, and that based on my inquiry of those individuals immediately responsible for obtaining the information, I believe that the submitted information is true, accurate, and complete. I am aware that there are significant penalties for submitting false information, including the possibility of fine and imprisonment.

A. PRINT OR TYPE NAME	B. SIGNATURE	C. DATE SIGNED

EPA Form 8700-13 (4-80)

PAGE ___1___ OF _____

Figure 8-4 / EPA format for manifest forms.

Federal Register / Vol. 45, No. 98 / Monday, May 19, 1980 / Rules and Regulations   332

Please print or type with ELITE type *(12 characters/inch)*.

GSA No. 12345-XX
Form Approved OMB No. 158-R00XX

⊕EPA

U.S. ENVIRONMENTAL PROTECTION AGENCY
FACILITY REPORT — PARTS B & C
(Collected under the authority of Section 3004 of RCRA.)

FOR OFFICIAL USE ONLY *(Items 1 & 3)*	1. DATE RECEIVED	XVI. TYPE OF REPORT *(enter an "X")*	XVII. FACILITY'S EPA I.D. NO.
	2. RECEIVED BY	☐ PART B ☐ PART C	

XVIII. GENERATOR'S EPA I.D. NO.

XX. GENERATOR ADDRESS *(street or P.O. box, city, state, & zip code)*

XIX. GENERATOR NAME *(specify)*

XXI. WASTE IDENTIFICATION

LINE NUMBER	A DESCRIPTION OF WASTE	B. EPA HAZARDOUS WASTE NUMBER *(see instructions)*	C. HANDLING METHOD *(enter code)*	D. AMOUNT OF WASTE	E. UNIT OF MEASURE *(enter code)*
1					
2					
3					
4					
5					
6					
7					
8					
9					
10					
11					
12					

XXII. COMMENTS *(enter information by line number — see instructions)*

EPA Form 8700-13B (5-80)

BILLING CODE 6560-01-C

PAGE ____ OF ____

Figure 8-4 / *(Continued)*

carrier to return the completed manifest to the shipper. This increases the carrier's role in the overall process, as illustrated in Figure 8–5. Other states have followed suit. A copy of the manifest proposal for use with extremely hazardous wastes in Washington is provided in Figure 8–6.

Differences in the forms employed by existing programs and the potential proliferation of new forms under EPA guidelines have raised concern throughout the industry. The scarcity of approved disposal sites and economies of scale have necessitated interstate transport of hazardous wastes. Hence, there is a growing movement to encourage the development of a standard manifest form for all states. Not only would such a move reduce the complexities for shippers and carriers faced with a variety of forms from different jurisdictions, it also would simplify regulation for state officials reviewing manifests and reports for out-of-state wastes. The first major move in this direction has been made in the northeastern states. The New York Division of Solid Waste is coordinating the development of the basic form, which will then be submitted for acceptance by 10 other states: Massachusetts, Connecticut, New Hampshire, Maine, Rhode Island, Vermont, New Jersey, Pennsylvania, Maryland, and Delaware. In addition to promoting compatibility of manifests, this group has expressed concern that EPA annual reporting requirements will not allow sufficient real-time turnaround to use the manifest as an enforcement tool. This is of vital interest to an area of the country where significant nefarious dumping has been reported. As a consequence, the standardized manifest is being designed to accommodate immediate copy coverage to regulators and automated data processing. Thus, regulators will receive copies of each manifest, as in the California system, rather than copies of only manifests that are thought not to have reached their designated destinations. With a similar system in New Jersey, it was determined that approximately 400 transactions a day were involved and that batch computing of the subsequent manifest could be accomplished for $70,000 per year (Battelle Memorial Institute, 1980).

Distribution of manifest copies in the proposed northeastern states system will be similar to that in California: retained copies for the generator, each transporter, and disposer; completed copy from disposer to the generator; copies sent by generator and disposer to generator state (and disposer's state, if different). All copies are to be sent by first-class mail. The transporter copy must accompany all shipments. However, state enforcement of this requirement is limited to having the disposer refuse receipt of shipments not so accompanied. The elimination of over-the-road compliance inspections reflects U.S. DOT policy to preempt regulations that would allow states to disrupt interstate commerce for reasons other than meeting minimal DOT manifest requirements. All intermediate processing stops would be defined as the disposer for the original manifest and would be required to initiate a new manifest when shipment proceeded.

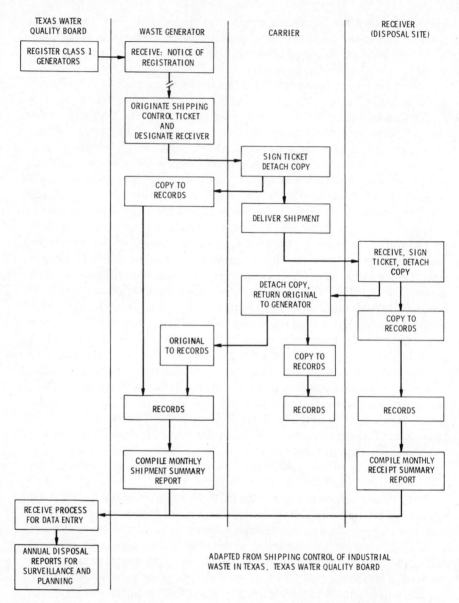

Figure 8-5 / Off-site disposal shipping control procedures in Texas. (Reprinted from Andres, 1977.)

State of
Washington
Department
of Ecology

EXTREMELY HAZARDOUS WASTE MANIFEST
Required by WAC 173-302

1)   (206) 753-6884

GENERATOR _____
            (see note 1 below)

ADDRESS _____ E.I.N. _____
                                                    (see note 2 below)

WASTE ORIGIN _____ PHONE _____
            (If different from address above)

WASTE DESCRIPTION (including hazardous constituents and range of concentration)	QUANTITY	CONTAINER	BULK SHIPMENTS		
			GAL.	LB.	CU.FT.

DOT SHIPPING NAME _____ DOT HAZARD CLASS _____
PHYSICAL STATE (CIRCLE): **SOLID   LIQUID   SLUDGE   OTHER** _____
MAJOR HAZARD (CIRCLE): **TOXID   CORROSIVE   IGNITABLE   REACTIVE   OTHER** _____
                                                              (Describe)
SPILL/SPECIAL INSTRUCTIONS _____
_____

IN THE EVENT OF A SPILL, CONTACT EITHER THE DEPARTMENT OF ECOLOGY AND/OR
THE NATIONAL RESPONSE CENTER, U.S. COAST GUARD 800-424-8802 FOR EMERGENCY ASSISTANCE.

This is to certify that the above-named materials are properly classified, described,
packaged, marked, labeled and are in proper condition for transportation according to the
applicable regulations of the U.S. Department of Transportation. Signed and dated by
the authorized agent of the generator:

GENERATOR SIGNATURE _____ DATE _____

2) COLLECTION SITE _____ E.I.N. _____
   ADDRESS _____ PHONE _____

3) TRANSPORTER _____ E.I.N. _____
   ADDRESS _____ PHONE _____
   TRANSPORTER SIGNATURE _____ DATE _____

4) TREATMENT OR DISPOSAL SITE _____ E.I.N. _____
   ADDRESS _____ PHONE _____
   TREATER/DISPOSER SIGNATURE _____ DATE _____

Note 1: Generator should also complete parts 2,3 and 4, except for signatures.
Note 2: EIN is the IRS employer identification number.

Distribution: Generator: **Green Copy** Transporter: **Pink Copy** Treater/Disposer: **Canary**
                                          (Original returned to Generator)   Copy

ECY 030-34   9/79

Figure 8-6 / Extremely hazardous waste manifest *Required by WAC 173-302.*

INFORMATION	FEDERAL	REQUIREMENTS STATE IDENTIFIED INFORMATION NEEDS
SPECIAL HANDLINE INSTRUCTIONS		•
ANALYSES	(NOT REQUIRED)	(NOT DEEMED NECESSARY)
DATES/SIGNATURE CERTIFICATION	•	
TRANSPORTER (FOR EACH)	•	
NAME INDIVIDUAL SIGNATURE	•	
EPA I.D. #	•	
STATE I.D. #		•
ROUTE	(NOT REQUIRED)	(NOT DEEMED NECESSARY)
HAULER'S LICENSE #	(NOT REQUIRED)	(NOT DEEMED NECESSARY)
FACILITY		
FACILITY NAME	•	
EPA I.D. #	•	
ADDRESS	•	
INDIVIDUAL SIGNATURE	•	
INDIVIDUAL TITLE		•
PHONE		•
REASONS - IF REJECTION OF SHIPMENT		•
OTHER		
NATURAL EMERGENCY RESPONSE PHONE		•
STATE EMERGENCY RESPONSE PHONE		•

Figure 8-7 / Proposed contents and origin of needs for recommended northeastern states manifest. (From Battelle Memorial Institute, 1980.)

    The proposed system for northeastern states includes a reduced reporting time for exceptions of 7 days from expected shipment delivery date, as opposed to the EPA time period of 30 days. This reduction is expected to enhance timely tracking and enforcement activities. Recommendations for manifest form content and origin of the requirement are indicated in Figure 8-7. The proposed form is illustrated in Figure 8-8.

    Among other things, the proposed northeastern states system includes conceptual design of an automated data processing unit for maintenance and inquisition of manifest records. A modular approach was developed by Battelle to accommodate construction of the system over time. This would allow for each module to be designed, once estimates of size are available, and still meet EPA requirements for the base program. The overall system is

INFORMATION	FEDERAL	REQUIREMENTS
		STATE IDENTIFIED INFORMATION NEEDS
MANIFEST DOCUMENT NUMBER	●	
GENERATOR		
GEN. NAME	●	
INDIVIDUAL NAME	●	
INDIVIDUAL TITLE		●
BUSINESS ADDRESS		●
PICK-UP ADDRESS	●	
PHONE	●	
EPA I.D. NUMBER	●	
WASTE		
SHIPPING NAME	●	
HAZARD CLASS	●	
TOTAL QUANTITY	●	
UNITS	●	
TYPE AND NUMBER OF CONTAINERS	●	
UN I.D. NUMBER	●	
EPA WASTE TYPE		●
EPA HAZARD CLASS		●
PHYSICAL STATE		●
HAZARDOUS CONSTITUENTS AND CONCENTRATIONS	(NOT REQUIRED)	(NOT DEEMED NECESSARY)

Figure 8-7 / (*Continued*)

conceptualized in Figure 8-9. Briefly, the master index module interfaces data bases such as the generator and disposer identification codes. The manifest module functions to process all manifest inputs and store relevant data for inquisition. Both audit and real-time tracking can be conducted as required. The emergency/administrative module covers several additional functions such as identification of emergency numbers and procedures and cataloging of the manifest system's operating guidelines. Such a system requires an investment in both the automatic data processing hardware and the software programs. Recent announcements suggest that firms may offer services to supply both and operate the system for states on a contract basis. Estimated costs of alternatives for system management are summarized in Table 8-8. It should be noted that a manual system was rejected on the basis

See reverse side for instructions
Please TYPE or PRINT clearly using
a ball point pen—PRESS HARD

Uniform State
HAZARDOUS WASTE MANIFEST

PART A:

DOCUMENT NO.

TO BE FILLED OUT BY GENERATOR

	NAME	SITE ADDRESS	PHONE NO.	EPA I.D. NO.
GENERATOR				
TRANSPORTER NO. 1				
TRANSPORTER NO. 2 (IF ANY)				
TREATMENT, STORAGE OR DISPOSAL FACILITY				

IF MORE THAN TWO TRANSPORTERS ARE TO BE UTILIZED, FILL OUT THE FOLLOWING AS APPROPRIATE.

THIS FORM IS NO ____ OUT OF A TOTAL OF ____ THE FIRST MANIFEST DOCUMENT NO. IS.

	PROPER US DOT SHIPPING NAME	US DOT HAZARD CLASS	UN NUMBER	FORM				QUANTITY	UNITS				CONTAINERS		EPA HAZ CODE	EPA WASTE TYP
				SOLID	LIQUID	GAS	SLUDGE		GALLONS	CU YDS	POUNDS	TONS	NO	TYPE		
1																
2																
3																
4																

SPECIAL HANDLING INSTRUCTIONS INCLUDING CONTAINER EXEMPTION (i.e., IDENTIFICATION OF ADDITIONAL WASTES INCLUDED IN SHIPMENT OF A NONHAZARDOUS NATURE WHICH DO NOT HAVE TO BE MANIFESTED)

GENERATOR'S CERTIFICATION. This is to certify that the above named materials are properly classified, described, packaged, marked and labelled and are in proper condition for transportation according to the applicable regulations of the Department of Transportation. U.S. EPA, and the State. The wastes described above were consigned to the Transporter named. The Treatment, Storage or Disposal Facility can and will accept the shipment of hazardous waste, and has a valid permit to do so. I certify that the foregoing is true and correct to the best of my knowledge.

GENERATOR'S SIGNATURE	TITLE	DATE TO BE SHIPPED	EXPECTED ARRIVAL DATE
		MONTH DAY YEAR	MONTH DAY YEAR

TRANSPORTER VEHICLE I.D. NO. STATE NUMBER	TRANSPORTER NO. 1 SIGNATURE AND CERTIFICATION OF RECEIPT OF SHIPMENT	DATE RECEIVED MONTH DAY YEAR

— — — — — — — TEAR AT THIS PERFORATION — — — — — — — — — — — — — — — —

TO BE FILLED OUT BY TRANSPORTER

PART B:

TRANSPORTER NO. 1 SIGNATURE AND CERTIFICATION OF DELIVERY AND NON-TAMPERING WITH SHIPMENT	DATE DELIVERED MONTH DAY YEAR

TRANSPORTER NO. 2 VEHICLE I.D. NO. STATE NUMBER	TRANSPORTER NO. 2 SIGNATURE AND CERTIFICATION OF RECEIPT OF SHIPMENT	DATE RECEIVED MONTH DAY YEAR

TRANSPORTER NO. 2 SIGNATURE AND CERTIFICATION OF DELIVERY AND NON-TAMPERING WITH SHIPMENT	DATE DELIVERED MONTH DAY YEAR

TO BE FILLED OUT BY TSD FACILITY

TREATMENT STORAGE OR DISPOSAL FACILITY INDICATION OF ANY DIFFERENCES BETWEEN MANIFEST AND SHIPMENT OR LISTING OF REASONS FOR AND DISPOSITION OF REJECTED MATERIALS	HANDLING METHOD
	1
	2
	3
	4

TREATMENT, STORAGE OR DISPOSAL FACILITY SIGNATURE AND CERTIFICATION	TITLE	DATE RECEIVED MONTH DAY YEAR

In case of an emergency or spill, immediately call The National Response Center (800) 424-8802 and the State (xxx) xxx-xxxx

DOCUMENT NO.

Figure 8-8 / Proposed uniform state hazardous waste manifest. (From Battelle Memorial Institute, 1980.)

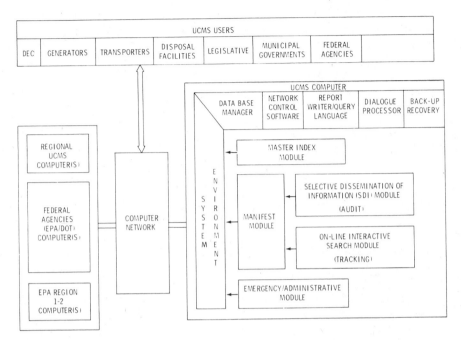

Figure 8-9 / Conceptual design of manifest management system. (From Battelle Memorial Institute, 1980.)

Table 8-8 / Estimated Development and Operation Costs for Manifest Data Processing (Assuming 500 Manifests/Shipment per Day.[a]

System	Support Package	Development and Operational Costs	
		Estimated One-Time Costs (Startup)	Estimated Recurring Costs per Year
I Dedicated mini-computer system	Program language	$588,700	$213,744
	Data base manager	$494,325	$201,396
II Shared computer facility system	Program language	$314,700	$202,344
	Data base manager	$220,325	$189,996
III Batch-fed shared computer facility	Program language	$175,000	$173,554
IV Manual manifest system	None	Minimal	$ 97,092

[a]Adapted from Battelle Memorial Institute (1980).

of its impracticality for tracking purposes. Similarly, a batch system was deemed too slow for the turnaround times required by participating northeastern states.

Implementation of manifest system controls is not accomplished without costs. A review of economic impacts was conducted prior to adoption of a manifest system in Illinois (Raufer and Croke, 1979). Direct cost impacts identified at that time included

1. Costs associated with completing and processing manifests
2. Costs associated with recordkeeping requirements
3. Costs associated with applying for and maintaining permits, identification numbers, and/or licenses

The Illinois study also looked at costs associated with permits and analysis for disposal and more sophisticated disposal alternatives generally attributed to hazardous waste management as a whole (as opposed to just the manifest system). The study team concluded that Illinois industry would be impacted at a level of $4.2 million annually. Current opinions (Raufer and Croke, 1979) rate this estimate as conservative. A similar study for the state of Minnesota concluded that manifest costs would average $0.10 per ton of waste (Battelle, 1977). If this value is inflated to $0.25 per ton of waste to accommodate recordkeeping, mailing, and reduced economies for small or mixed loads, the total cost of manifest systems for the 10.8 million wet metric tons of hazardous wastes in the United States (27% of total) currently available for contract disposal would be approximately $6 million.

The preceding includes only direct costs. Indirect costs, both monetary and otherwise, may be more significant. The manifest places a great amount of responsibility on the transporter. He is prohibited from accepting hazardous waste for shipment unless it is accompanied by a properly completed manifest. Whereas the generator is charged with completing the manifest, the transporter must be satisfied that the data on the form correspond with what is being shipped. All containers should be accounted for and properly classified, described, labeled, and packaged in undamaged, nonleaking containers. These assurances are needed for several reasons:

1. On signing the manifest, the carrier assumes custody of the wastes.
2. Should the disposal operator determine an inconsistency, the transporter may well find it difficult to prove he is not at fault, because the manifest certifies his initial receipt of the purported wastes.
3. In the event of a spill or "hazardous waste discharge," as the EPA has chosen to designate spills, the transporter has prescribed responsibility with respect to reporting, emergency response, and ultimate cleanup. Serious errors and liabilities would be involved if the transporter was unaware of the actual contents of a load.

The implications of these responsibilities are quite clear. The transporter cannot afford to rely entirely on the technical expertise and judgment of the shipper. Rather, he must be intimately familiar with the regulations and at least conversant in the emerging language of hazardous wastes. This necessitates education and training for supervisory staff and drivers. The costs of failure to do so may be extremely severe. In 1979, the average cost for cleanup of an accidental discharge was $100,000 (Fisher and Lamberton, 1980). Hence, in addition to the need for learning about the new regulations, there is increased incentive for driver safety programs and equipment inspection.

All of the foregoing increase the cost of operating, the risks and liability associated with operation, and consequently the cost of entry into the hazardous waste hauling industry. This is likely to reduce the current number of haulers and skew the industry profile more heavily to larger organizations. A resulting undercapacity would also encourage the further development of generator- and disposer-owned fleets.

Regulation outside the United States

The manifest system is by no means unique to regulation in the United States. The need for management of hazardous waste transportation is recognized throughout many of the industrialized nations of the world. The province of Ontario uses liquid industrial waste transfer forms such as that in Figure 8–10 to track the movement of waste materials. Member states of the European Community have been provided guidance on identification and accompanying documents through the directive of toxic and dangerous waste and its associated annexes (Working Party on Toxic and Dangerous Waste, 1979). Article 14(2) of the directive requires that

When toxic and dangerous waste is transported in the course of disposal it shall be accompanied by an identification form containing at least the following details:

- nature
- composition
- volume or mass of the waste
- name and address of the next holder or of the final disposer
- location of the site of final disposal where known.

Within this framework, member states are free to develop a shipping document or manifest system to meet their own needs.

Control in West Germany is maintained through an elaborate manifest

PLEASE PRINT   NUMBER

A.

COMPANY

ADDRESS

DESCRIPTION OF WASTE

QUANTITY kg (lb.), litres (gal.)	TIME OF TRANSFER			
	DAY	MONTH	YEAR	hr a.m./p.m.

TRANSFER LOCATION (IF DIFFERENT FROM COMPANY ADDRESS)

AUTHORIZED COMPANY OFFICIAL: NAME (PLEASE PRINT)	TITLE:

SIGNATURE	DATE:

B. DISPOSAL SITE OR WASTE MANAGEMENT SYSTEM

METHOD OF TREATMENT OR DISPOSAL

CERTIFICATE OF APPROVAL NO.	DATE OF DISPOSAL		
	DAY	MONTH	YEAR

C. CARRIER   D.

NAME	NUMBERS FROM SOURCE FORMS
ADDRESS	
VEHICLE LICENSE PLATE NO.	
DATE	
SIGNATURE	

Figure 8-10 / Transfer of liquid industrial waste.

system promulgated by the regulations of June 2, 1978 (Abfallnach Weisverordnung). A set (six copies) of detailed manifests is required of the generator. The copies are distributed as in the case of manifests in the United States, with two copies retained by the generator, two going to the regulatory body, and one each retained by the carrier and disposer. One of the copies retained by the generator is routed to him from the disposer as evidence of receipt of the wastes. The regulatory authorities' copies are provided by the generator and the disposer. The information flow thus generated is parallel to that of the California program. Retained copies are maintained in a register, which must be made available for inspection by the authorities. Special forms are required for wastes that are imported from outside the county. Similarly, customs inspection is employed for the export of any wastes. Once inspected, however, all tracking ceases. There is no mechanism to ensure that wastes that have crossed the frontier are in fact disposed of in a safe manner. Transporters are licensed for the specific carriage of hazardous wastes for (1) a specific consignment, (2) an unspecified number of consignments over a given period, (3) an unspecified number of consignments until further notice.

Regulation of hazardous waste transportation in the United Kingdom stems from the Poisonous Waste Act of 1972, which requires notification at least 3 days in advance of any plans to remove or deposit "toxic, dangerous or polluting wastes" within a local authority area. Although a sample form has been provided by the Department of the Environment, most local authorities have created their own, and as a consequence there is no standardization. Draft regulations are under discussion to further require the use of shipping documents and a register to assist authorities in assuring that wastes have been shipped to licensed sites. Without such a system, the lines of responsibility for wastes remain unclear during the management process. Presently, the U.K. system is designed for all wastes, not just toxic wastes.

Dutch authorities have drafted a ministerial order establishing a notification system for dispatch and receipt of toxic wastes similar to that in the United Kingdom. In Denmark, a "chemical waste information sheet" is employed to convey instructions on the storage, transport, and disposal of chemical waste. Carriers as well as disposers are held responsible to ensure that wastes do not cause pollution. Luxembourg requires maintenance of a register and use of documents for all waste collection, transport, import, export, storage, and treatment operations.

It should be noted, however, that the use of manifests and specific regulations are not universal throughout Europe. Ireland, for instance, had no provisions for regulating hazardous waste transportation as of 1979.

REFERENCES

Andres, D. R. 1977. Managing hazardous wastes—manifests destined for success in Texas. *Waste Age*, Vol. 8, No. 10, pp. 38–41.

Battelle Memorial Institute, Columbus Division. 1980. *A proposed uniform state hazardous waste system.* New York State Department of Environmental Conservation.

Battelle, Pacific Northwest Laboratory. 1977. *The impact of hazardous waste generation in Minnesota.* Minnesota Pollution Control Agency.

Booz Allen & Hamilton. 1979. *Options for establishing hazardous waste management facilities.* New York State Environmental Facilities Corporation.

Fisher, T. L.; and Lamberton, W. L. 1980. Hazardous waste transportation: are the RCRA regs workable? *Solid Wastes Management* Vol. 23, No. 7, pp. 58–59.

National Conference of State Legislature. 1982. *Hazardous Waste Management: A Survey of State Legislation 1982.* Denver.

Raufer, R.; and Croke, K. 1979. The hazardous waste manifest system: preview of things to come. *Pollution Engineering*

C. J. Touhill. 1980. Personal communications, letter dated August 8, 1980.

Working Party on Toxic and Dangerous Waste. 1979. *Toxic and dangerous wastes in the community.* Commission of the European Communities, Annex 3 ENV/324/79-EN, BR/rac.

9

Treatment Processes

Introduction

Effective management of hazardous wastes involves much more than careful burial. There are several options that should be considered before undertaking the disposal of a hazardous waste in a secure landfill. The application of treatment processes, the subject of this chapter, is one of the most widely used options for either eliminating a hazardous waste or reducing the volume and hazards associated with it. First, however, one should verify the need for treatment, because additional options frequently can minimize the potential for environmental damage or reduce the cost for treatment and disposal.

The U.S. Environmental Protection Agency has suggested a hierarchical structure of waste management options (U.S. EPA, 1977) based primarily on environmental concerns. The importance of economics in the management of hazardous wastes is recognized in this structure. The prioritized order of options is as follows:

Waste reduction

Waste separation and concentration

Waste exchange

Energy/materials recovery

Waste incineration/treatment

Secure ultimate disposal

Waste reduction merits first consideration because eliminating the waste or reducing the quantity of it produced at the source can provide substantial cost benefits as well as increased environmental protection. Waste reduction can be accomplished through improved housekeeping and process control or by changes in the process itself. Leaking equipment, for example, can create a hazardous waste stream where none would exist if proper maintenance programs had been followed. Poor process control may produce off-standard batches of product that become hazardous wastes. Changing the manufacturing process may simply involve substitution of a chemical reagent that eliminates a hazardous component of a waste, or it may involve complete redesign of the process. A cost analysis is usually performed to establish the feasibility of the process changes.

Separation and concentration of hazardous waste, the next option, can achieve a volume reduction, but not necessarily a reduction in the quantity of hazardous components. The separation or removal of water to produce a more concentrated waste is commonly undertaken to minimize transportation and ultimate disposal costs. Disposal costs are normally directly related to the quantity of waste, not the amount of hazardous components. Separation is often accomplished in plants where many waste streams are blended into a common waste stream, with only one or two of the streams being hazardous, and these can be isolated and processed separately to avoid dilution with the other streams. Dewatering of sludges (e.g., filtration) and evaporation of dilute wastes are examples of concentration techniques.

The third option for consideration is waste exchange, which has been discussed in Chapter 6.

Recovery of energy or valuable material from a hazardous waste is the fourth and last option before treatment. Through this option it is possible to return a profit in some instances, as well as avoid the disposal of a hazardous waste. Combustion of waste oil or other flammable materials to recover heat or energy is desirable both to conserve fuel and to reduce the potential for environmental damage by disposing of these materials to the land. Extraction of valuable materials from waste may require less energy than extraction from ores because of the more concentrated forms of these materials in wastes. A prime example of recovery and recycle that is widely practiced in the United States is provided by the waste solvent processing industry. The higher costs of secure landfill disposal relative to past practices of landfilling should provide industry with greater incentive for recovery of useful materials from hazardous wastes.

The following discussion of treatment processes will include the technology for recovering useful materials as well as the technology for eliminating hazardous components from a waste or rendering these components less hazardous for safer disposal. This chapter is divided into two sections, the first dealing with wastewater treatment and the second with treatment of nonaqueous wastes. Wastewater treatment technology has been well developed through the recognized need to preserve and enhance the quality of receiving surface waters. Increased emphasis is now being given to technology for treating solid wastes that is primarily directed toward protection of groundwater quality, because toxic components of these wastes may be leached into groundwater at the disposal site. A detailed discussion of the theory and application of the unit processes used for treatment of hazardous wastes is beyond the scope of this book.

Many excellent textbooks are available for those who wish to pursue this subject matter in greater depth. Some of these books will be cited in the text. The principal objectives of the discussion are to inform the reader of the types of treatment processes available for specific purposes and provide guidance on selection of the most cost-effective processes for a given waste.

Wastewater Treatment

Perspective

Large volumes of wastewater that could be defined as hazardous by RCRA guidelines are not classified as such because they are easily rendered nonhazardous by on-site treatment. Waste pickle liquor and other wastewaters produced by metal finishing operations are examples of this type of waste. For steel finishing operations alone, it has been estimated that 350 million gal of spent pickle liquor and at least 12 billion gal of rinse water would be generated by hydrochloric acid pickling of the more than 30 million tons of steel that are treated each year (National Commission on Water Quality, 1975) (a simplifying assumption, as other acids are also used). If all the rinse water is less than pH 2 (RCRA guideline for corrosivity), this volume of pickle liquor and rinse water will correspond to over 98 million tons of waste (U.S. EPA, 1971). By comparison, only 29 million tons of hazardous waste were reported as being produced in 1977 (see Chapter 4). The importance of wastewater treatment in minimizing the amount of hazardous waste produced cannot be overemphasized.

Treatment process selection is dependent on the nature of the wastewater and the quality of the effluent desired. Hazardous components of the wastewater may be either separated or converted to nonhazardous forms to permit disposal of the wastewater effluent by conventional methods. Conversion processes can frequently be accomplished in one step, as in the case of neutralization of strong acids and bases and oxidation/reduction reactions. On the other hand, separation processes often require multiple steps, because pretreatment or posttreatment is required either to prepare the wastewater for the separation process or to polish the effluent to meet discharge standards. Furthermore, the hazardous components separated from the wastewater must be disposed of, and that may require additional processing steps such as sludge dewatering. Multiple processing steps may also be required for a conversion process such as biological treatment to remove substances that interfere with biological degradation of the hazardous components or to dispose of sludge containing these substances.

A thorough characterization of the wastewater is very helpful for preparing a preliminary assessment of the types of treatment processes that may be used for the wastewater. Parameters that are most often used for assessing the potential of available treatment processes for a given hazardous wastewater are presented in Table 9–1, along with direct and secondary implications associated with the character and treatment of the waste.

Removal or conversion of the specific hazardous components is the primary objective of treatment; however, process selection is quite dependent on the other parameters. For example, the wastewater may contain a relatively small concentration of a hazardous organic compound that by itself could be readily removed by activated carbon adsorption. But a very high total organic carbon (TOC) concentration would indicate potential interference with the activated carbon adsorption process. These parameters also indicate the need for pretreatment steps, such as the amount of acid or base required to adjust the pH to near neutrality for biological treatment. Discussions of the individual treatment processes that follow will address the use of characterization data for assessing the potential of the processes.

Liquids–Solids Separation

General Considerations.   Separation of suspended matter from wastewater can be accomplished by a number of different processes, the selection of which depends on the quantity and characteristics of the suspended solids. Large heavy solids usually are easier to remove than finely divided light solids. The amount of residual solids that can be allowed in the treated effluent is also a consideration. Selection of a liquids–solids separation process may be directed at removal of a hazardous component associated with the solids, in which case nearly complete removal is desired. If liquids–solids separation is used as a pretreatment step to reduce organic loading to a biological treatment process, then nearly complete removal probably will not be required. Dewatering of sludge does not normally require a clear effluent, because this effluent usually is a small fraction of the raw wastewater stream and can be recycled to the head end of the treatment process train. Separation processes such as ion exchange and granular activated carbon in fixed beds require a feed stream that is relatively free of suspended solids; therefore, a liquids–solids separation step is required if the suspended solids exceed 15 to 25 mg/l.

There are several different devices that are commercially available for carrying out a particular unit operation such as gravity sedimentation or filtration. Additional information is available on the application of these devices and the theory associated with the unit operations (Culp et al., 1978; Weber, 1972; Rich, 1963; Metcalf and Eddy, 1979; Eckenfelder, 1966; U.S. EPA, 1975; Arthur D. Little, 1976; WPCF/ASCE, 1977).

Table 9-1 / Wastewater Parameters Commonly Used for Assessing Treatability.

Parameter	Direct Implications	Secondary Implications
pH	≤ 3 or ≥ 12 constitutes a hazardous waste pH control may reduce heavy-metal levels	pH control may be required as pretreatment for a number of different processes
Alkalinity/acidity	Determines reagent requirements for neutralization or pH adjustment	High reagent requirements for neutralization result in high salt concentrations in treated effluent
Nonfilterable residues 103°C and 550°C	Insoluble hazardous wastes in suspension	Pretreatment for -ion exchange, -activated carbon, -membrane processes
Settleable solids	Toxic chemicals sorbed onto particulates	
Filterable residues 103°C and 550°C	A measure of solute concentration in the form of stable residues at the specified temperature; residue at 550°C indicates stable salt concentrations that impact ion-exchange and membrane processes	High concentrations of dissolved salts may adversely affect biological treatment; discharge of salt waste to receiving waters may be restricted
Conductivity	Indicates dissolved electrolyte concentrations that may include volatile salts such as NH_4HCO_3 that escape residue determinations	Impacts ion-exchange and membrane processes
Chemical oxygen demand (COD)	May represent toxic organic chemicals	Governs biological treatment process design (by correlation with BOD)
Biochemical oxygen demand (BOD)	May represent toxic organic chemicals	Governs biological treatment process design
Total organic carbon	May respresent toxic organic chemicals	Indication of refractive organics in effluents
Heavy metals	Toxic constituents of a waste	Major component of sludges
Specific hazardous components	Hazardous constituent of a waste	May adversely affect biological treatment

Screening.   Screening devices are commonly used in conventional waste-water treatment to remove large pieces of solid matter that would interfere with subsequent processing operations or would cause damage to equipment such as pumps. Coarse screening devices may consist of parallel bars, rods or wires, perforated plates, gratings, or wire mesh with apertures greater than 0.6 cm. Self-cleaning units are available to eliminate manual cleaning of the screening devices (Metcalf and Eddy, 1979). Fine screening devices with apertures less than 0.6 cm are also used to a limited extent in wastewater treatment.

Gravity Sedimentation.   Gravity sedimentation is a well-established process that enjoys widespread use for removal of settleable and floatable matter from wastewater. The process involves quiescent containment of a wastewater for a sufficient period of time to allow some or all of the suspended matter to either settle out or float to the surface of the wastewater. In its simplest form as a batch process, a given volume of wastewater is transferred to a vessel or basin and held there until essentially all the settleable and floatable matter separates. Floating matter, if any, may then be skimmed and the clarified wastewater decanted for discharge or further treatment. Sludge may be allowed to collect in the vessel or basin. It is subsequently removed after several batches of wastewater have been processed. Removal of the sludge may be a manual operation using shovels or front-end loaders to transfer the sludge to containers for disposal. Transfer of sludge from a batch settling vessel can be facilitated by constructing the vessel with a conical bottom, as shown in Figure 9-1, to route the sludge through a valve into a container or transfer line. The conical bottom should have sufficient slope for drainage of the sludge; however, a scraper sweeping the bottom may be included to assure nearly complete removal of the sludge. Outlets equipped with valves are located vertically on the side of the vessel for decanting the supernatant liquid after settling is complete.

Large settling ponds may be constructed where sufficient land area is available. These are frequently sized to contain several days' output of wastewater, thereby achieving quiescent conditions for settling while waste-water flows continuously into and out of the ponds. The ponds are drained periodically to permit removal of sludge collected on the bottom of the ponds.

When sufficient land area is not available for a settling pond and relatively large volumes of wastewater must be treated, more sophisticated units called clarifiers usually are employed. These units are engineered to achieve maximum settling rates for a given volume and configuration. Many different configurations have been developed to increase the rate of settling per unit volume of clarifier. Figure 9-2 shows a diagram of a conventional circular clarifier that is one of the most frequently used designs in the

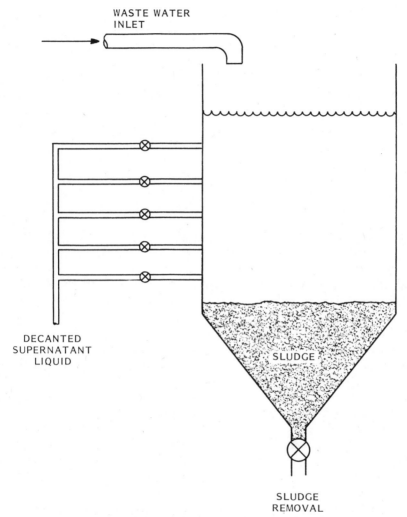

Figure 9-1 / Batch sedimentation tank.

industry. Influent flows continuously through the inlet to a well that is baffled to prevent excessive velocities and is located at the center of the clarifier. The wastewater flows radially outward from the well to a peripherally mounted serrated weir protected by a baffle to prevent stray currents from disturbing the quiescence of the wastewater. The clarifier is equipped with radial rotating collector arms moving along the bottom of the clarifier to plow settled sludge toward a center draw-off for continuous or periodic removal. A radial arm or skimmer that sweeps the surface to remove floating matter or scum may also be included.

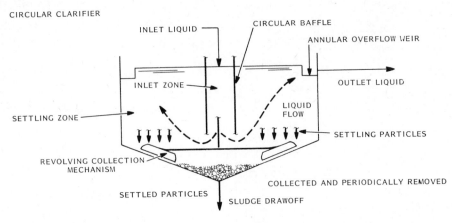

Figure 9–2 / Circular clarifier design (A.D. Little, Inc., 1976).

Sedimentation of a dilute suspension of discrete particles of known volume and density can be calculated for design purposes (Weber, 1972; Metcalf and Eddy, 1979), but this is not possible for flocculating particles commonly contained in chemically coagulated wastewater. The settling velocity of these particles changes continuously as the particles agglomerate during their descent in the clarifier. Some breaking apart of agglomerated particles may also occur as a result of eddy currents. Laboratory column batch settling tests (Zanoni and Blomquist, 1975) or continuous-flow pilot-plant studies are needed for design information if literature data are not available for the specific solids to be settled. Use of laboratory batch settling data also requires oversizing the clarifiers by factors of 25% to more than 100% to allow for nonideal conditions. Design safety factors probably are necessary for scale-up from small pilot-plant clarifiers to large open clarifiers that may be affected by wind.

High-rate clarifiers have been introduced commercially in the past two decades. These units employ the shallow-depth settling concept in slightly inclined or steeply inclined tube settlers (Culp et al., 1978). This type of clarifier is useful where space is limited or where it is desired to increase the capacity of an existing conventional clarifier.

Solids-contact or sludge-blanket clarifiers have been particularly useful in settling sludges that are flocculent and of low density. Several different equipment configurations are offered by manufacturers of these units. A major portion of the volume of the solids-contact clarifier may be taken up by mixing and reaction zones, as shown in Figure 9–3. These mixing and reaction zones coupled with the sludge blanket or slurry pool account for greater efficiency in solids removal. Flow enters the clarifier through an annular opening in the top of the primary mixing and reaction zone and proceeds from there to a secondary zone. Chemicals may be added to the incoming stream or directly into either mixing zone. The large circulation patterns in these zones provide opportunity for thorough mixing and particle contact to enhance flocculation. Treated wastewater leaves the mixing and reaction zone through two concentric draft tubes. Part of the flow is recycled back through the return-flow zone, and the remainder passes through the sludge blanket or slurry pool, where essentially all of the particulate matter is retained. Clarified water flows up from the sludge blanket and overflows to the effluent line. Sludge is collected in the concentrator and discharged at a rate designed to maintain an optimum solids concentration in the sludge blanket.

Solids-contact clarifiers have been reported to be particularly effective for enhancing sulfide precipitation of heavy metals using ferrous sulfide reagent (U.S. EPA, 1980). This reagent has limited solubility, which is advantageous to maintain low sulfide concentrations in the treated effluent, but the slow release of sulfide requires a long contact time in the clarifier to achieve low heavy-metal concentrations in the effluent.

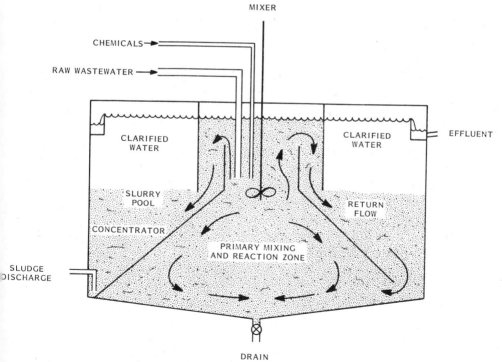

Figure 9–3 / Solids contact clarifier.

Gravity sedimentation is also employed for thickening or removal of additional water from sludges collected from wastewater clarifiers. Gravity thickeners operate on essentially the same principles as the clarifiers, except that the thickening action is relatively slow compared with settling in the initial clarification stage. The simple batch clarifier in Figure 9–1 could readily serve as a thickener by allowing more time for settling. Gravity thickening is frequently used as an inexpensive intermediate step for sludge dewatering prior to filtration or centrifugation.

Dissolved-Air Flotation. Suspended matter that does not sink or float in a reasonable period of time for separation by gravity sedimentation can frequently be separated efficiently by dissolved-air flotation. Separation is brought about by the introduction of finely divided gas bubbles, which become attached to the particulate matter, causing it to float to the surface, where it is removed by skimming. Introduction of the gas bubbles is usually accomplished by reducing the pressure of the wastewater, causing dissolved gases to be released as tiny bubbles on the particulate matter. A vacuum may be drawn on water saturated with air to release the bubbles, or air may be injected and dissolved into the water above atmospheric pressure, followed by release of the pressure. Dissolved-air flotation is commonly used to separate greasy or oily matter from industrial wastes. Chemical coagulation and flocculation are frequently used to enhance the removal of finely divided or colloidal solids.

Granular-Media Filters. Removal of small amounts of suspended matter from wastewater is typically the objective of granular-media or deep-bed filtration. This process serves as a polishing step in most wastewater treatment operations to produce a highly clarified water. Chemical coagulation/flocculation and sedimentation typically precede granular-media filtration, and the small portion of the particulate matter that escapes the clarifier is caught by the filter. If the particulate matter contains hazardous substances, the filter assures the removal of final traces of these materials. Filtration also assures an effluent that is suitable as a feed stream to other fixed granular-media beds, such as granular activated carbon and ion-exchange resins. Although the carbon and resin beds can function to a limited degree as filters, it is usually desirable to avoid this because of damage to the granules caused by frequent backwashing to remove the solids.

Granular-media filtration is a well-established, widely used process that appears in many different equipment configurations (Culp et al., 1978). Graded sand and pulverized anthracite coal are commonly used in filter beds, which vary in thickness from 1.5 to 3 ft. The filter medium is supported on a layer of gravel or a porous plate. The sand and coal may be used together to form a dual-media filter bed, which is generally superior to a single-medium bed. The difference in specific gravity between the sand (sp. gr. $\sim$ 2.6) and coal (sp. gr. $\sim$ 1.6) allows the placement of a layer of coarse coal granules (e.g., 0.8–2.0-mm diameter) over a layer of finer sand granules (0.4–1.0-mm diameter) for downflow filtration. Backwashing of the bed to remove filtered solids does not alter the layered structure of the bed, because the lighter coal remains on top after backwashing to present a more open structure and greater filter capacity. A third filter medium such as garnet sand (sp. gr. $\sim$ 4.2) can also be included in the filter bed to improve performance.

Conventional operation of high-rate granular filter beds is generally by downflow, either by gravity or in a pressurized vessel at flow rates typically in the 3- to 15-gpm/ft^2 range (Culp et al., 1978). When the filter capacity of the bed is fully utilized, as indicated by high differential pressure across the bed or breakthrough of turbidity, the service flow is terminated, and the bed is backwashed. Backwash water is pumped upflow to fluidize and scour the filter medium to release the filtered solids. The backwash water containing the filtered solids usually can be recycled to the head end of a chemical coagulation/sedimentation process if available. If sedimentation is not included, then the backwash water may be treated separately, as in a batch clarifier (see Figure 9-1). Backwash water is normally a small fraction of the total wastewater stream.

Granular-media filtration in deep beds is a key process in the production of highly clarified wastewater and should not be confused with surface filtration, which is more frequently used to dewater sludges rather than to produce highly clarified effluents. The ability of deep-bed granular-media filters to produce very clear effluents results not only from straining action, which physically blocks the passage of particulate matter through the bed, but also from adhesion, which removes particles finer than the pore spaces in the bed. Adhesion makes filtration of submicron particles possible. These small particles are attracted to the surface of the filter medium by physical forces but are removed by scouring during the backwashing step. The capacity of the filter bed to remove small particulate matter by adhesion is limited by the amount of surface area available. Surface filtering devices, as discussed in the next section, depend largely on straining action.

Surface Filters.   Surface filters typically make use of a fine medium such as cloth or close-mesh screen. A large number of surface filtering devices are commercially available. The most frequent application of surface filtration is for dewatering sludges or filtering large amounts of solids precipitated from wastewater, such as metal finishing wastes. The rotary vacuum filter illustrated in Figure 9-4 is representative of this type of filter. The filter medium, fabric or wire mesh, in the form of a continuous belt, rotates over a perforated drum that is partially submerged in the slurry to be filtered. Water is pulled through the filter cake that forms on the belt to the inside of the drum,

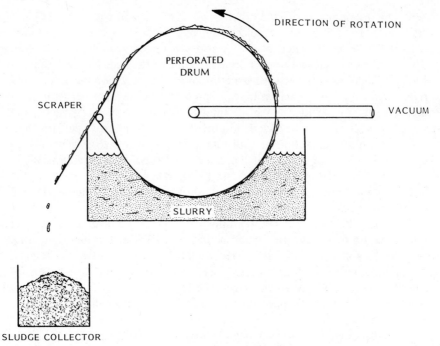

Figure 9-4 / Rotary vacuum filter.

where it is transferred to the vacuum system. The filter cake also acts as a filter medium to increase the efficiency of solids removal. The filter cake rotates around to the scraper, where it is removed and drops to a collector. Various appurtenances, such as spray washers to rinse the cake before the scraper and remove residue after the scraper, are often included with this equipment (Metcalf and Eddy, 1979).

Sludge characteristics are important to successful operation of rotary vacuum filters as well as other types of surface filtration devices. The sludge must be sufficiently coagulated to allow water to drain freely through the sludge. Chemical conditioning with ferric chloride or polymers, for example, is frequently necessary to effectively filter a sludge. Filter presses are also widely used for dewatering sludges (Arthur D. Little, 1976).

Other types of surface filters, such as microstrainers, precoat filters, disc filters, and filter leafs, are used to remove small to moderate amounts of solids from wastewater (Culp et al., 1978; Metcalf and Eddy, 1979). Application of a particular type of surface filter for a wastewater is dependent on the characteristics and concentration of the particulate matter contained therein and the quality of filtrate desired. Microstrainers, for example, may remove only a portion of the suspended matter to attain the desired effluent quality, whereas a precoat filter may remove essentially all of it, to produce a highly clarified effluent.

Centrifugation.   Centrifugation is a well-established process that has been used for many years in the chemical industry and in municipal sludge treatment applications. It is often a useful alternative to filtration, especially for sticky gelatinous sludges that do not dewater rapidly on a filter. Its most frequent use in waste treatment is for dewatering wet sludges that are typically 2% to 5% solids. These are dewatered to 15% to 50% solids in the cake removed from the centrifuge (Arthur D. Little, 1976). The liquid stream or centrate from the centrifuge is usually high in suspended matter and is commonly recycled to a coagulation/sedimentation process that precedes the centrifugation step. Basically, centrifuges operate by a rapid rotation of a liquid suspension, which induces a much greater force than gravity to hasten the separation of the suspended matter. Matter in aqueous suspension that is less dense than water can be separated, as well as matter that is more dense than water. A number of different types of centrifuges are commercially available (U.S. EPA, 1975).

Chemical Treatment

General Considerations.   Destruction or separation of hazardous constituents in wastewater by chemical treatment is an extremely important technology for processing hazardous wastes. Enormous quantities of potentially hazardous wastes are eliminated on site each day in manufacturing industries by chemical treatment. Chemical treatment is particularly suited for dilute wastewaters, where the effect of dilution has little or no impact on the reactions employed to render the waste nonhazardous. Neutralization of acidic and alkaline wastewaters is a prime example of this type of reaction. Furthermore, toxic components such as H^+ and OH^- in these wastewaters can be reduced well below hazardous levels in relatively uncomplicated equipment. Oxidation–reduction reactions can also be carried out with high efficiency in wastewaters.

Separation and concentration of dilute toxic heavy metals in wastewater are typically performed by chemical treatment that precipitates the metals to form sludges, which are processed by the liquids–solids separation techniques discussed earlier in this chapter. This method has wide application in the metal plating and metal finishing industries. Although the heavy metals cannot be destroyed by this treatment or any other type of treatment, the degree of hazard is greatly reduced through the formation of relatively insoluble sludge.

Neutralization. Neutralization of acidic or alkaline wastewater is one of the most common chemical treatment methods employed by industry. The process equipment can be very simple, as, for example, a batch operation to neutralize a small quantity of dilute acid waste. The waste is transferred to a small tank, where a neutralizing agent such as baking soda (sodium bicarbonate) is added with mixing (a paddle will suffice) until test paper indicates pH 8. Protective apparel, such as a face shield, rubber apron, and gloves, will also be needed to protect the operator against splashes of acid. Use of a weak basic compound such as sodium bicarbonate avoids converting the waste to a strongly alkaline solution by adding too much reagent. In addition, sodium bicarbonate is very safe to handle. Although this type of operation may be satisfactory for occasional batches of waste, it would be expensive for large volumes generated continuously or frequently.

Automated equipment for continuous-flow neutralization of acidic or alkaline wastewaters is generally used for large volumes generated frequently or continuously. An example of an automated continuous-flow system is shown in Figure 9-5. Influent acidic wastewater is fed to the neutralization tank, where a solution of sodium hydroxide is added at a rate controlled by a valve that responds to a pH meter. A sample pump circulates a small flow of neutralized wastewater through a cell containing the pH sensing electrodes. If the pH falls below some predetermined value, the meter signals to increase the valve opening to allow a greater flow of sodium hydroxide into the neutralization tank. A similar type of system can be constructed to neutralize alkaline wastes, where acid is added at a rate controlled by a pH meter.

Automated continuous-flow systems become more complex for highly variable wastewaters. A system such as that illustrated in Figure 9-5 would have difficulty responding to large fluctuations in acid content of the wastewater; therefore, an additional neutralization tank with pH-controlled addition of sodium hydroxide may be necessary to assure effluent in the desired pH range. Wastewaters that fluctuate from acidic to alkaline would require separate systems for adding the neutralizing acid and the base. If space is available, an equalizing tank or basin can be installed to provide a more consistent feed to the neutralizing system. Production facilities that are discharging both acidic and alkaline wastes usually have equipment for storing and blending the wastes to minimize consumption of neutralizing chemicals.

The choice of neutralizing chemical depends primarily on availability, cost, ease of handling, and compatibility with processing objectives. Commonly used alkaline chemicals for neutralizing acidic wastes include lime, limestone, sodium hydroxide, and soda ash. Lime is available in either the unslaked form (CaO) or the slaked form [Ca(OH)$_2$]. Unslaked lime is generally used at large installations that can take advantage of purchasing large quantities of this material at a low cost. The equipment needed to unload,

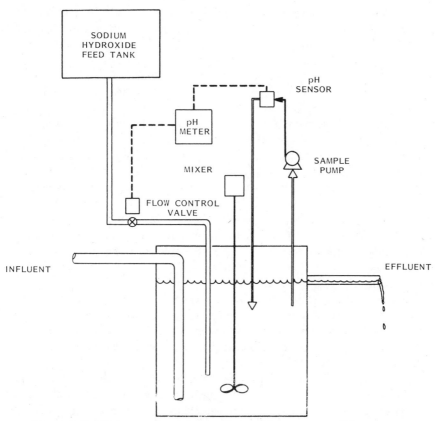

Figure 9-5 / Automated continuous–flow acid waste neutralizing system.

store, and slake (react with water) the lime can be quite costly; however, the cost can be justified on a unit basis of lime consumed. Although lime is generally much less expensive than sodium hydroxide or soda ash, it is more difficult to transfer to the neutralization tank because of its limited solubility. It must be added dry from a solid feeder or conveyer, either directly to the neutralization tank or to a slurry tank that feeds the neutralization tank. Maintenance problems are frequently encountered with these solids handling and slurry feed systems because of blockage in the solid feeder or slurry lines. Furthermore, lime neutralization may contribute to increased sludge production, which would be undesirable if the cost of disposing of the sludge exceeds the cost advantage of using lime in place of sodium hydroxide or soda ash. Lime neutralization of sulfuric acid or acidic wastes containing sulfates or sulfites can produce calcium sulfate and sulfite, which have limited solubilities in water. Rapid scaling of equipment in contact with the precipitating substances can also contribute to maintenance problems. In spite of the disadvantages of lime neutralization, lime is widely used in industry because of its low cost for treating acidic wastes. Limestone or calcium carbonate is also used extensively and can be purchased at a very low cost if the source of supply is nearby. It can be used in powdered form added to the neutralization tank, or in cases where precipitation does not interfere, the granular form can be used by percolating the wastewater through a bed of limestone. Neutralization with limestone is not as rapid as with lime, particularly for neutralizing acid wastes that form a layer of precipitated material [e.g., $Fe(OH)_3$] on the surface of the limestone particle. Access to the reactive surface area of the limestone is diminished by the formation of this layer.

The cost of chemicals for neutralization is governed by transportation costs as well as production costs. Furthermore, the neutralization capacity of the chemicals must be considered when performing cost comparisons. For example, 1.00 kg of pure CaO is equivalent to the following weights of pure compounds: 1.32 kg $Ca(OH)_2$, 1.43 kg of NaOH, 1.78 kg of $CaCO_3$, and 1.89 kg of Na_2CO_3. Impurities in commercial-grade chemicals must also be factored into the cost comparisons.

Selection of an acidic neutralization agent for alkaline wastewaters is usually between sulfuric acid and hydrochloric acid, with the former generally having a price advantage. The use of waste acids and alkalies from nearby industrial facilities should always be considered. These are frequently available through waste exchanges.

Alkalinity or acidity and pH are important analytical parameters for establishing the amount of neutralizing agent required. Alkalinity and acidity are usually expressed as mg/l $CaCO_3$, which has an equivalent weight of 50 g (50 mg per milliequivalent). A solution of 40 mg of NaOH in pure water would have an alkalinity of 50 mg/l as $CaCO_3$ and would require 48 mg/l of H_2SO_4 for neutralization. The presence of buffering agents (e.g., carbonate/bicarbonate) in wastewaters tends to complicate calculating the amount of neutralizing agent to attain a specified pH; however, conventional acidity or alkalinity data can still be used as a rough estimate of neutralization requirements. For example, the phenolphthalein alkalinity for an alkaline waste signifies the equivalent amount of acid that will be required to neutralize the waste to about pH 8.3. This will give an adequate safety factor if the discharge limit is pH 9.0, but some additional acid may be required for a safety factor if the discharge standard is pH 8.5.

In the case of an acidic waste, the phenolphthalein acidity can be used with the realization that the presence of buffering agents can indicate a greater neutralization requirement than actually necessary if the discharge standard is pH 6.0 or 6.5. If the wastewater is known to be relatively free of buffering agents, such as phosphate or heavy-metal salts that hydrolyze in the pH 6.0 to 8.3 range, then the phenolphthalein acidity provides a good estimate of the alkalinity required for neutralization. This fact is graphically illustrated by the titration curve for sulfuric acid given in Figure 9–6. The titration curve for phosphoric acid is also shown in Figure 9–6 to illustrate the effect of buffering. The titration curve for the sulfuric acid sample shows very little difference between the amount of standard base used to reach pH 6.5 and the amount to reach pH 8.3, whereas the amount of base used to reach pH 8.3 for the phosphoric acid titration is about 45% more than used to reach pH 6.5

Neutralization of acidic industrial wastewater frequently causes the precipitation of heavy metals contained in the wastewaters.

Precipitation/Coagulation/Flocculation.   Treatment of wastewater by precipitation/coagulation/flocculation is widely used throughout the industrial and municipal wastewater treatment field for removal of pollutants. It is particularly useful for removal of heavy metals from hazardous wastewater. Although the terms "precipitation," "coagulation," and "flocculation" describe different phenomena or processes, they are discussed collectively because of their close relationship. Precipitation in aqueous systems commonly refers to the formation of a solid phase, usually particulate matter suspended

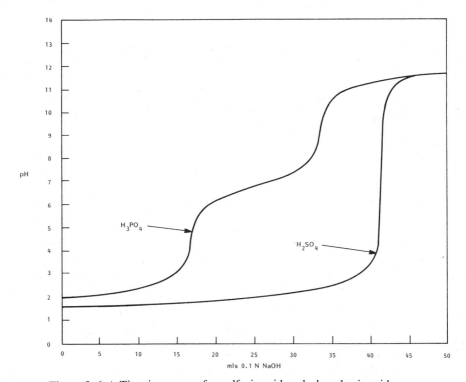

Figure 9-6 / Titration curves for sulfuric acid and phosphoric acid.

in a liquid phase. It is the first step in the separation sequence that ultimately removes the pollutant from the wastewater. Precipitation is most frequently accomplished by the addition of chemical reagents to a wastewater to cause the formation of a solid phase or precipitate containing the pollutant to be removed. The pollutant may be precipitated directly, through the formation of a sparingly soluble compound, or if the pollutant exists as finely divided particulate matter, it may be entrapped by the formation of another precipitate. Although the latter involves the formation of a precipitate, it is usually referred to as coagulation, because the principal objective is to collect the finely divided particles into larger particles to promote settling. In addition, coagulation includes the aggregation of colloids by reducing the repulsive charges on the colloidal particles or overcoming the effects of these charges that otherwise keep the particles in suspension. Flocculation also refers to particle aggregation, but it is generally used to describe the agglomeration of coagulating chemical throughout the wastewater. Coagulation usually takes place rapidly, whereas flocculation requires a longer time interval to allow particle contact for agglomeration. Agglomeration is hastened by gentle stirring of the treated wastewater to increase the collision rate of the coagulated particles, but not sufficient to shear the floc apart. A theoretical discussion of coagulation and flocculation is available (Weber, 1972).

Precipitation of metal hydroxides by addition of alkaline reagents to wastewater is commonly used for removal of hazardous metal ions. This is frequently accomplished coincidentally with neutralization treatment of plating wastes containing Cr^{3+} or Cd^{2+}, for example. Addition of alkaline reagents to these acidic wastewaters increases the hydroxyl ion content to a level where the metal hydroxide concentration exceeds its solubility limit. The equilibrium expression for solubility of a sparingly soluble substance such as $Cd(OH)_2$ (Weber and Posselt, 1974) is given in terms of the solubility product K_{sp} as follows:

$$K_{sp} = [Cd^{2+}][OH^-]^2 = 2.2 \times 10^{-14}$$

The concentration of the ions enclosed by brackets is given in molarities, which for a pure solution of $Cd(OH)_2$ would yield 1.8×10^{-5} M Cd^{2+} (2.0 mg/l) and 3.6×10^{-5} M $(OH)^-$. The simple relationship expressed by this equation is useful for illustrating the effect of pH on the solubility of heavy-metal hydroxide; however, many metals deviate from this model because of their ability to form soluble hydroxide complexes at the higher pH levels. A series of species may be formed at elevated pH levels; this is depicted for a divalent metal cation as follows:

$$M_T = M^{2+} + M(OH)^+ + M(OH)_2^0 + M(OH)_3^- + M(OH)_4^{2-} \ldots$$

The sum of these species, M_T, defines total metal solubility. Therefore, the theoretical solubility of many metals becomes a complex function with a characteristic minimum, as illustrated in Figure 9-7.

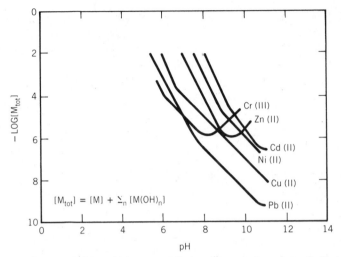

Figure 9-7 / The solubility of pure metal hydroxides as a function of pH. (Adapted from Nilsson, 1971.)

Metal removed by hydroxide precipitation may also be complicated by variances in ionic strength, temperature, the concentration of complexants, and the time required to attain equilibrium. These factors have frustrated attempts to match empirical data with theoretical expectations (Patterson and Minear, 1973).

Alkaline reagents commonly used for hydroxide precipitation of heavy metals are lime and sodium hydroxide, which are added to adjust the wastewater to an alkaline pH. The precipitate that forms is removed by settling, and the supernatant liquid may be further treated by granular-media filtration to remove small amounts of particulate matter that escaped the settling process. This treatment method is generally effective, but it has several problem areas, such as poor filterability of gelatinous hydroxide sludges, large volumes of sludge if lime is used on sulfate-containing wastewater, and high pH requirements (pH 10) for optimum precipitation of some metals, such as nickel, lead, and cadmium. High-pH treatment may require subsequent addition of acid for neutralization of the supernatant within acceptable limits for discharge to surface receiving water (Patterson et al., 1977).

The electroplating industry, which includes a large number of small shops across the country, has made extensive use of precipitation technology for removing hazardous metals from plating wastes. The "chemical destruct" waste treatment system consisting of cyanide destruction and hydroxide precipitation is widely used throughout this industry and serves as a basis for effluent standards promulgated by the EPA. Cyanide destruction is required not only to eliminate this hazardous component but also to free the heavy metals from the complexing action of cyanide. In spite of effective removal of cyanides from copper cyanide plating wastes, some complexing of the copper by other complexants (Yost and Maserich, 1977) appears to interfere with complete precipitation of the copper. This is readily evident by comparison of the pH–solubility curves given in Figures 9-8 and 9-9 for dilution of a pure copper sulfate solution and a copper cyanide waste that had been treated with hypochlorite to oxidize cyanide.

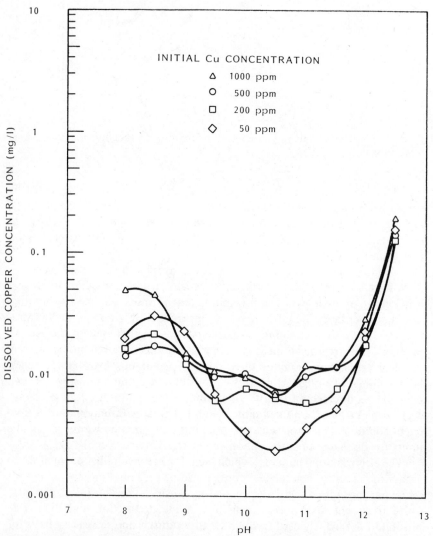

Figure 9–8 / Apparent solubility of copper versus solution pH in copper sulfate solution. (Adapted from Yost and Scarfi, 1979.)

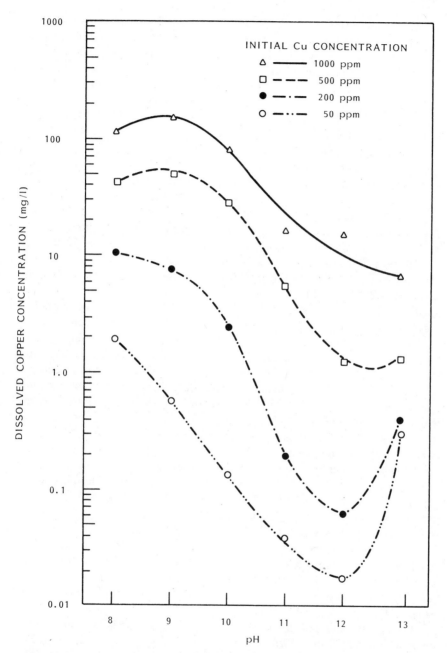

Figure 9-9 / Apparent solubility of copper versus solution pH in a copper cyanide plating waste dilution to 1000, 500, 200, and 50 mg/l copper. (Adopted from Yost and Scarfi, 1979.)

The dissolved copper concentrations for the pH-adjusted copper sulfate dilution in Figure 9–8 were essentially the same as would be expected. In the absence of complexation, initial copper concentrations have no effect on the final dissolved copper concentration. The pronounced concentration effect of the copper cyanide waste in Figure 9–9 indicates the presence of a complexant or complexants that may have been organic additives or carbonate (Yost and Scarfi, 1979). Pretreatment standards issued January 28, 1981, by EPA for existing sources of electroplating wastewater from common metal facilities discharging 38,000 l or more per day are given in Table 9–2. These data are indicative of the maxium concentrations of heavy metals that may be expected in precipitation process effluents from the treatment of electroplating wastewater.

The allowable copper concentration, for example, is more than 100 times the level reached for a pure copper sulfate solution shown in Figure 9–8, but less than the minimum attained for the copper cyanide wastes at 1000 mg/l Cu given in Figure 9–9. The pretreatment standards in Table 9–2 are below the limits issued under RCRA on May 19, 1980, for chromium, cadmium, and lead (see Table 3–3 for the extractant from the EP test described in these regulations.) However, the limits originally proposed were one-tenth the limits that were issued, and this would have caused some difficulty in attempting to exempt sludges of this type from the hazardous category. The interstitial liquid in the sludges would have exceeded the proposed limits without further leaching of metals from the solid phase of the sludges. Dynamic leaching tests on metal finishing waste sludges show that once the interstitial fluid is replaced with deionized water, the metal concentrations are substantially reduced in the leachates (Caulter, 1981). The results of these tests suggest that heavy-metal sludges can be safely disposed in landfills where organic wastes are excluded.

Coagulants such as alum, iron salts, and lime are effective for removing soluble heavy metals as well as heavy-metal particulates. The particulate forms of the metals are removed by entrapment in the flocs formed by the coagulants, and the soluble forms are removed by coprecipitation or by adsorption on the flocs. The results of alum coagulation experiments with tapwater containing heavy metals are presented in Table 9–3 to illustrate the effectiveness of this method for heavy-metal removal (Nilsson, 1971). Only nickel is not removed. Arsenic does not precipitate as a hydroxide, but the arsenate ion is known to coprecipitate with hydroxides with heavy metals such as Fe (Skripach et al., 1971).

Ferric hydroxide precipitation was reported to remove 82% of the arsenic present (influent 20 mg/l ferric chloride) (Shen, 1973). This was improved to 92% removal when the ferric chloride dosage was increased to 30 mg/l. The best results were obtained by first oxidizing the water with 20 mg/l chlorine and then coagulating with 50 mg/l ferric chloride. With this approach,

Table 9-2 / Pretreatment Standards for Existing Sources: Common Metals Facilities Discharging 38,000 l or More per Day.[a]

Pollutant or Pollutant Property	Maximum for Any 1 Day (mg/l)	Average of Daily Values for 4 Consecutive Monitoring Days Shall Not Exceed (mg/l)
CN, (cyanide-total)	1.9	1.0
Cu	4.5	2.7
Ni	4.1	2.6
Cr	7.0	4.0
Zn	4.2	2.6
Pb	0.6	0.4
Cd	1.2	0.7
Total metals	10.5	6.8

[a]From *Federal Register* (1981).

arsenic removal reached 98.7%. Lime coagulation was found to reduce the arsenic levels by about 20%. Other metals such as zinc, nickel, manganese, and mercury are also removed by ferric hydroxide precipitation (Kolthoff and Overholser, 1939; Ebersale, 1972).

Precipitation of heavy metals as sulfides is an alternative to metal hydroxide precipitation. The solubilities of heavy-metal sulfides are generally much less than the solubilities for heavy-metal hydroxides, as illustrated by the solubility curves in Figure 9–10. Operational difficulties in the past have prevented widespread use of sulfide precipitation for heavy-metal removal from wastewater, but recent developments have overcome these problems to a large extent. Addition of excessive soluble sulfide reagents such as sodium sulfide or sodium hydrosulfide can now be more carefully controlled by selective ion electrodes to minimize odor and toxicity problems associated with the soluble sulfide precipitation process. In addition, an insoluble sulfide precipitation process has been developed that supplies sulfide ion by addition of sparingly soluble ferrous sulfide to the wastewater (U.S. EPA, 1980). The

Table 9-3 / Heavy-Metal Removal from Tapwater by Alum Coagulation at pH 6.5–7.0.[a]

Metal	Heavy-Metal Concentrations (mg/l)	
	Added before Treatment	Present after Treatment
Arsenic(V)	21	1.2
Copper(II)	15	1.7
Chromium(III)	15	0.2
Lead(II)	17	1.3
Nickel(II)	16	17.0
Zinc(II)	17	0.3

[a]Data from Nilsson (1971).

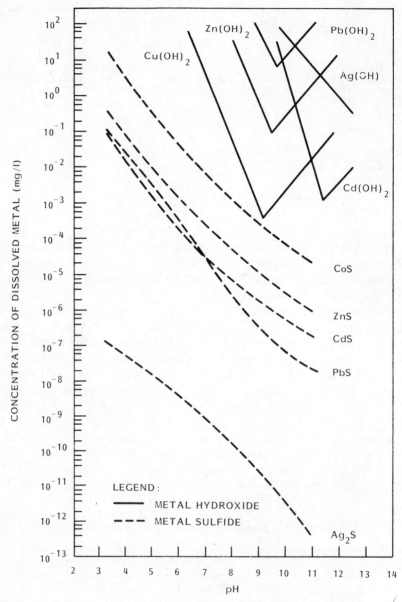

Figure 9–10 / Metal solubility as a function of pH (From U.S. EPA, 1980.)

ferrous sulfide dissociates into ferrous and sulfide ions, and because of the very low solubility of most heavy-metal sulfides, the sulfide ions are consumed by precipitation. Additional sulfide is released to maintain the equilibrium sulfide concentration. When all the heavy metals are precipitated, only a small excess of soluble sulfide remains in solution.

In addition to removal of toxic pollutants from wastewaters, coagulation/flocculation serves the needed function of removing suspended matter from feed streams to processes such as fixed-bed ion exchange and activated carbon adsorption columns that become inoperative due to plugging by the solids. Coagulation and flocculation are usually carried out in two steps, as illustrated in Figure 9-11. Coagulants such as alum, iron salts, lime, and polymers are added to the rapid-mix tank, where these materials are quickly dispersed to precipitate and collect the fine particulate matter into larger particles. Effluent from the rapid-mix tank flows to the flocculation tank, where gentle agitation allows the small coagulant particles to grow into larger particles for enhanced settling. The effluent from the flocculation tank is usually directed either to a filter, if the solids content is low, or to a sedimentation tank or basin, if the solids content is high.

Oxidation–Reduction.   Oxidation–reduction or "redox" processes are used for converting toxic pollutants to harmless or less toxic materials or to other forms that are more easily removed. These processes involve the addition of chemical reagents to wastewaters, causing changes in the oxidation states of substances both in the reagents and in the wastewaters. In order for one substance to be oxidized, another must be reduced. Oxidation occurs when the oxidation state of a substance is increased (e.g., $SO_3^{2-} \rightarrow SO_4^{2-}$), and

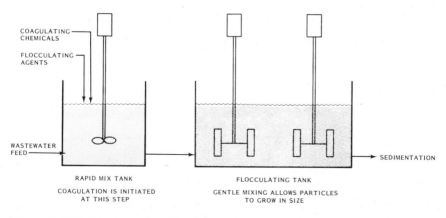

Figure 9-11 / Diagram of rapid mix and flocculating tanks for coagulation and flocculation.

reduction occurs when the oxidation state of a substance is reduced (e.g., $Cr_2O_7^{2-} \rightarrow Cr^{3+}$). Oxidation may be defined as a loss of electrons (e.g., $S^{4+} \rightarrow S^{6+}$) and reduction as a gain in electrons (e.g., $Cr^{6+} \rightarrow Cr^{3+}$). This latter definition is not directly applicable to reactions with organic substances. In the case of organic reactions, the earlier concept of oxidation, i.e., addition of oxygen and removal of hydrogen, is useful for explaining the reaction mechanism in most waste treatment applications.

The oxidation of cyanide and reduction of Cr^{6+} are widely used throughout the metal finishing industry for treatment of wastewater. Cyanide is typically oxidized with alkaline chlorine or hypochlorite solutions according to the following reactions:

$$CN^- + Cl_2 + 2OH^- = CNO^- + H_2O + 2Cl^-$$

or

$$CN^- + OCl^- = CNO^- + Cl^-$$

which on further addition of alkali and chlorine or hypochlorite proceeds to

$$2CNO^- + 3Cl_2 + 8OH^- = 2CO_3^{2-} + 6Cl^- + N_2 + 4H_2O$$

or

$$2CNO^- + 3OCl^- + H_2O = 2HCO_3^- + N_2 + 3Cl^-$$

The initial oxidation of cyanide to cyanate is rapid, requiring only a few minutes to complete, whereas the oxidation to carbonates and nitrogen gas is fairly slow and may require several hours for completion. Chemical usage for complete oxidation is approximately 7 lb of chlorine and 9 lb of sodium hydroxide per pound of cyanide.

Reduction of Cr^{6+} to Cr^{3+} is commonly performed with sulfur dioxide in large electroplating operations, but sulfite salts may be used, particularly for small operations. Reductions are carried out under acidic conditions (pH range 2 to 3) wherein the following reactions occur:

$$SO_2 + H_2O = H_2SO_3$$

$$2CrO_3 + 3H_2SO_3 = Cr_2(SO_4)_3 + 3H_2O$$

The approximate chemical usage is 1 lb of SO_2 per 1 lb of chromic acid (CrO_3).

The equipment used for redox reactions such as the foregoing is generally quite simple. Equipment for the process includes storage vessels for the reactants, metering devices for the feed stream and reactants, and a reaction vessel equipped with an agitator. Instrumentation may include pH and redox-potential monitoring systems.

Although oxidation of cyanide with chlorine and reduction of Cr^{6+} with sulfur dioxide or sulfite salts are the most widely used redox processes, there are a number of other reactants that are or have been used to detoxify these waste constituents (Howe, 1963; Battelle Memorial Institute, 1968; Patterson, 1975). Oxidation of cyanide has been accomplished in commercial waste treatment with hydrogen peroxide (*Products Finishing*, 1970), ozone (Arthur D. Little, 1976), and electrolysis (Easton, 1967). Electrolysis is particularly well adapted for treatment of concentrated cyanide wastes (Patterson, 1975).

Reduction of Cr^{6+} can be carried out with ferrous sulfate, but this is infrequently used in large operations because of the excessive amount of sludge produced when the added iron salts are precipitated during neutralization of the effluent.

Oxidation of toxic organic wastes with ozone is being used for treating certain types of industrial wastewaters. Oxidation with ozone has been effective for eliminating final traces of phenol from waste effluents (Arthur D. Little, 1976) and for treating concentrated cyanide wastes (Streebin et al., 1980).

Ozone is a powerful oxidizing agent that must be used immediately on being generated, because of its instability. Ozone is a gas formed by passing oxygen or air through an electrical discharge produced between two highly charged electrodes. The process for producing ozone suffers from low efficiency (only 10% of the electrical energy is used in converting O_2 to O_3), and only a small percentage of the oxygen is converted to ozone. Furthermore, the ozone is difficult to diffuse into a waste solution because of its low solubility. Nevertheless, the very high oxidizing power of ozone makes it a unique and useful oxidizing agent. It does not leave an undesirable residue when complete oxidation of organic compounds containing C, H, and O is achieved (water and CO_2 are the products).

As a general rule, ozone is used to eliminate traces of organics. The major cost item for ozonation is generating the ozone, which if used in large quantities makes the process very expensive.

Wet oxidation is a relatively new technology for treating hazardous wastes. This process has been reported to remove more than 99% of the cyanide during oxidation of acrylonitrile wastes (Wilhelmi and Knopp, 1979). Wet oxidation utilizes oxygen (pure or in air) at temperatures and pressures up to 350° C and 180 atm to treat organic wastes. The cost of oxygen is not a major factor in wet oxidation, particularly if air is used, and high concentrations of oxidizable organics in the waste can eliminate or reduce the energy requirement, which can be a major cost item in this process.

Numerous oxidizing and reducing agents are potentially applicable to waste treatment operations employing redox processes, but only relatively few have been used to an appreciable extent. Table 9–4 lists some of the common oxidants and types of water treated in industrial applications. Table 9–5 lists some of the common reductants used in industrial applications.

Ion Exchange. Ion exchange is sometimes viewed as a physical process (i.e., adsorption), but because it involves a change in chemical form (e.g., soluble salt to insoluble resin) it is considered more appropriate to classify it as a chemical process. This process involves the exchange of ions in solution with other ions held by fixed anionic or cationic groups or charges on insoluble polymers or minerals (e.g., zeolites). Typically, a waste solution is percolated through a granular bed of the ion exchanger, where certain ions in solution are replaced by ions contained in the ion exchanger. If the exchange involves cations, the exchanger is called a cation exchanger, and correspondingly the exchanger is called an anion exchanger, where anions are exchanged.

When a bed of ion-exchange material has been largely converted from one ionic form to another, it loses its ability to remove ions from the waste solution and becomes "exhausted." The ion-exchange bed is then "regenerated" with a solution of electrolyte that replaces ions removed from the waste solution with the desired ion originally on the exchanger. The regeneration cycle usually produces a more concentrated solution of the ions removed from the wastewater, which is an advantage if the recovered ions can be reused, but a disadvantage if they must be disposed of.

Industrial waste treatment applications have been rather limited except in the nuclear energy field, where this process is widely used for treating low-level radioactive wastewaters at nuclear power stations and fuel reprocessing facilities (Emility, 1967). When properly designed and operated, ion-exchange processes can achieve very high decontamination factors (D.F. = influent concentration/effluent concentration), which is especially useful for achieving the very low radionuclide concentrations required for release of effluents to the environment. Regeneration of the ion exchanger produces a concentrated radioactive regenerant waste that is typically evaporated and

Table 9–4 / Oxidants Used in Industrial Waste Treatment.

Oxidant	Waste	References
Chlorine or hypochlorite	Cyanide	Arthur D. Little, 1976; Patterson, 1975
Hydrogen peroxide	Cyanide	Patterson, 1975; *Products Finishing*, 1970
Nitrous acid	Benzidine	Arthur D. Little, 1976
Ozone	Phenol, cyanide	Arthur D. Little, 1976; Patterson, 1975; Streebin et al., 1980
Wet oxidation	Acrylonitrile, cyanide	Wilhelmi and Knapp, 1980
Electrolysis	Cyanide	Patterson, 1975; Easton, 1967

Table 9-5 / Reductants Used in Industrial Waste Treatment.

Reductant	Waste	References
SO_2, sulfites	Cr^{6+}	Arthur D. Little, 1976; Patterson, 1975
Ferrous sulfate	Cr^{6+}	Arthur D. Little, 1976; Patterson, 1975
Sodium borohydride	Tetraethyl lead	Arthur D. Little, 1976; Lores and Moore, 1974
Scrap iron	Copper	Patterson, 1975; Dean et al., 1972

fixed in the form of concrete for disposal. The exhausted resin may be discarded without generation. Inorganic ion exchangers such as aluminosilicate zeolites have been found effective for removing radioactive cesium from high-salt wastes. These exchangers are highly specific for the cesium ion and have also been used for removing ammonium ion from municipal effluents (Mercer and Ames, 1978).

Ion exchange is used in the electroplating industry to recover nickel and chromate from rinse waters (Arthur D. Little, 1976; Patterson, 1975; Gold, 1978). Ion exchange effectively reconcentrates the dilute nickel ion for recycle to the plating bath. The nickel is loaded into a cation resin bed from the waste rinse water, which is further treated by anion exchange to remove sulfate, chloride, and borate impurities. The treated rinse water can then be recycled. The cation resin bed loaded with nickel is regenerated with sulfuric acid, which produces a nickel sulfate solution suitable for recycle to the plating bath (Gold, 1978).

Biological Treatment

General Considerations.
Biological treatment is the workhorse of the municipal and industrial wastewater treatment field. Biological treatment is generally used for the removal of organic pollutants from wastewater, and its great popularity for serving this vital function results largely from its low cost and its effectiveness for removing many soluble and colloidal organics. Application of biological treatment to wastewaters containing toxic organic matter requires considerably more care than in the case of nontoxic wastes. The microorganisms used in biological treatment can easily be destroyed by shock loading or rapid increases in the amount of toxic material fed to the process. A considerable period of time may be required to reestablish an adequate population of microorganisms to treat the waste.

Many, but not all, toxic organic compounds are amenable to biological treatment. Very little information is available in the literature that is relevant to biological treatment of hazardous waste containing priority pollutants. Most of the information available concerns removal of priority pollutants from domestic waste at municipal treatment plants, and this information is not necessarily applicable to industrial wastewaters.

Biological treatment uses microorganisms, mainly bacteria that metabolize organic matter in wastewater to yield energy for synthesis, motility, and respiration. Biological utilization of organic compounds involves a series of enzyme-catalyzed reactions. Simple dissolved organic compounds are readily taken into the cells of microorganisms and oxidized. When microbial cells come into contact with complex insoluble organics, enzymes are released outside the cells to hydrolyze these materials (proteins and fats) into soluble fractions, enabling their transport through the cell wall for assimilation. Thus, the larger, more complex organic compounds are metabolized at a slower rate. There are some organic compounds that are not degraded by biological oxidation, and these are called "refractory" substances. Other compounds can be metabolized by the microorganisms at low concentrations but are toxic at high concentrations. A period of acclimation is frequently necessary in the case of toxic substances to allow the microorganisms to "adjust" to these materials. A different population of microorganisms may develop during the acclimation period for more effective treatment.

Relationships between metabolism, energy, and synthesis are important in understanding biological treatment systems. The primary product of metabolism is energy, and the chief use of this energy is for synthesis. Energy release and synthesis are coupled biochemical processes, where the maximum rate of synthesis occurs simultaneously with the maximum rate of energy yield (maximum rate of metabolism). The primary purpose of biological treatment is to convert soluble or colloidal organic substrates to CO_2, H_2O, and settleable matter that can be removed by sedimentation. In the case of hazardous or toxic substances, it is important to note that complete removal may not be achieved and that the metabolites from the process may be toxic in themselves. It is frequently necessary to depend on both biochemical conversion and dilution to achieve the desired reductions in concentrations of pollutants.

Efficient and successful biological oxidation of organic wastes requires a minimal quantity of nitrogen and phosphorus for the synthesis of new cells. In addition, trace quantities of several other elements such as potassium and calcium are required. These trace elements are usually present in natural waters in sufficient quantities to satisfy requirements for microbial metabolism. However, nitrogen and phosphorus are sometimes deficient in wastewater substrates, causing reductions in removal efficiencies of biological treatment systems. In such cases, nutrients must be added to supplement those in the wastewater substrate. Nitrogen should be added as a supplement in the form of ammoniacal nitrogen, because nitrogen as nitrite and nitrate is not readily available for microbial usage. Several soluble phosphorus salts that are readily assimilated by microorganisms are available. Generally, a BOD:N:P ratio of 100:5:1 is thought to be the optimum ratio of nutritional requirements for microorganisms utilized in biological waste treatment.

(BOD, or biochemical oxygen demand, is the term applied to signify the strength of organics in wastewater and is defined generally as the amount of oxygen required by microorganisms to biologically oxidize a given quantity of organics. The more concentrated the organic waste material, the higher the BOD.)

There are several different biological process configurations, including activated sludge, biological filters, aerated lagoons, oxidation ponds, land application systems, and anaerobic fermentation. Selection of a particular process is generally based on wastewater characteristics and volume, desired levels of pollutant removal, and location. Biological systems generally achieve 50% to 90% BOD removal, although higher removal can be attained under optimum conditions. Activated sludge, biological filters, and stabilization ponds are the most widely used biological treatment processes. These processes, along with land application, a widely used industrial process, are discussed next.

Activated Sludge Process. The activated sludge process involves the production of a suspended mass of microorganisms in a reactor to biologically degrade soluble and finely suspended organic compounds in wastewater to carbon dioxide, water, microorganisms, and energy. In operation of the activated sludge process, wastewater containing organic compounds is fed to the aerobic reactor (aeration tank), which furnishes (1) air required by microorganisms to biochemically oxidize the waste organics and (2) mixing to ensure intimate contact of microorganisms with the organic waste. The aerobic reactor contents are referred to as mixed liquor suspended solids (MLSS). In the vigorously mixed aerobic reactor, the organic wastes are metabolized to provide energy and growth factors for the production of more microorganisms, with the release of carbon dioxide and water as metabolic end products. The organic waste compounds are thus transformed into innocuous end products and microorganisms. The MLSS flow from the aeration tank to a sedimentation tank, which provides quiescent settling to allow separation of the biological solids from the treated wastewater. The treated and clarified water is collected and discharged as process effluent. Most of the settled biological solids are recycled as return activated sludge back to the aerobic reactor to provide an activated mass of microorganisms for continuous treatment of incoming wastewater. Some of the settled biological solids are wasted to maintain a proper balance in the population of microorganisms in the MLSS of the aerobic reactor.

The activated sludge process is very flexible and can be used for the treatment of almost any type of biodegradable waste. The original process configuration is called the conventional activated sludge process and has been modified in numerous ways. In the original conventional (or plug-flow) activated sludge process, wastewater with return activated sludge enters one end of a long narrow aeration tank and is mixed as it flows in a longitudinal direction down the length of the tank. The long rectangular aeration tanks are generally designed so that the total tank length is 5 to 50 times the width. Air is supplied via bubble-type diffusers, causing a spiral flow of the mixed liquor as it flows to the exit end of the tank. The spiraling flow down the length of the tank is a uniform straight-line flow pattern; hence the name "plug flow." Conventional and other activated sludge process variatons are discussed elsewhere (Rich, 1963; Metcalf and Eddy, 1979).

The complete-mix activated sludge system has become one of the more popular designs in recent years because of its greater ability to withstand shock loads and toxic substances. In the complete-mix system, influent wastewater is uniformly mixed throughout the entire aeration basin as rapidly as possible. The mixing tends to produce a uniform organic load through the entire content of the aeration basin. Figure 9–12 illustrates a typical flowsheet. Because the influent wastes are mixed throughout the aeration basin, the entire basin volume acts to buffer hydraulic surges and organic shock loads. For example, it has been shown that a phenol feed concentration of 100 mg/l is toxic to the conventional activated sludge process, whereas phenol at 2000 to 3000 mg/l is not toxic in the complete-mix system (McKinney and Ooten, 1969). This enables the establishment of equilibrium (or nearly so) conditions for stable operation.

Industrial application of the activated sludge process for removal of phenolic compounds from wastewater is well documented (Arthur D. Little, 1976). Phenol removals of 85% to 95% have been reported for petroleum refinery wastewater containing an average phenol concentration of 135 mg/l. Activated sludge is used by the steel industry for removing both phenol and cyanide. Examples of organic wastewaters that have been successfully treated by activated sludge to remove a wide variety of organic pollutants, including phenolic compounds, aromatic hydrocarbons, phthalic acid and esters, and acrylonitrile, are reported by Ottinger et al. (1973).

Trickling Filter Process.   The trickling filter process consists of a fixed bed of medium over which wastewater is intermittently or continuously distributed in a uniform manner by a flow distributor. Microorganisms grow on the surface of the filter medium, forming a biological or zoogloeal slime layer. Wastewater flows downward through the filter, passing over the slime layer containing the microorganisms. Organic material and nutrients in the wastewater are taken up by the slime layer for utilization by the microbial population. Oxidized end products are released to the liquid and collected in the underdrain system for discharge via the effluent channel. Aerobic conditions are maintained by air passing through the filter bed induced by the difference in specific weights of air on the inside and outside of the bed. A trickling filter will operate properly so long as the void spaces are not clogged by solids or

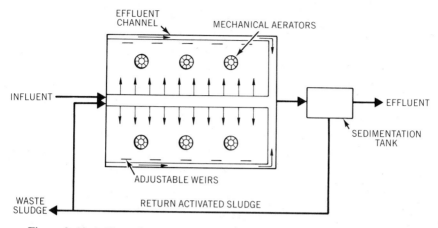

Figure 9-12 / Flow diagram of completely mixed activated sludge process.

excessive growth of the slime layer. The slime layer grows and gradually increases in thickness to the point that hydraulic shear force from the downward flow of wastewater causes portions of the film layer to slough off the filter medium. This solid matter is carried by the liquid stream to secondary clarification units, where it is separated as sludge.

The trickling filter process has the advantage of reliability over the activated sludge process. The reservoir of captive microorganisms that are readily adjustable to shock loadings are the basis of its dependability. The trickling filter achieves consistent BOD removals in the face of fluctuating hydraulic and organic demands. The plastic medium recently introduced results in a shortened detention time through the filter, but BOD removal is still limited to a maximum of 85%. A recently introduced innovation employs a design for recirculating biofloc from the system back through the filter, achieving high BOD/COD removal efficiency. The recirculated trickling filter is simlar to the activated sludge process and attains the same high BOD removal. Additional information concerning the design and operation of trickling filters is available (Rich, 1963; Metcalf and Eddy, 1979).

Trickling filters have been reported to be successfully applied to wastewaters containing acrolein, benzene, chlorinated hydrocarbons, and phenol (Arthur D. Little, 1976; Ottinger et al., 1973).

Stabilization Ponds.   Wastewaters may be effectively stabilized by the natural biological processes occurring in relatively shallow ponds. Stabilization is accomplished by the photosynthetic processes of algae and/or the oxidative processes of bacteria. Waste stabilization ponds (or lagoons, as they are sometimes called) are very popular for small operations in remote areas because their low construction and operating costs offer a significant financial advantage over other recognized treatment methods.

Waste stabilization ponds are generally classified according to the nature of the biological activity and environment within the pond. Thus, stabilization ponds are classified as aerobic-anaerobic (or facultative) and anaerobic. A waste stabilization pond system may include a single pond or a number of ponds in series or parallel. Also, the differently classified ponds may be utilized in series, i.e., aerobic followed by anaerobic, or vice versa. This is usually done to effect greater treatment efficiencies than can be achieved via a single pond type.

Aerobic ponds are generally separated into two categories, based on whether natural or artificial methods of supplying oxygen to the bacteria in the pond are utilized. Natural aeration supplies oxygen by natural surface aeration and by algal photosynthesis, and these ponds are generally termed "oxidation ponds." Mechanical aeration units can be used to artificially supply oxygen to the bacteria, and the process is essentially the same as the activated sludge process, without recycle of microorganisms. Mechanically aerated ponds are generally termed "aerated lagoons."

"Oxidation ponds" utilize algae and bacteria in a symbiotic relationship to stabilize waste organics. The oxygen released by the algae through the process of photosynthesis is utilized by bacteria in the aerobic degradation of organic matter. The nutrients and carbon dioxide released via respiration are, in turn, used by the algae. During the daylight hours of increased algal photosynthetic activity, it is possible for oxygen concentrations to reach supersaturation levels. Generally, solids will accumulate and settle in an oxidation pond because of the lack of mixing. These settled solids accumulate, forming an anaereobic sludge layer on the bottom. Oxidation ponds generally are relatively shallow, 3 to 5 ft in depth (McKinney, 1971).

Aerated lagoons are an outgrowth of the development of the completely mixed activated sludge process in that surface mechanical aerators are applied to overloaded oxidation ponds. Aerated lagoons are typically 8 to 15 ft in depth. Generally, no consideration is given to algae for supplying dissolved oxygen, because the turbulent pond surface inhibits the growth of algae.

Aerobic-anaerobic, or facultative, ponds were historically known as stabilization ponds. The symbiotic algae-bacteria relationship is utilized to its fullest in these ponds. The ponds are generally 3 to 8 ft in depth, and solids settle to the bottom to eventually decompose. This decomposition is anaerobic and results in the interchange of anaerobic decomposition by-products with aerobic oxidation by-products between the upper and lower portion of the pond.

Anaerobic ponds were the inevitable result of the widespread use of "stabilization" ponds (facultative) where the organic loading rates became excessive, causing anaerobic conditions throughout the pond. The symbiotic stabilization relationship failed, but was substituted for by an anaerobic stabilization process in which waste organics were stabilized for anaerobic methane-forming bacteria, in similarity to what occurs in anaerobic digesters.

Information concerning the use of waste stabilization ponds is rather limited in the published literature. Decomposition of phenolic compounds has been reported (Arthur D. Little, 1976). It is the authors' opinion that many so-called industrial waste stabilization ponds are actually just holding ponds, because the waste contained in these ponds is too toxic for biological activity to take place.

Anaerobic Digestion.   Biological decomposition of wastes in an enclosed vessel in the absence of molecular oxygen is widely used for stabilization of sewage sludges. Application to concentrated hazardous wastes is limited because of the generally high toxicity of these wastes.

Land Application.   Land application as a treatment and disposal method utilizes the interaction between plants and the soil surface to effectively stabilize many different types of wastes. The combination of plant and soil serves as a natural biological filter (Spyridakis and Welch, 1976), because most topsoils already contain the microorganisms needed for biochemical decomposition of organic matter. In addition, physical and chemical processes can occur within the soil to neutralize either strong acids or bases, remove inorganic constituents, and filter out suspended solids. Passage of the federal Water Pollution Control Act (PL 92-500) has focused attention on land application as an alternative for effective treatment and disposal of wastewaters and sludges to comply with zero-discharge requirements originally slated for 1985.

General criteria for judging the suitability of land disposal for a particular waste are as follows (Wallace, 1976):

1. The organic material must be biologically degradable at reasonable rates.
2. It must not contain materials in concentrations toxic to soil microorganisms. Because some toxic materials may accumulate through adsorption or ion exchange and approach toxic levels after prolonged operation, there must be reasonable assurance that this effect can be either prevented or mitigated.
3. It must not contain substances that will adversely affect the quality of the underlying groundwater. In many instances, decisions relative to this aspect of land disposal systems are difficult because of the uncertain nature of available estimating techniques.
4. It must not contain substances that cause deleterious changes to the soil structure, especially its infiltration, percolation, and aeration characteristics. An imbalance of sodium is a common problem, because sodium-form clays tend to swell and plug the soil.

Land treatment of refinery sludges dating back more than 20 yr has been reported (Synder, 1976; Matthews et al., 1981). Land treatment has not been used extensively for wastes other than petroleum refining wastes, but investigations are under way to identify other hazardous wastes that can be treated and disposed of in this manner (Berkovitch et al., 1981).

Separation of Dissolved Pollutants by Physical Methods

General Considerations.   The principal physical separation processes used for separating dissolved pollutants from wastewater are activated carbon adsorption, evaporation, reverse osmosis, solvent extraction, and steam stripping. There are several other physical separation processes that are potentially applicable to treatment of hazardous wastewaters (e.g., freeze crystallization, electrolysis, dialysis, and flotation); however, for various reasons these processes have not been applied to any significant extent in full-scale wastewater treatment plants (Arthur D. Little, 1976).

The composition of the wastewater is a very important consideration in selecting a physical separation process. As a general rule, activated carbon is economically feasible for dilute wastes only because rapid exhaustion of the carbon requires frequent replacement of the carbon, which is quite costly. The cost of activated carbon adsorption is usually directly related to the concentration of the pollutant to be removed from the wastewater, whereas the costs of evaporation, solvent extraction, reverse osmosis, and steam stripping are not as dependent on the pollutant concentration.

Carbon adsorption with thermal regeneration is the only process being addressed in this section that involves destruction of the separated pollutants. Additional methods are required for other processes if disposal of the pollutants is required. Reverse osmosis will generally produce the largest volume of concentrated wastewater for disposal or recycle.

The chemical and physical characteristics of the pollutants to be removed are also important to the selection of a physical removal method. Steam stripping is effective only for substances that have an appreciable vapor pressure at the boiling point of water, whereas evaporation is effective only for substances that do not have an appreciable vapor pressure at the boiling point of water. Highly soluble, small organic molecules usually are not adsorbed well by activated carbon. Large ions (hydrated) and organic molecules are separated more effectively by reverse osmosis than small ions and molecules.

Activated Carbon Adsorption.   The removal of organic and inorganic substances from water by activated carbon is largely accomplished through the process of adsorption, which is the accumulation of one substance on the surface of another (Culp et al., 1978; Weber, 1972). A large surface area per unit weight and a suitable pore structure to permit access to the surface are therefore very important factors in determining the effectiveness of an adsorbent. The chemical nature of the surface of an adsorbent is much less significant than the amount of surface area. The surface area of commercial activated carbon typically ranges from 580 to 1400 m^2/g. Variations in the behavior of commercially activated carbons are related to the amount of surface area, the pore structure, and the chemical nature.

The solubility of an organic compound in water is often a useful guide in determining how well the compound will be adsorbed by activated carbon. Insoluble (hydrophobic) organic compounds usually are good candidates for carbon adsorption. Similarly, adsorption will be high for compounds with a high octanol–water partition coefficient. The adsorption isotherm is the standard method used in obtaining quantitative data on the adsorption characteristics of an activated carbon. An adsorption isotherm relates the quantity of adsorbate (substances adsorbed) per unit weight of activated carbon to the concentration of the adsorbate in solution in contact with the carbon at equilibrium under a given set of conditions (e.g., temperature). A logarithmic plot of the isotherm data for a single adsorbate will usually yield a straight line. The empirical Freundlich equation is a useful formula for relating the amount of adsorbate in solution with that adsorbed by the carbon:

$$x/m = kC^{1/n}$$

where x is the amount of adsorbate adsorbed, m is the weight of activated carbon, C is the concentration of adsorbate in solution, and k and n are constants. In logarithmic form:

$$\log x/m = (\log k) + (1/n)(\log C)$$

where $1/n$ is the slope of a straight-line isotherm and k is the intercept at $C = 1$.

Adsorption isotherms are extremely valuable in a preliminary evaluation of the feasibility of using activated carbon for removal of a particular substance or of several substances from a waste solution. The EPA has published adsorption isotherms for 60 different toxic organic compounds (Dobbs and Cohen, 1978). These data are useful for comparison of pure compounds; however, wastewaters usually are mixtures, which complicates the evaluation of isotherm data. The individual compounds in a mixture are in competition for the available surface area, and the compounds with the strongest attraction will be favored at the expense of the compounds with weaker attractive forces.

Inspection of an adsorption isotherm will reveal whether or not a particular degree of removal can be achieved and will also show the approximate adsorptive capacity of the carbon. Adsorption isotherms are useful for determining the effects of pH and temperature on these systems. Comparison of different carbons under the same conditions is frequently performed to facilitate the selection of a particular carbon for treatment of a wastewater.

Adsorption isotherms for four different activated carbons are illustrated in Figure 9–13 to provide examples for interpretation. The high level and slight slope of carbon A reveals a high capacity over the entire range of

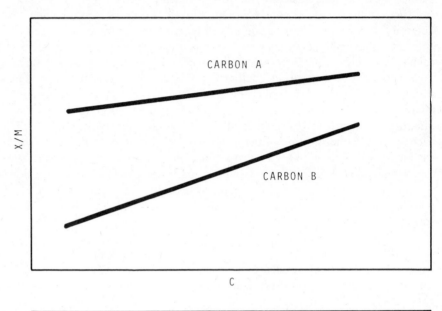

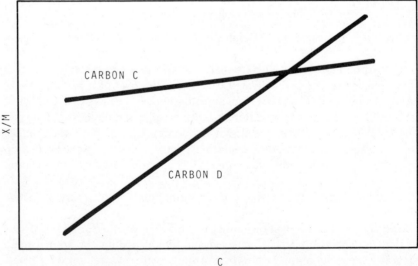

Figure 9-13 / Example absorption isotherms.

concentrations studied. Carbon B has essentially the same slope but proportionately less capacity; therefore with all other considerations being equal, Carbon A will be the choice for either batch or column operations. Carbon C will have a superior capacity in batch processes up to the point where the carbon D isotherm crosses over the carbon C isotherm. Carbon D will be better than carbon C for countercurrent column operation because of the higher capacity at the influent concentration. Column operation is generally favored by a steep isotherm slope.

Activated carbon is applied in a powdered form in batch contact processes and in granular form for column operation. Because adsorption involves diffusion of the adsorbate from the solution to the carbon particle and then inside the particle, consideration must be given to contact time. Commercial powdered carbons are very fine; therefore, equilibrium is attained rapidly, and most of the capacity can be utilized within a few minutes if adequate mixing occurs. Equilibrium is attained much more slowly with granular carbons exceeding 1 mm in particle diameter. The rate of adsorption increases with decreasing particle size; consequently the smaller mesh sizes are favored in this respect, but they suffer the disadvantage of high pressure differentials (and high pumping costs) when used in column operations. A compromise between adsorption rate and pressure differential is reached for most applications.

Column operation may be accomplished with either fixed beds or continuous columns. Fixed beds are generally used in small operations where the carbon is not regenerated on site. The carbon either is discarded when it becomes exhausted or is returned to the vendor for regeneration. Continuous column operation frequently employs on-site regeneration, as illustrated in Figure 9–14. The wastewater influent flows countercurrent to the carbon in the column; therefore the effluent has contacted the freshest carbon to attain a low concentration of adsorbate. The nearly exhausted carbon first contacts the raw influent, which maximizes the loading of adsorbates.

Exhausted carbon is withdrawn from the bottom of the column, transferred to a regeneration furnace, and then returned to the top of the column. The rate at which the carbon is regenerated and returned to the column is governed by the quality of effluent desired. Higher-quality effluent requires a higher regeneration rate. In a true continuous column operation, the effluent quality is constant, as opposed to a fixed-bed operation, where the effluent quality is high at the beginning but deteriorates as the column becomes exhausted. Pilot-plant testing is recommended for most column wastewater treatment applications to establish feasibility and costs and to optimize operating conditions.

Activated carbon adsorption is widely used in industrial applications for removing hazardous chemicals from wastewaters. A list of full-scale treatment systems is given by Lyman (1978), including both disposal and recovery applications.

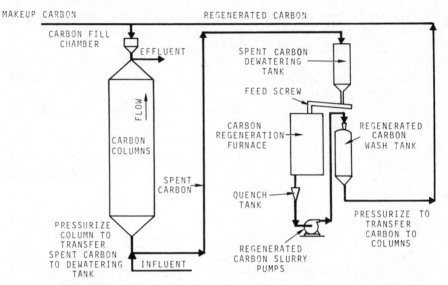

Figure 9-14 / Simplified flowsheet of a granular activated carbon column with a thermal regeneration system. (Adapted from Shuliger, 1978.)

Evaporation.   Evaporation is a process that employs the transfer of heat to a liquid, causing the vaporization of volatile components from the liquid. Application of this process to wastewater is usually directed at volatilizing water to separate it from nonvolatile pollutants. These pollutants are thus concentrated to produce a thick liquor or slurry that may facilitate recovery of valuable materials or disposal. This process is similar to distillation, except that no attempt is made to separate components of the vapor. Transfer of heat in older single-effect evaporators is generally accomplished by circulating steam through coils immersed in the liquid or through a jacket surrounding the liquid to cause the liquid to boil. The vapors may be either vented to the atmosphere or condensed to recover the wastewater. Single-effect evaporators are costly to operate where conventional fuels must be purchased to supply heat. The cost drops dramatically if a source of waste heat or steam is available. The cost of energy is substantially reduced in modern evaporators through the use of vapor recompression systems and multiple-effect evaporators.

Schematic diagrams of a vapor recompression evaporator and a three-effect evaporator are given in Figures 9-15 and 9-16, respectively. A vapor recompression evaporator operates much like a heat pump in that work is performed by a compressor to transfer heat from one source to another—in this case, from the vapor back to the liquid. The heat transfer is accomplished by compressing the vapor to elevate its temperature sufficiently to transfer the

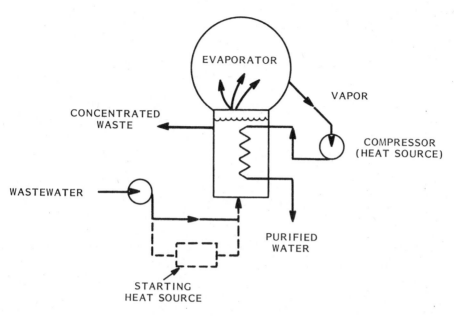

Figure 9-15 / Schematic diagram of a vapor recompression evaporator.

heat back to the evaporator through condenser coils or plates. The work required to effect the heat transfer is a small fraction of the energy required to evaporate water.

A multiple-effect evaporator, as in Figure 9–16, utilizes the energy for evaporation several times before it is wasted by operating each effect at a lower pressure and temperature than the previous effect. Steam supplies energy for the first effect, and condensing wastewater vapor supplies energy for the succeeding effects. Energy economy is gained by increasing the number of effects, but this also increases equipment costs to a point where additional effects offer no advantage.

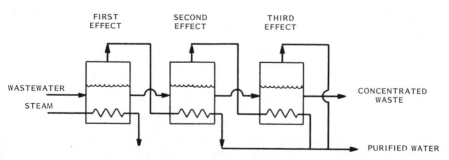

Figure 9-16 / Schematic diagram of a three-effect evaporator.

Corrosion, scaling, and heat transfer efficiency are important considerations in the design and operation of evaporators. Corrosion can be minimized through proper selection of materials that resist corrosion and through control of the chemistry of the wastewater (e.g., pH). Chemical control can also be important to avoid scaling. Scaling may be controlled through proper design of the heat transfer surfaces and by the seeding of crystals in the evaporating liquid. For example, maintaining a good flow of seeded liquid or slurry over the heat transfer surface (e.g., falling film) can avoid severe scale formation in many instances and permit high concentration factors, i.e., feed volume/concentrate volume.

When high decontamination factors are desired for the condensate, as in the case of radioactive wastewater evaporation, removal of entrained droplets of liquid or particulate matter from the vapor must be effected. This is accomplished through the use of de-entrainment devices, such as wire-mesh pads that remove the entrained matter from the vapor before the vapor is condensed.

Evaporation is a well-established process in the chemical manufacturing industry (Arthur D. Little, 1976). It is widely used for treatment of radioactive wastewaters generated by nuclear power plants and nuclear fuel reprocessing facilities (Godbee and Kibbey, 1975).

Recovery of cadmium and chromate from electroplating rinse water has been accomplished through the use of evaporation (Patterson, 1975). The treatment of concentrated solutions of pollutants by evaporation is generally favorable, because alternative processes such as ion exchange, carbon adsorption, and reverse osmosis are not applicable to this type of waste when it is desired to further concentrate it to a minimal volume.

Reverse Osmosis.   Osmosis may be defined as the spontaneous movement of a solvent from a dilute solution to a more concentrated solution across an ideal semipermeable membrane. This membrane allows the solvent (e.g., water) to pass, but not the dissolved solids. Flow occurs spontaneously from the dilute solution through the membrane to the concentrated solution. The head that develops in the concentrated solution is defined as the osmotic pressure (about 1 psi per 100 ppm TDS).

If a pressure exceeding the osmotic pressure is applied to the concentrated solution, the flow through the membrane will be reversed (Figure 9–17). Pure water will then flow through the membrane from the concentrated solution. The process may be run continuously by providing a mechanism for carrying away the remaining concentrate.

The magnitude of the osmotic pressure depends on the concentration of the salt solution and its temperature. By exerting pressure to the salt solution, the osmosis process can be reversed. When the pressure on the salt solution is greater than the osmotic pressure, fresh water is diffused through the mem-

OSMOSIS

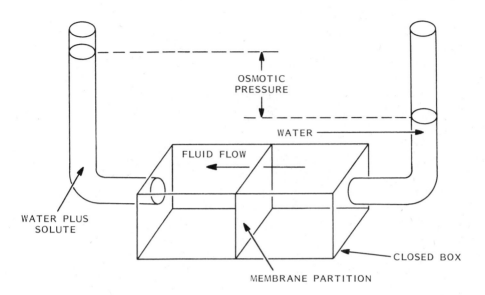

REVERSE OSMOSIS

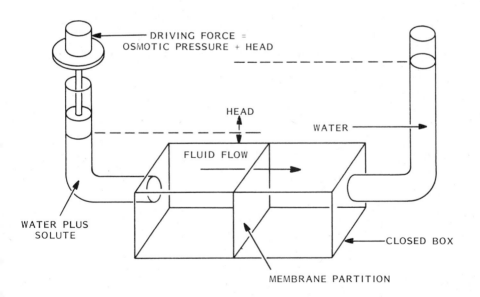

Figure 9–17 / Osmosis and reverse osmosis.

brane in the direction opposite to the normal osmotic flow—hence reverse osmosis.

Reverse osmosis (RO) membranes are prepared to have a very thin and very dense outer "skin" backed by a thicker and much more porous layer. The "skin" is in contact with the brine and is the principal zone of separation of salts from the water. Because of its density, it provides a barrier to the passage of dissolved salts and other impurities. The porous layer allows the diffusion of water through the fresh-water side of the membrane. Figure 9–18 shows a schematic representation of this structure for a hollow fine fiber membrane configuration.

Because there is a large amount of pressure transmitted to the membrane, a substantial mechanical support is necessary to hold the fragile membrane and prevent damage to it. This support must also be designed for compact packaging to accommodate its installation into piping systems. Several schemes have been explored for this membrane support configuration. The four principal types of membranes are (1) spiral wound, (2) hollow fine fiber, (3) tubular, and (4) plate and frame module. The first three types are in commercial production and are currently in use in operating plants.

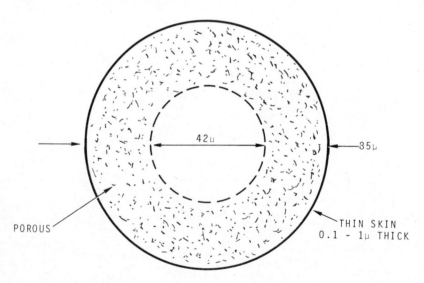

```
1μ = 1 MICRON = 0.00004 INCH
              = 4/100,000 INCH

42μ = 0.0016 INCH

85μ = 0.0033 INCH
```

Figure 9–18 / Hollow fine fiber: Asymmetric aromatic polyamide.

Spiral-wound RO Membranes. A sketch of the spiral-wound membrane and a module assembly that is a popular design for wastewater treatment are shown in Figure 9-19. The spiral-wound RO module was developed by Gulf Environmental Systems Company under contract to the U.S. Office of Saline Water. It was conceived as a method of obtaining a relatively high ratio of membrane area to pressure vessel volume. The membrane is supported on both sides of a backing material and sealed with glue on three of the four edges of the laminate. The laminate is also sealed to a central tube, which is drilled. The membrane surfaces are separated by a screen material that acts as a brine spacer. The entire package is then rolled into a spiral configuration and wrapped in a cylindrical form with tape for an outer wrap. Feed flow is parallel to the central tube. A reverse-osmosis flowsheet showing the position of the RO modules is presented in Figure 9-20.

Important operating parameters for a reverse-osmosis unit include temperature, pressure, and pH. Ideal operating pH ranges from 4 to 6 for cellulose acetate, the most commonly used membrane material. Increasing the temperature and pressure increases the water flow through the membrane. Maintaining high fluid flow at right angles across the membrane surface is also important to avoid the phenomenon of "concentration polarization," which occurs as water is removed from the salt solution, resulting in a layer of highly concentrated brine at the surface of the membrane. High fluid shear, which mixes the brine layer, can be accomplished by flowing the wastewater at high velocities in thin channels to promote laminar shear or wide channels to cause turbulence.

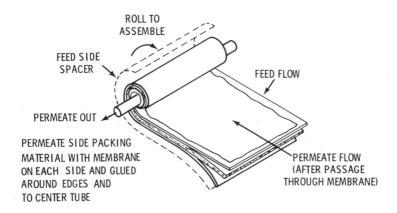

SPIRAL-WOUND REVERSE OSMOSIS MODULE

Figure 9-19 / Spiral-wound RO module. (From Office of Saline Water, 1965.)

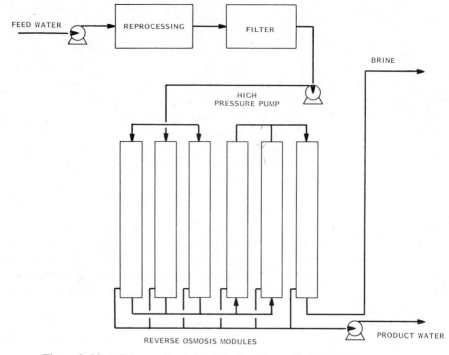

Figure 9-20 / Reverse Osmosis plant flowsheet. (A.D. Little, Inc., 1976.)

Reverse osmosis is becoming more popular for wastewater treatment applications as industry becomes better acquainted with the technology. Full-scale treatment facilities employing these processes are being used for recovering chemicals at electroplating operations (Patterson, 1975).

Solvent Extraction.   Solvent extraction or liquid–liquid extraction is a process whereby solutes are transferred from one liquid phase (e.g., water) to another (e.g., organic solvent) by intimate contact of the two liquid phases. As in the case of activated carbon adsorption, the affinity of a solute for an organic solvent is dependent on the hydrophilic nature of the solute. Solutes that are hydrophilic tend to be less extractable than hydrophobic solutes. The degree of extractability is measured by the partition coefficient, P, which is the ratio of the concentrations of solute in each liquid phase at equilibrium.

The effects of changes in chemical structure and substituents of an organic solute on the partition coefficient are revealed by the data given in Table 9–6.

A typical solvent extraction flowsheet is illustrated in Figure 9–21. Wastewater is intimately contacted with a suitable solvent by means of mixer-

Table 9-6 / Changes in Partition Coefficient[a] with Changes in Chemical Substituents and Structure.[b]

X	Aliphatic R $\log P_{RX} - \log P_{RH}$	Aromatic R $\log P_{C_6H_5X} - \log P_{C_6H_6}$
$-OC_6H_5$	1.61	2.08
$-I$	1.00	1.12
Intramolecular H bonding	0.65	—
$-Br$	0.60	0.86
$-CH_2$	0.50	0.50
$-SCH_3$	0.45	0.61
$-Cl$	0.39	0.71
Ring closure	−0.09	—
$-F$	−0.17	0.14
Branching in C chain	−0.20	—
Branching of functional group	−0.20	—
$-COOCH_3$	−0.27	−0.01
$-N(CH_3)_2$	−0.30	0.18
Double bond	−0.30	—
$-OCH_3$	−0.47	−0.02
Folding	−0.60	—
$-COOH$	−0.67	−0.28
$-COCH_3$	−0.71	−0.55
$-CN$	−0.84	−0.57
$-NO_2$	−0.85	−0.28
$-O-$	−0.98	—
$-OH$	−1.16	−0.67
$-NH_2$	−1.19	−1.23
$-C=O$	−1.21	—
$-CONH_2$	−1.71	−1.49

[a]A partition coefficient is the ratio of concentrations of a chemical in two immiscible liquid phases at equilibrium.
[b]From Leo et al. (1971).

settlers, packed columns, or centrifugal contactors. Contact of the solvent generally occurs countercurrent to the flow of wastewater through a series of contactors or through a packed column. The organic solvent containing the solute is transferred to the solvent regeneration unit, where the solvent is recovered by one of several processes given in Figure 9–21. The solute is separated for disposal or reused, and the solvent is returned to the extraction unit. The treated wastewater from the extraction unit is also processed for solvent recovery. The solubility of the solvent in the wastewater varies widely depending on the type of solvent. The solvent itself is usually considered a pollutant and must be removed to render the wastewater suitable for disposal.

A solute with a high partition coefficient is more efficiently extracted than a solute with a low partition coefficient and therefore requires less solvent. The amount of solvent used is one of the major factors for determining the cost of the process, particularly if an energy-consuming solvent recovery

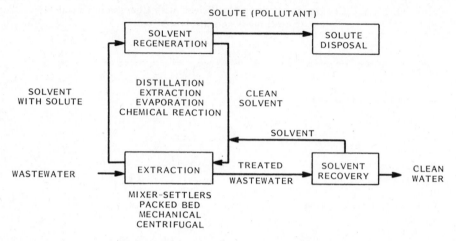

Figure 9-21 / Typical solvent extraction process.

process, such as distillation, is employed. Steam stripping to recover solvent from the extracted wastewater is also an energy-intensive process.

Removal of phenol from wastewaters is one of the primary applications of solvent extraction (Arthur D. Little, 1976).

Steam Stripping.   Steam stripping can be considered a form of gas stripping, where water vapor at elevated temperature is used as the medium for transfer and removal of volatile constituents from a liquid. Steam stripping is generally carried out in a continuous operation with a tower equipped with trays (e.g., bubble cap or sieve) or a suitable packing (e.g., saddles or rings) that enhances the contact of the gaseous phase with the liquid phase.

A typical steam stripper design is illustrated in Figure 9-22. The preheated feed from the heat exchanger is introduced at the top of the tower and flows by gravity countercurrent to the steam, flowing up from the bottom of the tower. As the liquid flows down the tower, it contacts steam containing progressively lower concentrations of volatile constituents stripped from the liquid. At the same time, the volatile constituents in the liquid are reduced to lower concentrations as the liquid flows to the bottom of the tower. The concentration of the volatile constituents is reduced to the lowest value at the bottom of the tower, where the liquid is first contacted by the incoming steam. Heat conservation is practiced by preheating the feed with the stripped bottoms in a heat exchanger.

Steam from the top of the tower either may be condensed or may be further processed in the gaseous phase (e.g., incinerated) before discharge. Condensation of the steam and organic vapors may produce an aqueous phase and an organic phase (e.g., water-immiscible organics) that can be

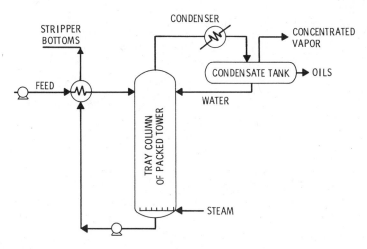

Figure 9–22 / Typical steam stripping system. (From A. D. Little, Inc., 1976.)

separated by decantation. Noncondensable vapor and gases may receive further treatment or may be released to the atmosphere if pollutant levels are low. Recycle or reflux of the condensate stream to the tower serves to increase the concentration of volatile constituents in the effluent stream from the tower and also eliminates a secondary wastewater stream that might require treatment before release.

Removal of volatile matter by a stripping tower may be increased by increasing the number of vapor/liquid contact stages in the tower. For example, the height of the tower can be increased to include additional trays or packing to increase the number of stages. Alternatively, a more effective packing may be substituted for the existing packing in the tower to increase the number of stages without altering the dimensions of the tower.

Steam stripper efficiency is a function not only of the number of vapor/liquid contact stages but also of temperature, steam/liquid reactions, and, in the case of the volatile acids and bases, the pH of the liquid. The effects of the steam-to-feed ratio and number of stages on ammonia removal from petroleum refinery sour water are illustrated in Figure 9–23. These data show that less stripping steam is needed as the number of stages is increased, but the potential for further steam savings diminishes. The terminology "theoretical stage" is used to denote the segment of a tower needed to achieve a liquid equilibrium concentration that corresponds to the vapor phase at a given zone in the tower. Although this concentration cannot actually be attained, in practice equilibrium can be approached so closely that the difference is unimportant. Several trays may be needed to achieve liquid equilibrium concentrations corresponding to the vapor concentrations entering the lower tray of a sequence of trays in the tower. Theoretical stages are used in

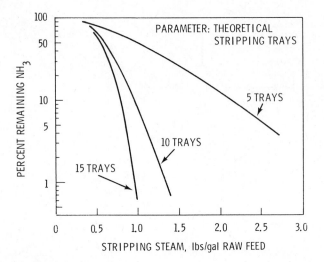

Figure 9-23 / Effect of steam-to-feed ratio and number of stages of ammonia removal. (Adapted from Melin et al., 1975.)

computational methods to describe the performance of a steam stripper under a variety of operating conditions.

The temperature of the feed to the tower is important to maintaining good stripper performance. Introduction of a cold feed stream to the tower results in substantial condensation of steam in the upper portion of the tower and therefore requires a greater steam input to maintain an outflow of steam from the tower.

Well-established wastewater treatment applications for steam stripping include ammonia and hydrogen sulfide removal from petroleum refinery sour waters (Beychok, 1967) and ammonia removal from coking liquors (Arthur D. Little Inc., 1976). Steam stripping is also used for removal of phenol from wastewater (Arthur D. Little Inc., 1976).

Cost of Wastewater Treatment

Cost Basis

Capital and operating costs for the principal treatment processes discussed in this chapter are provided for facilities of 0.1 and 1.0 million gallons per day (mgd). These costs were derived from curves given by the EPA (*Treatability Manual*, 1980). The cost data are generalized and are presented for comparative purposes only. Costs for specific locations and for different wastewaters may vary substantially from the data given here. Reference to the *Treatability Manual* is recommended for readers who desire additional information on the costing methods and other details. Except for land application of waste-water, land acquisition costs for the treatment facilities are not included in the cost data.

A breakdown of the cost components of the total capital investment is as follows:

Direct cost components:
    Purchased equipment and installation
    Instrumentation and controls
    Piping
    Electrical equipment and controls
    Buildings
    Yard improvements
    Service facilities
Indirect cost components:
    Engineering and supervision
    Construction expenses
    Contractor fee
    Contingency

Annual operating costs are broken down as follows:

Total direct operating cost:
    Labor
    Materials
    Chemicals
    Power
    Fuel
Total indirect operating cost:
    Plant overhead
    Taxes and insurance
    General and administrative expenses
    Depreciation
    Interest on working capital

The Engineering News Record (ENR) index for these cost data is 3119. The cost of energy and chemicals has a major impact on the cost of wastewater treatment; therefore, processes that use little energy and no chemicals (e.g., sedimentation, filtration, screening) are generally much less costly than processes that have high energy or chemical usage (e.g., evaporation, oxidation).

Liquids–Solids Separation

Capital and operating costs for six liquids–solids separation processes are presented in Table 9-7. The costs for vacuum filtration and centrifugation are not actually related to the volume of wastewater processed, but rather to the amount of sludge handled, which in this case is 4500 lb (2045 kg) of dry

Table 9-7 / Costs of Liquids–Solids Separation Processes (1000 gal = 3.8 m³).

Treatment Process	Costs for 0.1-mgd Facility			Costs for 1.0-mgd Facility		
	Total Capital ($1,000)	Total Annual Operating ($1,000)	Cost per 1000 gal ($)	Total Capital ($1,000)	Total Annual Operating ($1,000)	Cost per 1000 gal ($)
Screening (stationary)	48	24	0.66	67	27	0.07
Primary sedimentation	210	27	0.74	450	70	0.19
Dissolved-air flotation	840	85	2.32	2300	190	0.52
Granular-media filtration	270	45	1.23	1200	190	0.52
Vacuum filtration	620	85	2.33	850	180	0.49
Centrifugation	530	84	2.30	1000	150	0.41

Table 9-8 / Costs of Chemical Treatment Processes (1000 gal = 3.8 m³).

Treatment Process	Costs for 0.1-mgd Facility			Costs for 1.0-mgd Facility		
	Total Capital ($1,000)	Total Annual Operating ($1,000)	Cost per 1000 gal ($)	Total Capital ($1,000)	Total Annual Operating ($1,000)	Cost per 1000 gal ($)
Neutralization	300	300	8.57[a]	1500	1200	3.43[a]
Chemical clarification	260	50	1.37	500	130	0.36
Oxidation	4000	2500	68.4	—	—	—
Ion exchange	700	190	5.42[a]	3500	1000	2.86[a]

[a]Based on 350 days/yr operation.

solids (lime sludge) per million gallons (3.8×10^6 l). It is assumed that the sludge is removed from the wastewater by gravity sedimentation and fed to the vacuum filters or centrifuges as a 10% slurry. Considerable adjustment of the vacuum filtration and centrifugation costs will be required for sludge quantities that vary widely from these values.

The costs for gravity sedimentation are based on a circular clarifier operating at a surface loading rate of 600 to 1200 gpd/ft² (24,000 to 49,000 lpd/m²). The costs for granular-media filtration are based on the use of a dual-media gravity filter operated at 4 gpm/ft² (163 lpm/m²) filter surface and a filter run of 12 h.

Chemical Treatment

Costs for chemical treatment of wastewater are presented in Table 9–8. These costs will vary considerably with the composition of a specific waste. Neutralization costs are based on a 1% by volume sulfuric acid waste that is neutralized with quicklime in a three-reactor (series) system. The neutralization facility is equipped with a lime slaker, a clarifier for the treated wastewater, and a filter for the sludge. Chemical clarification involves 200 mg/l alum addition and gravity sedimentation. Oxidation costs are based on alkaline chlorine destruction of cyanide in a wastewater containing 1000 ppm copper cyanide and 1000 ppm sodium cyanide. Ion-exchange costs are based on treatment of a metal finishing wastewater containing 15 mg/l Zn^{2+}, 0.5 mg/l Cu^{2+}, 19 mg/l CN^-, and 22 mg/l Cr^{6+}.

Biological Treatment

Costs for three biological treatment processes are given in Table 9–9. The BOD_5 concentration in the wastewater feed to the activated sludge was assumed to be 1300 mg/l. The activated sludge process employed conventional diffused aeration and was loaded at a rate of 0.25 kg of BOD per kilogram of MLSS (MLSS = 2000 mg/l in aeration tank). The aerated lagoon was operated with a wastewater feed of 2100 mg/l BOD_5 at a hydraulic detention time of 7 days. Land application was used for disposal of digested sludge, which was generated at a rate of 900 lb of dry solids per million gallons of wastewater (0.11 kg/m³) and was applied to the land at 10 tons per acre per year (22 tonnes/hectare/yr).

Separation of Dissolved Matter

Costs for separation of dissolved matter from wastewater by five different processes are presented in Table 9–10. Of the five processes, only activated carbon adsorption is highly sensitive to the concentration of pollutant re-

Table 9-9 / Costs of Biological Treatment Processes (1000 gal = 3.8 m³).

Treatment Process	Costs for 0.1-mgd Facility			Costs for 1.0-mgd Facility		
	Total Capital ($1,000)	Total Annual Operating ($1,000)	Cost per 1000 gal ($)	Total Capital ($1,000)	Total Annual Operating ($1,000)	Cost per 1000 gal ($)
Activated sludge[a]	405	63	1.72	1,310	270	0.74
Aerated lagoon	210	30	0.82	380	57	0.16
Land application (sludge)	—	—	—	67	13	0.04

[a]Includes secondary clarification.

Table 9-10 / Costs of Physical Separation Processes (1000 gal = 3.8 m³).

Treatment Process	Costs for 0.1-mgd Facility			Costs for 1.0-mgd Facility		
	Total Capital ($1,000)	Total Annual Operating ($1,000)	Cost per 1000 gal ($)	Total Capital ($1,000)	Total Annual Operating ($1,000)	Cost per 1000 gal ($)
Activated carbon adsorption	360	110	3.01	1600	260	0.71
Evaporation[a]	—	—	—	1300	5342	8.99
Reverse osmosis	180	55	1.51	950	370	1.01
Solvent Extraction[b]	230	260	7.88	1300	770	2.33
Steam stripping[c]	130	90	2.57	600	520	1.49

[a]Costs are for 2 mgd, with a six-effect evaporation (Arthur D. Little, Inc., 1976).
[b]330 operating days/yr.
[c]350 operating days/yr.

moved. The costs for activated carbon adsorption are based on treatment of secondary effluent (20 to 100 mg/l COD) at a COD loading of 0.1 to 0.3 kg of COD per kilogram of activated carbon. Reverse-osmosis costs were based on treatment of a nickel plating rinse water to remove nickel salts for recycle. Solvent extraction costs were estimated for treatment of a 1.5% phenol solution and involve removing the solvent (toluene) by distillation of the spent solvent and by steam stripping of the aqueous effluent.

Steam stripping costs are based on treatment of refinery sour water. The major cost factor for steam stripping is energy consumption, which in this case is steam used at the rate of 2 lb per gallon of water treated.

REFERENCES

Battelle Memorial Institute. 1968. *A state-of-the-art review of metal finishing waste treatment.* U.S. Department of the Interior, Federal Water Quality Administration Report 12010 EIE 11/68.

Berkovitch, J. B.; Harris, J. C.; and Goodwin, B. 1981. Identification of hazardous waste for land treatment research. In *Proceedings of the seventh annual research symposium on land disposal: hazardous waste*, ed. D. W. Shultz. EPA-600/9-81-0026. U.S. Environmental Protection Agency, pp. 168–177.

Beychok, M. R. 1967. *Aqueous waste from petroleum and petrochemical wastes.* New York: Wiley.

Caulter, K. R. 1981. Report on the results of the AES/EPA sludge characterization project. In *Proceedings of the third conference on advanced pollution control for the metal finishing industry.* Eds. H. J. Schumacher, Jr. and G. S. Thompson, Jr. Report No. EPA-600/2-81-0028.

Culp, R. L.; Wesner, G. M.; and Culp, G. L. 1978. *Handbook of advanced wastewater treatment.* New York: Van Nostrand Reinhold.

Dean, J. R.; Bosqui, F. L.; and Lanouette, K. H. 1972. Removing heavy metals from wastewaters. *Environmental Science and Technology*, Vol. 6, 518.

Dobbs, R. A.; and Cohen, J. M. 1978. *Carbon adsorption isotherms for toxic organics.* EPA report, MERL, Office of Research and Development, Cincinnati.

Easton, J. K. 1967. Electrolytic decomposition of concentrated cyanide plating wastes. *Journal of the Water Pollution Control Federation*, Vol. 39, pp. 1621–1625, October 1967.

Ebersale, G. 1972. The removal of mercury from water by adsorption on aluminum and ferric hydroxide. Paper presented at the annual conference of the American Water Works Association, Chicago.

Eckenfelder, W. W. 1966. *Industrial water pollution control.* New York: McGraw-Hill.

Emility, L. A. 1967. *Operation and control of ion-exchange process for treatment of radioactive wastes.* Technical Report Series No. 78, International Atomic Energy Agency, Vienna.

Federal Register. 1981. Vol. 46, pp. 9467–74, January 28, 1981.

Godbee, H. W.; and Kibbey, A. H. 1975. *Application of evaporation to the treatment of liquids in the nuclear industry. Nuclear Safety* Vol. 16, p. 458.

Gold, H. 1978. Metal finishing waste. In *Ion exchange for pollution control,* ed. C. Calmon and H. Gold, Boca Raton, Fl.: CRC Press, pp. 173–190.

Howe, R. H. L. 1963. Recent advances in cyanide reduction practice. In *Proceedings of the 18th industrial waste conference,* pp. 690–704.

Kolthoff, I. M.; and Overholser, L. S. 1939. Studies on aging and coprecipitation with ortho ferric hydroxide in ammoniacal solution. *Journal of Physical Chemistry,* Vol. 43, p. 909.

Leo, A.; Hansch, C.; and Elkins, D. 1971. Partition coefficients and their uses. *Chemical Reviews,* Vol. 71, No. 6, pp. 525–626.

Arthur D. Little, Inc. 1976. *Physical, chemical, and biological treatment techniques.* U.S. Environmental Protection Agency report No. SW-148c.

Lores, C.; and Moore, R. B. 1974. U.S. patent 3,770,423.

Lyman, W. J. 1978. Applicability of carbon adsorption to the treatment of hazardous industrial wastes. In *Carbon adsorption handbook,* ed. P. M. Cheremismoff and F. Ellerbush, pp. 131–65. Ann Arbor: Ann Arbor Science Publishers.

Matthews, J. E.; Pfeffer, F. M.; and Weiner, L. A. 1981. Closure techniques at a petroleum land treatment site. In *Proceedings of the seventh annual research symposium on land disposal: hazardous waste,* ed. D. W. Shultz. EPA-600/9-81-0026. U.S. Environmental Protection Agency, pp. 240–245.

McKinney, R. E. 1971. *Waste treatment lagoon—state-of-the-art.* EPA Water Pollution Control Research Series No. 17090 EHA 07/71.

McKinney, R. E.; and Ooten, R. J. 1969. Concepts of complete mixing activated sludge. In *Transactions of the 19th annual conference on sanitary engineering.* Bulletin of Engineering and Architecture No. 60, University of Kansas, Lawrence.

Melin, G. A.; Niedzwielki, J. L.; and Goldstein, A. M. 1975. Optimum design of sour water strippers. *Chemical Engineering Progress* Vol. 71, p. 78.

Mercer, B. W.; and Ames, L. L. 1978. Zeolite ion exchange in radioactive and municipal wastewater treatment. In *Natural zeolites, occurrence, properties and use,* ed. L. B. Sand and F. R. Mumpton. New York: Pergamon.

Metcalf and Eddy, Inc. 1979. *Wastewater engineering: treatment, disposal, reuse.* 2nd ed. New York: McGraw-Hill.

National Commission on Water Quality. 1975. *Water pollution abatement technology: capabilities and cost, iron and steel industry.* Report PB-249 661, National Technical Information Service, Springfield, Va.

Nilsson, R. 1971. Removal of metals by chemical treatment of municipal wastewaters. *Water Research*, Vol. 5, p. 51.

Office of Saline Water, 1965. *1965 saline water conversion report*. U.S. Department of Interior, p. 193, U.S. Government Printing Office.

Ottinger, R. S.; Blumethal, J. L.; DalPorto, D. F.; Gruber, G. I.; Santy, M. J.; and Shih, C. C. 1973. *Recommended methods of reduction, neutralization, recovery or disposal of hazardous waste*. EPA-6x/2-73-053, U.S. Environmental Protection Agency.

Patterson, J. W. 1975. *Wastewater treatment technology*. Ann Arbor: Ann Arbor Science Publishers.

Patterson, J. W.; Allen, H. E.; and Scala, J. J. 1977. Carbonate precipitation for heavy metal pollutants. *Journal of the Water Pollution Control Federation* Vol. 49, p. 2351, December 1977.

Patterson, J. W.; and Minear, R. 1973. Physical chemical methods of heavy metals removal. Paper presented at a conference on heavy metals in the aquatic environment, Vanderbilt University, Nashville, Tennessee, December 4–7, 1973.

Products Finishing. 1970. New chemical destroys cyanide in zinc plate rinses. *Products Finishing* Vol. 34, pp. 130–133, December 1970.

Rich, L. G. 1963. *Unit operations of sanitary engineering*. New York: Wiley.

Shen, Y. S. 1973. A study of arsenic removal from drinking water. *Journal of the American Water Works Association*, Vol. 65, p. 543, August 1973.

Shuliger, W. G. 1978. Purification of industrial liquids with granular activated carbon: techniques for obtaining and interpreting data and selecting the type of commercial system. In *Carbon adsorption handbook*, ed. P. M. Cheremismoff and F. Ellerbush, pp. 55–85. Ann Arbor: Ann Arbor Science Publishers.

Skripach, T.; Kagan, V.; Ramanov, V.; et al. 1971. Removal of fluorine and arsenic from wastewater of the rare-earth industry. In *Proceedings of the 5th international water pollution research conference*, London: Pergamon.

Snyder, H. J., Jr. 1976. Disposal of waste oil re-refining residues by land farming. In *Proceedings of a research symposium at the University of Arizona, February 2–4, 1975, on residual management by land disposal*. EPA-600/9-76-015. ed. W. H. Fuller. U.S. Environmental Protection Agency, 195–205.

Spyridakis, P. E.; and Welch, E. B. 1976. Treatment process and environmental impacts of waste effluent disposal of land. In *Land treatment and disposal of municipal and industrial wastewaters*, ed. R. L. Sands and T. Asano, pp. 1–16. Ann Arbor: Ann Arbor Science Publishers.

Streebin, L. E.; Schornich, H. M.; and Washinski, A. M. 1980. Ozone oxidation of concentrated cyanide wastewater from electroplating operations. In *Proceedings of the 35th Industrial Waste Conference*, ed. J. M. Bell, Ann Arbor: Ann Arbor Science Publishers, pp. 665–676.

Treatability manual. 1980. *Vol. IV. Cost estimating*. EPA-600/8-80-042d.

U.S. EPA. 1980. *Summary report, control and treatment technology for metal finishing industry, sulfide precipitation.* Environmental Protection Agency–Technology Transfer, EPA 625/8-80-003.

U.S. EPA. 1977. *State decision-makers guide for hazardous waste management.* Guide No. SW-612, U.S. Government Printing Office.

U.S. EPA. 1975. *Process design manual for suspended solids removal.* U.S. Environmental Protection Agency–Technology Transfer, EPA 625/1-75-003a.

U.S. EPA. 1971. *Limestone treatment of rinse water from hydrochloric acid pickling of steel.* 1201D DUL2/71.

Wallace, A. T. 1976. Land disposal of industrial wastes. In *Land treatment and disposal of municipal and industrial wastewaters*, ed. R. L. Sands and T. Asano, Ann Arbor: Ann Arbor Science Publishers.

Weber, W. J., Jr. 1972. *Physicochemical processes for water quality control.* New York: Wiley-Interscience.

Weber, W. J., Jr.; and Posselt, H. S. 1974. Equilibrium models and precipitation reactions for cadmium II. In *Aqueous-Environmental Chemistry of Metals*, ed. A. J. Rubin, Ann Arbor: Ann Arbor Science Publishers, pp. 255–290.

Wilhelmi, A. R.; and Knopp, P. V. 1979. Wet air oxidation—an alternative to incineration. *Chemical Engineering Progress*, Vol. 75, No. 8, 46–52, August 1979.

WPCF/ASCE. 1977. *Wastewater treatment plant design.* Water Pollution Control Federation and American Society of Civil Engineers, MOP/8. Lancaster, Pa.: Lancaster Press.

Yost, K. J.; and Maserich, D. R. 1977. A study of chemical-destruct waste treatment systems in the electroplating industry. *Plating and Surface Finishing*, Vol. 64, p. 35, January 1977.

Yost, K. J.; and Scarfi, A. 1979. Factors affecting copper solubility in electroplating waste. *Journal of the Water Pollution Control Federation*, Vol. 51, p. 1887, July 1979.

Zanoni, A. E.; and Blomquist, M. W. 1975. Column settling tests for flocculent suspensions. *Journal of the Environmental Engineering Division*, *ASCE* Vol. EE3; pp. 309–318, June 1975.

10

Incineration

Introduction

The environmental problems associated with disposal of hazardous wastes in landfills (see Chapter 11) have intensified consideration of alternative methods that can quickly and safely convert these wastes to nonhazardous forms (*Hazardous Waste News*, 1981; *Toxic Materials News*, 1981; Serper, 1981; Scurlock et al., 1975). Incineration is the leading alternative method for disposal of hazardous organic wastes. Hazardous organic wastes constitute the greatest portion of all hazardous wastes, and incineration is applicable to essentially all forms of organic wastes (Ross, 1979).

Incineration is a controlled high-temperature oxidation process that converts the principal elements (carbon, hydrogen, and oxygen) in most organic compounds to CO_2 and H_2O. The toxic or hazardous nature of an organic molecule usually is due to the structure of the molecule, as opposed to the elements it contains. Therefore, destruction of the molecular structure usually eliminates the toxic or hazardous property (Scurlock et al., 1975). The existence of elements other than carbon, hydrogen, and oxygen in a waste may, on incineration, result in the production of gaseous or particulate pollutants that require removal in off-gas treatment systems.

In the last two decades, incinerator technology has moved from the highly polluting open pit and simple trash burners to modern efficient incinerators equipped with sophisticated off-gas treatment systems that emit only minimal amounts of pollutants. Although air quality in areas surrounding waste incinerators has greatly improved, capital and operating costs for the incinerators have substantially increased. Many waste incinerators across the country have therefore been phased out in favor of the less costly option of landfilling. Because of high short-term costs, incineration frequently does not appear to be the most economical disposal method. However, long-term costs should be considered when selecting a disposal method for wastes. The latter include potential liability for leakage and damages from landfills over time.

The objective of this chapter is to give the reader an understanding of some of the more important basic concepts of the chemistry and physics of combustion and how these can be applied to the incineration of hazardous wastes. The discussion of incinerator design and operation will focus largely on the two most frequently used types of hazardous waste incineration: rotary kiln and liquid injection.

Process Description

Combustion Chain Reactions

Knowledge of the chemistry and physics of combustion is helpful in understanding the reasons for the various designs of hazardous waste incinerators. This section might be entitled "Incineration Chemistry," to address the reactions that occur in a waste incinerator; however, "combustion" is generally considered the proper term to define a flame oxidation process. Incineration is the process that utilizes combustion to destroy unwanted material.

Very few combustion reactions are simple. The overall stoichiometric equations that represent combustion reactions seldom provide the detailed mechanisms involved. Several reactions may proceed simultaneously or in sequence, depending on the nature of the material undergoing combustion and the conditions that influence combustion, such as temperature, pressure, and oxygen supply. Combustion in a sequence of steps is described as a chain reaction. Oxidation of methane is a spectacular example to show how complex the seemingly simple reaction $CH_4 + 2O_2 \rightarrow CO_2 + 2H_2O$ can be (Kanury, 1977):

$$CH_4 + OH\bullet \rightarrow CH_3\bullet + O\bullet + H_2$$
$$CH_4 + O\bullet \rightarrow CH_3\bullet + OH\bullet$$
$$CH_3\bullet + O_2 \rightarrow H_2CO\bullet + OH\bullet$$
$$H_2CO\bullet + OH\bullet \rightarrow HCO\bullet + H_2O$$
$$HCO\bullet + OH\bullet \rightarrow CO + H_2O$$
$$CO + OH\bullet \rightarrow CO_2 + H\bullet$$
$$H\bullet + O_2 \rightarrow OH\bullet + O\bullet$$
$$H\bullet + H_2O \rightarrow OH\bullet + H_2$$
$$O\bullet + O\bullet + M \rightarrow O_2 + M^*$$
$$20H\bullet + M \rightarrow H_2O + O\bullet + M^*$$

M is a third body (which could be any other atom or molecule in the system or system walls) rendered active (designated with an asterisk) in the reaction.

Hydrocarbon combustion, as exemplified here, involves the reaction of free radicals such as $CH_3\bullet$, $HCO\bullet$, $OH\bullet$, etc., that play the role of chain carrier to sustain the chain reaction. The chain reaction is terminated whenever two or more of the unstable chain carriers collide with a third body, *M*, to form an inactive molecule. Combustion of complex organic substances can therefore produce a wide variety of intermediate compounds if the reaction is not allowed to go to completion. Incomplete combustion of a hazardous organic substance may release hazardous intermediates as well as the substance itself. Absolutely complete combustion is rarely, if ever, achieved in a practical incinerator, but it can be approached closely enough to achieve the designed objective of minimal impact due to emissions from the incinerator.

Heat of Combustion

Combustion of a mixture of fuel and oxygen begins when the temperature of the mixture is raised to a point where a flame reaction occurs. The flame reaction is manifested by a high rate of heat evolution and frequently, but not always, by the emission of light. The reaction begins when chemical bonds between the elements of the reactants (compounds to be burned) are broken to form free radicals, which combine or enter into a chain process to produce new compounds. Some or all of the new compounds will consist of combinations of elements initially in the fuel with oxygen (e.g., CO_2 and H_2O for complete combustion of hydrocarbons). The sum of the bond energies of the products of combustion will be less than the sum of the bond energies for the reactants—the difference being released energy (Kanury, 1977). The heat released by complete combustion of a mole of a compound is known as the "heat of combustion" or, more rigorously, as the "enthalpy of combustion" if combustion occurs at constant pressure. The heats of combustion for hazardous compounds are listed in Appendix L along with other data pertinent to combustion of these compounds.

The energy required to break a particular bond between two atoms is approximately the same regardless of the type of molecule containing the bonded pair of atoms. Different pairs of atoms and different types of bonds between the atoms result in different affinities for one another; thus the energies to break the atoms apart are different. Mean bond energies for atoms and bonds within various organic compounds are given in Table 10–1. The approximate heat of combustion for an organic compound may be calculated by determining the difference in the sum of the bond energies for the reactants (compound + oxygen) and the sum of the bond energies for the products of the combustion reaction (e.g., CO_2 + H_2O). This calculation can be useful for determining the approximate heat of combustion of a compound for which no literature values are available (e.g., complex waste organic compounds). An example calculation for estimating the heat of combustion of acrolein from bond energies is presented next.

Table 10-1 / Mean Bond Energies (kilocalories/mole). (Dean, 1973)

Bond	Energy	Bond	Energy
C—C	83	O—H	111
C=C	146	O—N	48
C≡C	200	N—H	93
C—H	99	Cl—Cl	57
C—O	86	Br—Br	46
C=O	192[a]	I—I	36
C—N	72	F—F	36
C=N	147	H—Cl	103
C—Cl	81	H—Br	88
C—Br	68	H—I	71
C—I	52	H—F	135
C—F	116	H—P	79
C—S	65	H—S	83
O—O	47	P—Cl	79
O=O	118	P—Br	64
N—N	52	S—Cl	61
N≡N	225		
H—H	104		

[a] Mean bond energy for CO_2 (Pimental and Spratley, 1970). Mean bond energy for aldehydes and ketones is 176 kcal/mole.

The stoichiometric equation for the combustion of acrolein in skeletal form is

$$
\begin{array}{ccc}
H & H & O \\
\end{array}
$$

$$
C = C - C \qquad + (4)\ O = O \rightarrow (3)\ O = C = O + (2)\ H - O - H
$$

$$
\begin{array}{ccc}
H & & H \\
\end{array}
$$

$$(C_3H_4O + 3.5O_2 \rightarrow 3CO_2 + 2H_2O)$$

which shows four C—H bonds, one C=C bond, three and one half O=O bonds, one C—C bond, and one C=O (aldehyde) bond being broken and six C=O (CO_2) bonds and four H—O bonds being formed. The energy required to break the bonds of the acrolein and oxygen is computed from values given in Table 10-1:

$$(4 \times 99) + 146 + (3.5 \times 118) + 83 + 176 = 1214$$

The bond energy in the products CO_2 and H_2O computed from values in Table 10-1 are

$$(6 \times 192) + (4 \times 111) = 1596 \text{ kcal}$$

The computed energy released by the reaction is $1596 - 1214 = 382$ kcal, which roughly approximates the value of 389.6 kcal/mole given for acrolein in Appendix K. Although heat-of-combustion values computed from mean bond energies are only rough approximations, these values can nevertheless be useful, for example, in determining preliminary supplemental fuel requirements for incineration of a waste for which no actual combustion data are available.

Changes in molecular structure that decrease the sum of the bond energies for the reactants or increase the sum of the bond energies of the products will lead to a greater heat of combustion. Substitution of an OH radical for H on the right-side carbon with the double-bonded O of the acrolein molecule gives acrylic acid, which has a heat of combustion of 327 kcal/mole. Substitutions of two hydrogen atoms for the double-bonded O on the acrolein molecule give propylene, which has a heat of combustion of 490.2 kcal/mole.

Halogenated organic compounds, such as chlorinated hydrocarbons, may have very low heats of combustion if the halogen content of the compound is very high. The effect of progressive substitution of chlorine atoms for hydrogen atoms on methane is illustrated by decreasing heats of combustion in Table 10-2.

Chlorinated hydrocarbons containing more than 70% chlorine by weight will generally require auxiliary fuel to maintain combustion. A "thermal vortex" burner is reported to be capable of maintaining stable combustion without auxiliary fuel, with wastes having heating values between 2500 and 3000 kcal/kg (4500–5400 Btu/lb) (Santoleri, 1973). Carbon tetrachloride, which was used to extinguish small fires prior to recognition of its high toxicity, will require a substantial addition of auxiliary fuel to sustain combustion.

Combustion of Halogenated Organics

Incineration of halogenated organics is also complicated by the formation of toxic products of combustion such as free halogens, halogen acids, and, in the case of chlorine, the deadly combination of carbon monoxide with chlorine to form phosgene, a poisonous gas used in World War I. The complete combustion of trichloroethylene with oxygen and no auxiliary fuel will result in the following reaction (Serper, 1981):

$$CHCl = CCl_2 + 2O_2 \rightarrow 2CO_2 + HCl + Cl_2$$

The Cl_2 forms as a result of the lack of hydrogen to produce HCl. Incomplete oxidation yields CO, which combines with the Cl_2 to make phosgene, $COCl_2$. The foregoing reaction probably will not sustain itself without some additional heat input, because the heat of combustion computed from bond energies is only 1600 kcal/kg. If additional heat input is obtained from the use

Table 10-2 / Heats of Combustion for Methane and
Chlorinated Products of Methane.[a]

Compound	Formula	Weight % Chlorine	Heat of Combustion (kcal/mole)	Heat of Combustion (kcal/kg)
Methane	CH_4	0	210.8	13,180
Methyl chloride	CH_3Cl	70	164.2	3,255
Methylene chloride	CH_2Cl_2	84	106.8	1,258
Chloroform	$CHCl_3$	89	89.2	747
Carbon tetrachloride	CCl_4	92	37.3	243

[a]Data from *CRC Handbook of Chemistry and Physics.*

of an auxiliary fuel such as natural gas (methane), then sufficient hydrogen is available to convert the free chlorine to hydrogen chloride. Conversion of chlorine to hydrogen chloride greatly improves the treatment of the off-gas from the reaction, because hydrogen chloride is readily removed by scrubbing with water, whereas chlorine gas is relatively insoluble and will largely pass through a wet scrubber. The chemical reaction for combustion of trichloroethylene with methane addition is illustrated by the following equation (Serper, 1981):

$$CHCl = CCl_2 + 3.5O_2 + CH_4 \rightarrow 3CO_2 + 3HCl + H_2O$$

Adsorption of HCl in scrubber water produces corrosive hydrochloric acid, which requires special corrosion-resistant materials for the off-gas treatment system or the use of alkaline reagents to neutralize the acid.

Combustion Products of Sulfur Compounds

Compounds containing sulfur either organically or inorganically bound in a reduced form (e.g., sulfides) will produce SO_2 on complete combustion in air. An example combustion reaction for methanethiol (methyl mercaptan) is illustrated by the following equation:

$$CH_3SH + 3O_2 \rightarrow CO_2 + 2H_2O + SO_2$$

Evolution of SO_2 and SO_3 from a waste incinerator may be restricted in certain areas of the country; therefore, scrubber systems may be required for removal of these substances. Alkaline scrubbers may be required if high SO_2 removal is needed, because SO_2 is not highly soluble in water.

Formation of NO_x

The formation of oxides of nitrogen (NO_x) may also be restricted, especially in regions where air quality problems exist (e.g., Southern California). Two possible sources of NO_x exist in combustion chambers: (1) formation of NO from oxygen and nitrogen present in the air supply to the chamber and (2) formation of NO from oxidation of compounds containing nitrogen (e.g., amines, nitrates) in the waste. Nitrogen oxide formed by process (1) is called "thermal NO," because elevated temperatures ($>1600°$ C) are required for the reaction to take place. Nitrogen oxide produced by process (2) is called "fuel NO" and its formation is essentially independent of combustion temperature over a broad range of practical interest (Argonne National Laboratory, 1977). Because minimum combustion temperatures are required for destruction of specific hazardous organic substances over a given time period, production of thermal NO is more or less fixed within practical design limitations for waste incinerators. Production of fuel NO is, however, controllable to some extent by controlling the air supply for combustion. Fuel NO increases markedly with increasing excess air. In a single combustion chamber, excess air cannot be reduced below a certain minimum (e.g., 3% excess O_2) for effective combustion of hazardous waste. The use of two consecutive combustion chambers can circumvent this limitation by operation of the first chamber under oxygen-starved conditions, which will produce N_2 rather than NO from organic nitrogen, and the second chamber with sufficient excess oxygen to complete combustion.

Based on data on the combustion of coal in utility boilers, increasing the fuel nitrogen content will not cause a commensurate increase in the fuel NO concentration of the off-gas (Argonne National Laboratory, 1977). Under equivalent combustion conditions, the percentage conversion of organic nitrogen to NO_x decreases as the nitrogen content increases. Furthermore, fuel NO production is a function of the type of nitrogen bonds as well as total bound nitrogen.

Nitrogen oxides are not effectively removed by conventional wet scrubber systems, and the high cost of catalytic and reductant systems for removing NO_x from off-gas renders these methods unattractive for most situations (Ottinger et al., 1973).

Combustion of Organic Phosphorus Compounds

Combustion of organic phosphorus compounds in excess oxygen produces phosphorus pentoxide (P_2O_5, actually a dimer P_4O_{10}) as the main product under most conditions (Hackman, 1978). If insufficient oxygen is available during combustion, some phosphorus trioxide (P_2O_3, actually a dimer P_4O_6) may be formed. Phosphorus pentoxide reacts readily with water to form phosphoric acid; therefore, a water or wet alkaline scrubber will effectively remove it. Phosphorus trioxide reacts with cold water to form phosphorous acid, but in hot water a number of compounds, including phosphine and elemental phosphorus, may be formed.

Incineration of Organometallic Wastes

Incineration of wastes containing hazardous heavy metals can be expected to produce secondary wastes such as incinerator ash and off-gas scrubbing liquors that may be classified as hazardous wastes. Furthermore, heavy metals that tend to volatilize during combustion (e.g., Hg, As, Se) may escape the off-gas treatment system in sufficient quantity to be of concern.

Combustion Rate

In the previous section we described the heat of combustion (the amount of energy released by flame oxidation of a substance), which is important in incinerator operation because a minimum amount is required to sustain combustion for a given set of conditions. Equally-important in the operation of an incinerator is the rate at which combustion takes place. Combustion rates are controlled by a number of different factors, some of which are related to inherent properties of the fuel and others to design and operational parameters of the incinerator. Factors that relate to the fuel (or waste) are

Heat of combustion
Activation energy
Heat of vaporization of liquid
Heat capacity of the fuel

Factors that relate to incinerator design and operation are

Ratio fuel:O_2
Degree of mixing
Heat input or removal

The temperature of combustion, which is critical to the reaction rate, is directly related to heat of combustion and heat input, whereas it is inversely related to heat of vaporization, heat capacity of fuel, heat removal, and degree of deviation from optimum reactant ratio (e.g., stoichiometric fuel: oxygen ratio). Thermodynamic relationships that are beyond the scope of this discussion can be established to show the effects of the foregoing factors on combustion temperature (Kanury, 1977; Eggers et al., 1964; Smith and Van Ness, 1975). The reactant ratio (i.e., excess air) is commonly adjusted to control the temperature of combustion. If the temperature in an incinerator is exceeding the desired level for a given fuel, increasing the amount of excess air will reduce the temperature of combustion.

The degree of reactant mixing is also critical to the reaction rate. A finite amount of time is required to bring the reactants together at the proper temperature for combustion. The speed with which this can be accomplished in a combustion chamber has a significant effect on reaction rate. Incinerators are usually designed to achieve a high degree of turbulence in the combustion zone to promote rapid mixing. Liquid fuels are introduced as sprays to hasten volatilization and mixing with air.

The reaction rate for a complex sequence of reactions in an incinerator can be simplified (Reed and Moore, 1980) to the first-order kinetic expression

(10.1)

$$dC/dt = kC$$

where C is the concentration of fuel (waste) oxidized, k is a kinetic rate constant, and t is time in seconds. The solution to this equation can be given as

(10.2)

$$\ln v = kt$$

where

$$v = \frac{\text{effluent organic concentration}}{\text{influent organic concentration}}$$

$$t = \text{incincerator residence time}$$

The Arrhenius expression can be used to determine the temperature dependence of the reaction rate constant of equation 10.2:

(10.3)

$$k = A \exp(-E/RT)$$

where A is the kinetic rate constant intercept, E is activation energy (cal/mol), T is temperature (°K), and R is the universal gas constant.

Kinetic rate constants have been determined for several toxic waste incinerators (Reed and Moore, 1980). These are plotted against the reciprocal of the temperature in Figure 10–1. The solid line in Figure 10–1 represents a least-squares correlation of the data, and the dotted lines indicate the 95% upper and lower confidence levels of the correlation. The least-squares equation for all the toxic waste incineration data is:

(10.4)

$$k = 86 \exp(-8580/RT)$$

and the correlation coefficient is 0.8. Equation 10.2 can then be combined with equation 10.4 to calculate the fraction of a combustible compound remaining at a certain temperature and time in an incinerator. (Reed and Moore, 1980).

Because the data in Figure 10–1 represent a wide variety of incinerator

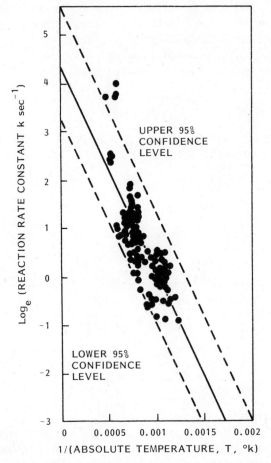

Figure 10-1 / Correlation of reaction rate constant with temperature. (Adapted from Reed and Moore, 1980.

designs, conservative estimates for determining the stability of given time–temperature combinations can be made by using the 95% lower confidence limit as follows:

(10.5)

$$k = 27.1 \; \exp(-8580/RT)$$

Time–temperature relationships for 99.9% destruction efficiency of organic waste material are illustrated in Figure 10–2 for the average of all incinerators (equation 10.4) and for the 95% lower confidence level (equation 10.5). The 99.9% destruction efficiency represents the minimum required for

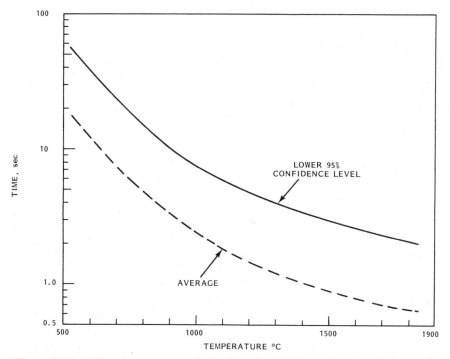

Figure 10-2 / Time–temperature relationships for 99.9% combustion efficiency of waste organic material.

combustion of toxic organic wastes at sea (IMCO, 1978). On the average, a temperature of 1060°C will be adequate to achieve 99.9% destruction efficiency at a 2-s residence time. At the conservative lower 95% confidence level, a temperature of 1820°C is indicated for the 2-s residence time. These data suggest that off-gas monitoring should be conducted for incineration of wastes below 1820°C at 2 s to assure adequate destruction efficiency.

Types of Incinerators

Liquid Injection

Liquid-injection incinerators operate by spraying the combustible waste mixed with air into a chamber where flame oxidation takes place. The purpose of spraying is to atomize the waste into small droplets (less than 40 microns). The small droplets present a large surface area for rapid heat transfer, thereby increasing the rate of vaporization and mixing with air to promote combustion. Air is supplied as a forced draft to provide the necessary mixing and turbulence.

Effective atomization is very important to the successful operation of a liquid-injection incinerator. The viscosity of the waste liquid or slurry should be less than 750 SSU to permit the shear forces available in the atomizer to break the liquid into small droplets (Ottinger et al., 1973). Waste liquids having viscosities higher than this value may be preheated or mixed with a lower-viscosity liquid to reduce viscosity. Devices used for atomization include rotary cups and various types of nozzles (Ottinger et al., 1973). Simple pressure nozzles are prone to plugging by solid particles frequently found in the liquid wastes.

Liquid-injection incinerators are widely used by industry for destruction of liquid organic wastes. These incinerators require minimal labor because of the relative ease of handling pumpable liquids compared with solid wastes. The maximum allowable viscosity for pumping is 10,000 SSU. A typical liquid waste incinerator system is shown in Figure 10–3. Liquid waste incinerator systems are generally equipped with transfer stations, storage tanks, and blending tanks, as shown in Figure 10–3, to ensure a reasonably steady and constant-composition waste flow (Hitchcock, 1979).

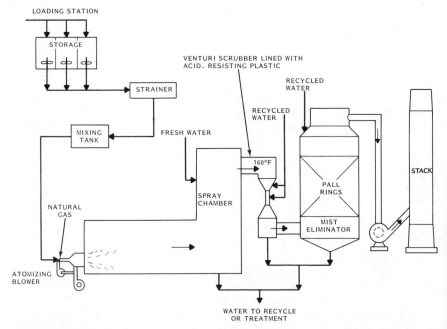

Figure 10-3 / Horizontally fired liquid waste incineration system. (From Ottinger et al., 1973.)

An off-gas treatment system consisting of a precooler, venturi scrubber, and cooling scrubber is also illustrated in Figure 10–3. Treatment of scrubber water (e.g., sedimentation of solid particles) is also provided.

Liquid-injection incinerators may be designed for horizontal firing, as shown in Figure 10–3, or for vertical firing, as shown in Figure 10–4, with the highly turbulent vortex design as an example (Lund, 1971; Witt, 1971). The vortex combustion chamber features tangential firing of liquid waste by a modified oil burner at the bottom of the chamber. A vortex of hot gas and

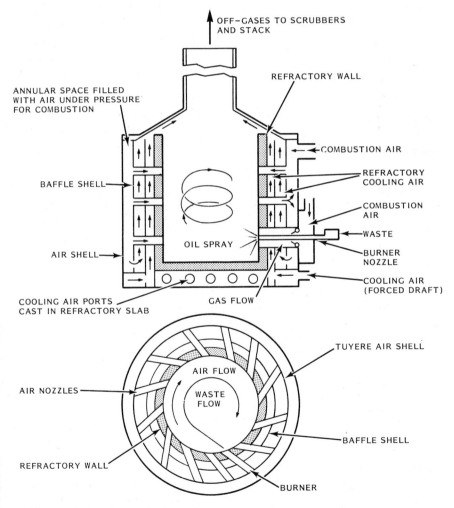

Figure 10-4 / Vortex liquid waste incinerators. (From Ottinger et al., 1973.)

primary air is created at this point and is maintained as it rises through the chamber by tangential injection of secondary air to complete the combustion process. Vortex combustion can achieve a heat release rate of 900,000 kcal/h/ m^3, which is about four times that of conventional liquid-injection combustion (Hitchcock, 1979). Generally, a good design avoids impingement of the flame on the refractory wall, and the unit operates at temperatures below the ash fusion point. Liquid-injection incinerators typically operate in the temperature range of 850°C to 1650°C.

Examples of full-scale liquid waste incineration include the following: (1) The Rollins Environmental Services unit at Logan Township, New Jersey, burns 3800 l/hr of waste at a temperature of 1000–1200°C, with a residence time of 2.5 s. (2) A liquid waste incinerator that has a capacity of 480 to 900 l/h at 980–1310°C and 1–12 s residence time is used by the General Electric Corporation at Pittsfield, Massachusetts (Cross, 1980).

Rotary-Kiln Incinerators

Rotary-kiln incinerators are highly versatile units that can accept virtually any type of combustible waste but are designed primarily to incinerate solids and tars that cannot be processed in liquid-injection incinerators. The rotary kiln is a cylindrical shell lined with refractory material that is horizontally mounted at a slight incline. The kiln typically has length-to-diameter ratios between 2:1 and 10:1 and is normally rotated at speeds between 1 and 5 rpm. Combustion temperatures are between 850°C and 1650°C, and residence times vary from several seconds to hours, depending on the type of waste being processed. The rotating kiln causes a tumbling action that effectively mixes combustible solids with air to promote complete burning of the waste. Noncombustible waste (e.g., ash, scrap metal) travels down the inclined kiln and is collected in drums after water quench cooling.

The 3M Company of St. Paul, Minnesota, incinerates chemical waste from manufacturing operations with a 23,000-kcal/h rotary kiln (Lewis et al., 1976). The kiln is 11 m long, with 0.3 m of super-heavy-duty refractory brick lining. Material is fed into the kiln in 210l (55-gal) drum quantities, having an average weight of 80 kg and ranging in weight from 70 to 230 kg. An example of a rotary kiln is illustrated in Figure 10–5.

The 3M rotary-kiln incinerator employs two waste handling systems— one for processing pumpable wastes and one for nonpumpable drummed wastes. The system for processing pumpable wastes consists of a pumping room, storage tanks, and blending tanks. All the waste is mechanically crushed to remove large agglomerations. Care must be exercised to avoid mixing wastes that react, solidify, or polymerize in the tanks, because these materials must be manually removed at great expense.

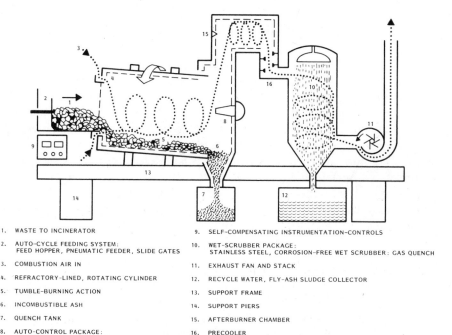

Figure 10-5 / Portable rotary kiln incineration units. (From Ottinger et al., 1973.)

1. WASTE TO INCINERATOR

2. AUTO-CYCLE FEEDING SYSTEM:
   FEED HOPPER, PNEUMATIC FEEDER, SLIDE GATES

3. COMBUSTION AIR IN

4. REFRACTORY-LINED, ROTATING CYLINDER

5. TUMBLE-BURNING ACTION

6. INCOMBUSTIBLE ASH

7. QUENCH TANK

8. AUTO-CONTROL PACKAGE:
   PROGRAMMED PILOT BURNER

9. SELF-COMPENSATING INSTRUMENTATION-CONTROLS

10. WET-SCRUBBER PACKAGE:
    STAINLESS STEEL, CORROSION-FREE WET SCRUBBER: GAS QUENCH

11. EXHAUST FAN AND STACK

12. RECYCLE WATER, FLY-ASH SLUDGE COLLECTOR

13. SUPPORT FRAME

14. SUPPORT PIERS

15. AFTERBURNER CHAMBER

16. PRECOOLER

Nonpumpable wastes are handled by a pack-and-drum feeder, a double air lock, and a drum conveyor. The conveyor moves the drums to the pack-and-drum feeder. Operating personnel remove the lids from the drums that are moving on the conveyor to visually inspect the contents of the drums. The drums are charged one at a time into the kiln at a predetermined rate commensurate with the heating value of the drummed waste. A vise-type device automatically grasps the drum as it enters the air lock. The operator then has the option of tipping the drum to discharge the contents into the kiln or releasing the drum itself into the kiln if the waste will not drain from it.

The rotary-kiln incinerator system is equipped with a secondary combustion chamber to allow for the oxidation of combustible solids suspended in the gas streams from the kiln. This chamber is operated at 820–890°C at a 1-s residence time, which is sufficient for complete oxidation of 1-micron combustible solids. Air flow through the kiln and secondary chamber is induced by a fan downstream of the wet scrubber air pollution equipment. After waste is introduced to the kiln, the only means of controlling temperature is by controlling the flow of air through the incinerator. Addition of waste must, therefore, be constant to maintain a steady input of energy to the incinerator.

The air emission control system of the rotary-kiln incinerator consists of five major components: a quench chamber, a venturi scrubber, a mist separator, an induced draft fan, and a 60-m stack. The quench chamber serves to cool the off-gas stream to about 80°C to eliminate the need for refractory linings in the remaining chambers. The quench tank itself is lined with acid-resistant brick and mortar. Halogen halides such as hydrogen chloride are present in the off-gas, which requires the use of corrosion-resistant materials.

The venturi scrubber is designed for removal of particulate matter down to 0.1 micron in size. A water-spray header with atomizing nozzles is added to the venturi throat for high-efficiency particulate removal. Fine water droplets entrained in the gas stream are removed with a demister. The incinerator is restricted to a particulate emission standard of 0.23 g per standard cubic meter of dry exhaust gas. This figure is adjusted to a 12% carbon dioxide concentration as required by regulation.

Miscellaneous Types of Incinerators

Multiple-Hearth Furnaces

The multiple-hearth furnace is the most widely used incineration system because of its simplicity, durability, and flexibility. This type of unit was initially designed to incinerate sewage plant sludges in 1934 and has been used quite successfully in this application (Ross, 1979; Ottinger et al., 1973; Lund, 1971). It is most suitable for wastes that are difficult to burn or wastes that have valuable metals that can be recovered.

The incinerator consists of a refractory-lined circular steel shell, with refractory hearths located one above the other. Solid waste or partially dewatered sludge is fed to the top of the unit, where a rotating rake plows it across the hearth to drop holes. The uncombusted material falls to the next hearth, and the process is repeated until eventually combustion is complete and ash is discharged at the bottom. Combustion air flows countercurrent to the sludge, and the exhaust gases exit at the top of the incinerator. Supplemental fuel is introduced at the bottom of the incinerator with air. In the upper zone of the incinerator the incoming solid waste or sludge is heated by the hot exhaust gases. Temperatures of around 540°C are typical in this zone. In the middle zone, gases and solids are burned at temperatures of 870°C to 980°C. In the lower zone, the ash is cooled by incoming combustion air.

Multiple-hearth furnaces are large and expensive and generally use large amounts of auxiliary fuel to process low-calorific wastes. Maximum temperatures are limited to 980°C to protect the hearths, but residence times up to several hours are possible. These units have seen only limited service for hazardous wastes.

Fluidized-Bed Incinerators

Fluid-bed technology from the petroleum and chemical processing industries has been adapted to incineration of wastes. The most common application involves the disposal of sludges or slurried wastes (Ross, 1979; Ottinger et al., 1973; Lund, 1971). The major processing steps are

1. Grit removal to protect unit from abrasion
2. Sludge thickening
3. Solids size reduction
4. Dewatering
5. Incineration
6. Exhaust gas treatment and ash disposal

The incinerator consists of a bed of hot sand or alumina into which air and the waste are injected. The incinerator operates at a pressure of about 2 psig and a temperature of 760°C to 810°C. Lower-temperature operation is avoided to ensure odor control. The sludge is fed at the bottom of the incinerator just above a distributor plate. Fluidizing air enters below the distributor plate. The sludge dries and oxidizes, with much of the heat being transferred to the sand or alumina bed. The combustion gases and the ash leave at the top of the reactor. An auxiliary burner is used to heat the bed to temperature prior to feeding sludge. Once the unit has reached the proper operating temperature, this auxiliary burner can be used to incinerate liquid or gaseous wastes.

Fluid-bed incinerators are relatively new and are becoming increasingly popular for sludge incineration. Some of the advantages of these units are (1) good mixing of sludge and air, (2) no moving parts (less maintenance), (3) heat exchange takes place within the sand bed, requiring fewer heat exchanges for efficient operation, and (4) the sand bed serves as a heat reservoir, permitting intermittent operation without excessive temperatures.

The fluidized bed has a high capital cost and should be used only for processing large quantities of waste that are difficult to incinerate by conventional systems. Its use for hazardous waste incineration has been limited.

Catalytic Incineration

Catalytic incinerators are used primarily for burning dilute, combustible, waste gas streams (Ross, 1979; Ottinger et al., 1973). The waste gas is preheated and then contacts a catalyst supported on porous media. Oxidation takes place on the surface of the catalyst. Most catalytic reactions can be carried out at lower temperatures, 320°C to 540°C, resulting in a significant fuel saving.

Because transporting gases over significant distances is not economical, gas incinerators are typically found at the site of waste gas production. Gas incinerators are common in the chemical process industries for incineration of solvents and destruction of odors. These are also found in petroleum refineries for the disposal of waste gases. Catalytic incinerators are generally sensitive to temperature (920° C maximum) and thus cannot be used for fuel-rich mixtures.

Pyrolysis

Organic materials can be transformed to different solid, liquid, or gaseous compounds by thermal decomposition in the absence of oxygen. Much work has been done to develop a pyrolysis system for converting refuse into useful fuel, but the results have been disappointing (Serper, 1981).

Molten-Salt Incineration

Hazardous organic wastes can be oxidized in a molten-salt medium that captures toxic combustion products such as halogen halides and sulfur oxides (Yosim et al., 1980). Molten alkali carbonates or hydroxides or mixtures thereof are useful for this purpose. A very high degree of destruction of chemical agents has been demonstrated, but as of the date of this writing, this process has not progressed beyond the demonstration stage.

Incineration at Sea

Incineration of hazardous organic wastes at sea by European vessels has been reported to be an environmentally acceptable and cost-effective means of disposal for these wastes. However, many people in Europe and the United States now claim that adequate demonstration of hazardous waste incineration at sea has not been performed. Several countries in Europe are currently planning to ban this method.

The cost of incineration at sea is less than for on-shore facilities, primarily because of elimination of expensive off-gas treatment systems. Toxic gases such as hydrogen halides are safely absorbed by the sea. The U.S. Environmental Protection Agency has recommended that a U.S. incinerator ship be constructed that is capable of destroying liquid wastes at sea and that can explore extending the capability of at-sea incineration to solid or semisolid materials (Johnson et al., 1981). Liquid-injection incineration and rotary-kiln incineration were recommended for use in this program. The first venture into at-sea incineration by a U.S. firm has begun recently with the purchase of the German ship *Vulcanus* by Chemical Waste Management, Inc. Subsequently, U.S. shipbuilders have planned construction of several new vessels. Controversy over permitting has forestalled all but a limited number of test runs. As a result, the Vulcanus has been operated out of Europe pending clarification of the U.S. position.

Cost of Incineration

Liquid-Injection Incinerators

A wide range of costs is possible, depending on the size of the incinerator and the type of off-gas treatment system employed. Operating costs in the range of $0.25 to $65.00 per 1000 l have been reported (Cross, 1980).

Rotary-Kiln Incineration

Capital and operating costs are also highly variable for rotary-kiln incinerators, for much the same reasons as liquid-injection incinerators. Uninstalled costs for a rotary-kiln incinerator are reported to range between $1100 and $2200/m³ of kiln, with installation costs being about 200% of these figures (Cross, 1980). Installed costs for small industrial systems vary between $2750 and $5500 per daily metric ton of capacity. Maintenance costs for the kiln average about 5–10% of the total installed cost. No operating cost data are available.

NOTES

1. Destruction efficiency is

$$\frac{W_{in} - W_{out}}{W_{in}} \times 100$$

where W_{in} is mass feed rate of principal toxic components fed into the incinerator and W_{out} is mass emission rate of principal toxic components in effluent from the combustion zone.

REFERENCES

Argonne National Laboratory. 1977. *Environmental control implication of generating electric power from coal.* Status report ANL/ECT-3, Appendix D.

Cross, F. L., Jr. 1980. Incineration of hazardous wastes. In *The handbook of hazardous waste management*, ed. A. A. Metry, pp. 310–22. Westport, CT: Technomic Publishing.

Dean, J. A. 1973. *Lange's handbook of chemistry*, 11th ed. New York: McGraw-Hill.

Eggers, D. F., Jr.; Gregory, H. W.; Halsey, G. D., Jr.; and Rabinovitch, B. S. 1964. *Physical chemistry*. New York: Wiley.

Hackman, E. E., III. 1978. *Toxic organic chemicals—destruction and waste treatment*, p. 73. Park Ridge, N.J.: Noyes Data Corporation.

Hazardous Waste News. 1981. Missouri hazardous waste task force recommends phasing out land disposal. *Hazardous Waste News*, Vol. 3, No. 41, p. 327. Business Publishers, Inc., Silver Spring, MD, October 19, 1981.

Hitchcock, D. A. 1979. Solid waste disposal: incineration. *Chemical Engineering*, p. 185, May 1979.

IMCO. 1978. *Regulations for the control of incineration of wastes and other matter at sea.* Addendum adopted 10-12-78 to *Convention of the prevention of marine pollution by dumping of wastes and other matter* (The London Convention), Intergovernmental Maritime Consultative Organization, completed 12-29-72 and entered into force 8-30-75.

Johnson, R. J.; Weller, P. J.; Oberacker, D. A.; and Neighbors, M. L. 1981. *Project summary: report on the interagency ad hoc work group for the chemical waste incineration ship program.* EPA-600/52-81-037.

Kanury, A. M. 1977. *Introduction to combustion phenomena.* New York: Gordon & Breach.

Lewis, C. R.; Edwards, R. E.; and Santora, M. A. 1976. Incineration of industrial wastes. *Chemical Engineering*, Vol. 83, No. 2, pp. 115–121, October 18, 1976.

Lund, H. F. 1971. *Industrial pollution control handbook.* New York: McGraw-Hill.

Ottinger, R. S.; Blumenthal, J. L.; Dal Porto, D. F.; Grubes, G. L.; Santy, M. J.; and Shik, C. C. 1973. *Recommended methods of reduction, neutralization, recovery or disposal of hazardous waste. Vol. III.* Report PB 224582, Springfield, VA: National Technical Information Service.

Pimental, G. C.; and Spratley, R. D. 1970. *Chemical bonding clarified through quantum mechanics.* San Francisco: Holden-Day.

Reed, J. C.; and Moore, B. L. 1980. Ultimate hazardous waste disposal by incineration. In *Toxic and hazardous waste disposal, Vol. 4, New ultimate disposal option*, ed. R. B. Pojasek. Ann Arbor: Ann Arbor Science Publishers. pp. 163–174.

Ross, R. D. 1979. The burning issue: incineration of hazardous wastes. *Pollution Engineering*, Vol. 11, No. 8, p. 25.

Santoleri, J. J. 1973. Chlorinated hydrocarbon waste disposal and recovery systems. *Chemical Engineering Progress*, Vol. 69, No. 1, pp. 68–80, January 1973.

Scurlock, A. C.; Lindsey, A. W.; Fields, T., Jr.; and Huber, D. R. 1975. *Incineration in hazardous waste management.* Environmental Protection Agency report EPA/530/SW-141.

Serper, A. 1981. Consider alternatives to landfilling for disposal of hazardous wastes. *Solid Wastes Management*, Vol. 24, No. 2, p. 62, February 1981.

Smith, J. M.; and Van Ness, J. C. 1975. *Introduction to chemical engineering thermodynamics*, 3rd ed. New York: McGraw-Hill.

Toxic Materials News. 1981. California begins program to end land disposal of toxic substances. *Toxic Materials News*, Vol. 8, No. 21, p. 323, October 21, 1981.

Witt, P. A., Jr. 1971. Disposal of solid wastes. *Chemical Engineering*, Vol. 78, No. 22, pp. 62–78, October 4, 1971.

Yosim, S. U.; Barsley, K. M.; Gray, R. L.; and Grantham, L. F. 1980. Disposal of hazardous wastes by molten salt combustion. In *Toxic and hazardous waste disposal*, ed. R. B. Pojasek, Ann Arbor: Ann Arbor Science Publishers. pp. 227–242.

11

Landfill Disposal

Introduction

Landfill disposal refers to the placement of waste on or beneath the surface of the ground with the intent of isolating the wastes from the environment. The waste is covered with soil or some other material to keep the waste in place and enhance the appearance of the site. Unfortunately, isolation or containment of wastes has been much less than satisfactory at many hazardous waste landfills in the United States. This has prompted legislation to correct deficiencies in the collection, transport, and disposal of hazardous wastes (Resource Conservation and Recovery Act) and to remedy unsafe conditions at operating and abandoned hazardous waste processing and disposal sites across the country (total estimated to be 8000) (Deland, 1981). At the outset of the Superfund program, the U.S. Environmental Protection Agency identified 114 hazardous waste sites requiring immediate remedial action to reduce the threat of escaping toxic materials (Inside EPA, 1981). Of the 114 sites listed, 33 were recorded as operating or abandoned landfills. Leaching of toxic components of the wastes from a landfill into groundwater is a common failure of poorly engineered hazardous waste landfills. Sometimes the primary failure is not the result of poor landfill engineering but is caused by inadequate control to protect the landfill from encroachment by the public, as in the case of the Love Canal incident (*Chemical Engineering*, 1979; *Solid Wastes Management*, 1980).

Landfills for hazardous wastes will be much more strictly regulated in the future by federal and state agencies than in the past. The increased cost of landfilling of hazardous wastes under these regulations is expected to reduce the amount sent to landfills for disposal. Although alternative methods of disposal, such as incineration, may be currently replacing landfilling in some instances, landfills are nevertheless anticipated to be the predominant method in the near term for disposal of the nearly 57 million metric tons per year of such waste produced in the United States (*Environmental Science*, 1981).

Over the long term, the goal is to implement waste destruction, reduction at the source, and reuse as the preferred methods of dealing with hazardous wastes (U.S. EPA, 1980). The state of California has initiated a program to phase out land disposal of millions of tons of toxic waste (*Toxic Materials News*, 1981). The cooperative efforts of state, private industry, and local governments are being solicited to help with the construction of new advanced facilities that will enable California to recycle, treat, and destroy almost 75% of all of the hazardous wastes that would otherwise be deposited in landfills. Similarly, 29 other states were found to have policies encouraging alternatives to landfilling as of 1982 (Table 11-1).

Table 11-1 / States with Policies Encouraging Alternatives to Landfilling of Hazardous Waste (National Conference of State Legislatures, 1982).

	AL	AZ	CA	CT	FL	GA	IL	IN	IA	KS	KY	ME	MD	MA	MI
Fee structure	x		x		x	x	x	x	x	x	x				x
Tax incentives															x
Bonds					x	x	x								
Definitional exclusions				x											x
Fast-track permitting								x		x	x				
Land burial restrictions			x			x								x	
Waste exchanges														x	
Research and development programs						x								x	
Other	x		x	x	x							x	x	x	

	MN	MS	MO	NH	NJ	NY	NC	OH	OK	OR	RI	SC	TN	WV	WI
Fee structure			x	x		x	x					x	x		x
Tax incentives	x					x	x			x			x		x
Bonds		x				x	x								
Definitional exclusions															
Fast-track permitting	x		x	x							x				
Land burial restrictions			x						x						
Waste exchanges			x			x	x		x					x	x
Research and development programs			x												
Other	x	x				x									

Although landfilling of hazardous wastes may be phased out to a large extent in the future, it will be necessary to rely on this method for much of these wastes for the next decade or so. As a consequence, landfilling must be carried out in a manner that will minimize the risk to public health and welfare. Key preventive measures for minimizing the risk include (Metry, 1980)

1. Siting waste disposal sites in environmentally favorable areas
2. Incorporating environmental and safety considerations in design, construction, and operation of ultimate disposal facilities
3. Fixation and encapsulation
4. Environmental surveillance and monitoring of water and air quality
5. Preparing contingency plans for counteracting spills, fires, explosions, and contamination of air, water, and land resources
6. Proper closure and perpetual care of completed disposal sites

The first item has been addressed in Chapter 5, which contains geologic and hydrologic information that is critical to the proper selection of a landfill disposal site. The remaining items from this list are the subjects of this chapter.

Landfill Design

In order to understand the important features of secure landfill design, it is best to first review the pathways by which contaminants move from facilities. Consider that if the cap is disturbed, contaminant movement may arise from surface runoff, vapor loss, wind-activated sediment suspension, or infiltration, followed by the generation of leachate. Of these, the last is most often encountered and usually results in problems that are difficult and costly to rectify. This is the focus of the examples provided here.

There are several modes in which chemical contamination can migrate to an aquifer from a disposal site. In the arid west or other areas where infiltration is limited, a release from a storage tank or impoundment constitutes a plug flow of a finite volume through the unsaturated zone. The plug will progress downward and, if the water table is relatively deep, expend its driving force until its advance is slowed to a rate of centimeters per year. Such an incident occurred at Hanford, where over 100,000 gal of waste in an underground tank leaked out over a period of a month. Both mathematical models and drilling programs confirmed that after the fluid had penetrated 80 ft, it hit a caliche layer and virtually stopped moving (Dawson and Cearlock, 1981).

In wetter climates, or where facilities hold much larger volumes of liquid, the driving force is not likely to diminish before the plume reaches the aquifer. This is the most common scenario, and the most difficult to mitigate.

The contaminants flow continuously in a column until entry as a ribbon in the aquifer. At that point, flow and attenuation mechanisms determine the fate of the chemicals involved. Contaminants may also be brought to the surface by deep-rooted plants or burrowing animals.

Because of the more serious nature of potential releases via any of the foregoing pathways, secure landfills for hazardous waste disposal require additional design features over conventional sanitary landfills to provide long-term protection of groundwater, surface water, air, and human health. Although the technology of secure landfilling is still in the developmental stage, a number of techniques are available for reducing the adverse effects of landfilling of hazardous wastes. Standards have been issued by the EPA on most of these techniques (e.g., control of surface water runon and runoff and liner design); however, "best engineering judgment" will be in effect for some time as exemptions are sought for existing interim status facilities.

Environmental Performance Standard

Interim standards were issued by the EPA on February 13, 1981, to go into effect on August 13, 1981, that included an environmental performance standard (40 CFR 267.10 Subpart B) for new hazardous waste landfills. This standard specified that all new hazardous waste landfills shall be located, designed, constructed, operated, and closed in a manner that will assure the protection of human health and the environment. Protection of human health and the environment would include but not be limited to the following:

A. Prevention of adverse effects on groundwater quality considering
   1. The volume and physical and chemical characteristics of the waste in the facility, including its potential for migration through soil or through synthetic liner materials
   2. The hydrogeologic characteristics of the facility and surrounding land
   3. The quantity, quality, and directions of groundwater flow
   4. The proximity and withdrawal rates of groundwater users
   5. The existing quality of groundwater, including other sources of contamination and their cumulative impact on the groundwater
   6. The potential for health risks caused by human exposure to waste constituents
   7. The potential damage to wildlife, crops, vegetation, and physical structures caused by exposure to waste constituents
   8. The persistence and permanence of the potential adverse effects
B. Prevention of adverse effects on surface-water quality, considering
   1. The volume and physical and chemical characteristics of the waste in the facility
   2. The hydrogeologic characteristics of the facility and surrounding land, including the topography of the area around the facility

3. The quantity, quality, and directions of groundwater flow
4. The patterns of rainfall in the region
5. The proximity of the facility to surface waters
6. The uses of nearby surface waters and any water quality standards established for those surface waters
7. The existing quality of surface water, including other sources of contamination and their cumulative impact on surface water
8. The potential for health risks caused by human exposure to waste constituents
9. The potential damage to wildlife, crops, vegetation, and physical structures caused by exposure to waste constituents
10. The persistence and permanence of the potential adverse effects

C. Prevention of adverse effects on air quality, considering
1. The volume and physical and chemical characteristics of the waste in the facility, including its potential for volatilization and wind dispersal
2. The existing quality of the air, including other sources of contamination and their cumulative impact on the air
3. The potential for health risks caused by human exposure to waste constituents
4. The potential damage to wildlife, crops, vegetation, and physical structures caused by exposure to waste constituents
5. The persistence and permanence of the potential adverse effects

D. Prevention of adverse effects due to migration of waste constituents in the subsurface environment, considering
1. The volume and physical and chemical characteristics of the waste in the facility, including its potential for migration through soil
2. The geologic characteristics of the facility and surrounding land
3. The patterns of land use in the region
4. The potential for migration of waste constituents into subsurface physical structures
5. The potential for migration of waste constituents into the root zone of food-chain crops and other vegetation
6. The potential for health risks caused by human exposure to waste constituents
7. The potential damage to wildlife, crops, vegetation, and physical structures caused by exposure to waste constituents
8. The persistence and permanence of the potential adverse effects

Compliance with this standard will be required for (1) the design of the impermeable liners, (2) the design and operation of leachate and runoff control systems, (3) closure and post-closure activities, and (4) any additional measures deemed necessary by the regulatory agency.

Subsequent guidance for landfill design, liner systems, and final cover were drafted in mid-1982 for use by permit writers and applicants. The regulations themselves required, at a minimum, a leachate collection-and-removal system and one liner. The former system is to reduce the head on the liner to one foot or less. The liner should prevent migration of liquids, allowing only *de minimis* infiltration of liquids. Synthetic membrane liners are required. If final cover will not be applied for 30 years or more, both primary and secondary leachate collection systems should be installed, and a secondary soil liner should be emplaced beneath the synthetic one.

The new RCRA amendments now make double liners mandatory, but there is some degree of discretion in the sidewall liner design.

Specific guidance for achieving a leachate head ≤ 1 foot includes the following:

≤ 30 cm drainage layer with hydraulic conductivity $\geq 1 \times 10^{-3}$ cm/sec and slope $\geq 2\%$.

Granular or fabric filter above the drainage layer (primary collection system only). Granular filter should meet the following criteria:

$$\frac{D15 \text{ (filter soil)}}{D85 \text{ (drainage layer)}} \leq 5$$

$$\frac{D50 \text{ (filter soil)}}{D50 \text{ (drainage layer)}} \leq 5$$

and

$$\frac{D15 \text{ (filter soil)}}{D15 \text{ (drainage layer)}} = 5\text{-}20$$

where $D15$ is the grain size, in millimeters, at which 15% of the filter soil used, by weight, is finer; $D85$ is the grain size, in millimeters, at which 85% of the filter soil or layer media, by weight, is finer; and $D50$ is the grain size, in millimeters, at which 50% of the filter soil or drainage media, by weight, is finer.

A drainage tile system and sump pump for removal of leachate.

Synthetic liners should be a minimum of 30 mil thick and consist of a material selected on the basis of compatibility with wastes anticipated at the site. Liners should be protected from above and below with at least 15 cm of sand bedding or finer material containing no large or sharp objects. Soil liners should consist of at least 60 cm of natural or recompacted emplaced soil with a saturated hydraulic conductivity $\leq 1 \times 10^{-7}$ cm/sec after contact with anticipated leachates.

In order to meet the regulatory requirement for caps or final cover to minimize infiltration, they should be installed as each cell is completed.

Where multiple lifts are employed, interim cover is encouraged. For large cells, cover should be applied progressively with completed fill. Caps should include:

A vegetative cover ≥60 cm thick utilizing flora that requires a minimum of maintenance and has no deep boring roots, having a final slope of 3 to 5%, and incorporating a surface drainage system that prevents erosion rills.

A middle drainage layer ≥30 cm with hydraulic conductivity $\geq 1 \times 10^{-3}$ cm/ sec, a bottom slope ≥2%, and a top-graded granular or fabric filter to promote lateral movement of fluids.

A low-permeability bottom layer incorporating an uppermost layer of synthetic membrane ≥ 20 mil thick, at least 15 cm of upper and lower bedding material (sand free of large or sharp objects), a final upper slope ≥2%, and a lower component ≥60 cm of recompacted soil with a hydraulic conductivity $\leq 1 \times 10^{-7}$ cm/sec emplaced at no more than 15 cm at a time before compaction.

Secure Landfill Design in High-Precipitation Regions

The design of a secure landfill is dependent on a number of factors, most of which are listed under the environmental performance standard given in the previous section. Climatic conditions can be added to the list because of the substantial input that the evapotranspiration/precipitation ratio can have on the landfill design requirements. Many arid or semiarid zones in the United States have evapotranspiration rates that greatly exceed precipitation; therefore, little or no water moves from the surface of the ground to the water table, which typically lies at a considerable distance below the surface. This effectively eliminates a leachate problem if only dry waste is disposed in the landfill and a suitable cover is placed over the landfill. Unfortunately, the major centers of hazardous waste production are for the most part far removed from these excellent landfill disposal sites.

Most of the eastern half of the United States, where the majority of the hazardous wastes are produced, has abundant precipitation that exceeds the evapotranspiration rate during the course of an average year. There is a high probability in this region that leachate will be generated in waste landfills, and control measures will be necessary to prevent the leachate from contaminating groundwater and surface-water supplies. Figure 11–1 illustrates a design of a secure landfill that has a cover to control the amount of leachate generated by precipitation and provides a backup system to collect and remove leachate if the cover should fail or be removed. The principal design features are as follows:

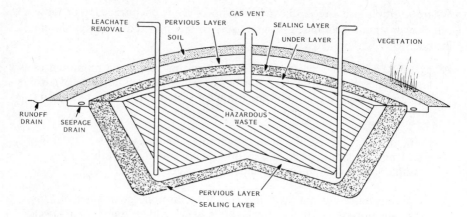

Figure 11-1 / Schematic of a secure landfill with a liner/drain-layer system in the cover and at the bottom.

1. A layer of topsoil over the landfill to provide vegetation for stabilization of the cover and to promote evapotranspiration of moisture that infiltrates the cover
2. A drainage system at the edges of the cover to transfer runoff away from the site; the topsoil cover is sloped to enhance runoff of precipitation
3. A highly permeable layer of sand or gravel between the soil cover and the sealing layer to divert infiltration to drains located at the sides of the landfill
4. A sealing layer such as fine clay or plastic membrane over the landfill to attenuate or prevent infiltration of precipitation into the waste
5. Buried waste that may be located in cells surrounded by fill material (e.g., soil) for segregation
6. An underlayer of material such as fine soil or sand to provide a base for the sealing layer
7. A venting system to remove gases such as methane and carbon dioxide that may be generated by biodegradation of organic waste in the landfill
8. A drainage layer of sand or gravel beneath the buried waste to collect leachate and divert it to drain at the edge of the landfill for removal to the surface
9. A bottom sealing layer to prevent the leachate from infiltrating to groundwater

Although prevention of leachate formation is a very desirable objective in a hazardous waste landfill design, this is very difficult to achieve for a broad spectrum of waste forms. In order to maintain the integrity of a sealed landfill cover, differential or uneven settling must be minimal, and this is simply not possible for poorly compacted waste, wet sludges, or any material that degrades to gases or liquids that flow out of the landfill. However, the long-range objective of hazardous waste management is to eliminate poorly com-pacted and degradable materials (which are mostly organic) from landfills through the use of alternative disposal technology. Landfills would then be used only for stable, solid, dry wastes that cannot be altered to safer forms by practical means.

If considerable subsidence of the landfill is anticipated, a well-sealed cover may not be practical; therefore, protection of groundwater from leachate contamination will depend on liner/drain-layer systems at the bottom of the landfill, as shown in Figure 11–2. The primary drainage and impervious layer diverts leachate to a collection system for removal and treatment at the surface. The secondary drainage and impervious layer acts as a backup and leak detection system. The cover of the landfill is constructed to provide a vegetative cover to promote evapotranspiration and is sloped to carry runoff to a drainage system. Such a cover should prevent most of the precipitation from entering the waste, but this will not reduce it to the levels possible with a sealed cover, as in Figure 11–1.

The double liner/drain-layer system illustrated in Figure 11–2 would be suitable for areas where there is a possibility that groundwater might inun-date the landfill. In a dual bottom liner system, one liner is the primary barrier, and the second liner is a backup barrier in case of failure of the primary barrier. The permeable layer of sand or gravel is placed between the barriers to remove infiltration water. This permeable layer is also sloped to collect the infiltration water for removal.

Recent work with uranium mill tailings has identified several key design changes that can greatly improve the efficiency of caps (Gee et al., 1981).

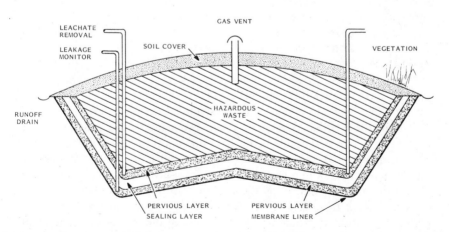

Figure 11-2 / Schematic of secure landfill with double liner/drain-layer systems at the bottom.

Whereas low-permeability materials such as clays can be effective barriers to infiltration, their characteristic high surface areas and small pore diameters may promote capillary movement and hence vapor loss. As a consequence, a multiple-layer design is more desirable. In the latter case, a layer of coarse gravel is inserted above the clay, but beneath the overburden, to form an effective barrier to capillary migration. It was also discovered that a layer of gellike material such as certain hygroscopic sludges could be employed with native soils in much smaller amounts as a replacement for clay. As a consequence, this moisture-holding layer stops infiltration at a lower cost than imported clays and is the alternate choice where clay cap material must be purchased and shipped. A comparison of the effectiveness of clay and gel caps can be seen in Table 11–2, based on the loss of radon from uranium mill tailings.

Efficiency of Liner/Drain-Layer Systems

Estimating the efficiency of a landfill to limit the amount of precipitation infiltrating the waste may be accomplished by several different methods. These methods generally involve computer models that can vary considerably with regard to complexity. A relatively simple model, HSSWDS, for computing approximate values of runoff/infiltration for landfills is described in a permit writer's guidance manual/technical resource document issued by the EPA (Perrier and Gibson, 1980). This model utilizes empirical data derived from measurements of runoff and drainage from a number of different soil types. Minimum input data include geographic location, site area and hydrologic length, and characteristics of final soil and vegetative cover. The model will simulate daily, monthly, and annual runoff, deep percolation, temperature (soil–water), and evapotranspiration.

More accurate estimates of runoff/infiltration are possible through the use of models that incorporate a theoretical approach to describing the unsaturated flow of precipitation through soils (Gupta et al., 1978; Hillel, 1977; Feddes et al., 1978) (e.g., under agricultural conditions). The accuracy of these unsaturated flow models is highly dependent on the quality and quantity of site-specific input data. These models require a greater degree of skill than the HSSWDS model but are being used on a routine basis by engineering firms to provide better data on runoff infiltration.

When inflow through the soil-vegetation cover of a landfill has been established, the amount diverted by liner/drain-layer systems (see Figure 11–1) and the amount infiltrating the waste or flowing to groundwater can be estimated. That flowing to groundwater represents leachate that may carry pollutants causing degradation of the quality of the groundwater. A simplified method of estimating the effectiveness of landfill liner/drain-layer systems is given by Wong (1977). This method may also be used for liner/drain-

Table 11-2 / Comparison of Radon Flux from Clay and Gel Caps.

Gel Layer (0.15 m)			Gel Layer (0.30 m)			Bentonite with Lime (0.40 m)			Bentonite (0.30 m)		
Plot No.	Pre	Post	Plot No.	Pre	Post	Plot No.	Pre	Post	Plot No.	Pre	Post
3	134.8	12.5	23	115.0	39.1	43	145.8	10.5	63	155.9	8.0
4	526.9	0.5	24	102.2	29.8	44	320.7	3.7	64	170.1	5.9
5	687.7	1.3	25	332.5	6.9	45	266.8	14.9	65	203.0	4.2
6	670.1	0.8	26	184.2	21.3	46	167.2	3.7	66	268.9	5.6
7	702.9	1.0	27	172.4	8.0	47	109.2	3.4	67	82.1	0.5
8	278.3	1.2	28	58.0	5.9	48	131.5	4.0	68	133.6	1.7
9	307.5	1.7	29	78.8	13.2	49	105.7	4.3	69	274.4	1.0
10	205.6	0.7	30	91.0	14.2	50	121.6	4.2	70	109.4	1.4
13	111.7	1.1	33	137.9	—	53	194.4	7.6			
14	558.6	25.8	34	493.5	16.4	54	316.8	2.1			
15	398.6	0.4	35	650.8	3.7	55	534.5	10.6			
16	485.8	1.0	36	460.5	1.1	56	731.9	4.2			
17	412.6	0.3	37	236.0	3.3	57	355.6	7.0			
18	203.0	0.8	38	201.5	3.8	58	119.3	0.4			
19	76.5	0.3	39	127.3	6.2	59	138.8	3.0			
20	106.4	1.1	40	171.8	8.2	60	86.2	7.1			
Average:											
336.7 ± 219.5		3.6 ± 6.7	225.8 ± 171.2		12.1 ± 10.7	240.4 ± 179.0		5.7 ± 3.7	174.7 ± 70.2		3.5 ± 2.7

layers in the cover of a landfill (Moore, 1980). The method assumes that a rectangular slug of liquid infiltrates instantaneously to the top of the clay liner illustrated in Figure 11-3 at a depth of H_0. The liquid then flows laterally through a pervious layer and vertically through the clay liner. The fraction of liquid that has flowed laterally at time t is given by

(11.1)

$$\frac{s}{s_0} = 1 - \frac{t}{t_1}$$

where s is length of saturated volume at time t (cm), s_0 is initial length of saturated volume that drains in time t_1 (cm), h is thickness of saturated volume at time t (cm), h_0 is initial thickness of saturated volume (cm), and t_1 is time in seconds for all the liquid to flow out of the drain layer. The fraction of liquid infiltrating the clay liner is given by

(11.2)

$$\frac{h}{h_0} = (1 + \frac{d}{h_0 \cos \theta}) e^{-Ct/t_1} - \frac{d}{h_0 \cos \theta} \quad (0 \le t \le t_1)$$

where

(11.3)

$$t_1 = \frac{s_0}{K_1 \sin \theta}$$

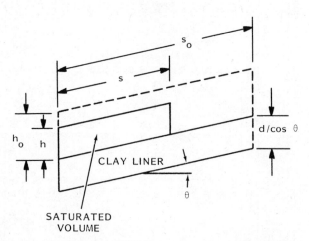

Figure 11-3 / Geometry for calculating efficiency of drain/liner systems. (Adopted from Wong, 1977.)

(11.4)

$$C = (\frac{s_0}{d}) (\frac{K_2}{K_1}) \cot \theta$$

and K_1 is the saturated permeability of material above clay liner (cm/s), K_2 is the saturated permeability of the clay liner (cm/s), θ is the slope angle of the liner (degrees), and d is the thickness of clay liner (cm).

The efficiency of the liner can be determined from a plot of h/h_0 versus s/s_0 or t/t_1, as shown in Figure 11–4 (curve can be approximated as a straight line). The efficiency of the system is given by the area, f, under the line, which can be computed from n (the value of h/h_0 at $t/t_1 = 1$),

(11.5)

$$n = (1 + \frac{d}{h_0 \cos \theta}) e^{-C} - \frac{d}{h_0 \cos \theta}$$

as follows:

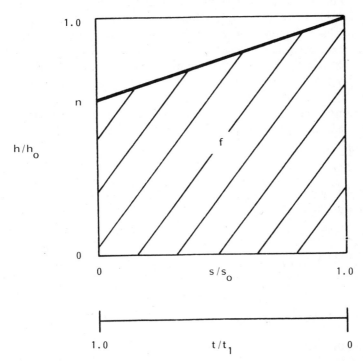

Figure 11-4 / Diagram for computing efficiency of drain/liner system. (From Moore, 1980.)

(11.6)

$$f = \frac{1 + n}{2} \quad \text{for} \quad n \geq 0$$

and

(11.7)

$$f = \frac{1}{2(1 - n)} \quad \text{for} \quad n \leq 0$$

The efficiency of the liner/drain system falls below 50% when n is less than zero. The area above the line in Figure 11–3 represents liquid infiltrating the liner.

The foregoing equations were derived on the basis that the clay liner is saturated and the liquid impinges on the liner instantaneously, which is not possible if porous material overlies the pervious drain layer and the precipitation event takes a finite amount of time. An unsaturated clay liner would have a slower rate of infiltration than a saturated clay liner and would divert more liquid to the drain. The error due to a finite rate of liquid impinging on the clay liner is small if this rate is much greater than the rate of lateral movement along the liner or the rate of infiltration into the liner. The thickness of the pervious drainage layer must also be sufficient to hold all the liquid penetrating the soil cover. Some mounding will actually occur in the drainage layer as the liquid moves laterally, and the mound must not penetrate into the overlying strata.

An example calculation for estimating the efficiency of a liner/drain layer in the cover of a landfill is as follows:

Conditions:
$h_0 = 2$ cm
$s_0 = 2000$ cm($\frac{\text{width of cover liner in figure 11–1}}{2}$)
$K_1 = 10^{-2}$ cm/s
$K_2 = 10^{-7}$ cm/s
$\theta = 5°$
$d = 100$ cm

From equation 11.3,

$$t_1 = \frac{2000}{(10^{-2})(\sin 5°)} = 2.29 \times 10^6 \text{ s} = 26 \text{ days}$$

From equation 11.4,

$$C = (\frac{2000}{100})(\frac{10^{-7}}{10^{-2}}) \cot 5° = 2.29 \times 10^{-3}$$

From equation 11.5,

$$n = (1 + \frac{100}{2 \cos 5°}) \, e^{-0.00229} - \frac{100}{2 \cos 5°} = 0.88$$

and from equation 11.6

$$f = \frac{1 + 0.88}{2} = 0.94$$

The effects of variable depth, d, of the liner and ratios of K_2/K_1 are illustrated in Figure 11-5, where h_0, s_0, and θ are held the same as in the example calculation earlier. These curves show that very little is gained by increasing linear depth above 10 cm. Liner thickness does not affect collection efficiency much as long as d is several times greater than h_0. The depth of the liner should be sufficient to assure the integrity of the liner under mechanical loading (e.g., heavy-equipment transport) and settling (Wong, 1977).

The ratio of K_2/K_1 has a major impact on the collection efficiency of a liner/drain-layer system, as seen in Figures 11–5 and 11–6. Collection efficiency remains high over a range of liner slopes from 1° (1.7%) to 6° (10.5%) for a K_2/K_1 ratio of 10^{-5}. Collection efficiency falls off rapidly with slope at K_2/K_1 ratios of 5×10^{-5} and 10^{-4}. Steep slopes on the curves of a landfill should be avoided to prevent erosion of the cover.

The height of liquid impinging on the clay liner also has a major impact on collection efficiency, as shown in Table 11–3, using the same parameters in the example calculation except for varying h_0 and setting K_2/K_1 at 1×10^{-4}. The collection efficiency increases from 0.42 at 2 cm of percolated liquid to 0.83 if an 8-cm depth of precipitation percolates through the vegetative-soil cover into the drain layer. The 8-cm depth would be equivalent to 2.8 cm (1.1 inches) of free-standing liquid without the porous medium (porosity = 0.35).

Estimating Impact of Leachate on Groundwater Quality

In areas of high precipitation it is reasonable to expect that some leachate will penetrate to the groundwater, because perfectly sealed liners have yet to be demonstrated. In most cases accurate forcasting of groundwater contamination is not possible because of the complexities of determining the amount of flow and flow paths of the leachate through the landfill. Furthermore, the physical chemistry of the leaching process under field conditions may not be understood well enough to give a good source term even if the flow path and rate are fully defined.

A worst-case approach is frequently used in a preliminary analysis to evaluate the potential impact of leachate infiltration to groundwater beneath

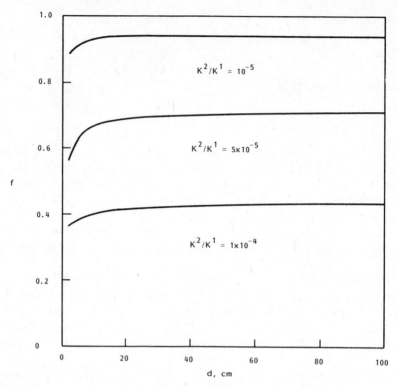

Figure 11–5 / Effect of liner depth, d, on efficiency, f (h_o, s_o, and Θ are given in example calculation).

a hazardous waste landfill. This approach is usually easy to accomplish because of the simplifying assumptions made in the analysis. To give an example: It has been determined through the use of simple models that a maximum of 1 cm of precipitation will infiltrate the sealed cover of a hazardous waste landfill over a period of 1 yr under the most adverse conditions apparent from historical precipitation records. If the 1 cm of water is evenly distributed throughout the year and the flow is impeded but not stopped by the bottom liner/drain-layer system, the waste and fill material, and the soil beneath the landfill, a constant flow of leachate to the groundwater may be assumed. Further, laboratory leach tests of the waste under the most severe conditions postulated for the landfill indicate that selenium is the controlling contaminant and might exist as a concentration as high as 1.0 mg/l in the leachate. Other conditions include (1) groundwater flow of 10 cm/day, (2) landfill width of 30 m and infinite length, (3) a producing well in the groundwater downgradient from the landfill with a 1500-cm screened intake that receives water uniformly throughout the length of the intake, (4) leach-

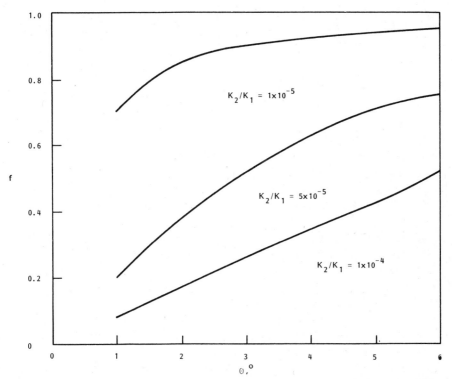

Figure 11-6 / Effect off cover slope, Θ on efficiency, f (h_o, s_o, and d are given in example calculation).

ate that is slightly more dense than the groundwater, (5) no attenuation of the selenium by adsorption or other means as it passes through the landfill and sediments below, and (6) a porosity of 0.35 in the strata containing the groundwater.

The problem may be viewed in two dimensions as in Figure 11-7. Although the leachate is not uniformly distributed in the aquifer, by the time it reaches the well it will be diluted by the uniform flow along the intake of the well. The dilution factor for the leachate plume to the well outlet is

Table 11-3 / **Effect of h_0 on Collection Efficiency of a Liner/Drain-Layer System with $K_2/K_1 = 10^{-4}$.**

Initial Liquid Depth, h_0 (cm)	Collection Efficiency, f
2	0.42
4	0.69
8	0.83

$$\text{dilution factor} = \frac{(365 \text{ days/yr}) (1500 \text{ cm}) (10 \text{ cm/day}) (0.35)}{(1 \text{ cm/yr}) (3000 \text{ cm})}$$
$$= 640$$

The selenium would be diluted to

$$\text{Se concentration (mg/l)} = \frac{1.0}{640} = 0.002 \text{ mg/l}$$

which would be below the primary drinking water standard of 0.01 mg/l for selenium.

The assumption of infinite length of the landfill maximizes the concentration of selenium by eliminating horizontal dispersion, which would tend to dilute the selenium, especially from a point source. The assumption of complete interception of the plume by the intake of the well also maximizes the selenium concentration as long as intake volume of uncontaminated water with contaminated water is a conservative estimate. The selenium concentration can be considerably less if the plume drops below the intake screen. If the leachate flow is not uniform throughout the year, a more conservative estimate can be made by assuming a shorter period (e.g., 3 months during winter) of flow.

Variations in groundwater flow due to changing precipitation and ground-water use patterns should also be considered in an impact analysis. Severe droughts and increased groundwater consumption, for example, can greatly decrease groundwater flows.

Secure Landfills in Regions of Low Precipitation

Penetration of precipitation through soils to groundwater in many areas of the western United States is either nonexistent or so limited that it is difficult to measure. This condition exists because of very light precipitation and evapotranspiration rates that greatly exceed precipitation. Moisture that collects in surface soils during periods of rain or melting snow is subsequently removed by direct evaporation from the soil or by the root systems of plants or grasses growing in the soil. If only dry waste is disposed into a secure

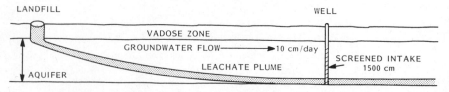

Figure 11-7 / Leachate plume intercepting a well.

landfill located in a region of low precipitation and a suitable vegetative cover is emplaced over the landfill, on final closure no leachate collection system will be required. The cover of these landfills should be shaped to promote runoff during periods of heavy precipitation (rare events) and should contain sufficient topsoil supporting natural plants or grasses to retain the moisture that collects during the wet periods. A layer of more permeable material such as sandy soil, sand, or gravel may be placed between the topsoil and the waste in the landfill to provide a barrier against moisture movement into the waste (Figure 11–8). Moisture will not move into this more permeable zone unless saturation is reached in the topsoil, which is a very infrequent event in many arid or semiarid regions. The HSSWDS model (Perrier and Gibson, 1980) may be used to determine the effectiveness of such a cover in a specific area of the United States.

Liner Materials

Although there are some areas in the eastern United States where natural barriers of highly impervious soils or sediments can be used for secure landfills, most areas will require some kind of man-made barriers to assure containment of the hazardous wastes. A wide variety of both natural and synthetic materials is available for constructing these barriers. Unfortunately, there is a lack of data concerning the effectiveness of these materials over long time periods (decades) when in contact with different types of hazardous wastes. Extensive laboratory testing of liner material has been undertaken under the sponsorship of the EPA (Haxo et al., 1977; Haxo, 1981), but at the present time laboratory data are useful only for comparing different types of liners with different wastes and cannot be used to predict the effectiveness of the liners under field conditions (Haxo, 1981). In the absence of field data for the liner materials, best engineering judgment will have to be used until such time as the field data become available.

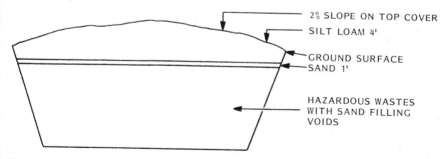

Figure 11-8 / Cross-sectional diagram of secure landfill for use in arid or semi arid areas.

The compatibility of a specific waste with a liner is the first consideration in the selection process. The principal waste properties that may have an adverse effect on a landfill liner are (Matrecon, 1980)

Acidic conditions, pH < 3.5
Alkaline conditions, pH > 10
Organic compounds
Oily wastes
Exchangeable cations such as Ca^{2+}

Characterization of a waste to be disposed in a lined secure landfill should therefore focus on these properties.

Materials available for the construction of liners include clay and clayey soils, admixes of asphalt or cement with soil or other materials, spray coatings, and polymeric membranes. A general compatibility matrix of waste types with various liner materials is given in Table 11–4. A rating of good (G) indicates that the combination is probably satisfactory, a rating of fair (F) means the combinations should be tested, and a rating of poor (P) indicates that the combination should be avoided. Descriptions of the different types of

Table 11–4 / Liner/Industrial Waste Compatibilities.[a]

Liner Material	Strong Caustic	Strong Acid	Organic Solvent	Oily Waste	Exchangeable Cations, e.g., Ca^{2+}
Soils					
Compacted clayey soils	F[b]	P	P-G[c]	G	P
Soil-bentonite	F	P	P-G[c]	G	P
Admixes					
Asphalt-concrete	G	F	P	P	G
Asphalt-membrane	G	G	P	P	G
Soil asphalt	F	P	P	P	F
Soil cement	F	P	P-G[c]	G	F
Polymeric membranes					
Butyl rubber	G	G	P	P	G
Chlorinated polyethylene	G	G	G	P	G
Polypropylene	G	G	G	G	G
Ethylene Propylene rubber	G	G	P	P	G
Polyethylene (low density)	G	P-G	F-G	G	G
Polyvinyl chloride	G	G	G	G	G

[a] Adapted from Stewart (1978).
[b] G = good; F = fair; P = poor.
[c] Clay liners will fail with some types of solvents (Green et al., 1981).

liners used for landfills and impoundments are given elsewhere (Matrecon, 1980).

Compatibility tests may be performed by several different methods. Cells may be constructed in the laboratory to simulate field conditions by exposing one side of a test liner to the waste under a hydraulic head while the other side overlies a crushed silica base. Leakage through the liners in the test cells is monitored over relatively long time periods (Haxo, 1981). Polymeric membranes have been tested by immersing samples of the membranes in waste mixtures for varying periods of time up to 12 months to determine the effect of exposure time. The membrane samples are subjected to elongation and swelling measurements after the exposure period to determine the change in these membrane properties compared with unexposed samples (Haxo, 1981). Pouch tests are also useful to determine the permeability of the membranes to wastes contained in a small pouch immersed in demineralized water over a period of time. Leakage to the demineralized water can be conveniently measured in this test (Haxo, 1981).

Some of the most recent work on liner–waste interactions was performed at Texas A&M (Anderson and Brown, 1981), where researchers evaluated the effects of leachate constituent on the permeability of clays. In that work, large increases in permeability were noted when two smectite clays were exposed to various neutral polar, neutral nonpolar, and basic organic fluids. In some cases, the final permeability will no longer meet design criteria for liners (Table 11–5).

Liner Installation

Proper installation of a liner for a secure landfill is very critical to achieving essentially complete isolation of the waste. The highest-quality liner materials cannot attain this goal if installation practices are faulty. Liner installation should be performed under the close supervision of an experienced engineer.

Preparation of the underlayer for the liner is the first task of construction. The underlayer should be placed over well-compacted subgrade material to

Table 11–5 / Final Permeabilities.

Contaminant	Initial Permeability (cm/s)	Permeability (cm/s)	At x Pore Volume
Acetic acid	10^{-8}	10^{-9}	0.5
Aniline	5×10^{-8}	5×10^{-6}	3.0
Acetone	10^{-9}	10^{-5}	1.5
Ethylene glycol	3×10^{-9}	3×10^{-6}	2.0
Heptane	3×10^{-9}	3×10^{-7}	1.0
Xylene	3×10^{-9}	10^{-5}	1.0

avoid settling. The primary purpose of the underlayer is to provide a firm smooth foundation for the liner. The underlayer should be free of large objects such as stones, roots, branches, and debris that could puncture the liner. Polymeric membranes are especially prone to punctures. Tests conducted to evaluate different methods of protecting membrane liners during installation indicate that fine sandy loam is preferable to coarse gravel on top of underlayers to minimize punctures caused by heavy-equipment traffic over the landfill (Gunkel, 1981). Organic matter that may degrade, causing gas pressure against the liner and ultimately creating a void below the liner, should not be incorporated in the underlayer.

Different types of liners will require different installation techniques. Polymeric membrane liners have essentially the same installation procedures except for welding at the seams. Some membranes, such as PVC, can be solvent-welded, and others require heat sealing, which is often difficult to accomplish in the field. Clay and clayey soils require careful compacting to assure a good seal. The moisture content of clay is critical to compaction. Paving asphalt can be applied by a conventional paving machine.

Protection of the liner by a drain layer of sand or a layer of soil over the liner is recommended to preserve its integrity. This cover should be at least 1 ft thick. The first lift of solid waste placed over the liner should not have protrusions, such as pipe that might penetrate the liner.

Groundwater Monitoring System

Groundwater monitoring around hazardous waste landfills is required to ensure that barrier systems that contain the hazardous wastes are functioning properly. Furthermore, if contamination is detected, early warning is given to initiate countermeasures before the contamination spreads very far. Federal regulations (40 CFR 265) set forth a minimum of four monitoring wells around a hazardous waste landfill. One of the wells should be located upgradient in the groundwater from the landfill and the other three downgradient around the limits of the landfill. The regulations set forth in detail how the wells should be implaced, screened, sealed, sampled, and located.

Although use of the "one-upflow, three-downflow" monitoring-well scheme will meet regulatory requirements, it will not necessarily provide the best information. If owner/operators wish to use monitoring data to avoid contaminant spread, they will be better served by individually designed monitoring programs that consider the specific geology of the site. It is naive to assume that a single layout plan will be optimal for all possible sites. Indeed, field studies have shown that in arid areas, limited fluid losses may move through tilted, unsaturated strata in directions far different than regional groundwater flow. Similarly, channels and discontinuities may allow plumes to bypass monitoring wells. Hence, well location is best determined after completion of site assessment work. The latter will provide, as one output, an analysis of the most likely models/locations of failures; these subsequent routes of contaminant migration are the optimum locations for monitoring wells. A final well should also be placed in an area that will not be affected by the site to help determine if contaminants observed in the monitoring wells are originating at another source. The owner/operator of the landfill is responsible for obtaining and analyzing the groundwater samples. Sample preservation and analysis are to be performed by established procedures (U.S. EPA, 1979). Groundwater quality must be maintained in keeping with the EPA's interim primary drinking water standards (Table 11–6). Additional analysis for chloride, sodium, ion, manganese, sulfate, and phenol must also be performed to monitor general water quality criteria. Other parameters for testing include pH, specific conductivity, total organic carbon, and total organic halogens, which are considered indications of potential pollution problems. Interim guidance waived monitoring requirements for facilities with double liners. However, a leachate collection system was required between the two liners with provisions to detect the presence of liquids.

Table 11-6 / Drinking Water Standards.

Parameter	Maximum Level (mg/l)
Arsenic	0.01
Barium	1.0
Cadmium	0.01
Chromium	0.05
Fluoride	1.4–2.4
Lead	0.05
Mercury	0.002
Nitrate (as N)	10
Selenium	0.01
Silver	0.05
Endrin	0.0002
Lindane	0.004
Methoxychlor	0.1
Toxaphene	0.005
2,4-D	0.1
2,4,5-TP Silvex	0.01
Radium	5 pCi/l
Gross Alpha	15 pCi/l
Gross Beta	4 mrem.yr
Turbidity[a]	1/TU
Coliform bacteria	1/100 ml

[a]Applicable to surface-water supplies only.

Overview

The objective of site design/engineering is to reduce risks associated with waste disposal. From the classic definition of risk, this can be accomplished in one of two ways: (1) Reduce the probability of a release. (2) Reduce the consequences of a release. However, experience suggests that for landfill or lagoon facilities, pursuit of the first of these objectives is deceptive. In the long run, the probability of failure is unity. This reflects the fact that natural forces, the limited life of fabricated barriers, and the likelihood of human intrusion will ultimately expose all buried materials. Engineered barriers can only hope to delay that exposure. If the disposed materials undergo a reduction of hazard over time, the delay will in fact achieve a reduction in risk. If not, the delay will be just that, a postponement of the same risk. It follows that major emphasis must be placed on reducing the consequences of a release.

One means of consequence reduction is through control of the source term. This may be accomplished by restricting the type of wastes placed in a repository or adding other materials to modify the properties of the waste. More risk-reduction options are available through measures to reduce release rates. These include engineered barriers, waste form changes (e.g., fixation), and designs to accelerate dilution. The underlying principle is that because failure is certain, a facility should be designed for failure in a way that will pose no more than acceptable risks. Slow programmed releases can be encouraged to allow the environment to assimilate contaminants at non-hazardous levels. This is particularly true for heavy metals that originate in the environment. Indeed, this is what is accomplished with fixation, which puts a waste in a less leachable form so that losses to the environment are slower. Hence, the final section of this chapter addresses fixation as an adjunct to landfilling.

Clearly, this approach flies in the face of political declarations that dilution is not a viable means of disposal. Yet, realization that all repositories will eventually fail highlights the impracticality of the foregoing position. If a repository is to fail, there is no question that it would be best if the failure were to occur in a dilute mode. By designing for dilution, that outcome can be assured.

Realization of ultimate repository failure should encourage such a change in landfill philosophy. As an alternative to controlled-release sites, landfill can be considered geologic storage sites employed until technology/economics make reclamation a viable alternative. Although this may seem a subtle point, the change in viewpoint will encourage maintenance of ownership and monitoring over longer periods of time. This will help avoid problems with abandoned sites, lost records, and the like, which foster sudden, often hazardous intrusions later in time.

One major hazardous waste disposal firm has taken a step in this direction with design of a complete above-ground landfill. The facility will sit on a concrete pad but use existing technology for the final cover and cap. The above-ground engineering affords several key advantages:

Ease of construction of a concrete underlining
Visibility to minimize future intrusion due to loss of records on use
Ease of monitoring
Ease of retrieval of wastes in the future

General Motors designed a similar facility for Rochester, N.Y. when they recognized that within 5 to 10 years an economic recovery process would likely be available for the metal sludge being managed.

Fixation and Encapsulation Methods

Fixation of a waste is herein defined as a method of rendering contaminants in the waste less mobile by reducing their solubility or by limiting their contact with a leaching solvent (e.g., water). Encapsulation of a waste is defined as a method of enclosing a bulk form (e.g., a block) of the waste with an impermeable membrane to limit contact with a leaching solvent. The terms "fixation" and "solidification" may be used interchangeably in many instances, although solidification implies the conversion of a liquid waste to a solid. Solidification also improves the physical characteristics of the waste to facilitate handling and transport. If the fixation method succeeds in reducing the mobility of the contaminants to the point where no serious stresses are exerted on the environment, then the wastes may be considered nonhazardous and may be disposed of in a conventional sanitary landfill.

Cement-based Fixation Processes

Processes employing cement are the most common type of fixation technique in use today. Cement technology is well understood, and the materials are readily available and inexpensive (Conner, 1979; Maugh, 1979). Cement fixation processes can be applied to wastes containing water, which is needed for the cementing reactions. When properly prepared, the solid product has a relatively low permeability and high compressive strength, although usually not as much as concrete (Bartos and Palermo, 1977). The disadvantages of cement fixation include an increase in waste volume and possible cement degradation under low-pH conditions.

Cement processes are used primarily for inorganic wastes, especially for fixation of heavy metals. These processes have been used extensively for radioactive wastes, metal plating and finishing wastes, electronic industry wastes, and flue gas desulfurization sludges (Christensen and Wakamiya,

1980). At the typically high pH levels in the cement mixtures (pH 9 to 11), heavy metals precipitate as hydroxides, carbonates, or silicates that are incorporated in the cement matrix (Lindsey, 1975). Minimum solubilities for many metals such as cadmium, copper, lead, nickel, and zinc occur between pH 9 and 11 (Malone et al., 1980). Cement processes are generally tolerant of chemical variations; additives can be added to adjust for problem compounds. Fixation with cement generally relies on reducing the solubility of compounds rather than limiting contact with water, because the porosity may be relatively high (e.g., > 35%). Contaminants that are not rendered insoluble by cement fixation techniques, such as organics, may be readily leached from the cement.

Soluble silicates are frequently added with cement to enhance the fixation of heavy metals. Lime may also be added to raise the pH (Coltharp et al., 1979). Portland cement has been successfully used for fixation of soda industry wastes containing mercury in Japan (Hiroaka and Takeda, 1979). Sodium sulfide is added to the cement in this process.

Lime-based Fixation Processes

A pozzolanic type of cement is used for waste fixation wherein lime is reacted with fine-grained silica and water at ambient temperatures. Sources of fine-grained silica include volcanic ash, burnt clay, diatomaceous earth, and granulated blast furnace slag. Lime-based processes are effective on inorganic wastes, particularly heavy metals, and the fixed product behaves similarly to portland cement products. At the typically high pH of the product, heavy metals precipitate as hydroxide or carbonates and are incorporated into the product matrix (Malone et al., 1980).

Thermoplastic Fixation Processes

Thermoplastic fixation processes may be viewed as a microencapsulation technique where individual waste particles are coated with an impermeable membrane such as asphalt or polymeric material (e.g., polyethylene). The process involves thorough mixing of the dry waste with heated molten plastic material, then allowing the mixture to cool and solidify in a solid block. The plastic material not only coats the individual particles but also fills in void spaces to prevent the passage of leaching solvent through the final product. A variation of this technique consists of mixing wet sludges with emulsified asphalt, then heating to evaporate the water. The final product generally contains 50% to 25% waste (on a dry basis), with the remainder being fixative (Hiroaka and Takeda, 1979). The thermoplastic fixative generally coats the waste particles rather than chemically combining with them.

Thermoplastic fixation techniques are effective for a broad range of contaminants. Contaminants not amenable to these techniques include organic solvents, anhydrous salts, strong oxidants, and thermally unstable wastes. Anhydrous salts tend to rehydrate when in contact with water and crack the final product. Strong oxidants and thermally unstable wastes may cause fires or violent reactions.

Advantages of thermoplastic fixation techniques include significantly lower leach rates for water-soluble components, potential volume reduction, and only a relatively small surface area available for leaching, because water cannot penetrate the product form to any significant degree. Disadvantages include higher costs, the need for more complicated equipment than for cement- or lime-based techniques, the risk of fire due to heated organic materials, large amounts of energy needed to dry wet wastes, and the possible production of noxious oil and odors during heating (Bell et al., 1981).

Organic Polymer Fixation Processes

Organic-polymer/waste mixtures can be prepared by mixing a suitable resin or monomer with the waste, then polymerizing the monomer or resins with a catalyst to form the fixed product. The waste usually is not bonded to the polymer matrix but is microencapsulated, as in the thermoplastic technique. Typical resins used in processing the wastes include urea/formaldehyde, vinyl-ester/styrene, and polyester. It is possible to apply this process to wet or aqueous wastes without dehydration by utilizing emulsified polyester resins and peroxide catalysts to produce a solid monolith composed of droplets of the waste encapsulated by polyester shells (Subramanian et al., 1977). The monolith is reported to have a compressive strength similar to that of concrete.

The advantages of organic polymer fixation include higher waste-to-fixative ratios and lower product densities than cement techniques and low processing temperatures. Disadvantages include possible production of noxious fumes during polymerization and the use of corrosive catalysts that require special mixtures and container liners. Urea/formaldehyde resins require low-pH conditions, which usually increase heavy-metal solubilities. Urea/formaldehyde resins also shrink with age and produce uncombined "weep water" during aging.

Glass Fixation Processes

Glass fixation processes were initially developed for highly radioactive wastes where extremely low leach rates were desired. Aqueous solutions containing radioactive metal ions are generally dried, calcined, mixed with glass-forming agents such as silica and borax, then fired at high temperatures to form glass. These glasses generally exhibit leaching rates several orders of magnitude less than thermoplastic waste mixtures. The principal advantage of the glass product is the very low leach rate, and the principal disadvantage is the high cost of the process.

Surface Encapsulation Techniques

Surface encapsulation involves coating a monolith of the waste with a thermoplastic material such as polyethylene to prevent contact of the waste with leaching solvent. This technique requires less effort than mixing the waste with the plastic material; however, if the coating cracks, the waste becomes subject to leaching.

Landfill Disposal Costs

Life-cycle average costs for land disposal of hazardous wastes have been estimated for the year 1978. These costs range from $110 to $340 per metric ton for facilities handling 2.3 and 0.45 metric tons per hour, respectively (Hansen and Rinkel, 1981). Chemical fixation with solids was estimated to add $198 per metric ton, and chemical fixation without solids was estimated to add $53 per metric ton to the cost of land disposal. Encapsulation costs were estimated to be $103 per metric ton at the rate of 0.45 metric ton per hour.

REFERENCES

Anderson, D.; and Brown, K. W. 1981. Organic leachate effects on the permeability of clay liners. In *Proceedings of the seventh annual research symposium—land disposal: hazardous wastes*, EPA-600/9-81-002b, March 1981.

Bartos, M. J., Jr.; and Palermo, M. R. 1977. *Physical and engineering properties of hazardous industrial wastes and sludges*. EPA-600/2-79-139.

Bell, N. E.; Halverson, M. A.; and Mercer, B. W. 1981. *Solidification of low-volume power plant sludges*. Electric Power Research Institute report No. EPRI-CS-2121, December 1981.

Chemical Engineering. 1979. Love Canal aftermath: learning from a tragedy. *Chemical Engineering*, Vol. 86, No. 23, pp. 86–92, October 22, 1979.

Christensen, D. C.; and Wakamiya, W. 1980. A solid future for solidification fixation processes. In *Toxic and hazardous waste disposal*, Vol. 4, ed. R. B. Pojasek, pp. 75–90. Ann Arbor: Ann Arbor Science Publishers.

Coltharp, W. M.; et al. 1979. *Review and assessment of the existing data base regarding flue gas cleaning wastes.* EPRI report FP-621, Vol. 1, prepared by Radian Corporation.

Conner, J. R. 1979. Considerations in selecting chemical fixation and solidification alternatives—the engineered approach. In *Proceedings of the second annual conference of applied research and practice in municipal and industrial waste,* pp. 129–147, Madison Waste Conference, Marine-on-St. Croix, Minn.

Dawson, G. W.; and Cearlock, D. B. 1981. The restoration process for chemically contaminated facilities. Presented at the Northeastern Conference on Hazardous Waste, Portsmouth, NH, October 26–27, 1981.

Deland, M. R. 1981. Superfund. *Environmental Science and Technology* Vol. 15, No. 3, p. 255, March 1981.

Environmental Science. 1981. Hazardous waste landfills. *Environmental Science and Technology,* Vol. 15, No. 3, March 1981.

Feddes, R. A.; Kowalik, P. J.; and Zaradny, H. 1978. *Simulation of field water, field water use and crop yield.* Center for Agricultural Publishing and Documentation, Wagenengin, The Netherlands.

Gee, G. W.; et al. 1981. Radon control by multilayer earth barriers: 2. Field tests. In *Proceedings of the fourth symposium on uranium mill tailings management,* October 26–27, 1981, Civil Engineering Dept., Colorado State University, pp. 289–308.

Green, W. J.; Lee, G. F.; and Jones, R. A. 1981. Clay-soils permeability and hazardous waste storage. *Journal of the Water Pollution Control Federation,* Vol. 53, No. 8, pp. 1347–1354, August 1981.

Gunkel, R. C. 1981. Membrane liner systems for hazardous waste landfills. In *Proceedings of the seventh annual research symposium—land disposal: hazardous waste,* Philadelphia, Pennsylvania, March 16–18, 1981, EPA-600/9-81-002b.

Gupta, S. K.; Tanji, K.; Nielsen, D.; Biggar, J.; Simmons, C.; and MacIntyre, J. 1978. *Field simulation of soil-water movement with crop water extraction.* Water science and engineering paper No. 4013, Department of Land, Air and Water Resources, University of California, Davis.

Hansen, W. G.; and Rinkel, H. L. 1981. *Cost comparisons of treatment and disposal alternatives for hazardous materials.* Project summary report EPA-600/52-80-188, February 1981.

Haxo, H. E., Jr. 1981. Durability of liner materials for hazardous waste disposal facilities. In *Proceedings of the seventh annual research symposium—land disposal: hazardous waste, Philadelphia, Pennsylvania, March 16–18, 1981.* EPA-600/9-81-002b.

Haxo, H. E., Jr.; Haxo, R. S.; and White, R. M. 1977. *Liner materials exposed to hazardous and toxic sludge.* First Interim Report, EPA-600-2-77-081.

Hillel, D. 1977. *Computer simulation of soil water dynamics: a compendium of recent work.* Ottawa: International Development Center.

Hiroaka, M.; and Takeda, N. 1979. *Behavior of hazardous substances in stabilization and solidification processes of industrial wastes.* Presented at the ACS 177th national meeting, Honolulu, Hawaii.

Inside EPA. 1981. EPA's top 114 superfund sites: 20 potentially worse than Love Canal. *Inside EPA*, Vol. 2, No. 44, p. 9, Inside Washington Publishers, Washington, D.C.

Lindsey, A. W. 1975. Ultimate disposal of spilled hazardous materials. *Chemical Engineering*, Vol. 82, No. 23, pp. 107-114.

Malone, P. G.; Jones, L. W.; and Larson, R. J. 1980. *Guideline to the disposal of chemically stabilized and solidified wastes.* EPA-SW-872.

Matrecon, Inc. 1980. *Lining of waste impoundment and disposal facilities.* EPA report No. SW-870, September 1980.

Maugh, T. H., II. 1979. Burial is last resort for hazardous waste. *Science*, Vol. 204, No. 22, pp. 1295-1298.

Metry, A. A. 1980. Minimizing risks in land disposal of hazardous waste. In *Toxic and hazardous waste disposal, Vol. 4*, ed. R. B. Pojasek, pp. 1-28, Ann Arbor: Ann Arbor Science Publishers.

Moore, C. A. 1980. *Landfill and surface impoundment performance evaluation manual.* EPA report No. SW-869, September 1980.

National Conference of State Legislatures. 1982. *Hazardous Waste Management: A Survey of State Legislation. 1982.* Denver.

Perrier, E. G.; and Gibson, A. C. 1980. *Hydrologic simulation on solid waste burial sites (HSSWDS).* EPA report No. SW-868, September 1980.

Solid Wastes Management. 1980. Drain system is installed to collect leachate from Love Canal bath tub. *Solid Wastes Management*, Vol. 23, No. 2, pp. 20-21, 34-38, February 1980.

Stewart, W. S. 1978. *State-of-the-art study of land impoundment techniques.* EPA-600/2-78-196.

Subramanian, R. V.; et al. 1977. Polymer encapsulation of hazardous industrial wastes. In *Proceedings of the national conference on treatment and disposal of industrial wastewaters and residues.* pp. 97-118. Hazardous Materials Control Research Institute, Rockville, MD.

Toxic Materials News. 1981. California begins program to end land disposal of toxic substances. *Toxic Materials News*, Vol. 8, No. 41, p. 323, October 21, 1981.

U.S. EPA. 1980. *Research summary: controlling hazardous wastes.* Report No. EPA-800/8-80-017.

U.S. EPA. 1979. *Methods for chemical analysis of water and wastes.* EPA-600/4-79-020, March 1979.

Wong, J. 1977. The design of a system for collecting leachate from a lined landfill site. *Water Resources Research*, Vol. 13, No. 2, pp. 404-410.

12

Ocean Dumping and Underground Injection

Introduction

Considerable controversy has been associated with ocean dumping and underground injection of wastes over the past decade or so. In response to public concern over actual and potential deterioration of the marine environment and underground water resources, government agencies have sought to severely restrict the use of these disposal methods. The advantages and disadvantages of ocean disposal and underground injection as opposed to other disposal methods have been vigorously debated by various sectors of society. In effect, restricted use of the ocean and deep subterranean strata for water disposal places a greater burden on other media such as land and the atmosphere. When the latter is also stressed, the basis for a raging controversy may be established. Such is the case for ocean disposal of sewage sludge from the city of New York. The EPA interpreted the 1977 amendment to the Marine Protection, Research, and Sanctuary Act (MPRSA) to mean termination of ocean dumping of sewage sludge by December 31, 1981. The city of New York undertook to comply with the deadline with the development of an interim solution that consisted of dewatering and composting sludge at two locations in the city. The finished compost was to be applied to underdeveloped parkland within the city. The annual operating cost for the city would have been almost $45 million, compared with $3 million for ocean dumping (Feliciano, 1981).

Prior to completing construction of the composting facilities, the city filed a lawsuit in March 24, 1980, alleging that the EPA had incorrectly implemented federal law. The lawsuit arose because of mounting concern over heavy-metal contamination in the sludge and possible adverse health effects from the composting sites. A lack of evidence was also indicated over whether land disposal of the sludge was less harmful than ocean disposal. The city argued that the intent of Congress in passing the MPRSA was to bar only disposal of wastes that would "unreasonably" degrade the marine environment and that all factors including costs and potential hazards of land-based alternatives should be considered. On April 14, 1981, a U.S. District Court ruled in favor of the city. At the date of this writing, the issue is far from being settled, but there are indications that sludge disposal in the ocean will again become a viable alternative.

The National Advisory Committee on Oceans and Atmosphere concluded in a report to Congress that ocean dumping of industrial wastes should continue where no unreasonable environmental damage occurs and where economic considerations suggest that this is the preferable option (*Hazardous Waste News*, 1981). The EPA was criticized for refusing to permit ocean dumping whenever land-based alternatives exist. The report contended that wastes should be disposed in a manner and medium that will minimize the risk to public health and the environment, and at a price this nation is prepared to pay.

Criticism of deep-well injection of hazardous wastes has largely centered on a purported lack of knowledge about the fate of injected wastes. Some critics are profoundly disturbed at the thought of carcinogenic, mutagenic, and other potent chemicals migrating below the earth's surface. Proponents argue, however, that general liquid movement in these deep formations is typically measured in inches per year; therefore, migration is negligible. Critics also point out the apparent triggering of earthquakes in the Denver, Colorado, area by injection of wastes at the Rocky Mountain Arsenal, but proponents counter that the wastes were injected into an unsuitable fractured granite rock instead of a sandstone formation (Maugh, 1979).

The EPA has been specifically directed by Congress to be very flexible in regulating deep-well injection so as not to impede the industry growth. This disposal method is therefore expected to remain a viable option for disposal of hazardous wastes in those areas having suitable formations to accept the wastes.

Ocean Dumping Regulations

The EPA has issued regulations and criteria (U.S. EPA, 1977) for ocean dumping of wastes under the authority provided by the Marine Protection, Research, and Sanctuary Act of 1972 (PL 92-532). The purpose of these regulations and criteria is to prevent or strictly limit ocean dumping of any material that would adversely affect the marine environment. Ocean dumping is controlled through a permit system administered by the EPA and the U.S. Army Corps of Engineers. District engineers of the latter agency are responsible for administering permits for dumping of dredged material, and the regional EPA administrators administer permits for dumping of other material. Ocean dumping of the following materials will not be approved by either the EPA or the Corps of Engineers:

High-level radioactive wastes

Materials in any form produced for radiological, chemical, or biological warfare

Materials insufficiently described by the permit applicant to determine the environmental impact of ocean dumping

Persistent inert synthetic or natural materials that float or remain suspended in the ocean and thus interfere materially with fishing, navigation, or other legitimate uses of the ocean.

The following types of permits are available under current ocean dumping regulations (U.S. EPA, 1977):

General permits are available for disposal of substances in emergency situations, or for substances generally disposed of in small quantities that will have a minimal adverse environmental impact.

Special permits valid for no more than 3 yr may be issued for dumping materials that satisfy the ocean dumping criteria.

Emergency permits may be available for dumping of certain materials posing unacceptable risk to human health, when there exists an emergency with no other feasible solution. The materials specifically listed in the regulations as "constituents prohibited as other than trace contaminants" are the ones singled out for concern in the Convention on the Prevention of Marine Pollution, Annex I; for these materials, permits may be issued only after consultation by the Department of State with the contracting parties. Most other materials may be permitted on approval by EPA.

Research permits, valid for 18 months, may be available for meritorious research projects.

Constituents prohibited as other than trace contaminants under the emergency permits described earlier include the following:

Organohalogen compounds

Mercury and mercury compounds

Cadmium and cadmium compounds

Oil of any kind and any form, including but not limited to crude oil and refined oil products and mixtures of oil with other materials such as oil sludge and oil refuse

Known carcinogens, mutagens, or teratogens or materials suspected of being in one or more of these categories by responsible scientific opinion.

Exclusions may be obtained for the foregoing materials if it can be demonstrated that the materials are inert or insoluble and thus are not available to marine organisms or that the materials will degrade to harmless substances in the marine environment.

Waste or dredged materials may be dumped only in areas designated for this purpose by the EPA or the Corps of Engineers. The need for ocean dumping must be satisfactorily demonstrated before permits will be granted. This need will be determined by evaluation of the following factors:

Relative environmental risk, impact, and costs of alternatives such as landfill, deep-well injection, incineration, land spreading, recycling, treatment, and storage

Modifications of manufacturing processes to eliminate the waste or render it less polluting

Alternative treatment schemes that will reduce the environmental impact of ocean disposal

Ocean Dumping Operations

Barge Disposal

Barge disposal is the principal method by which wastes are dumped into the ocean. The barges filled with liquid or slurried wastes are towed from loading stations to approved dump sites, and the wastes are pumped from the barges while in motion to enhance dilution of the wastes. Typical barge speeds are in the range of 5.6 to 11 km/h, and discharge rates vary from 3600 to 18,000 kg/min. The wastes are generally diluted in very large volumes of seawater within a few minutes. For example, titanium pigment manufacturing wastes containing 9% sulfuric acid are reported to be neutralized within a few minutes in the wake of a barge discharging these wastes (Peschiera and Freiherr, 1968). The acidity in this waste not only is diluted but also reacts with the alkality of the seawater to produce soluble products naturally present in the water (i.e., sulfate salts). The ferrous sulfate in the waste converts to light ferric hydroxide floc that is disposed and carried away by the currents. No harmful effects on fish and other marine life have been observed from the disposal of these wastes in the New York Bight (Peschiera and Freiherr, 1968).

Barge dispersion studies have been conducted with titanium dioxide manufacturing wastes containing 10% $FeSO_4$ and 8.5% H_2SO_4 to determine how much this waste dilutes and disperses with time after being pumped from a barge (Ketchum and Ford, 1952). Data from these studies were converted to iron concentrations in the wake of the barge at various times after discharge and plotted on a log-log graph (Ball and Reynolds, 1976), as illustrated in Figure 12–1. The waste was discharged at depths ranging from 2 to 4.6 m below the water surface while the barge was being towed at a speed

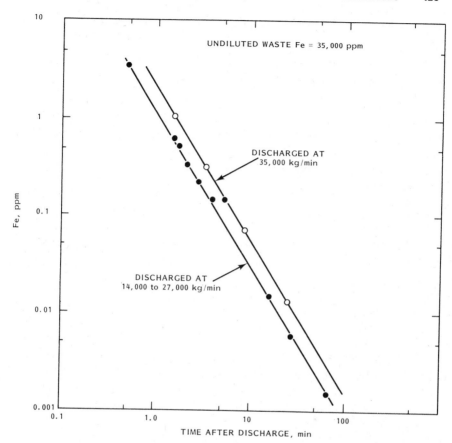

Figure 12-1 / Concentration of Fe versus Time after discharge of titanium dioxide pigment waste from a moving barge. (Adapted from Ball and Reynolds, 1976.)

of 11 km/h. The pumping rate varied from 14,600 to 35,000 kg/min. This waste was diluted by a ratio of 36,000 parts of seawater to one part of waste after 1 min. In another study with aniline production waste, a dilution of 1000 parts of seawater with one part of waste was measured after 1 min following discharge at a rate of 16,000 kg/min from a barge traveling 9.3 km/h (Ball and Reynolds, 1976).

Studies conducted on dispersion of a biosludge from a moving barge in the Gulf of Mexico indicate dilutions ranging from 1000 to 5000 parts of seawater to one part of waste after 20 min (Atlas et al., 1980). After 24 h, no trace of the sludge could be detected in the seawater. Dilution of wastes from moving barges appears to be related to the specific gravity of the waste as well as the speed of the barge and the discharge rate (Ball and Reynolds, 1976).

Environmental Impact of Ocean Dumping

A rapid buildup of bottom sediment due to disposal of solid wastes such as dredge spoils can result in the destruction of spawning areas, reductions in food supplies and vegetation cover, trapping of organic matter (with resultant development of anaerobic bottom conditions), and absorption or adsorption of organic matter (Smith and Brown, 1971). Suspended matter causing high turbidity may directly or indirectly affect fish and crustaceans. Direct effects cause an immediate response or even mortality by suffocation and reduced survival or growth in the larval stages of fish and shellfish. Indirect effects of turbidity include (1) reduced light penetration, resulting in reduced photosynthesis; (2) reduced visibility, rendering the search for food difficult by feeding organisms, and (3) reduction of vegetational cover.

Although acute toxic effects on fish and plankton have been observed immediately after dumping acid or chlorinated hydrocarbon wastes, these effects usually disappear within a few hours as the waste becomes dispersed (Smith and Brown, 1971). Dumping of wastes containing toxic persistent chemicals that concentrate in the food chain (mercury, cadmium, and polychlorinated biphenyls) is tightly controlled because of the potentially severe impact on the health of individuals who eat fish contaminated with these substances.

The classic example of bioconcentration problems related to ocean disposal took place in Japan. Mercury poisoning resulting from environmental contamination was first noted in Japan during the 1950s. The original outbreak occurred in a village near Minamata Bay, where at least 121 people were poisoned, including 46 who died (Furukawa et al., 1969). Studies of the victims and the surrounding bay revealed that methyl mercury from nearby acetaldehyde production facilities had been discharged to the bay. Subsequently, the organomercury was concentrated in resident fish and shellfish to levels of 1 to 50 ppm. Consumption of these marine organisms led to chronic mercury poisoning, which has since been referred to as Minamata disease. The symptoms are related to central nervous system disorders as described earlier, and the effects are irreversible. Cats, crows, and sea birds were also affected (Hartung and Dinman, 1972). A similar outbreak in Niigata, Japan, poisoned 47 persons, 6 of whom died in 1964–1965. More recently, an outbreak has been reported at Ariake, Japan, near Nagasaki (*Chemical and Engineering News*, 1973).

Based on these problems, it is clear that ocean disposal is most applicable to a given subset of hazardous wastes such as large-volume materials containing high brine concentrations and acid, base, or insoluble salts other than highly toxic heavy metals.

Underground Injection Regulations

In December 1974, the National Safe Drinking Water Act (PL 93-523) was passed by Congress. The purpose of this legislation was to assure that water supply systems meet minimum national standards for the protection of public health. The act was designed to achieve uniform safety and quality of drinking water in the nation by identifying contaminants and establishing maximum levels of acceptability. The major provision of the act that encompasses the underground injection program is the establishment of regulations to protect the underground drinking water sources by the control of subsurface injection of fluids, including wastes. Rules pertaining to the administration of the underground injection control (UIC) program were issued in 1980 as part of the consolidated permit program 40 CFR Parts 122, 123, and 124. Technical criteria and standards (40 CFR Part 146) for the UIC program were also issued in 1980. The states were given the options of (1) administering their own UIC programs on approval of the EPA or (2) allowing the EPA to do it. The UIC permit program established by the EPA regulates underground injection by five classes of wells:

Class I. Wells used by generators of hazardous wastes or operators of hazardous waste management facilities or other industrial and municipal entities that inject fluids beneath the lowermost formation containing an underground source of drinking water within one-quarter mile of the well bore

Class II. Wells used to inject fluids in connection with gas and oil recovery or storage operations

Class III. Wells used in connection with recovery of minerals or energy

Class IV. Wells used to inject hazardous or radioactive wastes into or above a formation that has a drinking-water source within one-quarter mile of the well bore

Class V. Injection wells not included in classes I, II, III, and IV

Class IV wells, which inject hazardous wastes directly into an underground drinking-water source, are to be phased out, and no permits will be issued for new wells in this class.

Disposal wells that do not discharge wastes within one-quarter mile of a drinking-water source are included in class V. At the date of this writing, class V wells are being inventoried and assessed, but regulatory action is to be taken at a later date. However, no class V well will be authorized by permit or rule if it results in the presence of any contaminant in an underground source of drinking water that may adversely affect human health.

EPA construction, operating, and monitoring criteria and standards applicable to class I wells (40 CFR Part 146) are as follows:

Construction Requirements

1. All class I wells shall be sited in such a fashion that they inject into a formation that is beneath the lowermost formation containing, within one-quarter mile of the well bore, an underground source of drinking water.
2. All class I wells shall be cased and cemented to prevent the movement of fluids into or between underground sources of drinking water. The casing and cement used in the construction of each newly drilled well shall be designed for the life expectancy of the well.
3. All class I injection wells, except those municipal wells injecting noncorrosive wastes, shall inject fluids through tubing with a packer set immediately above the injection zone, or tubing with an approved fluid seal as an alternative. The tubing, packer, and fluid seal shall be designed for the expected service.

The use of other alternatives to a packer may be allowed with the written approval of the director of the authorized regulatory agency. To obtain approval, the operator shall submit a written request to the director, which shall set forth the proposed alternative and all technical data supporting its use. The director shall approve the request if the alternative method will reliably provide a comparable level of protection to underground sources of drinking water. The director may approve an alternative method solely for an individual well or for general use.

Operating Requirements

1. Except during stimulation, injection pressure at the wellhead shall not exceed a maximum that shall be calculated so as to assure that the pressure in the injection zone during injection does not initiate new fractures or propagate existing fractures in the injection zone. In no case shall injection pressure initiate fractures in the confining zone or cause the movement of injection or formation fluids into an underground source of drinking water.
2. Injection between the outermost casing protecting underground sources of drinking water and the well bore is prohibited.
3. Unless an alternative to a packer has been approved, the annulus between the tubing and the long string of casing shall be filled with an approved fluid maintained at an approved pressure.

Monitoring Requirements

Monitoring requirements shall, at a minimum, include the following:

1. The analysis of the injected fluids with sufficient frequency to yield data representative of their characteristics

2. Installation and use of continuous recording devices to monitor injection pressure, flow rate and volume, and the pressure on the annulus between the tubing and the long string of casing
3. A demonstration of mechanical integrity once every 5 yr during the life of the well
4. The type, number, and location of wells within the area of review to be used to monitor any migration of fluids into the pressure in the underground sources of drinking water, the parameters to be measured, and the frequency of monitoring

Underground Injection Operations

A typical well consists of several pipes, as illustrated in Figure 12–2. The function of each of the pipes is as follows:

Surface Casing. This pipe extends from the surface to approximately 60 m below the base of the fresh water. This casing is cemented in place back to the ground level and inside the conductor pipe. Its function is to prevent the degradation of any fresh-water sands.

Protection Casing. This pipe extends from the surface to a point just above or below the disposal zone, depending on the type of completion intended. This casing is also cemented back to the surface and inside the surface casing. Its purpose is to seal off the formation above the disposal zone and below the fresh water.

Injection Tubing. This pipe is the conduit through which the effluent travels to the disposal zone. It is always sealed at the wellhead and usually just above the disposal zone. The annular space between the protecting casing and injection tubing is filled with a noncorrosive fluid. Pressure should be applied to this annulus approximately 35,000 kg/m² higher than the injection pressure. The higher annular pressure will prevent any flow of effluent into the casing tubing annulus.

Disposal-Zone Completion

If the well is completed in a sand formation, the gravel/packed-sand screen completion method should be used. If the well is completed in a limestone formation, open-hole or perforated completions are acceptable.

Wastewater Quality Requirements for Injection

The suitability of waste for underground injection depends on its volume and physical and chemical characteristics, and on the physical and chemical properties of the potential injection zones and their interstitial fluids. Wastewater that is desirable for injection must be (1) low in volume and high in concentration, (2) difficult to treat by surface methods, (3) free of any ad-

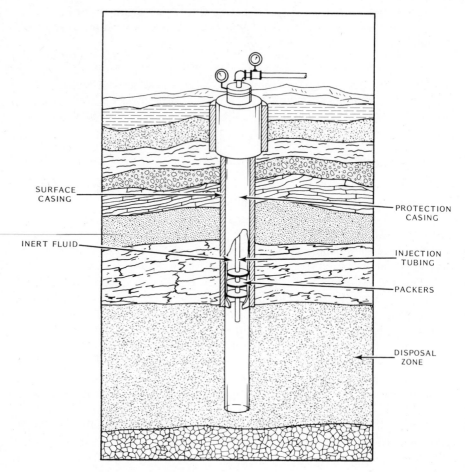

SURFACE CASING

INERT FLUID

PROTECTION CASING

INJECTION TUBING

PACKERS

DISPOSAL ZONE

Figure 12-2 / Disposal well. (From Ottinger et al., 1973.)

verse reaction with the formation fluid or the strata, (4) free of suspended solids, (5) biologically inactive, and (6) noncorrosive (Tofflemire and Brezner, 1971; Warner, 1968).

Waste disposal into underground aquifers constitutes the use of limited storage space, and only concentrated, very objectionable, relatively untreatable wastes should be considered for injection. The fluids injected into deep aquifers do not occupy empty pores, but displace the fluids that saturate the storage zone. The displaced fluids are frequently polluted (e.g., high salinity) and may migrate to fresh-water aquifers. Consequently, optimal use of the underground storage space will be realized by the use of underground injection only when more satisfactory alternative methods of waste treatment and disposal are not available.

Reaction of the wastewater with the formation water or the strata is important. Resulting problems could include dissolution of the formation, generation of a gas or precipitate in the formation, and clogging by biological growths. Walker and Stewart (1968) suggest a laboratory test to ensure compatibility of the wastewater with the formation. The wastewater can be mixed in a beaker with a formation water sample and held at formation temperature to see if there is any precipitate or adverse reaction. Pumping the wastewater through a core sample can reveal possible clogging problems. The wastewater should be free of suspended solids and biologically inactive to avoid reservoir clogging. The corrosiveness of the wastewater should be low to prevent tubing and pump corrosion.

Disposal Site Selection

Great care must be exercised in the selection of an underground disposal site for liquid wastes (National Industrial Pollution Control Council, 1971; Heikard, 1970; *Environmental Science*, 1968; NYS, 1968). The suitability of a specific location of a waste injection well must be evaluated by a detailed geologic subsurface investigation. However, regional geologic conditions can be used to evaluate general suitability of certain areas.

The regional favorability map (Figure 12–3) indicates that certain areas of the continental United States, such as the Rocky Mountains, are generally unsuitable for waste injection wells because igneous or metamorphic rocks lie at or near the ground surface. Such rocks do not have sufficiently high porosity or permeability to warrant their use as a disposal formation. Areas with extensive extrusive volcanic sequences exposed are also not suitable for waste disposal wells. Even though these rocks have porous zones, they usually contain fresh water. The waste disposal potentials of the basin and range areas (crosshatched areas on map) are largely unknown because the geologic conditions are complex.

The final appraisal of a disposal well site is usually determined by a two-phase geologic investigation. The first phase includes an evaluation of potential sites on the basis of available data. The second phase consists of a more detailed evaluation of subsurface conditions based on information obtained from drilling a pilot hole or the injection well.

Information sought during the first phase of the investigation and prior to the installation of an injection well includes the extent, thickness, depth, porosity, permeability, temperature, water quality, and piezometric pressure of potential injection zones. The presence of impermeable confining beds or lateral changes in rock properties, the existence of faults or joints, and the occurrence of any mineral resource in the area must also be evaluated. Existing wells in the area that may penetrate the potential injection zones

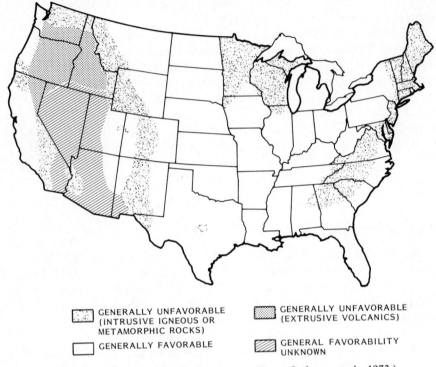

GENERALLY UNFAVORABLE
(INTRUSIVE IGNEOUS OR
METAMORPHIC ROCKS)

GENERALLY FAVORABLE

GENERALLY UNFAVORABLE
(EXTRUSIVE VOLCANICS)

GENERAL FAVORABILITY
UNKNOWN

Figure 12-3 / Deep-well disposal sites. (From Ottinger et al., 1973.)

must be located, because liquid wastes could escape through these wells if they are not properly plugged.

The second phase of the investigation is conducted during the drilling and testing of the injection well. Often the actual injection zone is not selected until the well has been drilled and a number of potential zones have been tested and until the chemical quality of water in the potential zones has been evaluated. Pumping tests are used to measure the permeability, and water samples are obtained for chemical analysis. Other important rock properties are measured by geophysical logging, drill-stem testing tools, or laboratory tests on core samples. The results of these geologic investigations are used not only in evaluating the feasibility of subsurface waste disposal but also to provide basic data for designing the injection well and the optimal rate of injection. Wastes may have different requirements relating to their extent of horizontal and vertical travel with time. For example, a chemically stable dilute waste may require only injection into, and dispersion in, a body of rapidly circulating groundwater that is recharged continually. However, a biochemically unstable effluent may require a residence time within the injection zone to permit further reaction. As an example, a very concentrated waste may require a long residence time without dispersion. To help in the selection of zones, the following classification system has been proposed:

Zone of Rapid Circulation.   The zone of rapid circulation extends from the land surface downward some tens or few hundreds of feet. Injection into this zone is normally precluded.

Zone of Delayed Circulation.   This zone is generally composed of fresh water that circulates continually and freely, but is retarded sufficiently that residence time varies from several to many decades or even a few centuries. Certain innocuous wastewaters have been injected into this zone successfully with suitable monitoring.

Subzone of Lethargic Flow.   In this subzone the native liquid is commonly saline and has very slow movement, measured in hundreds or even thousands of years. This subzone of lethargic flow is a primary zone for potential storage of the more concentrated wastes.

Stagnant Subzones.   These subzones are, with few exceptions, several thousand feet below land surface, and the fluid is hydrodynamically trapped. This zone would seem ideal for injection of very toxic wastes. However, the capabiltiy to accept and retain injected fluids needs to be assessed with extreme caution.

Dry Subzones.   A common type of dry subzone would be a salt bed or dome in which free water is virtually nonexistent, and it might be impermeable in a finite sense. Waste injected into such a zone would be wholly isolated from natural hydrodynamic circulation. However, because movement could occur through hydrofractures, performance of a dry subzone under injection should be assessed cautiously.

Well Operation and Monitoring

The operation of a disposal well should be closely controlled. Continuous monitoring of the injection pressure and flow should be made and checked for unusual variations. The annular pressure should also be monitored for variance. Biocides may have to be added to the wastes to prevent growths in the reservoir. To prevent operating problems, it is recommended to test the wastewater and annular fluid periodically for any change in composition. A uniform injection rate is more desirable than a widely varying or intermittent flow. Table 12-1 gives the relationship of injection pressure to well operation problems. Recommended disposal zone pressures are below 0.013 to

Table 12-1 / Problems Relative to Monitoring a Waste Injection Project.

Nature of Problem	Possible Cause	Preventive Design	Preventive Operations
High injection rates and low pressures at outset	1. Leak in system 2. Permeability and porosity exceed expectations	1. Use effective materials specifying adequate safety margins 2. Require whole-core analysis and more datum points, especially in questionable formations	Shutdown or slowdown until diagnosed. Back pressure on annulus of injection wells, if appropriate.
High injection rates and low pressures occurring suddenly during operations	1. Leak developed in system 2. Wastes attacking formation, increasing permeability or breaching reservoir	1. Anticorrosive materials and cathodic protection 2. pH adjustment and pressure limits	1. Locate leak and correct. 2. Alter waste treatment to optimize. A breached reservoir may have to be abandoned.
Low injection rates and high pressures at outset	1. Interaction of wastes with rock matrix or natural fluid 2. Performance wrongly predicted	1. Adequate geochemical data 2. Adequate data for accurate prediction of reservoir geometry, porosity, and permeability	1. Treat wastes 2. None
Declining injection rate and increasing pressures	1. Solids plugging formation 2. Precipitation in formation through interaction of wastes, rock and natural fluids 3. Naturally expected occurrence	1. Filtration of chemical, biological, and stabilization treatment 2. Reassessment of waste treatment 3. None	1. Flush bore hole or stimulate 2. Abandon project when pressure exceeds limits 3. None

0.017 kg/cm²/m (0.6 to 0.8 psi/ft) of depth and below 75% of the critical input pressure (NYS, 1968).

The critical input pressure is determined by conducting pumping tests in the well and plotting a curve of pressure versus flow rate. The point at which a sharp change in slope occurs is termed the critical input pressure (Figure 12–4). This is taken to be the point at which hydraulic fracturing occurs.

Salt Formation Disposal

As an outgrowth of nuclear waste repository studies, an alternate deep disposal method has been identified for hazardous wastes. The proposed approach centers around the use of natural salt formations (both bedded and domes) as vaults for solidified wastes.

As illustrated in Figure 12–5, salt deposits can be found in a number of locations throughout the Midwest and eastern United States. For over 30 years, various deposits have been used to store hydrocarbons including crude petroleum in the Strategic Petroleum Reserve. In recent years, comprehensive studies have been undertaken to determine the desirability of these deposits as repositories for high-level nuclear wastes. The very existence of such extensive masses of salt indicate a lack of local waters, which could dissolve wastes and transport them to potable aquifers. Hence, geohydrologic conditions will discourage waste migration. The salt is also strong. It has compressive strengths of 3000 to 7000 psi. These characteristics are desirable for disposal of hazardous wastes.

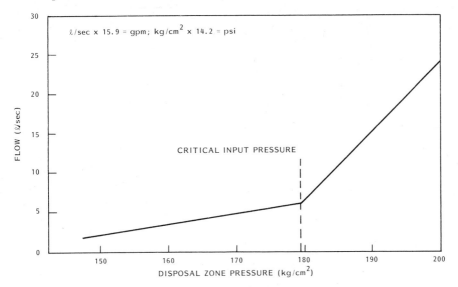

Figure 12-4 / A pressure–flow test of a deep well.

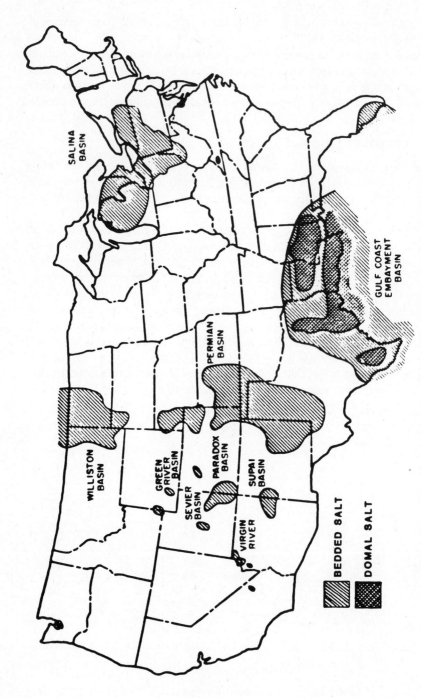

Figure 12-5 / Locations of major salt deposits in the United States (Pollution Engineering, 1984).

SALINA BASIN

GULF COAST EMBAYMENT BASIN

PERMIAN BASIN

WILLISTON BASIN

GREEN RIVER BASIN

SEVIER BASIN

PARADOX BASIN

SUPAI BASIN

VIRGIN RIVER

BEDDED SALT

DOMAL SALT

As currently proposed, the salt option would emplace wastes some 2000 to 5000 ft below land surface. Wells would be drilled into the salt formation and water injected to dissolve a cylindrical cavern (perhaps 2500 ft deep and 250 ft in diameter). Liquid wastes would then be injected until the cavern was filled and sealed with a cement cap. If desired, a solidification agent would be premixed with the waste so that the entire cylinder would set to a solid mass. Such a scheme would support disposal of 30 to 42 million gallons of waste (depending on the use of solidification agents) beneath an area of 2 acres.

Cost of Ocean Dumping and Deep-Well Injection

Ocean Dumping Costs

Estimated costs reported by NL Industries in 1977 for disposing of titanium dioxide production wastes at the New York Bight waste disposal site were $1.84 million per year (U.S. EPA, 1980a) or $2900 per trip (estimated 640 trips per year). A barge load is 3.7 million liters (about 4200 metric tons, sp. gr. 1.13), which is hauled from Sayreville, New Jersey, to the disposal site, which is located about 27 km off the New Jersey and Long Island coastlines. The time required for a round trip is about 12 h. The estimated cost per metric ton is $0.69.

Considerably higher per unit costs were reported by Allied Chemical Company for 12 barge trips per year at 1.6 million liters per load hauled from Elizabeth, New Jersey, to the New York Bight waste disposal site. The annual cost was $170,000 or $14,200 per trip (U.S. EPA, 1980a), which equates to $7.60 per metric ton (sp. gr. 1.17) for the acidic refrigerant production waste.

The cost of barging chemical wastes 106 miles off the New Jersey coastline was estimated to be in the range of $8.80 to $11.00 per metric ton in 1978 (U.S. EPA, 1980b). No data are available on the amount handled per trip, although 730,000 metric tons were reported to be dumped at this site in 1978.

The amount of waste disposed and the distance it is transported are important factors in the cost of ocean dumping.

Deep-Well Injection Cost

An estimated deep-well injection operating cost of $1.00 per thousand gallons for a 225,000-gpd operation was reported in 1973 (Battelle Memorial Institute, 1973). An update of these costs to 1981 yields a total operating cost of $2.40 per thousand gallons ($0.63/m³) based on construction and labor cost factors of 2.1 and 1.8, respectively, and an increase in capital recovery from 6.4% to 10% per annum. Only filtration was included in the surface treatment system. The injection well was drilled to 1850 m.

Smith reported that injection systems with wells drilled to 1630 m will generally cost from \$750,000 to \$1 million (Smith, 1979). At an injection rate of 530 l/min, maintenance costs are about \$60,000 per year. If neutralization is required prior to injection, an additional \$150,000 per year may be required.

REFERENCES

Atlas, E.; Brooks, J.; Trefry, J.; Sauer, T.; Schwab, C.; Bernard, B.; Schofield, J.; Giam, C. S.; and Meyer, E. R. 1980. Environmental aspects of ocean dumping in the western Gulf of Mexico. *Journal of the Water Pollution Control Federation*, Vol. 52, 329.

Ball, J.; and Reynolds, T. D. 1976. Dispersion of liquid waste from a moving barge. *Journal of the Water Pollution Control Federation* 48:2541.

Battelle Memorial Institute. 1973. *Program for management of hazardous wastes*. Environmental Protection Agency contract 68-01-0762. July 1973.

Chemical and Engineering News, 1973. Another outbreak of Minamata disease, Vol. 51, No. 23, p. 7.

Environmental Science. 1968. Deep well injection is effective for waste disposal. *Environmental Science and Technology*, Vol. 2, pp. 406–409.

Feliciano, D. V. 1981. Sludge: back into the ocean? *Journal of the Water Pollution Federation*, Vol. 53, No. 10, pp. 1442–1446, October 1981.

Furukawa, K.; Suzuki, T.; and Tonomura, K. 1969. Decomposition of organic mercurial compounds by mercury resistant bacteria. *Agricultural and Biological Chemistry*, Vol. 33, p. 128.

Hartung, R., and Dinman, B. D. 1972. *Environmental mercury contamination*. Ann Arbor Science Publishers, Inc. Ann Arbor, MI.

Hazardous Waste News. 1981. NACOA calls for continued use of oceans for industrial waste disposal. Business Publishers Inc., Silver Spring, MD, Vol. 3, No. 13, p. 98, March 30, 1981.

Heikard, J. 1970. *Deep well injection of liquid wastes*. Engineering bulletin No. 35, Dames and Moore, Los Angeles.

Ketchum, B. H.; and Ford, W. L. 1952. Rate of dispersion in the wake of a barge at sea. *Transactions of the American Geophysical Union*, Vol. 33, 680.

Maugh, T. A. 1979. Incineration, deep wells gain new importance. *Science*, Vol. 204, pp. 1188–1190, June 15, 1979.

National Industrial Pollution Control Council. 1971. *Waste disposal in deep wells*. U.S. Government publication 55-95, p. 34. February 1971.

NYS. 1968. *Confrence minutes, deep well injection conference*, New York State Department of Health, Albany.

Ottinger, R. S.; Blumenthal, J. L.; DalPorto, D. F.; Gruber, G. I.; Santy, M. J.; and Shih, C. C. 1979. Recommended methods of reduction, neutralization, recovery or disposal of hazardous waste. EPA-670/2-73-053c, U.S. Environmental Protection Agency.

Peschiera, L.; and Freiherr, F. H. 1968. Disposal of titanium pigment processing wastes. *Journal of the Water Pollution Control Federation*, Vol. 40, pp. 127–131.

Pollution Engineering Magazine. 1984. Pudvan Publishing.

Smith, D. D.; and Brown, R. P. 1971. *Ocean disposal of barge-delivered liquid and solid wastes from U.S. coastal cities.* EPA contract No. PH-86-68-203.

Smith, M. E. 1979. Solid-waste disposal: deep well injection. *Chemical Engineering*, Vol. 86, No. 8, p. 107, April 9, 1979.

Tofflemire, T. J.; and Brezner, G. P. 1971. Deep-well injection of wastewater. *Journal of the Water Pollution Control Federation*, Vol. 43, p. 1473.

U.S. EPA. 1980a. *Final environmental impact statement (EIS) for New York acid waste disposal site designation.* Environmental Protection Agency. September 1980.

U.S. EPA. 1980b. *Final environmental impact statement (EIS) for the 106-mile ocean waste disposal site designation.* Environmental Protection Agency. February 1980.

U.S. EPA. 1977. Ocean dumping: final revision of regulations and criteria. *Federal Register* Vol. 42, No. 7, January 11, 1977.

Walker, W. R.; and Stewart, R. C. 1968. Deep well disposal of waste. *Journal of the Sanitary Engineering Division, Proceedings of the American Society of Civil Engineers*, Vol. 94, p. 945.

Warner, D. L. 1968. Subsurface disposal of liquid industrial wastes by deep well injection. *American Association of Petroleum Geologists, Memoir*, No. 10, p. 16.

A

Summary and Conclusions of the 1973 Report to Congress on Hazardous Waste Management

The management of the nation's hazardous residues—toxic chemical, biological, radioactive, flammable, and explosive wastes—is generally inadequate; numerous case studies demonstrate that public health and welfare are unnecessarily threatened by the uncontrolled discharge of such waste materials into the environment.

Based on surveys conducted during this program, it is estimated that the generation of nonradioactive hazardous wastes is taking place at the rate of approximately 10 million tons yearly. About 40% by weight of these wastes are inorganic materials and 60% are organics; about 90% of the waste occurs in liquid or semiliquid form.

Hazardous waste generation is growing at a rate of 5% to 10% annually as a result of a number of factors: increasing production and consumption rates, bans and cancellations of toxic substances, and energy requirements (which lead to radioactive waste generation at higher rates).

Hazardous waste disposal to the land is increasing as a result of air and water pollution controls (which capture hazardous wastes from other media and transfer them to land) and denial of heretofore accepted methods of disposal such as ocean dumping.

Current expenditures by generators for treatment and disposal of such wastes are low relative to what is required for adequate treatment/disposal.

Ocean dumping and simple land disposal costs are on the order of $3 per ton, whereas environmentally adequate management could require as much as $60 per ton if all costs are internalized.

Federal, state, and local legislation and regulations dealing with the treatment and disposal of nonradioactive hazardous waste are generally spotty or nonexistent. At the federal level, the Clean Air Act, the Federal Water Pollution Control Act, and the Marine Protection, Research and Sanctuaries Act provide control authority over the incineration and water and ocean disposal of certain hazardous wastes, but not over the land disposal of residues. Fourteen other federal laws deal in a peripheral manner with the management of hazardous wastes, and approximately 25 states have limited hazardous waste regulatory authority.

Given this permissive legislative climate, generators of waste are under little or no pressure to expend resources for the adequate management of their hazardous wastes. There are few economic incentives (given the high costs of adequate management compared with costs of current practice) for generators to dispose of wastes in adequate ways.

Technology is available to treat most hazardous waste streams by physical, chemical, thermal, and biological methods, and for disposal of residues. Use of such treatment/disposal processes is costly, ranging from a low of $1.40/ton for carbon sorption, $10/ton for neutralization/precipitation, and $13.60/ton for chemical oxidation to $95/ton for incineration. Several unit processes are usually required for complete treatment/disposal of a given waste stream. Transfer and adaptation of existing technology to hazardous waste management may be necessary in some cases. Development of new treatment and disposal methods for some wastes (e.g., arsenic trioxide and arsenites and arsenates of lead, sodium, zinc, and potassium) is required. In the absence of treatment processes, interim storage of wastes on land is possible using methods that minimize hazard to the public and the environment (e.g., secure storage, membrane landfills).

A small private hazardous waste management industry has emerged in the last decade, offering treatment/disposal services to generators. The industry currently has capital investments of approximately $25 million and a capacity to handle about 2.5 million tons of hazardous materials yearly, or 25% of capacity required nationally. The industry's current throughput of hazardous waste is about 24% of installed capacity or 6% of the national total. The low level of utilization of this industry's services results from the absence of regulatory and economic incentives for generators to manage their hazardous wastes in an environmentally sound manner. This industry could respond over time to provide needed capacity if a national program for hazardous waste management, with strong enforcement capabilities, were created. This industry would, of course, be subject to regulation also.

The chief programmatic requirement to bring about adequate management of hazardous wastes is the creation of demand and adequate capacity for treatment/disposal of hazardous wastes. A national policy on hazardous waste management should take into consideration environmental protection, equitable cost distribution among generators, and recovery of waste materials.

A regulatory approach is best for the achievement of hazardous waste management objectives. A regulatory approach ensures adequate protection of public health and the environment. It will likely result in the creation of treatment/disposal capacity by the private sector without public funding. It will result in the mandatory use of such facilities. Costs of management will be borne by those who generate the hazardous wastes and their customers, rather than by the public at large, and thus cost distribution will be equitable. Private-sector management of the wastes in a competitive situation can lead to an appropriate mix of source reduction, treatment, resource recovery, and land disposal.

A regulatory program will not directly create a prescribed system of national disposal sites, however, because of uncertainties inherent in the private-sector response. EPA believes that the private sector will respond to a regulatory program. However, full assurance cannot be given that treatment/disposal facilities will be available in a timely manner for all regions of the nation nor that facility use charges will be reasonable in relation to cost of services. Also, private enterprise does not appear well suited institutionally to long-term security and surveillance of hazardous waste storage and disposal sites.

Based on analyses performed to date, EPA believes that no government actions to limit the uncertainties in private-sector response are appropriate at this time. However, if private capital flow were very slow and adverse environmental effects were resulting from the investment rate, indirect financial assistance in forms such as loans, loan guarantees, or investment credits could be used to accelerate investment. If facility location or user charge problems arose, the government could impose a franchise system with territorial limits and user charge rate controls. Long-term care of hazardous waste storage and disposal facilities could be assured by mandating use of federal or state land for such facilities.

EPA studies indicate that treatment/disposal of hazardous wastes at central processing facilities is preferable to management at each point of generation in most cases because of economies of scale, decreased environmental risk, and increased opportunities for resource recovery. However, other forces may deter creation of the "regional processing facility" type of system. For example, the pending effluent limitation guidelines now being developed under authority of the federal Water Pollution Control Act may force each generator to install water treatment facilities for both hazardous and nonhazardous aqueous waste streams. Consequently, the absolute volume of hazardous wastes requiring further treatment at central facilities may be reduced and the potential for economies of scale at such facilities may not be as strong as it is currently.

Given these uncertainties, several projections of future events can be made. Processing capacity required nationally was estimated assuming complete regulation, treatment, and disposal of *all* hazardous wastes at the earliest practicable time period. Estimates were based on a postulated scenario in which approximately 20 regional treatment/disposal facilities are constructed across the nation. Of these, 5 would be very large facilities serving major industrial areas treating 1.3 million tons yearly each, and 15 would be medium-size facilities each treating 160,000 tons annually. An estimated 8.5 million tons of hazardous wastes would be treated/disposed of away from the point of generation (off-site); 1.5 million tons would be pre-treated by generators on-site, with 0.5 million tons of residues transported to off-site treatment/disposal facilities for further processing. Each regional processing facility was assumed to provide a complete range of treatment processes capable of handling all types of hazardous wastes, and therefore each would be much more costly than existing private facilities.

Capital requirements to create the system described here are approximately $940 million. Average annual operating expenditures (including capital recovery and operating costs) of $620 million would be required to sustain the program. These costs are roughly estimated to be equivalent to 1% of the value of shipments from industries directly impacted. In addition, administrative expenses of about $20 million annually for federal and state regulatory programs would be necessary. For the reasons stated earlier, however, capacity and capital requirements for a national hazardous waste management system may be smaller than indicated, and more in line with the capacity and capital availability of the existing hazardous waste management industry.

B

EPA Listing of Hazardous Waste as a Part of the RCRA Regulations, Section 3001

§261.31 Hazardous waste from nonspecific sources.

The following solid wastes are listed hazardous wastes from non-specific sources unless they are excluded under

§§ 260.20 and 260.22 and listed in Appendix XI.

[261.31 introductory text added by 49 FR 37070, September 21, 1984]

Industry and EPA hazardous waste No.	Hazardous waste	Hazard code
Generic:		
F001	The following spent halogenated solvents used in degreasing: tetrachloroethylene, trichloroethylene, methylene chloride, 1,1,1-trichloroethane, carbon tetrachloride, and chlorinated fluorocarbons; and sludges from the recovery of these solvents in degreasing operations.	(T)
F002	The following spent halogenated solvents: tetrachloroethylene, methylene chloride, trichloroethylene, 1,1,1-trichloroethane, chlorobenzene, 1,1,2-trichloro-1,2,2-trifluoroethane, ortho-dichlorobenzene, and trichlorofluoromethane; and the still bottoms from the recovery of these solvents.	(T)
F003	The following spent non-halogenated solvents: xylene, acetone, ethyl acetate, ethyl benzene, ethyl ether, methyl isobutyl ketone, n-butyl alcohol, cyclohexanone, and methanol; and the still bottoms from the recovery of these solvents.	(I)
F004	The following spent non-halogenated solvents: cresols and cresylic acid and nitrobenzene; and the still bottoms from the recovery of these solvents.	(T)
F005	The following spent non-halogenated solvents: toluene, methyl ethyl ketone, carbon disulfide, isobutanol, and pyridine; and the still bottoms from the recovery of these solvents.	(I, T)
F006	Wastewater treatment sludges from electroplating operations except from the following processes: (1) sulfuric acid anodizing of aluminum; (2) tin plating on carbon steel; (3) zinc plating (segregated basis) on carbon steel; (4) aluminum or zinc-aluminum plating on carbon steel; (5) cleaning/stripping associated with tin, zinc and aluminum plating on carbon steel; and (6) chemical etching and milling of aluminum.	(T)
F019	Wastewater treatment sludges from the chemical conversion coating of aluminum.	(T)
F007	Spent Cyanide plating bath solutions from electroplating operations.	(R, T)
F008	Plating sludges from the bottom of plating baths from electroplating operations where cyanides are used in the process	(R, T)
F009	Spent stripping and cleaning bath solutions from electroplating operations where cyanides are used in the process.	(R, T)
F010	Quenching bath residues from oil baths from metal heat treating operations where cyanides are used in the process.	(R, T)
F011	Spent cyanide solutions from salt bath pot cleaning from metal heat treating operations.	(R, T)
F012	Quenching wastewater treatment sludges from metal heat treating operations where cyanides are used in the process.	(T)
F024	Wastes, including, but not limited to, distillation residues, heavy ends, fars, and reactor cleanout wastes from the production of chlorinated aliphatic hydrocarbons, having carbon content from one to five, utilizing free radical catalyzed processes. [This listing does not include light ends, spent filters and filter aids, spent dessicants, wastewater, wastewater treatment sludges, spent catalysts, and wastes listed in §261.32].	(T)
F020	Wastes (except wastewater and spent carbon from hydrogen chloride purification) from the production or manufacturing use (as a reactant, chemical intermediate, or component in a formulating process) of tri- or tetrachlorophenol, or of intermediates used to produce their pesticide derivatives. (This listing does not include wastes from the production of Hexachlorophena from highly purified 2,4,5-trichlorophenol.).	(H)
F021	Wastes (except wastewater and spent carbon from hydrogen chloride purification) from the production or manufacturing use (as a reactant, chemical intermediate, or component in a formulating process) of pentachlorophenol, or of intermediates used to produce its derivatives.	(H)
F022	Wastes (except wastewater and spent carbon from hydrogen chloride purification) from the manufacturing use (as a reactant, chemical intermediate, or component in a formulating process) of tetra-, penta-, or hexachlorobenzenes under alkaline conditions.	(H)
F023	Wastes (except wastewater and spent carbon from hydrogen chloride purification) from the production of materials on equipment previously used for the production or manufacturing use (as a reactant, chemical intermediate, or component in a formulating process) of tri- or tetrachlorophenols. (This listing does not include wastes from equipment used only for the production or use of Hexachlorophene from highly purified 2,4,5-trichlorophenol.).	(H)
F026	Wastes (except wastewater and spent carbon from hydrogen chloride purification) from the production of materials on equipment previously used for the manufacturing use (as a reactant, chemical intermediate, or component in a formulating process) of tetra-, penta-, or hexachlorobenzene under alkaline conditions.	(H)
F027	Discarded unused formulations containing tri-, tetra- or pentachlorophenol or discarded unused formulation containing compounds derived from these chlorophenols. (This listing does not include formulations containing Hexachlorophene synthesized from prepurified 2,4,5-trichlorophenol as the sole component.).	(H)
F028	Residues resulting from the incineration or thermal treatment of soil contaminated with EPA Hazardous Waste Nos. FO20, FO21, FO22, FO23, FO26, and FO27.	(T)

[261.31 amended by 45 FR 47833, July 16, 1980, revised by 45 FR 74890, November 12, 1980, 46 FR 4617, January 16, 1981, 46 FR 27476, May 20, 1981, 49 FR 5312, February 10, 1984; 50 FR 661, January 4, 1985; 50 FR 1999, January 14, 1985]

§ 261.32 Hazardous waste from specific sources.

The following solid wastes are listed hazardous wastes from specific sources unless they are excluded under

§§ 260.20 and 260.22 and listed in Appendix IX.

[261.32 introductory text added by 49 FR 37070, September 21, 1984]

[Sec. 261.32]

Industry and EPA hazardous waste No.	Hazardous waste	Hazard code
ood Preservation:		
K001	Bottom sediment sludge from the treatment of wastewaters from wood preserving processes that use creosote and/or pentachlorophenol	(T)
organic Pigments:		
K002	Wastewater treatment sludge from the production of chrome yellow and orange pigments	(T)
K003	Wastewater treatment sludge from the production of molybdate orange pigments	(T)
K004	Wastewater treatment sludge from the production of zinc yellow pigments	(T)
K005	Wastewater treatment sludge from the production of chrome green pigments	(T)
K006	Wastewater treatment sludge from the production of chrome oxide green pigments (anhydrous and hydrated)	(T)
K007	Wastewater treatment sludge from the production of iron blue pigments	(T)
K008	Oven residue from the production of chrome oxide green pigments	(T)
ganic Chemicals:		
K009	Distillation bottoms from the production of acetaldehyde from ethylene	(T)
K010	Distillation side cuts from the production of acetaldehyde from ethylene	(T)
K011	Bottom stream from the wastewater stripper in the production of acrylonitrile	(R, T)
K013	Bottom stream from the acetonitrile column in the production of acrylonitrile	(R, T)
K014	Bottoms from the acetonitrile purification column in the production of acrylonitrile	(T)
K015	Still bottoms from the distillation of benzyl chloride	(T)
K016	Heavy ends or distillation residues from the production of carbon tetrachloride	(T)
K017	Heavy ends (still bottoms) from the purification column in the production of epichlorohydrin	(T)
K018	Heavy ends from the fractionation column in ethyl chloride production	(T)
K019	Heavy ends from the distillation of ethylene dichloride in ethylene dichloride production	(T)
K020	Heavy ends from the distillation of vinyl chloride in vinyl chloride monomer production	(T)
K021	Aqueous spent antimony catalyst waste from fluoromethanes production	(T)
K022	Distillation bottom tars from the production of phenol/acetone from cumene	(T)
K023	Distillation light ends from the production of phthalic anhydride from naphthalene	(T)
K024	Distillation bottoms from the production of phthalic anhydride from naphthalene	(T)
K093	Distillation light ends from the production of phthalic anhydride from ortho-xylene	(T)
K094	Distillation bottoms from the production of phthalic anhydride from ortho-xylene	(T)
K025	Distillation bottoms from the production of nitrobenzene by the nitration of benzene	(T)
K026	Stripping still tails from the production of methy ethyl pyridines	(T)
K027	Centrifuge and distillation residues from toluene diisocyanate production	(R, T)
K028	Spent catalyst from the hydrochlorinator reactor in the production of 1,1,1-trichloroethane	(T)
K029	Waste from the product steam stripper in the production of 1,1,1-trichloroethane	(T)
K095	Distillation bottoms from the production of 1,1,1-trichloroethane	(T)
K096	Heavy ends from the heavy ends column from the production of 1,1,1-trichloroethane	(T)
K030	Column bottoms or heavy ends from the combined production of trichloroethylene and perchloroethylene	(T)
K083	Distillation bottoms from aniline production	(T)
K103	Process residues from aniline extraction from the production of aniline	(T)
K104	Combined wastewater streams generated from nitrobenzene/aniline production	(T)
K085	Distillation or fractionation column bottoms from the production of chlorobenzenes	(T)
K105	Separated aqueous stream from the reactor product washing step in the production of chlorobenzenes	(T)
organic Chemicals:		
K071	Brine purification muds from the mercury cell process in chlorine production, where separately prepurified brine is not used	(T)
K073	Chlorinated hydrocarbon waste from the purification step of the diaphragm cell process using graphite anodes in chlorine production	(T)
K106	Wastewater treatment sludge from the mercury cell process in chlorine production	(T)
esticides:		
K031	By-product salts generated in the production of MSMA and cacodylic acid	(T)
K032	Wastewater treatment sludge from the production of chlordane	(T)
K033	Wastewater and scrub water from the chlorination of cyclopentadiene in the production of chlordane	(T)
K034	Filter solids from the filtration of hexachlorocyclopentadiene in the production of chlordane	(T)
K097	Vacuum stripper discharge from the chlordane chlorinator in the production of chlordane	(T)
K035	Wastewater treatment sludges generated in the production of creosote	(T)
K036	Still bottoms from toluene reclamation distillation in the production of disulfoton	(T)
K037	Wastewater treatment sludges from the production of disulfoton	(T)
K038	Wastewater from the washing and stripping of phorate production	(T)
K039	Filter cake from the filtration of diethylphosphorodithioic acid in the production of phorate	(T)
K040	Wastewater treatment sludge from the production of phorate	(T)
K041	Wastewater treatment sludge from the production of toxaphene	(T)
K098	Untreated process wastewater from the production of toxaphene	(T)
K042	Heavy ends or distillation residues from the distillation of tetrachlorobenzene in the production of 2,4,5-T	(T)
K043	2,6-Dichlorophenol waste from the production of 2,4-D	(T)
K099	Untreated wastewater from the production of 2,4-D	(T)
xplosives:		
K044	Wastewater treatment sludges from the manufacturing and processing of explosives	(R)
K045	Spent carbon from the treatment of wastewater containing explosives	(R)
K046	Wastewater treatment sludges from the manufacturing, formulation and loading of lead-based initiating compounds	(T)
K047	Pink/red water from TNT operations	(R)
etroleum Refining:		
K048	Dissolved air flotation (DAF) float from the petroleum refining industry	(T)
K049	Slop oil emulsion solids from the petroleum refining industry	(T)
K050	Heat exchanger bundle cleaning sludge from the petroleum refining industry	(T)
K051	API separator sludge from the petroleum refining industry	(T)
K052	Tank bottoms (leaded) from the petroleum refining industry	(T)
on and Steel:		
K061	Emission control dust/sludge from the primary production of steel in electric furnaces	(T)
K062	Spent pickle liquor from steel finishing operations	(C, T)
econdary Lead:		
K069	Emission control dust/sludge from secondary lead smelting	(T)
K100	Waste leaching solution from acid leaching of emission control dust/sludge from secondary lead smelting	(T)

[Sec. 261.32]

Industry and EPA hazardous waste No.	Hazardous waste	Hazard code
Veterinary Pharmaceuticals:		
K084	Wastewater treatment sludges generated during the production of veterinary pharmaceuticals from arsenic or organo-arsenic compounds	(T)
K101	Distillation tar residues from the distillation of aniline-based compounds in the production of veterinary pharmaceuticals from arsenic or organo-arsenic compounds.	(T)
K102	Residue from the use of activated carbon for decolorization in the production of veterinary pharmaceuticals from arsenic or organo-arsenic compounds.	(T)
Ink Formulation:		
K086	Solvent washes and sludges, caustic washes and sludges, or water washes and sludges from cleaning tubs and equipment used in the formulation of ink from pigments, driers, soaps, and stabilizers containing chromium and lead.	(T)
Coking:		
K060	Ammonia still lime sludge from coking operations	(T)
K087	Decanter tank tar sluge from coking operations	(T)

[261.32 amended by 45 FR 47833, July 16, 1980; 45 FR 72039, October 30, 1980; revised by 45 FR 74980, November 12, 1980; 46 FR 4617, January 16, 1981; 46 FR 27476, May 20, 1981]

§ 261.33 Discarded commercial chemical products, off-specification species, container residues, and spill residues thereof.

[261.33 revised by 45 FR 78541, November 25, 1980]

The following materials or items are hazardous wastes when they are discarded or intended to be discarded as described in § 261.2(a)(2)(i), when they are burned for purposes of energy recovery in lieu of their original intended use, when they are used to produce fuels in lieu of their original intended use, when they are applied to the land in lieu of their original intended use, or when they are contained in products that are applied to the land in lieu of their original intended use.

[261.33 introductory text amended by 49 FR 37070, September 21, 1984; 50 FR 661, January 4, 1985]

(a) Any commercial chemical product, or manufacturing chemical intermediate having the generic name listed in paragraph (e) or (f) of this section.

(b) Any off-specification commercial chemical product or manufacturing chemical intermediate which, if it met specifications, would have the generic name listed in paragraph (e) or (f) of this section.

(c) Any container or inner liner removed from a container that has been used to hold any commercial chemical product or manufacturing chemical intermediate having the generic names listed in paragraph (e) of this section, or any container or inner liner removed from a container that has been used to hold any off-specification chemical product and manufacturing chemical intermediate which, if it met specifications, would have the generic name listed in paragraph (e) of this section, unless:

(1) The container or inner liner has been triple rinsed using a solvent capable of removing the commercial chemical product or manufacturing chemical intermediate; or

(2) The container or inner liner has been cleansed by another method that has been shown in the scientific literature, or by tests conducted by the generator, to achieve equivalent removal; or

(3) In the case of a container, the inner liner that prevented contact of the commercial chemical product or manufacturing chemical intermediate with the container, has been removed.

[261.33(c) revised by 45 FR 78541, November 25, 1980; 46 FR 27476, May 20, 1981]

(d) Any residue or contaminated soil, water or other debris resulting from the cleanup of a spill into or on any land or water of any commercial chemical product or manufacturing chemical intermediate having the generic name listed in paragraph (e) or (f) of this section, or any residue or contaminated soil, water or other debris resulting from the cleanup of a spill, into or on any land or water, of any off-specification chemical product and manufacturing chemical intermediate which, if it met specifications would have the generic name listed in paragraph (e) or (f) of this section.

[Comment: The phrase "commercial chemical product or manufacturing chemical intermediate having the generic name listed in . . ." refers to a chemical substance which is manufactured or formulated for commercial or manufacturing use which consists of the commercially pure grade of the chemical, any technical grades of the chemical that are produced or marketed and all formulations in which the chemical is the sole active ingredient. It does not refer to a material, such as a manufacturing process waste, that contains any of the substances listed in paragraphs (e) or (f

[Sec. 261.33(d)]

Where a manufacturing process waste is deemed to be a hazardous waste because it contains a substance listed in paragraphs (e) or (f), such waste will be listed in either §§ 261.31 or 261.32 or will be identified as a hazardous waste by the characteristics set forth in Subpart C of this Part.]

[261.33(d) amended by 46 FR 27476, May 20, 1981]

(e) The commercial chemical products, manufacturing chemical intermediates or off-specification commercial chemical products or manufacturing chemical intermediates referred to in paragraphs (a) through (d) of this section, are identified as acute hazardous wastes (H) and are subject to be the small quantity exclusion defined in § 261.5(e).

[Comment: For the convenience of the regulated community the primary hazardous properties of these materials have been indicated by the letters T (Toxicity), and R (Reactivity). Absence of a letter indicates that the compound only is listed for acute toxicity.]

These wastes and their corresponding EPA Hazardous Waste Numbers are:

261.33(e) amended by 46 FR 27476, May 20, 1981]

Hazardous waste No.	Substance
P023.............	Acetaldehyde, chloro-
P002.............	Acetamide, N-(aminothioxomethyl)-
P057.............	Acetamide, 2-fluoro-
P058.............	Acetic acid, fluoro-, sodium salt
P066.............	Acetimidic acid, N-[(methylcarbamoyl)oxy]thio-, methyl ester
P001.............	3-(alpha-acetonylbenzyl)-4-hydroxycoumarin and salts, when present at concentrations greater than 0.3%.
	[P001 amended by 49 FR 19923, May 10, 1984]
P002.............	1-Acetyl-2-thiourea
P003.............	Acrolein
P070.............	Aldicarb
P004.............	Aldrin
P005.............	Allyl alcohol
P006.............	Aluminum phosphide
P007.............	5-(Aminomethyl)-3-isoxazolol
P008.............	4-aAminopyridine
P009.............	Ammonium picrate (R)
P119.............	Ammonium vanadate
P010.............	Arsenic acid
P012.............	Arsenic (III) oxide
P011.............	Arsenic (V) oxide
P011.............	Arsenic pentoxide
P012.............	Arsenic trioxide
P038.............	Arsine, diethyl-
P054.............	Aziridine
P013.............	Barium cyanide
P024.............	Benzenamine, 4-chloro-
P077.............	Benzenamine, 4-nitro-
P028.............	Benzene, (chloromethyl)-
P042.............	1,2-Benzenediol, 4-[1-hydroxy-2-(methylamino)ethyl]-
P014.............	Benzenethiol
P028.............	Benzyl chloride
P015.............	Beryllium dust
P016.............	Bis(chloromethyl) ether
P017.............	Bromoacetone

Hazardous waste No.	Substance
P018.............	Brucine
P021.............	Calcium cyanide
P123.............	Camphene, octachloro-
P103.............	Carbamimidoselenoic acid
P022.............	Carbon bisulfide
P022.............	Carbon disulfide
P095.............	Carbonyl chloride
P033.............	Chlorine cyanide
P023.............	Chloroacetaldehyde
P024.............	p-Chloroaniline
P026.............	1-(o-Chlorophenyl)thiourea
P027.............	3-Chloropropionitrile
P029.............	Copper cyanides
P030.............	Cyanides (soluble cyanide salts), not elsewhere specified
P031.............	Cyanogen
P033.............	Cyanogen chloride
P036.............	Dichlorophenylarsine
P037.............	Dieldrin
P038.............	Diethylarsine
P039.............	O,O-Diethyl S-[2-(ethylthio)ethyl] phosphorodithioate
P041.............	Diethyl-p-nitrophenyl phosphate
P040.............	O,O-Diethyl O-pyrazinyl phosphorothioate
P043.............	Diisopropyl fluorophosphate
P044.............	Dimethoate
P045.............	3,3-Dimethyl-1-(methylthio)-2-butanone, O-[(methylamino)carbonyl] oxime
P071.............	O,O-Dimethyl O-p-nitrophenyl phosphorothioate
P082.............	Dimethylnitrosamine
P046.............	alpha, alpha-Dimethylphenethylamine
P047.............	4,6-Dinitro-o-cresol and salts
P034.............	4,6-Dinitro-o-cyclohexylphenol
P048.............	2,4-Dinitrophenol
P020.............	Dinoseb
P085.............	Diphosphoramide, octamethyl-
P039.............	Disulfoton
P049.............	2,4-Dithiobiuret
P109.............	Dithiopyrophosphoric acid, tetraethyl ester
P050.............	Endosulfan
P088.............	Endothall
P051.............	Endrin
P042.............	Epinephrine
P046.............	Ethanamine, 1,1-dimethyl-2-phenyl-
P084.............	Ethenamine, N-methyl-N-nitroso-
P101.............	Ethyl cyanide
P054.............	Ethylenimine
P097.............	Famphur
P056.............	Fluorine
P057.............	Fluoroacetamide
P058.............	Fluoroacetic acid, sodium salt
P065.............	Fulminic acid, mercury(II) salt (R,T)
P059.............	Heptachlor
P051.............	1,2,3,4,10,10-Hexachloro-6,7-epoxy-1,4,4a,5,6,7,8,8a-octahydro-endo,endo-1,4:5,8-dimethanonaphthalene
P037.............	1,2,3,4,10,10-Hexachloro-6,7-epoxy-1,4,4a,5,6,7,8,8a-octahydro-endo,exo-1,4:5,8-demethanonaphthalene
P060.............	1,2,3,4,10,10-Hexachloro-1,4,4a,5,8,8a-hexahydro-1,4:5,8-endo, endo-dimeth- an-onaphthalene
P004.............	1,2,3,4,10,10-Hexachloro-1,4,4a,5,8,8a-hexahydro-1,4:5,8-endo,exo-dimethanonaphthalene
P060.............	Hexachlorohexahydro-exo,exo-dimethanonaphthalene
P062.............	Hexaethyl tetraphosphate
P116.............	Hydrazinecarbothioamide
P068.............	Hydrazine, methyl-
P063.............	Hydrocyanic acid
P063.............	Hydrogen cyanide
P096.............	Hydrogen phosphide
P064.............	Isocyanic acid, methyl ester
P007.............	3(2H)-Isoxazolone, 5-(aminomethyl)-
P092.............	Mercury, (acetato-O)phenyl-
P065.............	Mercury fulminate (R,T)
P016.............	Methane, oxybis(chloro-
P112.............	Methane, tetranitro- (R)
P118.............	Methanethiol, trichloro-
P059.............	4,7-Methano-1H-indene, 1,4,5,6,7,8,8-hep-tachloro-3a,4,7,7a-tetrahydro-
P066.............	Methomyl

Hazardous Waste No.	Substance
P067.............	2-Methylaziridine
P068.............	Methyl hydrazine
P064.............	Methyl isocyanate
P069.............	2-Methyllactonitrile
P071.............	Methyl parathion
P072.............	alpha-Naphthylthiourea
P073.............	Nickel carbonyl
P074.............	Nickel cyanide
P074.............	Nickel(II) cyanide
P073.............	Nickel tetracarbonyl
P075.............	Nicotine and salts
P076.............	Nitric oxide
P077.............	p-Nitroaniline
P078.............	Nitrogen dioxide
P076.............	Nitrogen(II) oxide
P078.............	Nitrogen(IV) oxide
P081.............	Nitroglycerine (R)
P082.............	N-Nitrosodimethylamine
P084.............	N-Nitrosomethylvinylamine
P050.............	5-Norbornene-2,3-dimethanol, 1,4,5,6,7,7-hex-achloro, cyclic sulfite
P085.............	Octamethylpyrophosphoramide
P087.............	Osmium oxide
P087.............	Osmium tetroxide
P088.............	7-Oxabicyclo[2.2.1]heptane-2,3-dicarboxylic acid
P089.............	Parathion
P034.............	Phenol, 2-cyclohexyl-4,6-dinitro-
P048.............	Phenol, 2,4-dinitro-
P047.............	Phenol, 2,4-dinitro-6-methyl-
P020.............	Phenol, 2,4-dinitro-6-(1-methylpropyl)-
P009.............	Phenol, 2,4,6-trinitro-, ammonium salt (R)
P036.............	Phenyl dichloroarsine
P092.............	Phenylmercuric acetate
P093.............	N-Phenylthiourea
P094.............	Phorate
P095.............	Phosgene
P096.............	Phosphine
P041.............	Phosphoric acid, diethyl p-nitrophenyl ester
P044.............	Phosphorodithioic acid, O,O-dimethyl S-[2-(methylamino)-2-oxoethyl]ester
P043.............	Phosphorofluoric acid, bis(1-methylethyl)-ester
P094.............	Phosphorothioic acid, O,O-diethyl S-(ethylthio)methyl ester
P089.............	Phosphorothioic acid, O,O-diethyl O-(p-nitro-phenyl) ester
P040.............	Phosphorothioic acid, O,O-diethyl O- pyrazinyl ester
P097.............	Phosphorothioic acid. O,O-dimethyl O-[p-((di-methylamino)-sulfonyl)phenyl]ester
P110.............	Plumbane, tetraethyl-
P098.............	Potassium cyanide
P099.............	Potassium silver cyanide
P070.............	Propanal, 2-methyl-2-(methylthio)-, O-[(methylamino)carbonyl]oxime
P101.............	Propanenitrile
P027.............	Propanenitrile, 3-chloro-
P069.............	Propanenitrile, 2-hydroxy-2-methyl-
P081.............	1,2,3-Propanetriol, trinitrate- (R)
P017.............	2-Propanone, 1-bromo-
P102.............	Propargyl alcohol
P003.............	2-Propenal
P005.............	2-Propen-1-ol
P067.............	1,2-Propylenimine
P102.............	2-Propyn-1-ol
P008.............	4-Pyridinamine
P075.............	Pyridine, (S)-3-(1-methyl-2-pyrrolidinyl)-, and salts
P111.............	Pyrophosphoric acid, tetraethyl ester
P103.............	Selenourea
P104.............	Silver cyanide
P105.............	Sodium azide
P106.............	Sodium cyanide
P107.............	Strontium sulfide
P108.............	Strychnidin-10-one, and salts
P018.............	Strychnidin-10-one, 2,3-dimethoxy-
P108.............	Strychnine and salts
P115.............	Sulfuric acid, thallium(I) salt
P109.............	Tetraethyldithiopyrophosphate
P110.............	Tetraethyl lead
P111.............	Tetraethylpyrophosphate
P112.............	Tetranitromethane (R)
P062.............	Tetraphosphoric acid, hexaethyl ester
P113.............	Thallic oxide

Hazardous waste No.	Substance
P113	Thallium(III) oxide
P114	Thallium(I) selenite
P115	Thallium(I) sulfate
P045	Thiofanox
P049	Thioimidodicarbonic diamide
P014	Thiophenol
P116	Thiosemicarbazide
P026	Thiourea, (2-chlorophenyl)-
P072	Thiourea, 1-naphthalenyl-
P093	Thiourea, phenyl-
P123	Toxaphene
P118	Trichloromethanethiol
P119	Vanadic acid, ammonium salt
P120	Vanadium pentoxide
P120	Vanadium(V) oxide
P001	Warfarin, when present at concentrations greater than 0.3%.
	[P001 amended by 49 FR 19923, May 10, 1984]
P121	Zinc cyanide
P122	Zinc phosphide, when present at concentrations greater than 10%.
	[P122 amended by 49 FR 19923, May 10, 1984]

(f) The commercial chemical products, manufacturing chemical intermediates, or off-specification commercial chemical products referred to in paragraphs (a) through (d) of this section, are identified as toxic wastes (T) unless otherwise designated and are subject to the small quantity exclusion defined in § 261.5 (a) and (f).

[Comment: For the convenience of the regulated community, the primary hazardous properties of these materials have been indicated by the letters T (Toxicity), R (Reactivity), I (Ignitability) and C (Corrosivity). Absence of a letter indicates that the compound is only listed for toxicity.]

These wastes and their corresponding EPA Hazardous Waste Numbers are:

[261.33(f) amended by 46 FR 27476, May 10, 1984; 50 FR 1999, January 14, 1985]

Hazardous Waste No	Substance
U005	Acetamide, N-9H-fluoren-2-yl-
U112	Acetic acid, ethyl ester (I)
U144	Acetic acid, lead salt
U214	Acetic acid, thallium(I) salt
U002	Acetone (I)
U003	Acetonitrile (I,T)
U004	Acetophenone
U005	2-Acetylaminofluorene
U006	Acetyl chloride (C,R,T)
U007	Acrylamide
U008	Acrylic acid (I)
U009	Acrylonitrile
U150	Alanine, 3-[p-bis(2-chloroethyl)amino] phenyl-, L-
U011	Amitrole
U012	Aniline (I,T)
J248	3-(alpha-Acetonylbenzyl)-4-hydroxycoumarin and salts, when present at concentrations of 0.3% or less.

[U248 added by 49 FR 19923, May 10, 1984]

U014	Auramine
U015	Azaserine
U010	Azirino(2',3':3,4)pyrrolo(1,2-a)indole-4,7-dione, 6-amino-8-[((aminocarbonyl) oxy)methyl]-1\1a,2,8,8a,8b-hexahydro-8a-methoxy-5-methyl-,
U157	Benz[j]aceanthrylene, 1,2-dihydro-3-methyl-
U016	Benz[c]acridine
U016	3,4-Benzacridine
U017	Benzal chloride
U018	Benz[a]anthracene
U018	1,2-Benzanthracene
U094	1,2-Benzanthracene, 7,12-dimethyl-
U012	Benzenamine (I,T)
U014	Benzenamine, 4,4'-carbonimidoylbis(N,N-dimethyl-
U049	Benzenamine, 4-chloro-2-methyl-
U093	Benzenamine, N,N'-dimethyl-4-phenylazo-
U158	Benzenamine, 4,4'-methylenebis(2-chloro-
U222	Benzenamine, 2-methyl-, hydrochloride
U181	Benzenamine, 2-methyl-5-nitro-
U019	Benzene (I,T)
U038	Benzeneacetic acid, 4-chloro-alpha-(4-chlorophenyl)-alpha-hydroxy, ethyl ester
U030	Benzene, 1-bromo-4-phenoxy-
U037	Benzene, chloro-
U190	1,2-Benzenedicarboxylic acid anhydride
U028	1,2-Benzenedicarboxylic acid, [bis(2-ethylhexyl)] ester
U069	1,2-Benzenedicarboxylic acid, dibutyl ester
U088	1,2-Benzenedicarboxylic acid, diethyl ester
U102	1,2-Benzenedicarboxylic acid, dimethyl ester
U107	1,2-Benzenedicarboxylic acid, di-n-octyl ester
U070	Benzene, 1,2-dichloro-
U071	Benzene, 1,3-dichloro-
U072	Benzene, 1,4-dichloro-
U017	Benzene, (dichloromethyl)-
U223	Benzene, 1,3-diisocyanatomethyl- (R,T)
U239	Benzene, dimethyl-(I,T)
U201	1,3-Benzenediol
U127	Benzene, hexachloro-
U056	Benzene, hexahydro- (I)
U188	Benzene, hydroxy-
U220	Benzene, methyl-
U105	Benzene, 1-methyl-1-2,4-dinitro-
U106	Benzene, 1-methyl-2,6-dinitro-
U203	Benzene, 1,2-methylenedioxy-4-allyl-
U141	Benzene, 1,2-methylenedioxy-4-propenyl-
U090	Benzene, 1,2-methylenedioxy-4-propyl-
U055	Benzene, (1-methylethyl)- (I)
U169	Benzene, nitro- (I,T)
U183	Benzene, pentachloro-
U185	Benzene, pentachloro-nitro-
U020	Benzenesulfonic acid chloride (C,R)
U020	Benzenesulfonyl chloride (C,R)
U207	Benzene, 1,2,4,5-tetrachloro-
U023	Benzene, (trichloromethyl)-(C,R,T)
0234	Benzene, 1,3,5-trinitro- (R,T)
U021	Benzidine
U202	1,2-Benzisothiazolin-3-one, 1,1-dioxide
U120	Benzo[j,k]fluorene
U022	Benzo[a]pyrene
U022	3,4-Benzopyrene
U197	p-Benzoquinone
U023	Benzotrichloride (C,R,T)
U050	1,2-Benzphenanthrene
U085	2,2'-Bioxirane (I,T)
U021	(1,1'-Biphenyl)-4,4'-diamine
U073	(1,1'-Biphenyl)-4,4'-diamine, 3,3'-dichloro-
U091	(1,1'-Biphenyl)-4,4'-diamine, 3,3'-dimethoxy-
U095	(1,1'-Biphenyl)-4,4'-diamine, 3,3'-dimethyl-
U024	Bis(2-chloroethoxy) methane
U027	Bis(2-chloroisopropyl) ether
U244	Bis(dimethylthiocarbamoyl) disulfide
U028	Bis(2-ethylhexyl) phthalate
U246	Bromine cyanide
U225	Bromoform
U030	4-Bromophenyl phenyl ether
U128	1,3-Butadiene, 1,1,2,3,4,4-hexachloro-
U172	1-Butanamine, N-butyl-N-nitroso-
U035	Butanoic acid, 4-[Bis(2-chloroethyl)amino]benzene-
U031	1-Butanol (I)
U159	2-Butanone (I,T)
U160	2-Butanone peroxide (R,T)
U053	2-Butenal
U074	2-Butene, 1,4-dichloro- (I,T)

U031	n-Butyl alchohol (I)
U136	Cacodylic acid
U032	Calcium chromate
U238	Carbamic acid, ethyl ester
U178	Carbamic acid, methylnitroso-, ethyl ester
U176	Carbamide, N-ethyl-N-nitroso-
U177	Carbamide, N-methyl-N-nitroso-
U219	Carbamide, thio-
U097	Carbamoyl chloride, dimethyl-
U215	Carbonic acid, dithallium(I) salt
U156	Carbonochloridic acid, methyl ester (I,T)
U033	Carbon oxyfluoride (R,T)
U211	Carbon tetrachloride
U033	Carbonyl fluoride (R,T)
U034	Chloral
U035	Chlorambucil
U036	Chlordane, technical
U026	Chlornaphazine
U037	Chlorobenzene
U039	4-Chloro-m-cresol
U041	1-Chloro-2,3-epoxypropane
U042	2-Chloroethyl vinyl ether
U044	Chloroform
U046	Chloromethyl methyl ether
U047	beta-Chloronaphthalene
U048	o-Chlorophenol
U049	4-Chloro-o-toluidine, hydrochloride
U032	Chromic acid, calcium salt
U050	Chrysene
U051	Creosote
U052	Cresols
U052	Cresylic acid
U053	Crotonaldehyde
U055	Cumene (I)
U246	Cyanogen bromide
U197	1,4-Cyclohexadienedione
U056	Cyclohexane (I)
U057	Cyclohexanone (I)
U130	1,3-Cyclopentadiene, 1,2,3,4,5,5-hexa- ch
U058	Cyclophosphamide
U240	2,4-D, salts and esters
U059	Daunomycin
U060	DDD
U061	DDT
U142	Decachlorooctahydro-1,3,4-metheno-2H-cyclobuta[c,d]-pentalen-2-one
U062	Diallate
U133	Diamine (R,T)
U221	Diaminotoluene
U063	Dibenz[a,h]anthracene
U063	1,2:5,6-Dibenzanthracene
U064	1,2:7,8-Dibenzopyrene
U064	Dibenz[a,i]pyrene
U066	1,2-Dibromo-3-chloropropane
U069	Dibutyl phthalate
U062	S-(2,3-Dichloroallyl) diisopropylthiocarbam
U070	o-Dichlorobenzene
U071	m-Dichlorobenzene
U072	p-Dichlorobenzene
U073	3,3'-Dichlorobenzidine
U074	1,4-Dichloro-2-butene (I,T)
U075	Dichlorodifluoromethane
U192	3,5-Dichloro-N-(1,1-dimethyl-2-propynyl) benzamide
U060	Dichloro diphenyl dichloroethane
U061	Dichloro diphenyl trichloroethane
U078	1,1-Dichloroethylene
U079	1,2-Dichloroethylene
U025	Dichloroethyl ether
U081	2,4-Dichlorophenol
U082	2,6-Dichlorophenol
U240	2,4-Dichlorophenoxyacetic acid, salts esters
U083	1,2-Dichloropropane
U084	1,3-Dichloropropene
U085	1,2:3,4-Diepoxybutane (I,T)
U108	1,4-Diethylene dioxide
U086	N,N-Diethylhydrazine
U087	O,O-Diethyl-S-methyl-dithiophosphate
U088	Diethyl phthalate
U089	Diethylstilbestrol
U148	1,2-Dihydro-3,6-pyradizinedione
U090	Dihydrosafrole
U091	3,3'-Dimethoxybenzidine
U092	Dimethylamine (I)
U093	Dimethylaminoazobenzene
U094	7,12-Dimethylbenz[a]anthracene

[Sec. 261.33(f)]

Hazardous Waste No.	Substance
U095	3,3'-Dimethylbenzidine
U096	alpha,alpha-Dimethylbenzylhydroperoxide (R)
U097	Dimethylcarbamoyl chloride
U098	1,1-Dimethylhydrazine
U099	1,2-Dimethylhydrazine
U101	2,4-Dimethylphenol
U102	Dimethyl phthalate
U103	Dimethyl sulfate
U105	2,4-Dinitrotoluene
U106	2,6-Dinitrotoluene
U107	Di-n-octyl phthalate
U108	1,4-Dioxane
U109	1,2- Diphenylhydrazine
U110	Dipropylamine (I)
U111	Di-N-propylnitrosamine
U001	Ethanal (I)
U174	Ethanamine, N-ethyl-N-nitroso-
U067	Ethane, 1,2-dibromo-
U076	Ethane, 1,1-dichloro-
U077	Ethane, 1,2-dichloro-
U114	1,2-Ethanediylbiscarbamodithioic acid
U131	Ethane, 1,1,1,2,2,2-hexachloro-
U024	Ethane, 1,1'-[methylenebis(oxy)]bis[2-chloro-
U003	Ethanenitrile (I, T)
U117	Ethane,1,1'-oxybis- (I)
U025	Ethane, 1,1'-oxybis[2-chloro-
U184	Ethane, pentachloro-
U208	Ethane, 1,1,1,2-tetrachloro-
U209	Ethane, 1,1,2,2-tetrachloro-
U218	Ethanethioamide
U247	Ethane, 1,1,1,-trichloro-2,2-bis(p-methoxyphenyl).
U227	Ethane, 1,1,2-trichloro-
U043	Ethene, chloro-
U042	Ethene, 2-chloroethoxy-
U078	Ethene, 1,1-dichloro-
U079	Ethene, trans-1,2-dichloro-
U210	Ethene, 1,1,2,2-tetrachloro-
U173	Ethanol, 2,2'-(nitrosoimino)bis-
U004	Ethanone, 1-phenyl-
U006	Ethanoyl chloride (C,R,T)
U112	Ethyl acetate (I)
U113	Ethyl acrylate (I)
U238	Ethyl carbamate (urethan)
U038	Ethyl 4,4'-dichlorobenzilate
U114	Ethylenebis(dithiocarbamic acid)
U067	Etylene dibromide
U077	Ethylene dichloride
U115	Ethlene oxide (I,T)
U116	Ethylene thiourea
U117	Ethyl ether (I)
U076	Ethylidene dichloride
U118	Ethylmethacrylate
U119	Ethyl methanesulfonate
U139	Ferric dextran
U120	Fluoranthene
U122	Formaldehyde
U123	Formic acid (C,T)
U124	Furan (I)
U125	2-Furancarboxaldehyde (I)
U147	2,5-Furandione
U213	Furan, tetrahydro- (I)
U125	Furfural (I)
U124	Furfuran (I)
U206	D-Glucopyranose, 2-deoxy-2(3-methyl-3-nitro-soureido)-
U126	Glycidylaldehyde
U163	Guanidine, N-nitroso-N-methyl-N'-nitro-
U127	Hexachlorobenzene
U128	Hexachlorobutadiene
U129	Hexachlorocyclohexane (gamma isomer)
U130	Hexachlorocyclopentadiene
U131	Hexachloroethane
U132	Hexachlorophene
U243	Hexachloropropene
U133	Hydrazine (R,T)
U086	Hydrazine, 1,2-diethyl-
U098	Hydrazine, 1,1-dimethyl-
U099	Hydrazine, 1,2-dimethyl-
U109	Hydrazine, 1,2-diphenyl-
U134	Hydrofluoric acid (C,T)
U134	Hydrogen fluoride (C,T)
U135	Hydrogen sulfide
U096	Hydroperoxide, 1-methyl-1-phenylethyl- (R)
U136	Hydroxydimethylarsine oxide
U116	2-Imidazolidinethione
U137	Indeno[1,2,3-cd]pyrene
U139	Iron dextran
U140	Isobutyl alcohol (I,T)
U141	Isosafrole
U142	Kepone
U143	Lasiocarpine
U144	Lead acetate
U145	Lead phosphate
U146	Lead subacetate
U129	Lindane
U147	Maleic anhydride
U148	Maleic hydrazide
U149	Malononitrile
U150	Melphalan
U151	Mercury
U152	Methacrylonitrile (I,T)
U092	Methanamine, N-methyl- (I)
U029	Methane, bromo-
U045	Methane, chloro- (I,T)
U046	Methane, chloromethoxy-
U068	Methane, dibromo-
U080	Methane, dichloro-
U075	Methane, dichlorodifluoro-
U138	Methane, iodo-
U119	Methanesulfonic acid, ethyl ester
U211	Methane, tetrachloro-
U121	Methane, trichlorofluoro-
U153	Methanethiol (I,T)
U225	Methane, tribromo-
U044	Methane, trichloro-
U121	Methane, trichlorofluoro-
U123	Methanoic acid (C,T)
U036	4,7-Methanoindan, 1,2,4,5,6,7,8,8-octa-chloro-3a,4,7,7a-tetrahydro-
U154	Methanol (I)
U155	Methapyrilene
U247	Methoxychlor.
U154	Methyl alcohol (I)
U029	Methyl bromide
U186	1-Methylbutadiene (I)
U045	Methyl chloride, (I,T)
U156	Methyl chlorocarbonate (I,T)
U226	Methylchloroform
U157	3-Methylcholanthrene
U158	4,4'-Methylenebis(2-chloroaniline)
U132	2,2'-Methylenebis(3,4,6-trichlorophenol)
U068	Methylene bromide
U080	Methylene chloride
U122	Methylene oxide
U159	Methyl ethyl ketone (I,T)
U160	Methyl ethyl ketone peroxide (R,T)
U138	Methyl iodide
U161	Methyl isobutyl ketone (I)
U162	Methyl methacrylate (I,T)
U163	N-Methyl-N'-nitro-N-nitrosoguanidine
U161	4-Methyl-2-pentanone (I)
U164	Methylthiouracil
U010	Mitomycin C
U059	5,12-Naphthacenedione, (8S-cis)-8-acetyl-10-[(3-amino-2,3,6-trideoxy-alpha-L-iyxo-hexopyranosyl)oxy]-7,8,9,10-tetrahydro-6,8,11-trihydroxy-1-methoxy-
U165	Naphthalene
U047	Naphthalene, 2-chloro-
U166	1,4-Naphthalenedione
U236	2,7-Naphthalenedisulfonic acid, 3,3'-[(3,3'-di-methyl-(1,1'-biphenyl)-4,4-diyl)]-bis (azo](bis(5-amino-4-hydroxy)-,tetrasodium salt
U166	1,4-Naphthaquinone
U167	1-Naphthylamine
U168	2-Naphthylamine
U167	alpha-Naphthylamine
U168	beta-Naphthylamine
U026	2-Naphthylamine, N,N'-bis(2-chlorobenzyl)-
U169	Nitrobenzene (I,T)
U170	p-Nitrophenol
U171	2-Nitropropane (I)
U172	N-Nitrosodi-n-butylamine
U173	N-Nitrosodiethanolamine
U174	N-Nitrosodiethylamine
U111	N-Nitroso-N-propylamine
U176	N-Nitroso-N-ethylurea
U177	N-Nitroso-N-methylurea
U178	N-Nitroso-N-methylurethane
U179	N-Nitrosopiperidine
U180	N-Nitrosopyrrolidine
U181	5-Nitro-o-toluidine
U193	1,2-Oxathiolane, 2,2-dioxide
U058	2H-1,3,2-Oxazaphosphorine, 2-[bis(2-chloro-ethyl)amino]tetrahydro-, oxide 2-
U115	Oxirane (I,T)
U041	Oxirane, 2-(chloromethyl)-
U182	Paraldehyde
U183	Pentachlorobenzene
U184	Pentachloroethane
U185	Pentachloronitrobenzene
See F027	Pentachlorophenol
U186	1,3-Pentadiene (I)
U187	Phenacetin
U188	Phenol
U048	Phenol, 2-chloro-
U039	Phenol, 4-chloro-3-methyl-
U081	Phenol, 2,4-dichloro-
U082	Phenol, 2,6-dichloro-
U101	Phenol, 2,4-dimethyl-
U170	Phenol, 4-nitro-
See F027	Phenol, pentachloro-
Do	Phenol, 2,3,4,6-tetrachloro-
Do	Phenol, 2,4,5-trichloro-
Do	Phenol, 2,4,6-trichloro-
U137	1,10-(1,2-phenylene)pyrene
U145	Phosphoric acid, Lead salt
U087	Phosphorodithioic acid, O,O-diethyl-, S-methylester
U189	Phosphorous sulfide (R)
U190	Phthalic anhydride
U191	2-Picoline
U192	Pronamide
U194	1-Propanamine (I,T)
U110	1-Propanamine, N-propyl- (I)
U066	Propane, 1,2-dibromo 3-chloro-
U149	Propanedinitrile
U171	Propane, 2-nitro- (I)
U027	Propane, 2,2'-oxybis[2-chloro-
U193	1,3-Propane sultone
U235	1-Propanol, 2,3-dibromo-, phosphate (3:1)
U126	1-Propanol, 2,3-epoxy-
U140	1-Propanol, 2-methyl- (I,T)
U002	2-Propanone (I)
U007	2-Propenamide
U084	Propene, 1,3-dichloro-
U243	1-Propene, 1,1,2,3,3,3-hexachloro-
U009	2-Propenenitrile
U152	2-Propenenitrile, 2-methyl- (I,T)
U008	2-Propenoic acid (I)
U113	2-Propenoic acid, ethyl ester (I)
U118	2-Propenoic acid, 2-methyl-, ethyl ester
U162	2-Propenoic acid, 2-methyl-, methyl ester (I,T)
See F027	Propionic acid, 2-(2,4,5-trichlorophenoxy)-
U194	n-Propylamine (I,T)
U083	Propylene dichloride
U196	Pyridine
U155	Pyridine, 2-[(2-(dimethylamino)-2-thenyla-mino]-
U179	Pyridine, hexahydro-N-nitroso-
U191	Pyridine, 2-methyl-
U164	4(1H)-Pyrimidinone, 2,3-dihydro-6-methyl-2-thioxo-
U180	Pyrrole, tetrahydro-N-nitroso-
U200	Reserpine
U201	Resorcinol
U202	Saccharin and salts
U203	Safrole
U204	Selenious acid
U204	Selenium dioxide
U205	Selenium disulfide (R,T)
U015	L-Serine, diazoacetate (ester)

[Sec. 261.33(f)]

Hazardous Waste No.	Substance
See FO27	Silvex
U089	4,4'-Stilbenediol, alpha,alpha'-diethyl-
U206	Streptozotocin
U135	Sulfur hydride
U103	Sulfuric acid, dimethyl ester
U189	Sulfur phosphide (R)
U205	Sulfur selenide (R,T)
See FO27	2,4,5-T
U207	1,2,4,5-Tetrachlorobenzene
U208	1,1,1,2-Tetrachloroethane
U209	1,1,2,2-Tetrachloroethane
U210	Tetrachloroethylene
See FO27	2,3,4,6-Tetrachlorophenol
U213	Tetrahydrofuran (I)
U214	Thallium(I) acetate
U215	Thallium(I) carbonate
U216	Thallium(I) chloride
U217	Thallium(I) nitrate
U218	Thioacetamide
U153	Thiomethanol (I,T)
U219	Thiourea
U244	Thiram
U220	Toluene
U221	Toluenediamine
U223	Toluene diisocyanate (R,T)
U222	O-Toluidine hydrochloride
U011	1H-1,2,4-Triazol-3-amine
U226	1,1,1-Trichloroethane
U227	1,1,2-Trichloroethane
U228	Trichloroethene
U228	Trichloroethylene
U121	Trichloromonofluoromethane
See FO27	2,4,5-Trichlorophenol
Do.	2,4,6-Trichlorophenol
Do.	2,4,5-Trichlorophenoxyacetic acid
U234	sym-Trinitrobenzene (R,T)
U182	1,3,5-Trioxane, 2,4,5-trimethyl-
U235	Tris(2,3-dibromopropyl) phosphate
U236	Trypan blue
U237	Uracil, 5[bis(2-chloroethyl)amino]-
U237	Uracil mustard
U043	Vinyl chloride
U248	Warfarin, when present at concentrations of 0.3% or less.
	[U248 added by 49 FR 19923, May 10, 1984]
U239	Xylene (I)
U200	Yohimban-16-carboxylic acid, 11,17-dimeth-oxy-18-[(3,4,5-trimethoxy-benzoyl)oxy]-, methyl ester.
U249	Zinc phosphate, when present at concentrations of 10% or less.
	[U249 added by 49 FR 19923, May 10, 1984]

Appendix I—Representative Sampling Methods

The methods and equipment used for sampling waste materials will vary with the form and consistency of the waste materials to be sampled. Samples collected using the sampling protocols listed below, for sampling waste with properties similar to the indicated materials, will be considered by the Agency to be representative of the waste.

Extremely viscous liquid—ASTM Standard D140–70 Crushed or powdered material—ASTM Standard D346–75 Soil or rock-like material—ASTM Standard D420–69 Soil-like material—ASTM Standard D1452–65 Fly Ash-like material—ASTM Standard D2234–76 [ASTM Standards are available

from ASTM, 1916 Race St., Philadelphia, PA 19103]

Containerized liquid wastes—"COLIWASA" described in "Test Methods for the Evaluation of Solid Waste, Physical/ Chemical Methods," [1] U.S. Environmental Protection Agency, Office of Solid Waste, Washington, D.C. 20460. [Copies may be obtained from Solid Waste Information, U.S. Environmental Protection Agency, 26 W. St. Clair St., Cincinnati, Ohio 45268]

Liquid waste in pits, ponds, lagoons, and similar reservoirs.—"Pond Sampler" described in "Test Methods for the Evaluation of Solid Waste, Physical/ Chemical Methods." [1]

This manual also contains additional information on application of these protocols.

---
[1] These methods are also described in "Samplers and Sampling Procedures for Hazardous Waste Streams," EPA 600/2–80–018, January 1980.

Appendix II— EP Toxicity Test Procedure

[Revised by 46 FR 35247, July 7, 1981]

A. Extraction Procedure (EP)

1. A representative sample of the waste to be tested (minimum size 100 grams) shall be obtained using the methods specified in Appendix I or any other method capable of yielding a representative sample within the meaning of Part 260. [For detailed guidance on conducting the various aspects of the EP see "Test Methods for the Evaluation of Solid Waste, Physical/Chemical Methods" (incorporated by reference, see § 260.11).]

2. The sample shall be separated into its component liquid and solid phases using the method described in "Separation Procedure" below. If the solid residue [1] obtained using this method totals less than 0.5% of the original weight of the waste, the residue can be discarded and the operator shall treat the liquid phase as the extract and proceed immediately to Step 8.

3. The solid material obtained from the Separation Procedure shall be evaluated for its particle size. If the solid material has a surface area per gram of material equal to, or greater than, 3.1 cm² or passes through a 9.5 mm (0.375 inch) standard sieve, the operator shall proceed to Step 4. If the surface area is smaller or the particle size larger than specified above, the solid material shall be

---
[1] 1. The percent solids is determined by drying the filter pad at 80 C until it reaches constant weight and then calculating the percent solids using the following equation:

$$100 = \% \text{ solids } \frac{(\text{weight of pad} + \text{solid}) - (\text{tare weight of pad}) \times}{\text{initial weight of sample}}$$

prepared for extraction by crushing, cutting or grinding the material so that it passes through a 9.5 mm (0.375 inch) sieve or, if the material is in a single piece, by subjecting the material to the "Structural Integrity Procedure" described below.

4. The solid material obtained in Step 3 shall be weighed and placed in an extractor with 16 times its weight of deionized water. Do not allow the material to dry prior to weighing. For purposes of this test, an acceptable extractor is one which will impart sufficient agitation to the mixture to not only prevent stratification of the sample and extraction fluid but also insure that all sample surfaces are continuously brought into contact with well mixed extraction fluid.

5. After the solid material and deionized water are placed in the extractor, the operator shall begin agitation and measure the pH of the solution in the extractor. If the pH is greater than 5.0, the pH of the solution shall be decreased to 5.0 ± 0.2 by adding 0.5 N acetic acid. If the pH is equal to or less than 5.0, no acetic acid should be added. The pH of the solution shall be monitored, as described below, during the course of the extraction and if the pH rises above 5.2, 0.5N acetic acid shall be added to bring the pH down to 5.0 ± 0.2. However, in no event shall the aggregate amount of acid added to the solution exceed 4 ml of acid per gram of solid. The mixture shall be agitated for 24 hours and maintained at 20°–40°C (68°–104°F) during this time. It is recommended that the operator monitor and adjust the pH during the course of the extraction with a device such as the Type 45-A pH Controller manufactured by Chemtrix, Inc., Hillsboro, Oregon 97123 or its equivalent, in conjunction with a metering pump and reservoir of 0.5N acetic acid. If such a system is not available, the following manual procedure shall be employed:

(a) A pH meter shall be calibrated in accordance with the manufacturer's specifications.

(b) The pH of the solution shall be checked and, if necessary, 0.5N acetic acid shall be manually added to the extractor until the pH reaches 5.0 ± 0.2. The pH of the solution shall be adjusted at 15, 30 and 60 minute intervals, moving to the next longer interval if the pH does not have to be adjusted more than 0.5N pH units.

(c) The adjustment procedure shall be continued for at least 6 hours.

(d) If at the end of the 24-hour extraction period, the pH of the solution is not below 5.2 and the maximum amount of acid (4 ml per gram of solids) has not been added, the pH shall be adjusted to 5.0 ± 0.2 and the extraction continued for an additional four hours, during which the pH shall be adjusted at one hour intervals.

6. At the end of the 24 hour extraction period, deionized water shall be added to the extractor in an amount determined by the following equation:

[Appendix II]

C

Proposed California Hazardous Waste Listing, February 1975 (in Chapter 3, see California Department of Health, 1975)

Acetylene sludge
Acid and water (corrosive)
Acid sludge (corrosive)
Alkaline caustic liquids (corrosive)
Alkaline cleaner (corrosive)
Alkaline corrosive battery fluid (corrosive)
Alkaline corrosive liquids (corrosive)
Asbestos waste (toxic)
Battery acid (corrosive)
Beryllium waste (toxic)
Catalyst (toxic)
Caustic wastewater (corrosive)
Chemical cleaners (corrosive or irritant)
Cleaning solvents (flammable)
Data processing fluid (flammable)
Electrolyte, acid (corrosive)
Etching acid liquid or solvent (corrosive)
Lime and water (corrosive or irritant)

Lime sludge (corrosive or irritant)
Lime wastewater (corrosive or irritant)
Liquid cement (flammable)
Liquid cleaning compounds (corrosive or irritant)
Obsolete explosives
Oil of Bergamot and products containing 2% or more of oil of Bergamot (strong sensitizer)
Paint (or varnish) remover or stripper (flammable)
Paint waste (or slops)—except water-based (flammable or toxic)
Petroleum waste (flammable)
Pickling liquor (corrosive)
Powdered orris root and products containing it (strong sensitizer)
Printing ink (flammable)
Refinery waste (flammable)
Retrograde explosives
Sludge acid (corrosive)
Solvents (flammable)
Spent acid (corrosive)
Spent caustic (corrosive)
Spent (or waste) cyanide solutions (toxic)
Spent mixed acid (corrosive)
Spent plating solution (toxic)
Spent sulfuric acid (corrosive)
Sulfonation oil (flammable)
Toxic chemical toilet wastes (toxic)
Toxic tank sediment (toxic)
Unrinsed pesticide containers (toxic)
Unwanted or waste pesticides—an unusable portion of active ingredient or undiluted formulation (toxic)
Waste chemicals—where the chemical is a substance like tetraethyl lead (toxic)
Waste epoxy (strong sensitizer)
Waste (or slop) oil (flammable)
Wyandotte cleaner (corrosive or irritant)

D

Minnesota State Survey Forms

MINNESOTA ASSOCIATION OF
COMMERCE AND INDUSTRY

July 28, 1976

Dear Sir:

The Minnesota Pollution Control Agency (MPCA) is adopting regulations to control
hazardous wastes "from the cradle to the grave." The statutory authority under
which these regulations will be adopted require consideration of variations in
population density, transportation, economic impact and other factors which vary
throughout the state.

The Minnesota Association of Commerce and Industry (MACI) is trying to determine
if the regulations are justified and reasonable. However, we need facts regarding
1) the types and quantities of hazardous wastes in the various areas of the state;
and 2) how alternative regulatory proposals affect the economics of large and small
industries in these areas. We also are concerned that there presently are virtually
no satisfactory disposal facilities in the state. To attract capital investment in
such facilities, the types, quantities and general locations of hazardous wastes must
be known.

It is anticipated that disposal costs will soar under new regulations. Thus, MACI
is considering the establishment of a voluntary waste exchange referral service.
This service would put producers of wastes in direct contact with users of wastes,
and vice versa. The net result should be that some hazardous wastes would become
an economic asset instead of a liability.

It is for these purposes we are conducting this extensive survey of more than 5,000
Minnesota firms. We understand and respect the delicate and technical nature of the
information we are asking you to supply. To assure you that the information you
supply will not "come back to haunt you", the following safeguards are guaranteed:

1. The covering page identifying the respondent will be separated from the
    rest of the form. No one except MACI staff will have access to the
    covering page. It will be used only to make direct contact to clarify
    survey information, or to obtain more detailed information through a
    follow-up confidential interview.

2. Data will be compiled and summarized on a regional basis (six separate
    regions) showing gross totals or amounts for each region (see page 6
    of the survey). Thus, it will be impossible to trace amounts or types
    of wastes back to a single source.

MACI is conducting this survey with the knowledge, support and understanding of
the MPCA. We believe it is in your interest to cooperate fully, whether or not
you pay dues to our 1,700-member association of Minnesota businesses, industries,
local chambers of commerce and other business-oriented associations.

Please complete and return the enclosed survey no later than August 13, 1976. If
you have any questions, you may call me at 612/227-9591.

Sincerely,

James T. Shields

James T. Shields
Director of Environmental Affairs

JTS:djd
Encl.

(page 1)

<u>1976 Industrial Waste Management Survey</u>
(Please type or print)

1. Company Name _____

    Plant Address _____

    City _____ Zip Code _____

    County _____

2. Name of person responding to survey _____

    Title _____ Telephone _____/_____
                                      (Area   (Number)
                                      Code)

Return completed survey form to:

    MACI
    Industrial Waste Survey
    Hanover Building
    480 Cedar Street
    St. Paul, Minnesota 55101

454

(page 2)

3. Please circle the major Standard Industrial Classification (SIC) group(s) which best describes your plant. (If your plant does not fit in any of the listed groups, please describe it under the space titled "other".)

SIC #   SIC DESCRIPTION

(20)   Food and Kindred Products
(22)   Textile Mill Products
(23)   Apparel and Other Finished Products Made From Fabrics and Similar Materials
(24)   Lumber and Wood Products, except Furniture
(25)   Furniture and Fixtures
(26)   Paper and Allied Products
(27)   Printing, Publishing, and Allied Industries
(28)   Chemicals and Allied Products
(29)   Petroleum Refining and Related Industries
(30)   Rubber and Misce-laneous Plastic Products
(31)   Leather and Leather Products
(32)   Stone, Clay, and Glass Products
(33)   Primary Metal Industries
(34)   Fabricated Metal Products, except Machinery and Transportation Equipment
(35)   Machinery, except Electrical
(36)   Electrical Machinery, Equipment, and Supplies
(37)   Transportation Equipment
(38)   Measuring, Analyzing, and Controlling Instruments; Photographic Medical, and Optical Goods; Watches and Clocks
(39)   Miscellaneous Manufacturing Industries
( )   Other (Please specify) _____

4. For statistical purposes, please indicate the number of employees in your plant (please convert part-time and temporary employees to their equivalent in full time employees). _____

   If you circled more than one SIC in question 3, please indicate next to each circled SIC the number of employees engaged in each endeavor.

5. Does your plant produce any of the industrial wastes included in the following waste category list? Yes _____ No _____

Waste Category List

Oils
Solvents
Flammables (other than solvents, but not including sawdust, paper, and other such materials)
Oxidizers (including peroxides, oxides, permanganates, nitrates, chlorates, and persulfates)
Explosives (including compressed gases)
Irritants and Corrosives (materials causing skin or eye irritation, burns, or corrosion to metals)
Wastewater Sludges (residues remaining after wastewater treatment)
Pesticides (including insecticides, herbicides, rodenticides, and fungicides)
Paints (including paint rejects, residues, and sludges)
Heavy Metals (wastes other than paints, which contain metals such as Cu, Cr, Ni, and Pb)
Other points (including cyanide, arsenic, and selenium compounds)
Other Similar Materials (please specify)

(over)

(page 3)

* * * * *

If your answer to question 5 is yes, please complete the remainder of the survey. If your answer is no, ignore the remaining questions, but please return the questionnaire completed through question 5.

* * * * *

6. Would a statewide waste exchange referral service, under which you might find markets for your firm's wastes or purchase other firm's wastes for your own use, be of value to your plant?
   Yes _____ No _____

7. Please complete the Industrial Waste Management Table on Page 4 in accordance with the following instructions (note example shown in table):

In Section A
   1. List the appropriate waste category(ies) from question 5 (select the single most appropriate category for a particular waste-the choice of waste category may sometimes be arbitrary when a waste fits equally well into two categories).
   2. Specify waste components (for example if the category is solvents, the specific waste component might be xylene or cleaning solvents).
   3. Indicate the SIC # most important in producing that waste (see list in question 3).

In Section B and C
   Check the appropriate boxes under "Principal Disposal Method" and "Principal Disposal Location". (If, under a single category, you must check more than one "Principal Disposal Method" or "Principal Disposal Location", please consider that waste category as if it were two categories, using two lines of the table).

In Section D
   List the approximate one way haul distance from point of generation of the waste to final disposal (if applicable).

In Section E
   1. For each waste category, estimate the total gallons, tons, or cubic feet of that waste category generated, indicating the unit which is used.
   2. If the unit is gallons or cubic feet and you know the approximate weight per gallon or cubic foot, please indicate this.

In Section F
   1. Estimate your net savings (if any) resulting from recycling.
   2. If there is not a net savings from recycling, estimate your net total cost of waste management (exclude municipal sewer or NPDES permitted facility costs and subtract the value of recycled materials).
   3. Estimate the percentages of net total cost attributable to the various phases of waste management.

(page 4)

INDUSTRIAL WASTE MANAGEMENT TABLE
(see instructions in question 7 on page 3)

EXAMPLE

A.1. Waste Category (as listed in question 5 on page 2)	Wastewater Sludge	
2. Specific Waste Components	Ni, Zn, Cd, Cr	
3. Applicable SIC # (see question 3 on page 2)	28	
B. Principal Disposal Method (check one)		
1. Municipal Sewer		
2. NPDES Permit		
3. Incineration		
4. Sanitary Landfill		
5. Landspreading		
6. Lagooning		
7. Recycling		
8. Other (Specify)		
C. Principal Disposal Location (check one)		
1. On company property		
2. Off company property		
D. Miles Hauled to Final Disposal	25	
E. 1. Quantity Per Year (indicate tons, gallons, or cubic feet)	5,000 gal.	
2. Weight per gallon or cubic foot (if applicable and known)	11 lbs./gal.	
F. Waste Management Costs		
1. Recycling Savings	none	
2. Cost	$2,500	
3. Percent of Cost Attributable to:		
storage	0	
transport	40%	
disposal	60%	

(continue on reverse side)

(page 5)

INDUSTRIAL WASTE MANAGEMENT TABLE
(see instructions in question 7 on page 3)

A.1. Waste Category (as listed in question 5 on page 2)				
2. Specific waste components				
3. Applicable SIC # (see question 5 on page 2)				
B. Principal Disposal Method (check one)				
1. Municipal Sewer				
2. NPDES Permit				
3. Incineration				
4. Sanitary Landfill				
5. Landspreading				
6. Lagooning				
7. Recycling				
8. Other (Specify)				
C. Principal Disposal Location (check one)				
1. On company property				
2. Off company property				
D. Miles Hauled to Final Disposal				
E.1. Quantity Per Year (indicate tons, gallons, or cubic feet)				
2. Weight per gallon or cubic foot (if applicable and known)				
F. Waste Management Costs				
1. Recycling Savings				
2. Cost				
3. Percent of Cost Attributable to:				
storage				
transport				
disposal				

EXAMPLE OF THE MANNER IN WHICH THE INDUSTRIAL WASTE MANAGEMENT SURVEY DATA WILL BE SUMMARIZED BY REGION

WASTE DISPOSAL METHODS

Waste Category	Total Volume (Tons)	Municipal Sewer	NPDES Permit	Incineration	Sanitary Landfill	Land-Spread	Lagooned	Treated	Recycled
Oils	2,000								2,000
Solvents	600								500
Flammables	800	100		400				100	100
Oxidizers	50	10		30				200	
Explosives	30			10	20	10			
Irritants and Corrosives	400							350	50
Wastewater Sludges	1,200			200	300	400	200	100	
Pesticides	1,000	100				900			
Paints	100	100							
Heavy Metals	500		400					100	
Other Poisons	200					100		100	
Other Similar Materials	10	2			2		3	3	
TOTALS	6,890	312	400	640	322	1,410	203	953	2,650

WASTE DISPOSAL COSTS

Waste Category	Avg. Waste Mgmt. Cost Per Gallon	Storage	Collection Transport	Disposal	Average Haul Distance (miles)
Oils	None: Recyled at Profit				
Solvents	$ .05	25%	70%	5%	70
Flammables	.15	20%	50%	30%	40
Oxidizers	.15	15%	55%	30%	40
Explosives	.10	20%	60%	20%	10
Irritants and Corrosives	.20	10%	30%	60%	30
Wastewater Sludges	.10	30%	50%	20%	10
Pesticides	.05	5%	80%	15%	10
Paints	.10			100%	0
Heavy Metals	.60	10%	10%	80%	60
Other Poisons	.55	10%	15%	75%	200
Other Similar Materials	.10	30%	45%	25%	15

E

Model Industrial Survey Forms (in Chapter 4, see U.S. EPA 1977)

460

DATA COLLECTION GUIDE
FOR AN
INDUSTRIAL WASTE SURVEY

A.   General information (to be obtained from each facility).

Facility name _____

Facility location _____

_____

Facility owner _____

Facility mailing address _____

_____

Facility manager _____   Telephone no. _____

Facility contact _____   Telephone no. _____

SIC group name and four digit number.   Primary _____

Secondary _____

Time period for which data is representative _____

Number of employees _____   Facility area _____

Either obtain a plat of the facility showing the location of onsite process waste storage, treatment, and disposal from the facility personnel or sketch a diagram of the facility on the back of this page.

B.   Waste characterization (applicable to generator, treatment, and incinerator facilities).

Process waste			
Process origin			
Quantity of waste			
Annual rate			
Average hourly rate			
Maximum hourly rate			
Waste stream composition (weight basis)			
Process products			
Quantity			

Attach flow diagrams of each process showing product and waste streams, if available.

DATA COLLECTION GUIDE FOR AN INDUSTRIAL WASTE SURVEY

C. Storage methodology (applicable to generators, treatment and disposal facilities, and collectors and haulers).

Process wastes stored			
Quantity			
Type of storage			
Frequency of transfer to the storage area			
Frequency of transfer from the storage area			
Methods of transfer to and from storage			
Safety procedures			
Emergency plans			

D. Transportation methodology (applicable to generator, storage, and treatment facilities and collectors and haulers).

Wastes transported			
Quantity			
Destination			
Waste composition			
Special handling procedures			
Emergency plans			

E. Treatment methodology (applicable to generator and treatment facilities).

Wastes treated			
Quantity			
Composition of wastes treated			
Treatment methods			
Equipment used to treat wastes			
Products			

Describe the wastes from the treatment facility using the waste characterization portion of the guide.

F. Disposal methodology (applicable to generator, treatment, and disposal facilities).

Land disposal

Waste			
Quantity			
Composition			
Type of disposal			
Liner type			
Thickness			
Leachate collection			
Depth of facility			
Distance to ground water			
Site security			
Leachate treatment			
Burial methods			
Types of leachate analysis			

Describe methods used to identify and mark the location of hazardous wastes.

DATA COLLECTION GUIDE FOR AN INDUSTRIAL WASTE SURVEY

F.   Disposal methodology (continued)

Incineration

Wastes			
Quantity			
Composition			
Type of incinerator			
Rated capacity			
Auxiliary fuel used			
Quantity			
Design specifications			
Temperature			
Dwell-time			
Air pollution controls			
Air pollution permits			
Residue disposal			
Waste storage prior to incineration			

F.   Disposal methodology (continued)

Other disposal methods

Wastes			
Quantity			
Ocean dumping			
Reclaimer			
Well injection			
Other			

F

Site Selection Criteria Areas as Delineated in the 1973 Report to Congress

Waste Considerations

A. Geographic distribution and dispersion of sources
B. Intrinsic hazard—type and severity
C. Volume or other capacity factor
D. Uniqueness and frequency
E. Regulatory status

Process Considerations

A. Basic approach
   1. Recover and recycle
   2. Render inert and disperse
   3. Package and store
B. Anticipated site effluents—air, water, solid
C. Hazard inventory
D. Utility needs

Geologic considerations

A. Rock type (related to ultimate disposal/storage method)
   1. Stratigraphic sequence

    2. Structural relationships
    3. Mineral potential (economic)
    4. Historical potential (paleontological and archeological)
B. Geologic hazards
    1. Earthquake risk
    2. Tectonics (subsidence, faults and folds)
    3. Landslide potential
    4. Volcanic potential

Physiographic Considerations

A. Topographic
    1. Elevation
    2. Slope
    3. Accessibility
B. Soil types
    1. Primary disposal (storage of ultimate waste product)
      a. Erodability (texture, structure, permeability)
      b. Depth to bedrock and groundwater
      c. Sorption (suitability/capacity for hazardous materials)
    2. Secondary disposal (low-level or dilute process waste streams)
      a. Depth to bedrock and groundwater
      b. Sorption (suitability/capacity)
      c. Leaching potential
    3. Accidental spills of process feed or waste streams
      a. Soil texture and structure
      b. Depth to bedrock and groundwater
    4. Engineering limitations

Hydrologic Considerations

A. Confined/unconfined aquifers
    1. Depth
    2. Extent
    3. Porosity, permeability, transmissivity
    4. Chemistry
    5. Present and future uses
B. Waterways
C. Floodplains and swamps
D. Tidal basins
E. Seiche potential
F. Tsunami potential

Climatologic Considerations

A. Diffusion characteristics
B. Extreme conditions
    1. Frequency and severity of hurricanes, tornadoes, thunderstorms, other local storms and winds
C. Potential evaporation
D. Seasonal climatology
    1. Air temperature
    2. Relative humidity
    3. Fog frequency
    4. Solar radiation
E. Annual precipitation

Transportation Considerations

A. Economics and safety (risk) with respect to
    1. Carrier (mode)
    2. Distance
    3. Routing
    4. Modal splits
    5. Type of material transported
    6. Source of waste production
    7. Volume of material transported
B. Accessibility of transportation
    1. Rail
        a. Mainline
        b. Secondary
    2. Highway
        a. Interstate and major highways
        b. Secondary roads
    3. Waterway access

Water, Land, and Air Resources Considerations

A. Land ownership and prior association with hazardous wastes
B. Security and public access
C. Regulatory status
D. Livestock and grazing
E. Crops and timber
F. Industrial
    1. Mining
    2. Light and heavy manufacturing
    3. Commercial

G. Recreational
H. Air transportation—corridors and airports
 I. Water
   1. Irrigation
   2. Commercial and domestic water supply
   3. Surface water transportation
 J. Unique natural or historical areas

Human Environment Considerations

A. Demography
   1. Population—distribution and density
B. Public acceptance
C. Aesthetics
   1. Unique landforms, fauna or flora compositions
   2. Frequency and proximity of unique areas to view

Biological Considerations

A. Birds and wildfowl
   1. National, state, and local refuges
   2. Major flyways, including species and densities
   3. Rare and endangered species areas
B. Terrestrial wildlife
   1. National, state, and local refuges
   2. Ecosystem quality and stability
      a. Species diversity and density of producers, herbivores, and carnivores
   3. Rare and endangered species areas
C. Aquatic life
   1. National, state, and local refuges
   2. Ecosystem quality and stability
      a. Species diversity and densities of producers, benthos, and fish
   3. Rare and endangered species areas

G

Criteria for PCB Disposal: Excerpts from 40 CFR 761

Subpart E—List of Annexes

ANNEX I

§ 761.40   Incineration.

(a) *Liquid PCB's.* An incinerator used for incinerating PCB chemical substances or liquid PCB mixtures shall be approved by the Agency Reginal Administrator pursuant to paragraph (d) of this section. Such incinerator shall meet all of the requirements specified in paragraphs (a)(1) through (9) of this section, unless a waiver from these requirements is obtained pursuant to paragraph (d)(5) of this section. In addition, the incinerator shall meet any other requirements which may be prescribed pursuant to paragraph (d)(4) of this section.

(1) Combustion criteria shall be either of the following:

(i) Maintenance of the introduced liquids for a 2-second dwell time at 1200°C ($\pm$ 100°C) and 3 percent excess oxygen in the stack gas, or

(ii) Maintenance of the introduced liquids for a 1½-second dwell time at 1600°C ($\pm$ 100°C) and 2 percent excess oxygen in the stack gas.

(2) Combustion efficiency shall be at least 99 percent computed as follows:

$$\text{Combustion efficiency} = \frac{Cco_1 - Cco}{Cco_1} \times 100.$$

where
Cco_1 = Concentration of carbon dioxide.
Cco = Concentration of carbon monoxide.

(3) The rate and quantity of PCB's which are fed to the combustion system shall be measured and recorded at regular intervals of no longer than 15 minutes.

(4) The temperatures of the incineration process shall be continuously measured and recorded. The combustion temperature of the incineration process shall be based on either direct (pyrometer) or indirect (wall thermocouple-pyrometer correlation) temperature readings.

(5) The flow of PCB's to the incinerator shall stop automatically whenever the combustion temperature drops below the temperatures specified in paragraph (a)(1) of this section.

(6) Monitoring of stack emission products shall be conducted:

(i) When an incinerator is first used for the disposal of PCB's under the provisions of this regulation, and

(ii) When an incinerator is first used for the disposal of PCB's after the incinerator has been modified in a manner which may effect the characteristics of the stack emission products.

(iii) At a minimum such monitoring shall be conducted for the following parameters: (a) O_2; (b) CO; (c) CO_2; (d) Oxides of Nitrogen (NO_2); (e Hydrochloric Acid (HCL); (f) Total Chlorinated Organic Content (RCL); (g) PCB Chemical Substances; (h) Total Particulate Matter.

(7) At a minimum, continuous monitoring and recording of combustion products and incineration operations shall be conducted for the following parameters whenever the incinerator is incinerating PCB's: (i) O_2; (ii) CO; (iii) CO_2.

(8) Incinerator operations shall be immediately suspended when any one or more of the following conditions occur:

(i) Failure of monitoring operations specified in paragraph (a)(7) of this section.

(ii) Failure of the PCB rate and quantity measuring and recording equipment specified in paragraph (a)(3) of this section, or

(iii) Combustion temperature, dwell time, or excess oxygen fall below those specified in paragraph (a)(1) of this section.

(9) Water scrubbers shall be used for HCl control during PCB incineration and shall meet any performance requirements specified by the appropriate EPA Regional Administrator. Scrubber effluent shall be monitored and shall comply with applicable effluent or pretreatment standards, and any other State and Federal laws and regulations. An

alternate method of HCl control may be used if the alternate method has been approved by the Regional Administrator.

(b) *Non-liquid PCB's.* An incinerator used for incinerating non-liquid PCB mixtures, PCB articles, PCB equipment, or PCB containers shall be approved by the Agency Regional Administrator pursuant to paragraph (d) of this section. Such incinerator shall meet all of the requirements specified in paragraphs (b)(1) through (3) of this section, unless a waiver from these requirements is obtained pursuant to paragraph (d) (5) of this section. In addition, the incinerator shall meet any other requirements which may be prescribed pursuant to paragraph (d)(4) of this section.

(1) The mass air emissions from the incinerator shall be no greater than 0.001g PCB chemical substances/Kg of PCB chemical substance introduced into the incinerator.

(2) Such incinerator shall comply with the provisions of § 761.40(a) (2), (3), (4), (6), (7), (8) (i) and (ii) and (9).

(3) The flow of PCB's to the incinerator shall stop automatically whenever the combustion temperature falls below the temperatures specified in any approvals issued by the Regional Administrator pursuant to paragraph (d) of this section. Incinerator operations shall stop immediately whenever the excess oxygen measurements fall below those specified in any approvals issued by the Regional Administrator pursuant to paragraph (d) of this section.

ANNEX II

§ 761.41 Chemical waste landfills.

(a) *General.* A chemical waste landfill used for the disposal of PCB's shall be approved by the Agency Regional Administrator pursuant to paragraph (c) of this section. Such landfill shall meet all of the requirements specified in paragraph (b) of this section, unless a waiver from these requirements is obtained pursuant to paragraph (c)(4) of this section. In addition, the landfill shall meet any other requirements which may be prescribed pursuant to paragraph (c)(3) of this section.

(b) *Technical requirements.* Requirements for chemical waste landfills used for the disposal of PCB's are as follows:

(1) *Soils.* The landfill site shall be located in thick, relatively impermeable formations such as large-area clay pans. Where this is not possible, the soil shall have a high clay and silt content with the following parameters:

(i) In-place soil thickness, 4', or compacted soil liner thickness. 3'.

(ii) Permeability (cm/sec), 1×10^{-7}.

(iii) Percent soil passing No. 200 Sieve, "30.

(iv) Liquid Limit, "30.

(v) Plasticity Index, "15.

(vi) Artificial Liner Thickness, "30 mil.

NOTE: In the event that an artificial liner is used at a landfill site, special precautions shall be taken to insure that its integrity is maintained and that it is chemically compatible with PCB's. Soil underlining shall be provided as well as a soil cover.

(2) *Hydrology.* The bottom of the landfill shall be substantially above the historical high groundwater table. Floodplains, shorelands, and groundwater recharge areas shall be avoided. There shall be no hydraulic connection between the site and standing or flowing surface water. The site shall have monitoring wells and leachate collection and shall be at least fifty feet from the nearest groundwater.

(3) *Flood protection.* (i) If the landfill site is below the 100-year floodwater elevation, the operator shall provide surface water diversion dikes around the perimeter of the landfill site with a minimum height equal to two feet above the 100-year floodwater elevation.

(ii) If the landfill site is above the 100-year floodwater elevation, the operators shall provide diversion structures capable of diverting all of the surface water runoff from a 24-hour, 25-year storm.

(4) *Topography.* The landfill site shall be located in an area of low to moderate relief to minimize erosion and to help prevent landslides or slumping.

(5) *Monitoring Systems*—(i) *Water Sampling.* (a) The ground and surface water from the disposal site area shall be sampled for use as baseline operations.

(b) Defined water sources shall be sampled at least monthly when the landfill is being used for disposal operations.

(c) Defined water sources shall be sampled indefinitely on a frequency of no less than once every six months after final closure of the disposal area.

(ii) *Groundwater Monitor Wells.* (a) If underlying earth materials are homogeneous, impermeable, and uniformly sloping in one direction, only three sampling points shall be necessary. These three points shall be equally spaced on a line through the center of the disposal area and extending from the area of highest water table elevation to the area of the lowest water table elevation on the property.

(b) All monitor wells shall be cased and the annular space between the monitor zone (zone of saturation) and the surface shall be completely back-filled or plugged with portland cement to effectively prevent percolation of surface water into the well bore. The well opening at the surface shall have a removable cap to provide access and to prevent entrance of rainfall or stormwater runoff. The well shall be pumped to remove the volume of liquid initially contained in the well before obtaining a sample for analysis. The discharge shall be treated to meet applicable State or Federal discharge standards or recycled to the chemical waste landfill.

(iii) *Water analysis.* As a minimum, all samples shall be analyzed for the following parameters, and all data and records of the sampling and analysis shall be maintainted as required in Annex VI. Sampling methods and analytical procedures for these parameters shall be as specified in 40 CFR Part 136 as amended in 41 FR 52779 of December 1, 1976.

(*a*) PCB's.

(*b*) pH.

(*c*) Specific conductance.

(*d*) Chlorinated Organics.

(6) *Leachate Collection.* A leachate collection monitoring system shall be installed beneath the chemical waste landfill. Leachate collection systems shall be monitored monthly for quantity and quality of leachate produced. The leachate should be either treated to acceptable limits for discharge in accordance with a State or Federal permit or disposed of by another State or Federal approved method. Water analysis shall be as provided in paragraph (b)(5)(iii) of this section. Acceptable leachate collection monitoring/collection systems shall be one of the following designs unless a waiver is obtained pursuant to paragraph (c)(4) of this section.

(i) *Simple Leachate Collection.* This system consists of a gravity flow drainfield installed under the waste disposal facility liner. This design is recommended for use when semi-solid or leachable solid wastes are placed in a lined pit excavated into a relatively thick, unsaturated, homogeneous layer of low permeability soil.

(ii) *Compound Leachate Collection.* This system consists of a gravity flow drainfield installed under the waste disposal facility liner and above a secondary installed liner. This design is recommended for use when semiliquid or leachable solid wastes are placed in a lined pit excavated into relatively permeable soil.

(iii) *Suction Manometers.* This system consists of a network of porous "stones" connected by hoses/tubing to a vacuum pump. The porous "stones" or suction manometers are installed along the sides and under the bottom of the waste disposal facility liner. This type of system works best when installed in relatively permeable unsaturated soil immediately adjacent to the disposal facility's bottom and/or sides.

H

Texas Site Qualification Guidelines

Information Guidelines

The type of information needed by the Texas Water Quality Board to determine the adequacy of a disposal site and its facilities (pit, pond, landfill, and landfarm areas) depends to a large extent on (1) the chemical composition and degree of hazard of the materials to be disposed of and (2) the proposed or existing construction details/specifications of the disposal facilities. Information pertaining to the types of waste should be prepared and submitted in accordance with the Texas Water Quality Board's *Technical Guideline on Waste Classification*. Information developed for the purpose of determining the adequacy of proposed or existing waste disposal facilities themselves (pits/trenches, ponds/lagoons, landfills, landfarms, etc.) should be prepared in accordance with the following guidelines, which are keyed to the type of construction/preparation techniques utilized at a particular facility to *prevent ground and surface water pollution*. The information suggestions below apply to any waste disposal facility accepting liquid or leachable waste materials capable of degrading the quality of ground or surface waters in the vicinity of the disposal site (this includes classes IA, IB, and II).

A. Natural/In-Place Soils

If the in-place soils (without reworking/compaction) are used as the protective barrier to inhibit vertical and lateral movement of materials to be retained within the facility, then the following information should be obtained.

1. Soil Borings. An adequate number of soil borings (minimum of four) should be completed in the immediate vicinity of the facility in order to accurately determine the subsurface conditions existing at the site. To a certain extent, the number of borings depends on the size of the area being investigated. The following table can be used as a guide in determining the minimum number of borings to be completed.

Number of Borings	Size of Pit Area(s) in Acres
4	1 to 5
6	5 to 10
10	10 to 20
16	20 to 50
16 to 24	over 50

As a general rule, the number of borings necessary to adequately define the subsurface conditions at a site will decrease if the soil/sediment/substrate is relatively homogeneous and will increase if the soil/sediment/substrate is heterogeneous. Those bore holes that are not converted to monitor wells should be properly plugged to prevent their functioning as conduits for movement of fluid to the subsurface. Bore holes should be cemented bottom to top or filled bottom to top with a drilling mud mixture containing a minimum of 30% bentonite or montmorillonite clay.

The borings should be completed to a depth of at least 40 ft below natural grade or a minimum of 30 ft below the deepest part of an excavation. Logs describing the soil lithology encountered in each of the borings should be prepared/completed as the borings are made. The date(s) the borings were made should be recorded on the logs.

2. Soil Testing. Representative samples of all naturally occurring in-place soils that are to be utilized as barriers to vertical and lateral seepage should be collected during boring operations. These soil samples should be tested to determine the following: percent passing No. 200 sieve, liquid limit, plastic limit, plasticity index, unified soil class, and coefficient of permeability.

3. Permeability Testing. Permeability tests should be performed using both deionized water and the liquid material or leachate from the material to be retained as pore fluids during testing. Permeability tests should be performed on undisturbed soil samples obtained from the boreholes. In the event undisturbed soil samples cannot be obtained, samples remolded and recompacted to the same density and moisture conditions as are characteristic of the undisturbed soil at the site may be used to estimate the in-place permeability of the undisturbed soil. Every effort should be made to obtain undisturbed soil samples, because they will more convincingly demonstrate the impermeability of any material that is being depended on to halt the migration of any liquid disposed of or any leachate that may form at the disposal site. Where soil conditions permit, it is strongly recommended that "field permeability" tests be conducted in order to provide a more complete understanding of the behavior of the soil/sediment/substrate when subjected to a hydraulic head similar to that which may be encountered during actual operation of the facility. Permeability test methods should be included with test results.

4. Groundwater Data.   The initial and static depth (24 h) of the groundwater should be accurately recorded whenever groundwater is encountered in the course of the soil boring operations. The date(s) on which groundwater level measurements were taken should be recorded on the boring log. When possible, groundwater samples should be obtained for analysis to establish background quality data. Any information relating to water well location and groundwater use in the vicinity of the site should be submitted along with the other groundwater data. In addition, obtain information relating to the geologic formations or rock units underlying the site. For example, describe the depth and thickness of the units, the character (lithology) of the rock making up these units, and the water-bearing properties of these geologic units.

5. Specifications/Installation.   The operator/company should submit information pertaining to the construction of the facility (depth below grade, dikes, slope of walls, etc.).

B. Reworked/Reconstructed/Compacted Soil Liner

If the protective barrier utilized to inhibit vertical and lateral movement of materials to be retained within the facility consists of (1) reworked in-place soils with soil amendments or additives, (2) recompacted in-place soils without additives, or (3) reworked or compacted soil imported to the site, then the following information should be obtained:

1. Soil Borings.   No soil borings willl normally be necessary if the in-place soils are relatively permeable and a liner or protective barrier will be utilized to prevent seepage. However, in the event that a monitoring/leachate collection system is necessary, because of the hazardous nature of the waste(s) or adverse local geologic or hydrologic conditions, soil borings and soil testing may be recommended by the Texas Water Quality Board.

2. Soil Testing.   Those natural/in-place soils that are to be utilized (after reworking/amendment/compacting) as barriers to vertical and lateral seepage should be collected and tested as described in section A.2. In situations where natural/in-place soils are not suitable for reworking and imported soils are utilized for liner construction, samples of these imported soils should also be tested as described in section A.2.

3. Permeability Testing. Permeability tests should be performed using both deionized water and the liquid material or leachate from the material to be retained as pore fluids during testing. Permeability tests should be performed on soil samples that have been prepared (reworked, amended, compacted) in the same manner as the constructed soil liner of the retention facility in question. Permeability test methods should be included with test results.

4. Groundwater Data. Groundwater data as described in section A.4 should be submitted.

5. Specifications/Installation. The operator/company should submit information pertaining to the construction of the soil liner utilized (thickness, additives, compaction density, etc.).

C. Artificial Liners

If an artificial liner (PVC, CPE, concrete, gunnite, butyl rubber, etc.) is utilized to inhibit vertical and lateral movement of materials to be retained within the facility, then the following information should be obtained:

1. Soil Borings. No soil borings will normally be necessary if the in-place soils are relatively permeable and a liner or protective barrier will be utilized to prevent seepage.

2. Soil Testing. No soil testing will normally be necessary. However, in the event that a monitoring/leachate collection system is necessary because of the hazardous nature of the waste(s) or adverse local geologic or hydrologic conditions, soil borings and soil testing may be recommended by the Texas Water Quality Board.

3. Permeability/Chemical Resistance Testing. The operator/company should supply information pertaining to the permeabilty/chemical resistance of the artificial lining material to the chemicals/materials that the lining is expected to come in contact with. Where such information is not available, the liner materials permeability/chemical resistance must be tested using the chemicals/materials to be retained in facility as test fluids during such testing.

4. Groundwater Data. Groundwater data as described in section A.4 should be submitted.

5. Specifications/Installation. The operator/company should submit information pertaining to the composition and thickness of the artificial lining material, along with details of the installation methods (covered with soil, field seamed, etc.).

Flood Protection

In addition to the foregoing informational requirements, data should be submitted on flood protection. In order to prevent inundation of a disposal site area and to direct rainfall runoff around a disposal site, it is necessary in many cases (where favorable grade does not exist) to construct surface-water diversion structures. These diversion structures may be of many different designs or types, including, but not limited to, dikes, berms, ditches, contour plowing, etc.

At those sites accepting hazardous or class IA wastes, located at elevations above the 50-yr floodwater elevation, diversion structures employed should be capable of diverting all of the rainfall from a 24-h, 25-yr storm. Those sites accepting hazardous or class IA wastes, below the 50-yr floodwater elevation, should provide surface-water diversion dikes, with a minimum height equal to 2 ft above the 50-yr floodwater elevation, around the perimeter of the disposal site.

When submitting information on flood protection and floodplain elevations, land subsidence must be taken into consideration. Where land subsidence is known to occur, the expected extent of the floodplain must be considered at all sites by employing the United States Geological Survey 5-yr subsidence projection.

Ia

November 1981, List of 114 EPA Priority Superfund Sites

Identifier	Location	State	Content
Commencement Bay	Tacoma	WA	Heavy metals, organics
Keefe Environmental Services	Epping	NH	Metals, solvents, pesticides
Lipari Landfill	Pitman	NJ	Mercury, organics
Mark Phillip Trust	Woburn	MA	Metals
McAdoo Associates	McAdoo	PA	Organic solvents
Nyanza Chemical Waste Dump	Ashland	MA	Mercury, dyes, sludges
Pollution Abatement Services	Oswego	NY	Metals, organics
Prices Pit	Pleasantville	NJ	Solvents, oils
Tar Creek	Ottawa Co	OK	Lead, Zinc
Tybouts Corner	New Castle	DE	Solvents, inorganics
Northwest 58th St. Landfill	Hialeah	FL	Metals, solvents, phenols
Miami Drum Services	Miami	FL	Metals, solvents, phenols
Varsol Spill Site	Miami	FL	Petroleum solvent
Bruin Lagoon	Bruin	PA	Metals, sludges, organics
Burnt Fly Bog	Marlboro Township	NJ	PCB, lead, solvents
Llangollen Army Creek Landfills	New Castle	DE	Chlorinated solvents
Goose Farm	Plumstead Township	NJ	Solvents
Lone Pine Landfill	Freehold Township	NJ	Metals, organics
Motco	LaMargue	TX	Metals, styrene, VC
Pijack Farm	Plumstead Township	NJ	Metals, organics, oil, explosives
Vertac, Inc.	Jacksonville	AR	2,4,5-T, dioxin
Bridgeport Rental & Oil Service	Bridgeport	NJ	Organic solvents, oil
Spence Farm	Plumstead Township	NJ	Solvents, PCBs, oil, phenols
D'Imperio Property	Hamilton Township	NJ	Solvents
French Limited Disposal Site	Crosby	TX	Sludge
Love Canal	Niagara Falls	NY	Pesticide wastes
Old Bethpage Landfill	Oyster Bay	NY	PCBs, organics
Picketville Road Landfill	Jacksonville	FL	Hazardous materials
Reeves Southeastern Corp.	Tampa	FL	Metals
Seymour Recycling Corp.	Seymour	IN	Toxics, flammables
Sikes Disposal Pits	Crosby	TX	Chemical wastes

Name	City	State	Contaminants
South Carolina Recycling & Disposal	Richland Co.	SC	Toxics, flammables, reactives
Aerojet General Corp.	Sacramento	CA	Solvents
American Creosote Works	Pensacola	FL	Creosote, PCP
Charles George Land Reclamation Trust	Tynesborough	MA	Organics, inorganics
Iron Mountain Mines	Keswick	CA	Metals
Kin-Buc Landfill	Edison	NJ	PCBs, oil
Oakdale Dump Sites	Oakdale	MN	Toxic chemicals
Olean Well Fields	Olean	NY	Halogenated solvents
Picillo Farm Site	Coventry	RI	Laboratory reagents, explosives
Stauffer Chemical	Delaware City	DE	PVC component chemicals
Taylor Road Landfill	Tampa	FL	Volatile organics
Andover Sites	Andover	MN	Solvents, paints, oils
Broward County Solid Waste Disposal Facility	Davie	FL	Metals, ammonia
Butler Tunnel	Pittson	PA	Cyanide, solvents
Facet Enterprises, Inc.	Elmira	NY	Metals, oils, solvents
Fulbright Landfill	Springfield	MO	Cyanides, metals, solvents
Ottati Goss/Kingston Steel Drum	Kingston	NM	Solvents
Pioneer Sand Co.	Warrington	FL	Chromium
Timber Lake Battery Disposal	Tampa	FL	Lead
Whitehouse Waste Oil Pits	Whitehouse	FL	Metals, PCBs
Whitewood Creek	Deadwood	SD	Arsenic
Chem-Dyne Corp.	Hamilton	OH	Toxics, flammables
Chemical Control	Elizabeth	NJ	Chemicals, PCBs
Coleman-Evans Wood Preserving Co.	Whitehouse	FL	PCP
Davis Liquid Chemical Waste Disposal Site	Smithfield	RI	Metals, oils, solvents, pesticides
Fritt Industries	Walnut Ridge	AR	Metals, fertilizer
Hollingsworth Solderless Terminal Co.	Ft. Lauderdale	FL	Solvents, oil, copper
Re-Solve, Inc.	N. Dartmouth	MA	Chlorinated solvents
Laurel Park Landfill	Naugatuck	CT	Industrial Wastes
Reilly Tar & Chemical Co.	St. Louis Park	MN	Coal tars, creosote
Stringfellow Acid Pits	Glen Avon	CA	Metals, acids
Allen Transformer	Fort Smith	AR	PCBs
Alpha Chemical Corp.	Galloway	FL	Organics, dioxane

Fields Brook	Ashtabula	OH	Toxics
Koppers Gas & Coke Plant	St. Paul	MN	Toxics
Mid-South Wood Products	Mena	AR	PCP, arsenic, creosote
Neal's Landfill	Bloomington	IN	PCBs
United Nuclear Corp.	Churchrock	NM	Uranium tailings
Upper Freehold	Upper Freehold Township	NJ	Solvents
Zellwood Groundwater Contamination Site	Zellwood	FL	Metals
19th Avenue Landfill	Phoenix	AZ	Metals, solvents, pesticides
Batavia Landfill	Batavia	NY	Metal sludges, solvents
Gold Coast Oil Corp.	Miami	FL	Solvents, paints
Homestake Mining	Milan	NM	Uranium tailings
Uranica Landfill	Buffalo Township	PA	PCBs, metals, solvents
Valley of the Drums	Brooks	KY	Solvents, toxics
Lord-Shope Landfill	Girard Township	PA	Solvents, oils, inorganics
National Lead-Taracorp Site	St. Louis Park	MN	Metal slag
Outboard Marine Corp.	Waukegan	IL	PCBs
Sapp Battery Salvage	Jackson Co.	FL	Metals, acid
Tower Chemical Corp.	Clermont	FL	Pesticides
ABM-Wade	Chester	PA	Mercury, solvents, cyanides
Ellisville Area Sites	Ellisville	MO	Pesticides, solvents
Chemicals & Minerals Reclamation	Cleveland	OH	Solvents, acids, sludges
Gratiot County Landfill	St. Louis	MI	PBBs
Lehigh Electric & Engineering Co.	Old Forge	PA	PCBs
Marathon Battery Corp.	Philipston	NY	Cadmium
Mathews Electroplating	Roanoke Co.	VA	Chromium, cyanide
Silvestor's	Nashua	NH	Sludges, liquids
Western Sand & Gravel Site	Burrillville	RI	Liquids
Taputium Farm	American	Somoa	Pesticides
Fort Lincoln Barrel Site	Washington	D.C.	Solvents, inks, dyes
Arsenic Trioxide Disposal Site	Southeastern	ND	Arsenic
Chemical Metals Industries	Baltimore	MD	Cyanides, ammonia, inorganics

Site	Location	State	Waste
Walcott Chemical Co. Warehouse	Greenville	MS	Formic Acid, liquids
PCB Wastes	Trust Territories of the Pacific		PCBs
Rose Park	Salt Lake City	UT	Petroleum by-products
Ordot Landfill		Guam	Wastes
West Virginia Ordnance	Point Pleasant	WV	TNT, explosives
Atchison, Tokeka & Santa Fe Railroad	Clovis	NM	Diesel fuel
Bioecology Systems, Inc.	Grand Prairie	TX	Solvents, metals, inorganics
Chisman Creek Disposal	York Co.	VA	Selenium, vanadium
Criner Waste Disposal Site	Criner	OK	Asbestos, cyanide, sludges
Denver Radium Sites	Denver	CO	Radium
Lindane Dump	Harrison Township	PA	Lindane
Niagara County Refuse Site	Wheatfield	NY	Organics
Summit National Liquid Disposal Services	Deerfield	OH	Organics
Triana-Redstone Arsenal	Triana	AL	DDT
Winthrop Town Landfill	Winthrop	ME	Commercial wastes
Aidex Corp.	Council Bluffs	IA	Pesticides
Arkansas City Dump Site	Arkansas City	KS	Asphaltics
North Hollywood Dump	Memphis	TN	Metals, pesticides
PCB Spills	Multiple	NC	PCBs
PCB Warehouse	Siapan Northern Mariann Islands		PCBs
Luminous Processes, Inc.	Athens	GA	Radium

States with More Than One Site

Florida—16	Texas—4	Rhode Island—3
New Jersey—12	Arkansas—4	New Mexico—3
New York—8	Ohio—4	Oklahoma—2
Pennsylvania—8	New Hampshire—3	Indiana—2
Minnesota—5	Delaware—3	Missouri—2
Massachusetts—4	California—3	Virginia—2

Ib

Expanded December 1982 List of Proposed National Priorities Sites and Status

Group 1

EPA Region

State	City/County	Site Name	Response Status +

	State	City/County	Site Name	V	R	E	D
05	MN	FRIDLEY	FMC#				D
03	DE	NEW CASTLE COUNTY	TYBOUTS CORNER# *		R	E	
03	PA	BRUIN BORO	BRUIN LAGOON#		R		
01	MA	WOBURN	INDUSTRI-PLEX#	V	R	E	
02	NJ	PITTMAN	LIPARI LANDFILL#	V	R	E	
02	NY	WELLSBILLE	SINCLAIR REFINERY#				D
02	NJ	PLEASANTVILLE	PRICE LANDFILL# *		R	E	
02	NY	OSWEGO	POLLUTION ABATEMENT SERVICES# *		R	E	
07	IA	CHARLES CITY	LABOUNTY SITE	V			
02	NJ	MANTUA	HELEN KRAMER LANDFILL#				D
03	DE	NEW CASTLE	ARMY CREEK#				D
02	NJ	OLD BRIDGE TOWNSHIP	CPS/MADISON INDUSTRIES			E	
01	MA	ASBLAND	NYANZA CHEMICAL#			E	
02	NJ	GLOUCESTER TOWNSHIP	GEMS LANDFILL#			E	
01	RI	COVENTRY	PICILLO COVENTRY# *		R	E	
05	MI	SWARTZ CREEK	BERLIN & FARRO#		R	E	
07	KS	CHEROKEE COUNTY	TAR CREEK, CHER. CO.				D
01	MA	HOLBROOK	BAIRD & MCGUIRE			E	
02	NJ	FREEHOLD	LONE PINE LANDFILL#		R	E	
01	NH	SOMERSWORTH	SOMERSWORTH LANDFILL				D
03	PA	MCADOO	MCADOO# *				D
01	NH	EPPING	KES - EPPING#		R	E	
06	AR	JACKSONVILLE	VERTAC, INC.#	V		E	
08	MT	SILVER BOW/DEER LODGE	SILVER BOW CREEK				D
06	TX	CROSBY	FRENCH, LTD.#		R		
05	MI	UTICA	LIQUID DISPOSAL, INC.#		R		
01	NH	NASHUA	SYLVESTER/NASHUA# *		R	E	
06	TX	LA MARQUE	MOTCO# *		R		
05	OH	ARCANUM	ARCANUM IRON & METAL			E	
06	TX	CROSBY	SIKES DISPOSAL PITS#		R		
04	AL	LIMESTONE & MORGAN	TRIANA, TENNESSEE RIVER#			E	
09	CA	GLEN AVON HEIGHTS	STRINGFELLOW# *		R		
01	ME	GRAY	MCKIN COMPANY		R	E	
06	TX	HOUSTON	CRYSTAL CHEMICAL#		R	E	
02	NJ	BRIDGEPORT	BRIDGEPORT RENT. & OIL#	V	R	E	
05	IN	GARY	MIDCO I		R	E	
08	SD	WHITEWOOD	WHITEWOOD CREEK# *	V			
01	MA	ACTON	W R GRACE			E	
01	MA	EAST WOBURN	WELLS G&H				D
02	NJ	MARLBORO TOWNSHIP	BURNT FLY BOG#		R	E	
04	FL	PLANT CITY	SCHUYLKILL METALS				D
05	MN	NEW BRIGHTON/ARDEN	NEW BRIGHTON#				D
05	MN	ST. LOUIS	REILLY TAR# *		R	E	
02	NY	OYSTER BAY	OLD BETHPAGE LANDFILL#			E	
04	FL	JACKSONVILLE	PICKETTVILLE RD LANDFILL#				D
08	MT	ANACONDA	ANACONDA - ANACONDA	V			
03	PA	GROVE CITY	OSBORNE#				D
05	MN	BRAINERD/BAXTER	BURLINGTON NORTHERN#				D
02	NJ	FAIRFIELD	CALDWELL TRUCKING				D
06	OK	OTTAWA COUNTY	TAR CREEK#		R		

+: V = Voluntary or Negotiated Response, R = Federal and State Response,
 E = Federal and State Enforcement, D = Actions to be Determined.
# = IPL/EEL.   * = States' Designated Top Priority Sites.

Group 2

EPA Region

State	City/County	Site Name	V	R	E	D
05 IN	SEYMOUR	SEYMOUR# *	V	R	E	
02 NJ	BRICK TOWNSHIP	BRICK TOWNSHIP LANDFILL			E	
05 MI	CADILLAC	NORTHERNAIRE PLATING#				D
10 WA	VANCOUVER	FRONTIER HARD CHROME			E	
04 FL	DAVIS	DAVIE LANDFILL#				D
04 FL	MIAMI	GOLD COAST OIL#				D
09 AZ	TUSCON	TUSCON INT'L AIRPORT#				D
02 NY	BRANT	WIDE BEACH DEVELOPMENT				D
09 CA	REDDING	IRON MOUNTAIN MINE#				D
02 NJ	CARLSTADT	SCIENTIFIC CHEMICAL PROCESSING				D
02 NJ	HAMILTON TOWNSHIP	D'IMPERIO PROPERTY#		R		
05 MN	OAKDALE	OAKDALE#				D
04 FL	GALLOWAY	ALPHA CHEMICAL#				D
05 IL	GREENUP	A & F MATERIALS#		R	E	
03 PA	DOUGLASVILLE	DOUGLASVILLE DISPOSAL				D
02 NJ	HILLSBOROUGH	KRYSOWATY FARM#				D
05 MN	ST. PAUL	KOPPER'S COKE#				D
01 MA	PLYMOUTH	PLYMOUTH HARBOR/CORDAGE			E	
10 ID	SMELTERVILLE	BUNKER HILL				D
10 WA	TACOMA	COM. BAY, TACOMA CHANNEL#		R	E	
02 NJ	EAST RUTHERFORD	UNIVERSAL OIL PRODUCTS			E	
09 CA	RANCHO CORDOVA	AEROJET#			E	
09 AZ	PHOENIX	19TH AVENUE LANDFILL			E	
05 MI	ST. LOUIS	GRATIOT COUNTY LANDFILL# *	V	R	E	
01 MA	NEW BEDFORD	NEW BEDFORD# *		R	E	
06 LA	DARROW	OLD INGER# *		R		
05 OH	HAMILTON	CHEM DYNE# *	V	R	E	
04 SC	COLUMBIA	SCRDI BLUFF ROAD# *	V	R	E	
01 CT	NAUGATUCK	LAUREL PARK INC.# *			E	
05 IL	WAUKEGAN	OUTBOARD MARINE COPR.# *		R	E	
08 CO	BOULDER	MARSHALL LANDFILL# *				D
01 ME	WINTHROP	WINTHROP LANDFILL# *		R		
01 VT	BURLINGTON	PINE STREET CANAL# *				D
03 WV	POINT PLEASANT	WEST VA ORDNANCE# *		R		
06 NM	ALBUQUERQUE	SOUTH VALLEY# *				D
07 MO	ELLISVILLE	ELLISVILLE SITE# *		R		
08 ND	SOUTHEASTERN	ARSENIC TRIOXIDE SITE# *		R		
03 VA	ROANOKE COUNTY	MATTHEWS# 8		R		
07 IA	COUNCIL BLUFFS	AIDEX CORP.# *		R	E	
09 AS	AMERICAN SAMOA	TAPUTIMU FARMS# *		R		
09 AS	GLOBE	MT. VIEW MOBILE HOME# *				D
04 KY	BROOKS	A. L. TAYLOR# *		R		
04 TN	MEMPHIS	NORTH HOLLYWOOD DUMP# *		R		
04 NC	210 MILES OF ROADS	PCB SPILLS# *		R	E	
09 GU	GUAM	ORDOT LANDFILL# *		R		
04 MS	GULFPORT	PLASTIFAX# *		R		
08 UT	SALT LAKE CITY	ROSE PARK SLUDGE PIT# *	V			
07 KS	ARKANSAS CITY	ARKANSAS CITY DUMP# *		R		
09 CM	NORTH MARIANAS	PCB WAREHOUSE# *		R		

†: V = Voluntary or Negotiated Response, R = Federal and State Response,
E = Federal and State Enforcement, D = Actions to be Determined.
\# = IPL/EEL. * = States' Designated Top Priority Sites.

Group 3

EPA Region	State	City/County	Site Name	V	R	E	D
02	NY	OYSTER BAY	SYOSSET LANDFILL				D
04	AL	GREENVILLE	MOWBRAY ENGINEERING				D
05	MI	BRIGHTEN	SPIEGELBURG LANDFILL				D
04	FL	MIAMI	MIAMI DRUM#		R		
02	NJ	DOVER TOWNSHIP	REICH FARMS			E	
02	NJ	SOUTH BRUNSWICK	SOUTH BRUNSWICK LANDFILL	V			
04	FL	TAMPA	KASSAUF-KIMERLING#				D
05	IL	WAUCONDA	WAUCONDA SAND & GRAVEL#			E	
05	MI	MUSKEGON	OTT/STORY/CORDOVA#				D
01	NH	KINGSTON	OTTATI & GOSS#		R	E	
03	VA	SALTVILLE	SALTVILLE WASTE DISPOSAL				D
02	NJ	RINGWOOD	RINGWOOD MINES/LANDFILL				D
02	NY	NIAGARA FALLS	HOOKER - S AREA			E	
04	FL	WHITEHOUSE	WHITEHOUSE OIL PITS#		R		
05	OH	DEERFIELD	SUMMIT NATIONAL#	V		E	
02	NY	NIAGARA FALLS	LOVE CANAL#		R	E	
05	IN	KINGSBURY	FISHER CALO	V		E	
05	MI	PLEASANT PLAINS TWP	WASH KING LAUNDRY				D
04	FL	WARRINGTON	PIONEER SAND#			E	
04	FL	TAMPA	REEVES SE GALVANIZING#				D
05	MI	DAVISBURG	SPRINGFIELD TOWNSHIP DUMP				D
05	MI	FILER CITY	PACKAGING CORP. OF AMERICA				D
03	PA	BUFFALO	HRANICA#			E	
08	CO	LEADVILLE	CALIFORNIA GULCH				D
04	NC	CHARLOTTE	MARTIN MARIETTA, SODYECO				D
04	FL	ZELLWOOD	ZELLWOOD GROUNDWATER CONTAM#				D
05	OH	CIRCLEVILLE	BOWERS LANDFILL				D
05	OH	ASHTABULA	FIELDS BROOK#		R		
03	PA	HARRISON TOWNSHIP	LINDANE DUMP#			E	
04	PL	SERMNER	TAYLOR ROAD LANDFILL#			E	
01	RI	BURRILLVILLE	WESTERN SAND & GRAVEL#		R	E	
02	NJ	MAYWOOD & ROCHELLE PK	MAYWOOD CHEMICAL SITES				D
06	OK	CRINER	CRINER/HARDAGE#			E	
05	MN	ST. LOUIS PARK	NATIONAL LEAD TARACORP#				D
05	MI	ROSE TOWNSHIP	ROSE TOWNSHIP DUMP#				D
05	MN	ANOKA COUNTY	WASTE DISPOSAL ENGINEERING#				D
02	NJ	EDISON	KIN-BUC LANDFILL#	V	R	E	
05	MN	LEHILLIER/MANKATO	LEHILLIER#				D
05	MI	GRAND RAPIDS	BUTTERWORTH #2 LANDFILL				D
02	NJ	BOUND BROOK	AMERICAN CYANAMID			E	
02	NY	SOUTH GLENE FALLS	GE MOREAU SITE			E	
02	NJ	PEDRICKTOWN	N.L. INDUSTRIES			E	
01	RI	NORTH SMITHFIELD	L & RR - N SMITHFIELD			E	
04	FL	HIALEAH	NW 58TH STREET#			E	
04	FL	TAMPA	62ND STREET DUMP				D
05	MI	UTICA	G&H LANDFILL#		R		
02	NJ	FRANKLIN TOWNSHIP	METALTEC/AEROSYSTEMS			E	
02	NJ	PEMBERTON TOWNSHIP	LANG PROPERTY				D
02	NJ	PARSIPPANY, TROY HLS	SHARKEY LANDFILL				D
06	LA	SORENTO	CLEVE REBER				D

†: V = Voluntary or Negotiated Response, R = Federal and State Response,
   E = Federal and State Enforcement, I = Actions to be Determined.
# = IPL/EEL. * = States' Designated Top Priority Sites.

Group 4

EPA Region

State	City/County	Site Name	Response Status *

05	IL	MARSHALL	VELSICOL ILLINOIS				D
05	MI	ST. LOUIS	VELSICOL MICHIGAN	V		E	
05	MI	MANCELONA	TAR LAKE				D
10	OR	ALBANY	TELEDYNE WAH CHANG				D
02	NY	SOUTH CAIRO	AMERICAN THERMOSTAT			E	
01	MA	DARTMOUTH	RE-SOLVE#		R	E	
02	NJ	PLUMSTEAD TOWNSHIP	GOOSE FARM#		R		
04	TN	TOONE	VELSICOL CHEMICAL CO.	V			
02	NY	MOIRA	YORK OIL COMPANY#		R		
04	FL	COTTONDALE	SAPP BATTERY#		R		
07	KS	HOLIDAY	DOEPKE DISPOSAL HOLIDAY				D
01	RI	SMITHFIELD	DAVIS LIQUID#		R	E	
01	MA	TYNGSBORO	CHARLES-GEORGE#			E	
02	NJ	WINSLOW TOWNSHIP	KING OF PRUSSIA				D
03	VA	YORK COUNTY	CHISMAN#				D
05	OH	SALEM	MEASE CHEMICAL				D
02	NJ	ELIZABETH	CHEMICAL CONTROL#		R	E	
05	OH	IRONTON	ALLIED CHEMICAL				D
05	MI	PENNFIELD TOWNSHIP	VERONA WELL FIELD#				D
01	CT	BEACON FALLS	BEACON HEIGHTS			E	
03	PA	MALVERN	MALVERN TCE SITE				D
02	NY	ELMIRA HEIGHTS	FACET ENTERPRISES#				D
03	DE	NEW CASTLE	DELAWARE SAND & GRAVEL#				D
08	CO	IDAHO SPRINGS	CENTRAL CITY, CLEAR CREEK#				D
03	PA	PALMERTON	PALMERTON ZINC PILE				D
05	IN	BOONE COUNTY	ENVIROCHEM				D
04	TN	LAWRENCEBURG	MURRAY OHIO DUMP				D
04	FL	WHITEHOUSE	COLEMAN EVANS#			E	
04	FL	INDIANTOWN	FLORIDA STEEL				D
09	AZ	GOODYEAR	LITCHFIELD AIRPORT AREA				D
02	NJ	PLUMSTEAD	SPENCE FARM#		R		
02	NJ	DOVER TOWNSHIP	TOMS RIVER CHEMICAL				D
04	FL	LIVE OAK	BROWN WOOD				D
02	NY	PORT WASHINGTON	PORT WASHINGTON LANDFILL			E	
06	AR	MENA	MID-SOUTH#				D
02	NJ	CHESTER	COMBE FILL SOUTH LANDFILL				D
02	NJ	SOUTH BRUNSWICK TWP	JIS LANDFILL			E	
08	CO	COMMERCE CITY	WOODBURY CHEMICAL#				D
01	MA	WESTBOROUGH	HOCOMOCO POND				D
02	NY	RAMAPO	RAMAPO LANDFILL			E	
05	MI	ALBION	MCGRAW EDISON				D
02	NY	ALBANY	MERCURY REFINING				D
04	FL	FORT LAUDERDALE	HOLLINGSWORTH#				D
02	NJ	ROCKAWAY TOWNSHIP	ROCKAWAY TOWNSHIP WELLS				D
02	NY	OLEAN	OLEAN WELLFIELD#		R		
04	FL	MIAMI	VARSOL SPILL#				D
02	NY	BATAVIA	BATAVIA LANDFILL#				D
09	CA	UKIAH	COAST WOOD PRESERVING				D
08	CO	DENVER	DENVER RADIUM SITE#		R		
08	MT	MILLTOWN	MILLTOWN				D

*: V = Voluntary or Negotiated Response, R = Federal and State Response,
E = Federal and State Enforcement, D = Actions to be Determined.
# = IPL/EEL. * = States' Designated Top Priority Sites.

Group 5

EPA Region

State	City/County	Site Name	V	R	E	D
07 MO	VERONA	SYNTEX FACILITY	V		E	
02 NJ	PLUMSTEAD	PIJAX FARM#		R		
02 NJ	SOUTH KEARNY	SYNCON RESINS#			E	
09 CA	RICHMOND	LIQUID GOLD				D
09 CA	FRESNO	PURITY OIL SALES, INC.				D
02 NJ	HOWELL TOWNSHIP	BOG CREEK FARM				D
05 IN	BLOOMINGTON	NEAL'S LANDFILL#				D
01 MA	LOWELL	SILRESIM#		R	E	
01 NE	LONDONDERRY	TINKHAM SITE				D
02 NJ	PISCATAWAY	CHEMSOL			E	
02 NJ	MARLBORO TOWNSHIP	IMPERIAL OIL			E	
02 NJ	FAIR.LAWN	FAIR LAWN WELLFIELD				D
05 IN	ELKHART	MAIN STREET WELL FIELD				D
02 NJ	MT. OLIVE TOWNSHIP	COMBE FILL NORTH LANDFILL				D
02 PR	JUANA DIAZ	GE WIRING DEVICES				D
02 NJ	MONROE TOWNSHIP	MONROE TOWNSHIP LANDFILL			E	
02 NJ	ROCKAWAY BORO	ROCKAWAY BORO WELLFIELD				D
05 IN	COLUMBIA CITY	WAYNE WASTE OIL				D
06 NM	MILAN	HOMESTAKE#				D
02 NJ	BERKLEY	BEACHWOOD/BERKLEY WELLS				D
02 NJ	DOVER	DOVER MUNICIPAL WELL 4				D
02 NY	VESTAL	VESTAL WATER SUPPLY			E	
10 WA	TACOMA	COM. BAY, NEAR SHORE TIDE FLAT#				D
	PEMBROKE	CROSS BROS/PEMBROKE				D
		FLYNN QUARRY CO.				D
03 PA	WEST ORMROD	EELEVA LANDFILL			E	
10 WA	SEATTLE	HARBOR ISLAND LEAD				D
09 CA	FULLERTON	MCCOLL	V			
10 WA	MEAD	KAISER MEAD	V			
02 PR	RIO ABAJO	FRONTERA CREEK				D
09 CA	FRESNO	SELMA PRESSURE TREATING				D
02 PR	FLORIDA AFUERA	BARCELONETA LANDFILL				D
03 MD	ELKTON	SAND, GRAVEL AND STONE			E	
05 MI	WYOMING	SPARTAN CHEMICAL COMPANY				D
02 NJ	FLORENCE	ROEBLING STEEL CO.				D
05 MI	GREILICKVILLE	GRAND TRAVERSE OVERALL SUPPLY CO				D
02 NJ	VINELAND	VINELAND STATE SCHOOL				D
03 PA	PHILADELPHIA	ENTERPRISE AVENUE				D
07 MO	SPRINGFIELD	FULBRIGHT LANDFILL#				D
04 SC	CAYCE	SCRDI DIXIANA#				D
02 NJ	SWAINTON	WILLIAMS PROPERTY		R		
02 NJ	EDISON	RENORA				D
04 FL	PENSACOLA	AMERICAN CREOSOTE#			E	
05 OH	IRONTON	E.H. SCHILLING LANDFILL				D
02 NJ	BAYVILLE	DENZER & SCHAFER X-RAY			E	
02 NJ	GIBBSTOWN	HERCULES				D
05 IN	GARY	NINTH AVE. DUMP	V		E	
05 MI	ST. LOUIS	GRATIOT CO GOLF COURSE	V		E	
01 RI	CUMBERLAND	PETERSON/PURITAN				D
01 MA	GROVELAND	GROVELAND WELLS				D

†: V = Voluntary or Negotiated Response, R = Federal and State Response,
E = Federal and State Enforcement, D = Actions to be Determined.
# = IPL/EEL. * = States' Designated Top Priority Sites.

Group 6

EPA Region				Response Status †	
	State	City/County	Site Name		
10	WA	SPOKANE	COLBERT LANDFILL	R	
09	AZ	SCOTTSDALE	INDIAN BEND WASH AREA		D
09	AZ	KINGMAN	KINGMAN AIRPT INDUSTRIAL AREA		D
02	NY	WHEATFIELD	NIAGARA COUNTY REFUSE#		D
04	FL	DELAND	SHERWOOD MEDICAL		D
05	MI	PARK TOWNSHIP	SOUTHWEST OTTAWA LANDFILL		D
02	NY	HORSEHEADS	KENTUCKY AVE. WELLFIELD#		D
01	ME	WASHBURN	PINETTE'S SALVAGE YARD		D
02	NJ	MILLINGTON	ASBESTOS DUMP		D
04	KY	LOUISVILLE	LEE'S LANE LANDFILL#		D
03	PA	STATE COLLEGE	CENTRE COUNTY KEPONE	E	
05	OH	BYESVILLE	FULTZ LANDFILL		D
05	AR	WALNUT RIDGE	FRITT INDUSTRIES#		D
05	OH	COSHOCTON	COSHOCTON CITY LANDFILL		D
03	PA	GIRARD TOWNSHIP	LORD SHOPE#	E	
05	IL	WAUKEGAN	JOHNS-MANVILLE		D
01	MA	PALMER	PSC RESOURCES	R	
05	MI	OTISVILLE	FOREST WASTE PRODUCTS		D
04	FL	CLERMONT	TOWER CHEMICAL#	E	
03	PA	LOCK HAVEN	DRAKE CHEMICAL INC.#	R	
03	MD	ANNAPOLIS	MIDDLETON ROAD DUMP	E	
03	DE	NEW CASTLE	TRIS SPILL SITE		D
03	PA	HAVERFORD	HAVERTOWN PCP SITE	E	
05	IN	GARY	LAKE SANDY JO		D
05	MI	GRAND RAPIDS	CHEM CENTRAL		D
01	MA	BRIDGEWATER	CANNON ENGINEERING	E	
05	MI	TEMPERANCE	NOVACO INDUSTRIES		D
06	LA	BAYOU SORREL	BAYOU SORREL#		D
02	NJ	JACKSON TOWNSHIP	JACKSON TOWNSHIP LANDFILL	E	
02	MI	KALAMAZOO	K & L AVE. LANDFILL		D
06	AR	EDMONDSEN	GURLEY PIT		D
05	MI	WHITEHALL	WHITEHALL WELLS		D
05	MI	IONIA	IONIA CITY LANDFILL		D
02	NJ	MONTGOMERY TOWNSHIP	MONTGOMERY HOUSING DEV		D
02	NJ	ROCKY HILL	ROCKY HILL MUNICIPAL WELL		D
02	NY	BREWSTER	BREWSTER WELL FIELD		D
02	NJ	ORANGE	US RADIUM		D
08	MT	LIBBY	LIBBY GROUND WATER		D
03	PA	JEFFERSON	RESIN DISPOSAL	E	
06	TX	HIGHLANDS	HIGHLANDS ACID PIT#	R	
04	KY	NEWPORT	NEWPORT DUMP		D
03	PA	LOWER PROVIDENCE TWP	MOYERS LANDFILL		D
04	KY	WEST POINT	DISTLER BRICKYARD	R	
01	CT	SOUTHINGTON	SOLVENTS RECOVERY SYSTEM	E	
03	PA	ERIE	PRESQUE ISLE		D
02	NJ	SAYREVILLE	SAYREVILLE LANDFILL		D
08	CO	COMMERCE CITY	SAND CREEK		D
08	WY	LARAMIE	BAXTER/UNION PACIFIC	E	
01	NH	DOVER	DOVER LANDFILL		D
06	AR	FT. SMITH	INDUSTRIAL WASTE CONTROL		D

†: V = Voluntary or Negotiated Response, R = Federal and State Response,
  E = Federal and State Enforcement, D = Actions to be Determined.
# = IPL/EEL. * = States' Designated Top Priority Sites.

494

Group 7
EPA Region

State		City/County	Site Name	Response Status †			
				V	R	E	D
02	NY	CLAYVILLE	LUDLOW SAND & GRAVEL				D
07	MO	IMPERIAL	ARENA 2: FILLS 1 & 2				D
06	LA	SLIDELL	BAYOU BONFOUCA				D
03	WV	LEETOWN	LEETOWN PESTICIDE PILE				D
01	CT	CANTERBURY	YAWORSKI			E	
05	OH	DODGEVILLE	NEW LYME LANDFILL				D
02	NJ	OLD BRIDGE	EVOR PHILLIPS				D
03	PA	CHESTER	WADE (ABM)#		R	E	
03	PA	OLD FORGE	LACKAWANNA REFUSE				D
02	NJ	GALLOWAY TOWNSHIP	MANNHEIM AVENUE DUMP				D
02	NY	FULTON	FULTON TERMINALS				D
05	MI	MUSKEGON	SCA INDEPENDENT LANDFILL				D
01	NH	LONDONDERRY	AUBURN RD LANDFILL			E	
03	WV	NITRO	FIKE CHEMICAL	V			
10	WA	KENT	WESTERN PROCESSING#			E	
05	MI	PETOSKEY	PETOSKEY MUNICIPAL WELLS				D
05	OH	ROCK CREEK	ROCK CREEK/JACK WEBB		R		
05	OH	JEFFERSON	POPLAR OIL#		R	E	
07	KS	WICHITA	JOHN'S SLUDGE POND				D
02	NJ	PENNSAUKEN	SWOPE OIL AND CHEMICAL#				D
05	MI	KENTWOOD	KENTWOOD LANDFILL				D
05	MN	ANDOVER	SOUTH ANDOVER SITE#				D
06	AR	NEWPORT	CECIL LINDSEY				D
05	IN	MARION	MARION (BRAGG) DUMP				D
05	OH	READING	PRISTINE				D
04	KY	CALVERT CITY	AIRCO				D
05	OH	ST. CLAIRSVILLE	BUCKEYE RECLAMATION				D
06	TX	GRAND PRAIRIE	BIO-ECOLOGY#		R		
04	FL	MOUNT PLEASANT	PARRAMORE SURPLUS				D
01	VT	SPRINGFIELD	OLD SPRINGFIELD LANDFILL				D
02	NY	LINCKLAEN	SOLVENT SAVERS				D
03	VA	FINEY RIVER	US TITANIUM			E	
05	IL	GALESBURG	GALESBURG/KOPPERS				D
05	OH	KINGSVILLE	BIG D CAMPGROUNDS				D
02	NY	NIAGARA FALLS	HOOKER - HYDE PARK	V		E	
05	MI	MARQUETTE	CLIFF/DOW DUMP				D
05	MI	MUSKEGON	DUELL & GARDNER LANDFILL				D
02	NJ	EVESHAM	ELLIS PROPERTY				D
04	KY	JEFFERSON COUNTY	DISTLER FARM#				D
09	CA	CLOVERDALE	MGM BRAKES				D
05	MI	LUDINGTON	MASON COUNTY LANDFILL				D
05	MI	ROSE TOWNSHIP	CEMETARY DUMP SITE				D
01	RI	NORTH SMITHFIELD	FORESTDALE				D
06	TX	HOUSTON	HARRIS (FARLEY ST)#		R		
03	PA	SEVEN VALLEYS	OLD CITY OF YORK LANDFILL			E	
05	IL	OGLE COUNTY	BYRON SALVAGE YARD			E	
03	PA	KING OF PRUSSIA	STANLEY KESSLER			E	
02	NJ	FREEHOLD TOWNSHIP	FRIEDMAN PROPERTY#		R		
02	NJ	FRANKLIN TOWNSHIP	MYERS PROPERTY				D
02	NJ	BOONTON	PEPE FIELD				D

†: V = Voluntary or Negotiated Response, R = Federal and State Response,
E = Federal and State Enforcement, D = Actions to be Determined.
# = IPL/EEL. * = States' Designated Top Priority Sites.

Group 8

EPA Region State	City/County	Site Name	Response Status †
05 MI	SOUTH OSSINEKE	OSSINEKE	D
05 MI	NILES	U.S. AVIEX	D
06 NM	CLOVIS	ATSF/CLOVIS#	E
10 WA	YAKIMA	PESTICIDE PIT, YAKIMA	D
05 TN	LEWISBURG	LEWISBURG DUMP	D
01 ME	SACO	SACO TANNING	D
03 PA	PHILADELPHIA	METAL BANKS	E
06 AR	MARION	CRITTENDEN CO. LANDFILL	D
05 MI	GRANDVILLE	ORGANIC CHEMICALS	D
10 OR	PORTLAND	GOULD, INC.	D
02 PR	JUNCOS	JUNCOS LANDFILL	D
04 FL	NORTH FLORIDA	MUNISPORT	D
05 MI	CLARE	CLARE WATER SUPPLY	D
02 NJ	ASBURY PARK	M&T DELISA LANDFILL	D
10 WA	YAKIMA	FMC YAKIMA	D
05 MI	ODEN	LITTLEFIELD TOWNSHIP DUMP	D
05 MI	KALAMAZOO	AUTO ION	D
04 SC	FORT LAWN	CAROLAWN, INC.	R E
05 MI	SPARTA	SPARTA LANDFILL	D
05 IL	WINNEBAGO	ACME SOLVENT/MORRISTOWN#	D
05 MI	CHARLEVOIX	CHARLEVOIX MUNICIPAL WELL	D
03 WV	FOLLANSBEE	FOLLANSBEE SLUDGE FILL	D
01 ME	AUGUSTA	O'CONNOR SITE	D
03 PA	WESTLINE	WESTLINE	D
05 MI	BRIGHTON	RASMUSSEN'S DUMP	D
05 MI	OSCODA	BEDELUM INDUSTRIES	D
02 PR	BARCELONETA	RCA DEL CARIBE	D
05 IN	LESANON	WEDZEB INC	D
04 KY	CALVERT CITY	B.F. GOODRICH	D
03 PA	STROUDSBURG	BRODHEAD CREEK	R E
05 MI	ADRIAN	ANDERSON DEVELOPMENT	D
05 MI	LIVINGSTON COUNTY	SHIAWASSEE RIVER	E
05 IL	LA SALLE	LASALLE ELECTRIC UTILITIES	E
04 TN	GALLOWAY	GALLOWAY PONDS	D
03 DE	KIRKWOOD	HARVEY KNOTT DRUM SITE#	R
03 DE	DOVER	WILDCAT LANDFILL	D
03 PA	WEST CHESTER TWP	BLOSENSKI LANDFILL	E
03 DE	DELAWARE CITY	DE CITY PVC PLANT#	D
03 MD	CUMBERLAND	LIMESTONE ROAD SITE	E
02 NY	NIAGARA FALLS	HOOKER - 102ND STREET	E
03 DE	NEW CASTLE	NEW CASTLE STEEL SITE	D
06 NM	CHURCHROCK	UNITED NUCLEAR CORP.#	D
09 CA	HOOPA	CELTOR CHEMICAL	D
04 AL	PERDIDO	PERDIDO GRDWATER CONTAMINATION	D
02 NY	COLD SPRINGS	MARATHON BATTERY#	D
03 PA	OLD FORGE	LEHIGH ELECTRIC#	R E
04 TN	CHATTANOOGA	AMNICOLA DUMP	D
05 OR	WEST CHESTER	SKINNER LANDFILL	D
07 MO	MOSCOW HILLS	ARENA 1 (DIOXIN)	D
04 NC	SWANNANOA	CHEMTRONICS, INC.	D

†: V = Voluntary or Negotiated Response, R = Federal and State Response,
E = Federal and State Enforcement, D = Actions to be Determined.
# = IPL/EEL. * = States' Designated Top Priority Sites.

Group 9

EPA Region

State	City/County	Site Name	Response Status †

07	NE	BEATRICE	PHILLIPS CHEMICAL	D
05	MI	BUCHANAN	ELECTROVOICE	D
03	PA	KIMBERTON	KIMBERTON	D
05	IN	BLOOMINGTON	LEMON LANE LANDFILL	D
10	ID	RATHDRUM	ARRCOM (DREXLER ENTERPRISES)	D
03	PA	WARMINSTER	FISCHER & PORTER	E
10	WA	LAKEWOOD	LAKEWOOD	D
05	OH	ZANESVILLE	ZANESVILLE WELL FIELD	D
09	CA	SACRAMENTO	JIBBOOM JUNKYARD	D
02	NJ	SPARTA	A. O. POLYMER	R
07	IA	DES MOINES	DICO	D
06	TX	ORANGE COUNTY	TRIANGLE CHEMICAL	R E
02	NC	JERSEY CITY	PJP LANDFILL	D
05	OH	MARIETTA	VAN DALE JUNKYARD	D
03	PA	PARKER	CRAIG FARM DRUM SITE	D
03	PA	UPPER SAUCON TWP	VOORTMAN	D
05	IL	BELVIDERE	BELVIDERE	D
05	IN	ALLEN COUNTY	PARROT ROAD	D

---

†: V = Voluntary or Negotiated Response, R = Federal and State Response,
   E = Federal and State Enforcement, D = Actions to be Determined.
\# = IPL/EEL.   * = States' Designated Top Priority Sites.

J

Unit Costs for Remedial Actions

Table J-1 / Remedial Action Unit Process: Landfill Capital Cost Components.[a]

Component	Subcomponent	Definition	English Units	Source Cost	Low	High	Newark
Apply stabilized waste	Installation	Hauling and spreading, 10 miles (16 km.), round trip	$/yd.³	6.66	4.20	8.72	7.60
Area preparation	Labor	Rake and cleanup, average	$/acre	300	190	390	340
Area preparation	Equipment	Rake and cleanup, average	$/acre	90	90	90	90
Backfill	Labor	Dozer and sheepsfoot roller	$/yd.³	0.39	0.24	0.49	0.44
Backfill	Equipment	Dozer and sheepsfoot roller	$/yd.³	1.10	1.10	1.10	1.10
Bentonite, delivered	Materials/shipping	Shipment of bentonite by rail	$/ton	28.8	19.8	64.8	63.0
Berm construction	Materials/installation	Use scraper	$/yd.³	0.38	0.27	0.46	0.40
Blower	Materials/installation	Blower, air	$/each	1,150	800	1,360	1,200
Butterfly valves, 6 in.	Materials/installation	PVC valve	$/each	192	135	230	200
Butterfly valves, 8 in.	Materials/installation	PVC valve	$/each	304	213	360	317
Cement pipe, 4 in. perforated	Labor	Asbestos, Class 4000 underdrain	$/yd.	2.73	1.73	2.89	2.57
Cement pipe, 4 in. perforated	Materials	Asbestos, Class 4000 underdrain	$/yd.	4.62	3.56	4.90	4.34
Cement pipe, 6 in. perforated	Labor	Asbestos, Class 4000 underdrain	$/yd.	2.92	1.77	3.69	3.21
Cement pipe, 6 in. perforated	Materials	Asbestos, Class 4000 underdrain	$/yd.	7.28	5.61	7.72	6.84
Cement pipe, 6 in. perforated	Labor	Cement pipe, non-perforated	$/yd.	3.22	2.03	4.21	3.68
Cement pipe, 6 in. perforated	Materials	Cement pipe, non-perforated	$/yd.	5.67	4.37	6.01	5.32
Chemicals	Materials	Treatment chemical, sodium hypochlorite (NaClO)	&/gal.	0.61	0.45	0.64	0.57
Deep wells, 6 in.	Materials/labor	Drilled and cased	$/ft.	6.19	3.91	8.10	7.10
Dewatering system	Materials/labor	For cost breakdown, see "Discharge pipe", "Submersible pump", "Deep wells"					

Item	Type	Description	Unit				
Discharge pipe, 4 in.	Labor	PVC plastic, Schedule 40	$/yd.	20.0	12.6	26.2	22.8
Discharge pipe, 4 in.	Materials	PVC plastic, Schedule 40	$/yd.	8.33	5.42	8.83	7.86
Discharge pipe, 8 in.	Labor	PVC plastic, Schedule 40	$/yd	31.0	19.5	40.6	35.4
Discharge pipe, 8 in.	Materials	PVC plastic, Schedule 40	$/yd.³	24.9	19.2	26.4	23.4
Diversion ditch, construction	Installation	Construction and maintenance repair	$/yd.³	2.11	1.33	2.76	2.38
Drilled holes, 2.5 in.	Materials/installation	Drilled and cased with pipe	$/yd.	16.7	10.6	21.9	19.1
Drilled holes, 6 in.	Materials/installation	Drilled and cased with pipe	$/yd.	20.3	14.2	24.1	21.1
Drill rig	Rental (Equip./labor)	Crew and light duty rig, and grading	$/day	400	280	470	420
Excavation, drainage trench	Labor	Use backhoe loader	$/yd.³	2.01	1.29	2.66	2.28
Excavation, drainage trench	Materials	Use backhoe loader	$/yd.³	1.22	0.91	1.29	1.14
Excavation/grading, soil	Labor	Excavation/grading, soil, common borrow, 1000 ft (305 m.) haul	$/yd.³	0.17	0.11	0.22	0.19
Excavation/grading, soil	Equipment	Excavation/grading, soil, common borrow, 1000 ft (305 m.) haul	$/yd.³	0.58	0.58	0.58	0.58
Excavation/grading waste	Labor	300 ft (90 m.) haul, dozer and truck	$/yd.³	0.28	0.17	0.36	0.32
Excavation/grading waste	Equipment	300 ft (90 m.) haul, dozer and truck	$/yd.³	1.03	1.03	1.03	1.03
Exploratory boring	Materials/installation	Tests strata in or below landfill, and to apply chemical injection, 4 in. (0.03 m.) diameter holes	$/yd.	17.9	12.5	21.1	18.6
Flow meters, 6 in.	Materials/installation	Measures rate of flow of landfill gas to blower	$/each	880	550	1,150	1,003
Flow meters, 8 in.	Materials/installation	Measures rate of flow of landfill gas to blower	$/each	1,040	650	1,360	1,180
Geotechnical investigation	Labor/materials	Includes exploratory holes,	$/site	5,500	3,850	6,520	5,720

Table J-1 / (Continued)

Component	Subcomponent	Definition	English Units	Source Cost	Low	High	Newark
		surveying, mobilization, drilling addition, pump test, and report					
Gravel	Labor	One dozer operator, one truck driver	$/yd.^{3}	1.88	1.19	2.47	1.96
Gravel	Materials	3/4 in. screened gravel	$/yd.^{3}	5.47	4.21	5.80	5.70
Grout curtain	Labor	Two grid—pherolic resin (also for grout bottom seal)	$/yd.^{3}	199	125	261	227
Grout curtain	Materials	Two grid—pherolic resin (also for grout bottom seal)	$/yd.^{3}	42.7	32.9	45.3	40.1
Header pipe, 8 in.	Materials	PVC Schedule 40	$/yd.	24.9	19.2	26.4	23.4
Header pipe, 8 in.	Installation	PVC Schedule 40	$/yd.	31.0	19.5	40.6	35.4
Hydroseeding	Labor	Includes seed and soil supplements	$/acre	69.1	43.5	90.5	78.8
Hydroseeding	Materials	Includes seed and soil supplements	$/acre	355	273	376	333
Hydroseeding	Equipment	Includes seed and soil supplements	$/acre	50	50	50	50
Liner	Labor/materials	30 mil., bracketed with heavyweight geotextile fabric, Hypalon	$/yd.^{2}	5.22	3.65	6.19	5.43
Materials testing	Materials/installation	For costing, see "Exploratory Boring"					
Moisture traps	Materials/installation	Removes water from gas control system	$/each	480	335	570	500
Monitoring equipment	Equipment	Gas detection instrumentation to monitor	$/each	500	500	500	500

gas control systems (MSA Model 53 Gascope)

Item	Type	Description	Unit				
Monitoring wells, gas	Materials	0.5 in. (1.3 cm.), 12 ft (3.6 m.) deep, for landfill gas monitoring	$/yd.	1.22	0.94	1.29	1.15
Monitoring wells, gas	Installation	0.5 in. (1.3 cm.), 12 ft (3.6 m.) deep, for landfill gas monitoring	$/yd.	8.03	5.05	10.51	9.15
Mulching	Labor	Hay mulching	$/acre	34.5	21.7	45.2	39.3
Mulching	Materials	Hay mulching	$/acre	85.00	65.4	90.1	79.9
Mulching	Equipment	Hay mulching	$/acre	23.2	23.2	23.2	23.2
Pipe, PVC (elbows), 6 in.	Materials	90° fitting	$/each	27.0	20.8	28.6	25.4
Pipe, PVC (elbows), 6 in.	Installation	90° fitting	$/each	28.0	17.6	36.7	31.9
Pipe, PVC (elbows), 8 in.	Materials	90° fitting	$/each	52.0	40.0	55.1	48.9
Pipe, PVC (elbows), 8 in.	Installation	90° fitting	$/each	39.0	24.6	51.1	44.5
Pipe, PVC (Tees), 6 in.	Materials	T-fittings for gas wells	$/each	37.0	28.5	39.2	34.8
Pipe, PVC (Tees), 6 in.	Installation	T-fittings for gas wells	$/each	46.0	29.0	60.3	52.4
Pipe, PVC (Tees), 8 in.	Materials	T-fittings for gas wells	$/each	75.0	57.7	79.5	
Pipe, PVC (Tees), 8 in.	Installation	T-fittings for gas wells	$/each	59.0	37.2	77	
Pipe, PVC, laterals, 8 in.	Materials	For gas extraction wells, Schedule 40	$/yd.	24.9	19.2		
Pipe, PVC, laterals, 8 in.	Installation	For gas extraction wells, Schedule 40	$/yd.	31.0	19.5	40.6	35.4
Pipe, PVC, laterals, 12 in.	Materials	For gas extraction wells, Schedule 80	$/yd.	112	86.1	118	105
Pipe, PVC, laterals, 12 in.	Installation	For gas extraction wells, Schedule 80	$/yd.	14.1	10.9	14.9	13.2
Pipe, PVC, risers, 4 in.	Materials	For gas extraction wells, Schedule 40	$/yd.	8.33	6.42	8.83	7.83
Pipe, PVC, risers, 4 in.	Installation	For gas extraction wells, Schedule 40	$/yd.	20.0	12.6	26.2	22.8
Pipe, PVC, risers, 6 in.	Materials	For gas extraction wells, Schedule 40	$/yd.	14.9	11.5	15.8	14.6

Table J-1 / (Continued)

Component	Subcomponent	Definition	English Units	Source Cost	Low	High	Newark
Pipe, PVC, risers, 6 in.	Installation	For gas extraction wells, Schedule 40	$/yd.	23.4	14.8	30.7	26.7
Pump, centrifugal	Equipment/installation	3/4 HP pump	$/each	1,600	1,600	1,600	1,600
Pump, submersible	Labor	1 HP—4 in. submersible	$/each	344	220	450	390
Pump, submersible	Materials	1 HP—4 in. submersible	$/each	424	330	450	400
Recharge trench	Materials/installation	Excavation of trench	$/yd.^{3}	1.22	0.85	1.44	1.26
Sand	Materials/installation	For well point casing backfill	$/bag	5.80	4.06	6.87	6.03
Sheet piling	Materials	Steel sheet, PMA-22 (22 lbs/ft^2)	$/ton	460	354	488	433
Sheet piling	Installation	Install steel sheet, PMA-22 (22 lbs/ft^2)	$/ton	90.8	57.2	119	103
Slurry trench excavation	Materials/installation	Includes installation of bentonite slurry	$/yd.^{3}	35.0	24.5	41.4	36.4
Spread excavated material	Labor	One equipment operator, five laborers	$/yd.^{3}	0.16	0.10	0.19	0.17
Spread excavated material	Equipment	Spread by dozer, no compaction	$/yd.^{3}	0.39	0.39	0.39	0.39
Surface seal, bituminous concrete	Installation	3 in. (0.08 m.) thick cap	$/yd.^{2}	0.88	0.55	1.15	1.01
Surface seal, bituminous concrete	Materials	3 in. (0.08 m.) thick cap	$/yd.^{2}	2.52	1.94	2.67	2.37
Surface seal, clay cap	Materials/installation	6 in. (15-cm.) clay cap, includes 18 in. (46-cm.) soil cover	$/yd.^{2}	3.98	2.89	5.29	4.53
Surface seal, clay cap	Materials/installation	18 in. (46-cm.) clay cap, includes 18 in. (46-cm.) soil cover	$/yd.^{2}	5.11	3.73	6.79	5.83
Surface seal, fly ash cap	Materials/installation	12 in. (30-cm.) fly ash, includes 18 in. (46-cm.) soil cover	$/yd.^{2}	3.84	2.81	5.11	4.38

Item	Category	Description	Unit				
Surface seal, fly ash cap	Materials/installation	24 in. (60-cm.) fly ash cap, includes 18 in. (46-cm.) soil cover	$/yd.^{2}	5.20	3.71	6.91	5.93
Surface seal, lime-stabilized cap	Materials/installation	5 in. (13-cm.) lime-stabilized cap, includes 18 in. (46-cm.) soil cover	$/yd.^{2}	4.99	4.33	6.64	5.69
Surface seal, PVC membrane cap	Materials/installation	30-mil PVC membrane cap, includes 18 in. (46-cm.) soil cover	$/yd.^{2}	8.94	8.03	11.	10.2
Surface seal, soil-cement cap	Materials/installation	5 in. (13-cm.) soil-cement cap, includes 18 in. (46-cm.) soil cover	$/yd.^{2}	4.99	4.33	6.64	5.69
Surveying	Labor	Labor cost/day, establish surface topographic profile	$/day	220	150	260	230
Tipping fees	Unit costs	Dumping and grading wastes at new site	$/ton	100	100	100	100
Transportation	Labor/equipment	30-ton dump truck/driver, based on one-way hauling distance, return trip included	$/ton-mi.	0.070	0.044	0.093	0.080
Transportation	Labor/equipment	15-ton dump truck/driver, based on one-way hauling distance, return trip included	$/ton-mi.	0.080	0.050	0.106	0.091
Treatment system	Unit costs	Typical costs—interpolated at 0.12 MGD (440,000 l./d)	$/each	580K	406K	687K	603K
Trench excavation	Labor	One equipment operator/one laborer	$/yd.^{3}	0.17	0.13	0.27	0.24
Trench excavation	Equipment	Tractor or hydraulic backhoe to excavate trench, sloped 1/2 to 1	$/yd.^{3}	0.88	0.88	0.88	0.88
Well points, 2.5 in. (6.4 cm.) diameter	Materials/installation	PVC, 25 ft (7.6 m.)	$/yd.	54.0	37.8	64.0	56.1
Wellpoint fittings	Materials/installation	Fittings and accessories	$/well	12.0	8.40	14.2	12.5

[a]From Rishel et al. (1981), as cited in Chapter 7.

Table J-2 / Remedial Action Unit Process: Landfill O&M Cost Components.[a]

Component	Subcomponent	Definition	English Units	Source Cost	Low	High	Newark
Chemicals	Materials	Wastewater/leachate treatment plant chemicals	$/gal./day (influent)	0.095	0.095	0.095	0.095
Electricity	Power costs	For water treatment plant, extraction, injection, and gas control well/pumps	$/kwh	0.05	0.05	0.05	0.05
Grubbing	Labor/equipment	Assume annual grubbing (clearing) of brush	$/yd.^{2}	0.16	0.10	0.21	0.18
Maintenance/repair, diversion ditch	Installation	Assume diversion ditch needs rebuilding 2 times a year after major storms	$/yd.^{3}	2.10	1.32	2.75	2.40
Monitoring	Labor	For gas monitoring at active and passive gas control installations	$/hr.	12.5	7.88	16.6	14.2
Monitoring (analysis)	Laboratory costs	For ground water/leachate monitoring from monitoring wells	$/sample	330	330	330	330
Monitoring (sampling)	Labor	For ground water/leachate monitoring from monitoring wells	$/hr.	12.5	7.88	16.6	14.2
Grass mowing	Labor/materials	Use 58" power ride mower, one operator	$/acre	38	27	45	40
Operating cost	Labor	For water treatment plant operating personnel	$/hr.	10.0	6.30	13.10	11.4
Operating cost	Labor	For gas collection system operating personnel	$/hr.	15.0	9.45	19.9	17.1
Refertilization	Labor/materials	Assume refertilization once per year	$/acre	138	100	160	140
Water	Materials	Industrial process water	$/103 gal.	3.50	3.50	3.50	3.50

[a] From Rishel et al. (1981), as cited in Chapter 7.

Table J-3 / Remedial Action Unit Process: Surface Impoundment Capital Cost Components.[a]

Component	Subcomponent	Definition	English Units	Source Cost	Low	High	Newark
Area preparation	Labor	For area preparation, rake and clean up, average	$/yd.²	0.062	0.039	0.081	0.070
Area preparation	Equipment	For area preparation, rake and clean up, average	$/yd.²	0.019	0.019	0.019	0.019
Bentonite, delivered	Materials/shipping	Shipment of bentonite by rail near job site, includes materials and delivery	$/ton	28.8	19.8	64.8	63.0
Cement pipe, 6 in.	Materials	Class 4000, perforated, asbestos	$/yd.	7.28	5.61	7.72	6.84
Cement pipe, 6 in.	Installation	Class 4000, perforated, asbestos	$/yd.	2.82	1.77	3.69	3.21
Discharge trench	Labor	Including backfill 3 ft (1 m.) deep	$/yd.³	1.50	0.94	1.96	1.69
Discharge trench	Equipment	Including backfill 3 ft (1 m.) deep	$/yd.³	1.19	1.19	1.19	1.19
Diversion ditch, construction	Installation	Construction and maintenance/repair	$/yd.³	2.11	1.33	2.76	2.38
Drilled holes, 6 in.	Materials/installation	Drilled and cased with pipe	$/yd.	20.3	14.2	24.0	21.1
Drill rig	Rental (Equipment/labor)	Crew and light-duty rig	$/day	400	280	470	420
Excavation	Labor	One equipment operator	$/yd.³	0.12	0.08	0.16	0.14
Excavation	Equipment	Front end loader	$/yd.³	0.49	0.49	0.49	0.49
Excavating/grading soil	Labor	Common borrow (earth), 1000 ft (305 m.) haul	$/yd.³	0.17	0.11	0.22	0.19
Excavating/grading, soil	Equipment	Common borrow (earth), 1000 ft (305 m.) haul	$/yd.³	0.58	0.58	0.58	0.58
Geotechnical investigation	Unit costs	Includes surveying, test borings, equipment, mobilization, monitoring wells, pump tests, report	$/site	14,500	9,500	19,500	16,900

Table J-3 / (Continued)

Component	Subcomponent	Definition	English Units	Source Cost	Low	High	Newark
Geotechnical investigation	Unit costs	Slurry wall testing	$/each	2,000	1,260	2,620	2,260
Gravel	Labor/installation	3/4 in. screened gravel, one dozer operator, one truck driver	$/yd.³	1.88	1.19	2.47	1.96
Gravel	Materials	3/4 in. screened gravel, one dozer operator, one truck driver	$/yd.³	5.47	4.21	5.80	5.70
Grout curtain	Labor	Chemical grout, phenolic resin, 2-grid	$/yd.³	199	125	261	227
Grout curtain	Materials	Chemical grout, phenolic resin, 2-grid	$/yd.³	47.7	32.9	45.3	40.1
Header and discharge pipe, 8 in.	Materials	PVC class 150 pipe, laid in trench	$/yd.	22.0	17.0	23.3	20.7
Header and discharge pipe, 8 in.	Installation	PVC class 150 pipe, laid in trench	$/yd.	8.75	5.51	11.5	9.88
Hydroseeding	Labor	Seed and soil supplements/amendments	$/yd.²	0.0143	0.0090	0.0187	0.0163
Hydroseeding	Materials	Seed and soil supplements/amendments	$/yd.²	0.0733	0.0565	0.0777	0.0689
Hydroseeding	Equipment	Hydroseeding	$/yd.²	0.0103	0.0103	0.0103	0.0103
Mulching	Labor	Mulching hay	$/yd.²	0.0071	0.0045	0.0093	0.0081
Mulching	Materials	Mulching hay	$/yd.²	0.0175	0.0135	0.0186	0.0165
Mulching	Equipment	Mulching hay	$/yd.²	0.0048	0.0048	0.0048	0.0048
Pump, centrifugal	Equipment/installation	3/4 HP pump	$/each	1600	1600	1600	1600
Pump, submersible	Labor/installation	1 HP, 4 in. including wiring	$/each	344	220	450	390
Pump, submersible	Materials	1 HP, 4 in. including wiring	$/each	424	330	450	400
Sheet piling	Labor/equipment/installation	PMA-22 steel sheet piling (22 lbs/ft²)	$/ton	90.8	57.2	119	103

Item	Cost type	Description	Unit				
Sheet piling	Materials	PMA-22 steel sheet piling (22 lbs/ft²)	$/ton	460	354	488	433
Slurry wall, installation	Installation	Install slurry compound in excavated trench	$/yd.³	35.1	24.6	41.7	36.5
Slurry wall testing	Unit cost	See "Geotechnical Investigation, Slurry wall testing"					
Soil compacting	Labor	With sheepsfoot roller	$/yd.³	0.48	0.30	0.63	0.54
Soil compacting	Equipment	With sheepsfoot roller	$/yd.³	0.65	0.65	0.65	0.65
Sump	Labor/equipment	16 ft (5 m.) deep × 8 in. (20 cm.) thick concrete, cast in place	$/each	820	517	1,074	926
Sump	Materials	16 ft (5 m.) deep × 8 in. (20 cm.) thick concrete, cast in place	$/each	750	577	795	705
Surface seal, bituminous concrete	Labor/equipment	3 in. (0.08 m.) thick cap	$/yd.²	0.88	0.60	1.04	0.92
Surface seal, bituminous concrete	Materials	3 in. (0.08 m.) thick cap	$/yd.²	2.52	1.94	2.67	2.37
Surface seal, clay cap	Materials/installation	6 in. (15-cm.) clay cap, includes 18 in. (46-cm.) soil cover	$/yd.²	3.98	2.89	5.29	4.54
Surface seal, clay cap	Materials/installation	18 in. (46-cm.) clay cap, includes 18 in. (46-cm.) soil cover	$/yd.²	5.11	3.73	6.79	5.83
Surface seal, fly ash cap	Materials/installation	12 in. (30-cm.) fly ash cap, includes 18 in. (46-cm.) soil cover	$/yd.²	3.84	2.81	5.11	4.38
Surface seal, fly ash cap	Materials/installation	24 in. (60-cm.) fly ash cap, includes 18 in. (46-cm.) soil cover	$/yd.²	5.20	3.71	6.91	5.93
Surface seal, lime-stabilized cap	Materials/installation	5 in. (13-cm.) lime-stabilized cap, includes 18 in. (46-cm.) soil cover	$/yd.²	4.99	4.33	6.64	5.69

Table J-3 / (Continued)

Component	Subcomponent	Definition	English Units	Source Cost	Low	High	Newark
Surface seal, PVC membrane cap	Materials/installation	30-mil PVC membrane cap, includes 18 in. (46-cm.) soil cover	$/yd.²	8.94	8.03	11.9	10.2
Surface seal, soil-cement cap	Materials/installation	5 in. (13-cm.) soil-cement cap, includes 18 in. (46-cm.) soil cover	$/yd.²	4.99	4.33	6.64	5.69
Surveying	Labor	Labor costs/day	$/day	220	150	260	230
Tipping fee	Unit costs	Fee paid at secure landfill	$/ton	100	100	100	100
Transportation	Labor/equipment	30-ton dump truck/driver, based on one-way hauling distance, return trip included	$/ton-mi.	0.070	0.044	0.093	0.080
Transportation	Labor/equipment	15-ton dump truck/driver, based on one-way hauling distance, return trip included	$/ton-mi.	0.080	0.050	0.106	0.091
Treatment plant	Unit costs	Costs interpolated from Dodge Guide 1980, with SCS estimate	$/gpd	3.38	2.65	4.49	3.94
Trench excavation	Labor	16 ft (5 m.) deep × 3 ft (1 m.) wide, one equipment operator, one laborer	$/yd.³	0.17	0.13	0.27	0.24
Trench excavation	Equipment	16 ft (5 m.) deep × 3 ft (1 m.) wide, backhoe excavator, sloped 1/2:1	$/yd.³	0.88	0.88	0.88	0.88
Well fittings, 8 in.	Materials	PVC	$/well	127	97.8	135	119
Well fittings, 8 in.	Installation	PVC	$/well	98.0	61.7	128	111
Well points, 2.5 in.	Materials/installation	16 ft (4.9 m.) long	$/yd.	53.8	36.3	63.5	56.3
Wellpoint fittings	Unit costs	Fittings and accessories	$/well	12.0	8.40	14.2	12.5

[a]From Rishel et al. (1981), as cited in Chapter 7.

Table J–4 / Remedial Action Unit Process: Surface Impoundment O&M Cost Components.[a]

Component	Subcomponent	Definition	English Units	Source Cost	Low	High	Newark
Chemicals	Materials	Wastewater/leachate treatment plant chemicals	$/gpd (influent)	0.095	0.095	0.095	0.095
Electricity	Power costs	For water treatment plant or extraction, injection, and gas control wells/pumps	$/kwh	0.05	0.05	0.05	0.05
Refertilizing	Labor/materials	Assume fertilizing once/year	$/yd.^{2}	0.0050	0.0039	0.0054	0.0047
Grass mowing	Labor/equipment	Assume grass mowing 6 times/year, minimum $10 per visit	$/yd.^{2}	0.0013	0.00081	0.0017	0.0014
Grubbing	Labor/equipment	Assume annual grubbing (clearing) of brush	$/ydd.^{2}	0.13	0.10	0.21	.018
Maintenance repair/diversion ditch	Installation	Assume twice annual ditch repair	$/yd.^{3}	2.11	1.33	2.76	2.38
Monitoring (analysis)	Laboratory costs	For ground water/leachate monitoring from monitoring wells	$/sample	330	330	330	330
Monitoring (sampling)	Labor	For ground water/leachate monitoring from monitoring wells	$/hr.	12.5	7.88	16.6	14.2
Operator personnel	Labor	For operation of water treatment plant and sampling for monitoring	$/hr.	12.6	7.92	16.3	14.3

[a]From Rishel et al. (1981), as cited in Chapter 7.

Table J-5 / Average U.S. Low and High Costs of Unit Operations for Medium-Size Sites.[a]

Unit Operations	Typical English Units	Average U.S. Low Cost Per Unit			Average U.S. High Cost Per Unit		
		Initial Capital ($)	First Year O&M ($)	Life Cycle Cost ($)	Initial Capital ($)	First Year O&M ($)	Life Cycle Cost ($)
1. Contour grading and surface water diversion	Site area, ac.	6,210	46.3	6,600	7,240	97.2	8,060
2. Surface sealing	Site area, ac.	27,200	0	27,200	37,500	0	37,500
3. Revegetation	Site area, ac.	5,440	42.6	5,800	6,680	74.7	7,310
4. Bentonite slurry trench cutoff wall	Wall face area, ft.^{2}	5.08	0.075	5.71	9.21	0.081	9.90
5. Grout curtain	Wall face area, ft.^{2}	55.8	0.075	87	112	0.081	175
6. Sheet piling cutoff wall	Wall face area, ft.^{2}	6.81	0	6.80	10.1	0	10.1
7. Bottom sealing	Site area, ac.	2,136,000	10.3	2,141,000	4,128,000	708	4,134,000
8. Drains	Pipe length, ft.	22.1	9.26	109	32.3	11.2	127
9. Well point system	Intercept face area, ft.^{2}	5.81	0.486	9.97	9.72	0.524	14.2
10. Deep well system	Intercept face area, ft.^{2}	1.04	0.190	2.66	1.70	0.206	3.46
11. Injection	Intercept face area, ft.^{2}	7.18	18.0	164	8.38	18.7	166
12. Leachate handling by subgrade irrigation	Site area, ac.	2,130	688	7,960	3,380	748	9,720
13. Chemical fixation	Site area, ac.	27,900	647	33,400	52,700	708	58,700
14. Chemical injection	Landfill volume, ft.^{3}	1.69	0.081	2.38	3.24	0.089	3.98
15. Excavation and reburial	Landfill volume, ft.^{3}	10.8	0.001	10.8	11.1	0.001	11.1
16. Ponding	Site area, ac.	262	0	262	416	0	416
17. Trench construction	Trench length, ft.	3.72	0.105	4.61	4.37	0.216	6.20
18. Perimeter gravel trench	Trench length, ft.	30.2	0.036	30.5	43.8	0.078	44.5
19. Treatment of contaminated waste	Contaminated water, gal./d	5.75	0.446	9.52	9.74	0.810	16.6

No.	Item	Units						
20.	Gas migration control—passive	Site perimeter, ft.	228	0.710	234	334	1.50	346
21.	Gas migration control—active	Site perimeter, ft.	81.8	5.56	129	142	10.9	234
22.	Pond closure and contour grading of surface	Site area, ac.	10,900	431	14,500	14,200	879	21,600
23.	Surface sealing of closed impoundments	Site area, ac.	19,600	0	19,600	28,500	0	28,500
24.	Revegetation	Site area, ac.	1,030	69.0	1,600	1,540	77.6	2,200
25.	Slurry trench cutoff wall	Wall face area, ft.^{2}	5.58	0	5.58	10.10	0	10.10
26.	Grout curtain	Wall face area, ft.^{2}	30.4	0.180	31.9	58.7	0.197	60.4
27.	Sheet piling cutoff wall	Wall face area, ft.^{2}	7.14	0.196	8.80	10.7	0.215	12.5
28.	Grout bottom seal	Site area, ac.	350,000	7,470	413,250	654,000	8,160	723,000
29.	Toe and underdrains	Pipe length, ft.	96.1	44.4	471	185	48.6	595
30.	Well point system	Intercept face area, ft.^{2}	5.50	2.69	28.3	10.3	2.93	35.2
31.	Deep well system	Intercept face area, ft.^{2}	3.09	0.893	10.65	5.61	0.972	13.9
32.	Well injection system	Intercept face area, ft^2	2.92	0.859	10.16	5.17	0.938	13.1
33.	Leachate treatment	Contaminated water, gal./d	4.37	1.48	17	7.40	2.76	30.8
34.	Berm reconstruction	Replaced berm, yd.^{3}	2.28	0.093	3.06	2.91	0.187	4.48
35.	Excavation and disposal at secure landfill	Impoundment volume, gal.	1.47	0	1.47	1.52	0	1.52

[a]From Rishel et al. (1981), as cited in Chapter 7.

K

Combustion Properties of Hazardous Materials

Contaminant	Physical State	Flash Point °F	Auto Ignition Temperature °C	Explosive Carries Own Oxygen	Heat of Combustion Btu/lb (kg cal/gm)(a)	HCT/COCl2	SOx	NOx	Metal	Other
P001 3-(α-Acetonylbenzyl)-4-hydroxy coumarin	Solid				---					
P002 1-Acetyl-2-thiourea	solid						X	X		
P003 Acrolein	liquid	-15	455	explosive polymerization	12,507		X			
P004 Aldrin	solid	--nonflammable--			---					
P005 Allyl Alcohol	liquid	70	713		13,720		X			
P006 Aluminum phosphide	solid	--nonflammable--			---			X		POx, p
P007 5-(Aminomethyl)-3-isoxyzole	solid							X		
P008 Aminopyridine	Solid							X		
P009 Ammonium picrate	Solid	Explodes 423°C		May explode				X		
P010 Arsenic Acid	solid	--nonflammable--							X	
P011 Arsenic pentoxide	solid	--nonflammable--							X	
P012 Arsenic trioxide	solid	--nonflammable--							X	
P013 Barium Cyanide	Solid	--nonflammable--			---			X	X	Sulfides
P014 Benzenethiol	liquid	158					X			
P015 Beryllium dust	solid	--nonflammable--								Be cmpds
P016 Bis(chloromethyl) ether	liquid					X				
P017 Bromoacetone	liquid	--nonflammable--								HBr
P018 Brucine	solid				(2933)			X		
P019 2-Butanone peroxide	liquid			may explode						
P020 2-sec-Butyl-4,6-dinitrophenol	solid	--nonflammable--							X	
P021 Calcium cyanide	solid	--nonflammable--								sulfides
P022 Carbon disulfide	liquid	-22	194		5,814	X	X			
P023 Chloroacetaldehyde	liquid	190				X				
P024 p-Chloroaniline	solid							X		
P025 1-(p-chlorobenzoyl)-5-methoxy-2-methylindole-3-acetic acid	solid									
P026 1-(o-chlorophenyl)thiourea	solid					X	X	X		
P027 3-Chloropropionitrile	liquid	168				X		X		
P028 α-Chlorotoluene	liquid	153	1085			X		X		
P029 Copper cyanide	solid	--nonflammable--						X	X	
P030 Cyanides	solids	--nonflammable--						X		
P031 Cyanogen	gas				(25813)			X		
P032 Cyanogen bromide	solid	--nonflammable--				X		X		HBr
P033 Cyanogen chloride	liquid	--nonflammable--						X		
P034 2-Cyclohexyl-4,6-dinitrophenol	solid				---					
P035 2,4-Dichlorophenoxyacetic acid (2,4-D)	solid	--nonflammable--				X				
P036 Dichlorophenylarsine	solid	--nonflammable--				X				
P037 Dieldrin	solid				---	X				
P038 Diethylarsine	liquid									
P039 O,O-Diethyl-B-(2-(ethylthio) ethyl),phosphoro thioate	liquid				---		X			POx

(a) Combustion values in parentheses are estimates.

Contaminant	Physical State	Flash Point °F	Auto Ignition Temperature °C	Explosive Carries Own Oxygen	Heat of Combustion Btu/lb (kg cal/gm)(a)	HCT/COC12	SOx	NOx	Metal	Other
P040 0,0-Diethyl-0-(2-pyrazinyl) phosphorothioate	liquid				--					POx
P041 3,4-Dihydroxy-α-(methylamino)	solid				--			X		
P042										
P043 Di-isopropylfluorophosphate	liquid				--					HF,POx
P044 Dimethate	solid				--		X	X		POx
P045 3,3-Dimethyl-1-(methylthio)-2-butanone-0-[(methylamino) carbonyl]oxime	solid				--		X	X		POx
P046 α,α Dimethylphenylamine (dimethylaniline)	liquid	145	700		(1143)			X		
P047 4,6-Dinitro cresol	solid			may explode	--			X		
P048 2,4-Dinitrophenol	solid			may explode	(648)			X		
P049 2,4-Dithiobiuret	solid				--		X	X		
P050 Endosulfan	solid				--		X			
P051 Endrin	solid	--nonflammable--			--					
P052 Ethylcyanide (Propionitrile)	liquid	36			12,290			X		
P053 Ethylenediamine	liquid	110	725		15,930			X		
P054 Ethyleneimine	liquid	12	608		--			X		
P055 Ferric Cyanide	solid	--nonflammable--			--			X	X	HF
P056 Fluorine	gas	--nonflammable--			--					HF
P057 2-Fluoroacetamide	solid				--					HF
P058 Sodium Fluoracetate	solid				--					
P059 Heptachlor	solid				--	X				
P060 1,2,3,4,10,10-Hexachloro-1,4,4a,5,8,8a,-P060 hexahydro-1,15,8,-endo-endo-dimethanonaphthalene	solid	--nonflammable--			--	X				
P061 Hexachloropropene	liquid				--					Cl2
P062 Hexaethyltetraphosphate	liquid				--	X				POx
P063 Hydrocyanic acid	liquid	0	1000		10,560			X		
P064 Isocyanic acid, methylester	liquid				--			X		
P065 Mercury fulminate	solid			may explode	--			X		Hg
P066 Methomyl	solid				--		X			
P067 2-Methylaziride	liquid	<80			--			X		
P068 Methylhydrazine					--			X		
P069 2-Methyllactonitrile				may explode						
P070 2-Methyl-2-(methylthio) propinalidehyde-0-(methyl)- carbonyl) oxine										
P071 Methyl parathion	solid	--nonflammable--			--		X	X		POx,P
P072 1-Naphthyl-2-thiourea	solid				--		X	X		
P073 Nickel carbonyl	liquid	<-4	-31		--				X	
P074 Nickel cyanide	solid	--nonflammable--			--				X	
P075 Nicotine	liquid	471			(1428)			X		
P076 Nitric oxide	gas	--nonflammable--			--			X		
P077 p-Nitroniline	solid	329			(761)			X		
P078 Nitrogen dioxide	gas	--nonflammable--			--			X		

(a) Combustion values in parentheses are estimates.

Contaminant	Physical State	Flash Point °F	Auto Ignition Temperature °C	Explosive Carries Own Oxygen	Heat of Combustion Btu/lb (kg cal/gm)(a)	HCT/COCl2	SOx	NOx	Metal	Other
P079 Nitrogen pentoxide	gas							x		
P080 Nitrogen tetroxide	gas							x		
P081 Nitroglycerine	liquid	428	518	explodes	(368)			x		
P082 N-nitrosodimethylamine	liquid							x		
P083 N-Nitrosodiphenylamine	solid							x		
P084 N-Nitrosomethylvinylamine	liquid							x		
P085 Octamethylpyrophosphoramide	liquid				---			x		POx
P086 Oleyl alcohol (condensed with 2 moles ethylenoxide)	liquid									
P087 Osmiumtetroxide	solid	nonflammable							x	
P088 7-Oxabicyclof2.2.1]heptane-2,3-dicarboxylic acid	solid									
P089 Parathion	liquid	--nonflammable--			---		x	x		POx
P090 Pentachlorophenol	solid	--nonflammable--			---	x				Cl2
P091 Phenyl dichloroasine	liquid				---	x			x	
P092 Phenylmercury acetate	solid	>100	--nonflammable--		---		x		x	Hg
P093 N-Phenylthiourea	solid	--nonflammable--			---		x	x		
P094 Phorate	liquid				---					POx
P095 Phosgene	gas	--nonflammable--				x				
P096 Phosphine	gas		212							
P097 Phosphorothiou acid,0,0-dimethyl ester, 0-ester with N,N-dimethyl benzene sulfonamide										
P098 Potassium cyanide	solid	--nonflammable--						x		
P099 Potassium silver cyanide	solid	--nonflammable--						x	x	
P100 1,2-Propanediol (propylene glycol)	liquid	210	700		10,310					
P101 Propionitrile	liquid	36			(456)			x		
P102 2-Propyn-1-01	liquid				---					
P103 Selenourea	solid							x		Se cmpds
P104 Silver cyanide	solid	--nonflammable--						x	x	
P105 Sodium azide	solid							x		Na,NaOH,NOx
P106 Sodium cyanide	solid	--nonflammable--						x		
P107 Strontium sulfide	solid									
P108 Strychnine	solid				(2685)			x		
P109 Tetraethyldithiopyrophosphate	liquid	200					x			POx
P110 Tetraethyl lead	liquid	--nonflammable--			7,870				x	
P111 Tetraethyl pyrophosphate	liquid	266		may explode	---					POx
P112 Tetranitromethane	liquid	--nonflammable--						x		
P113 Thallic oxide	solid	--nonflammable--								
P114 Thallium selenite	solid	--nonflammable--			---		x	x	x	Se cmpds
P115 Thallium sulfate	solid	--nonflammable--			---		x		x	
P116 Thiosemicarbazide	solid							x		
P117 Thiuram	solid						x	x		
P118 Trichloromethenethiol	liquid					x				Cl2
P119 Ammonium Vanadate	solid	--nonflammable--						x	x	
P120 Vanadium pentoxide	solid	--nonflammable--							x	
P121 Zinc cyanide	solid	--nonflammable--						x	x	
P122 Zinc phosphide	solid	--nonflammable--							x	POx

(a) Combustion values in parentheses are estimates.

Contaminant	Physical State	Flash Point °F	Auto Ignition Temperature °C	Explosive Carries Own Oxygen	Heat of Combustion Btu/lb (kg cal/gm)(a)	HCl/COCl2	SOx	NOx	Metal	Other
U001 Acetaldehyde	liquid	-36	347		10,600					
U002 Acetone	liquid	0	869		12,250					
U003 Acetonitrile	liquid	42	975		13,360			x		
U004 Acetophenone	liquid	180	1058		14,850			x		
U005 2-Acetylaminofluorene	solid				---					
U006 Acetyl chloride	liquid	40	734		---	x				
U007 Acrylamide	solid				---					
U008 Acrylic acid	liquid	130	376		8,100			x		
U009 Acrylonitrile	liquid	32	898		9,900			x		
U010 Mitomycin C	solid				---			x		
U011 Amitrole	solid				---			x		
U012 Aniline	liquid	158	1139		14,980			x		
U013 Asbestos	solid	--nonflammable--								Particulates
U014 Auramine	solid				---					
U015 Azaserine	solid				---					
U016 Benz[c]acridine	solid				---			x		
U-17 Benzol chloride	liquid	153	1085		---	x				
U-18 Benz[a]anthracene	solid				---					
U019 Benzene	liquid	12	1040		17,460	x				
U020 Benzenesulfonyl chloride	liquid				(1,561)	x	x			MF
U021 Benzidine	solid				---			x		
U022 Benzo[a]pyrene	solid				---					
U023 Benzotrichloride	liquid				---	x				
U024 Bis(2-chloroethoxy)ethane	liquid				---	x				
U025 Bis(2-chloroethyl)ether	liquid	131			---	x				
U026 N,N-Bis(2-chloroethyl)-2-naphthylamine	solid				---			x		
U027 Bis(2)chloroisopropyl ether	liquid	185			---	x				
U028 Bis(2-ethylhexyl)phthalate	liquid				---	x				
U029 Bromomethane	liquid	practically nonflammable	999		3,188					HBr
U030 4-Bromophenyl phenylether	liquid				14,230					
U031 n-Butylalcohol	liquid	84	689		---					
U032 Calcium chromate	solid	--nonflammable--			---				x	
U033 Carbonyl fluoride	gas			can explode	---					HF
U034 Chloral	liquid				---	x				Cl2
U035 Chlorambucil	solid				---	x		x		
U036 Chlordane	liquid				---	x				Cl2
U037 Chlorobenzene	liquid	84	1184		12,000	x				
U038 Chlorobenzilate	liquid				---	x				
U039 p-Chloro-m-cresol	solid				---	x				
U040 Chlorodibromomethane					---					
U041 1-Chloro-23-epoxypropene (epichlorohydrin)	liquid	92	804		8,100	x				Cl2,HBr
U042 Chloroethyl vinyl ether	liquid	80			---	x				

(a) Combustion values in parentheses are estimates.

Contaminant	Physical State	Flash Point °F	Auto Ignition Temperature °C	Explosive Carries Own Oxygen	Heat of Combustion Btu/lb (kg cal/gm)(a)	Off Gas Problems: HCT/COCl2	SOx	NOx	Metal	Other
U043 Chloroethene (vinyl chloride)	liquid	-58	966		8,136	X				
U044 Chloroform	liquid	--nonflammable--			(89.2)	X				
U045 Chloromethane	gas		1170		---	X				
U046 Chloromethyl methyl ether	liquid				---	X				
U047 2-Chloronaphthalene	liquid	270	>1036		---	X				
U048 2-Chlorophenol	solid	147			---	X				
U049 4-Chloro-o-toluidine hydrochloride	solid				---	X		X		
U050 Chrysene	solid				---					
U051 Cresote	liquids	165	637				X	X		
U052 Cresols	liquid	178-202	1038-1110		14,720					
U053 Crotonaldehyde	liquid	55	450		13,700					
U054 Cresylic acid	liquid	178	1110		14,720					
U055 Cumene	liquid	111	797		17,710					
U056 Cyclohexane	liquid	-4	473		18,684					
U057 Cyclohexanone	liquid	111	788		15,430					
U058 Cyclophosphamide	solid				---	X		X		POx
U059 Paunomycin	solid				---			X		
U060 DDD	solid				---	X				
U061 DDT	solid	162			---	X				
U062 Diallate	liquid				---	X	X	X		
U063 Dibenz[a,h]anthracene	solid				---			X		
U064 Dibenza[a,i]pyrene	solid				---			X		
U065 Dibromochloromethane	liquid				---	X				Cl_2,HBr
U066 1,2-Dibromo-3-chloropropane	liquid	186	nonflammable		---	X				HBr
U067 1,2-Dibromo ethane (ethylene dibromide)	liquid	--nonflammable--			---					HBr
U068 Dibromomethane (methylene bromide)	liquid				---					HBr
U069 Di-n-butyl phthalate	liquid	315	757		13,200					
U070 1,2-Dichlorobenzene	liquid	151	1198		(671)	X				
U071 1,2-Dichlorobenzene	solid	150	1198		(671)	X				
U072 1,4-Dichlorobenzene	solid	150			(671)	X				
U073 3,3'-Dichlorobenzidine	solid				---	X		X		
U074 1,4-Dichloro-2-butene	liquid	80			---	X				
U075 Dichlorodifluoromethane	gas	--nonflammable--			---	X				Cl_2,HF
U076 1,1-Dichloroethane	liquid	22			(267)	X				
U077 1,2-Dichloroethane (ethylene dichloride)	liquid	56	775		3400	X				
U078 1,1-Dichloroethylene	liquid	0	1058		---	X				
U079 1,2-trans-Dichloroethylene	liquid	43			---	X				
U080 Dichloromethane (methylene chloride)	liquid	practically nonflammable	23	1139	---	X				Cl_2
U081 2,4-Dichlorophenol	solid	237			---	X				
U082 2,6-Dichlorophenol	solid				---	X				
U083 1,2-Dichloropropane	liquid	60	1035		---	X				

(a) Combustion values in parentheses are estimates.

Contaminant	Physical State	Flash Point °F	Auto Ignition Temperature °C	Explosive Carries Own Oxygen	Heat of Combustion Btu/lb (kg cal/gm)(a)	HCT/COCl2	SOx	NOx	Metal	Other
U084 1,3-Dichloropropene	liquid	95			---	X				
U085 Diepoxy butane	liquid				---					
U086 1,2-Diethylhydrazine					---			X		
U087 O,O-Diethyl-S-methyl phosphorodithioate	liquid				---					POx
U088 Diethyl phthalate	liquid	325			---		X			
U089 Diethylstilbesterol	solid				---					
U090 Dihydrosafrole					---					
U091 3,3'-Dimethoxybenzidine	solid				16,800			X		
U092 Dimethylamine	gas	20	752		---			X		
U093 p-Dimethylaminoazobene	solid				---			X		
U094 7,12-Dimethylbenz[a]anthracene	solid				---					
U095 3,3'-Dimethylbenzidine	solid				---			X		
U096 α,α-Dimethylbenzylhydroperoxide	liquid	34	452		14,170	X				
U097 Dimethylcarbamoyl chloride	liquid	175		May explode	---			X		
U098 1,1-Dimethylhydrazine	liquid	5	480		---			X		
U099 1,2-Dimethylhydrazine	liquid				---			X		
U100 Dimethylnitrosamine	liquid				---					
U101 2,4-Dimethylphenol (xylenol)	solid				---					
U102 Dimethylphthalate	liquid	295	1032		---		X			
U103 Dimethylsulfate	liquid	182	370		---					
U104 2,4-Dinitrophenol	solid	404			---			X		
U105 2,4-Dinitrotoluene	solid			can explode	(648)			X		
U106 2,6-Dinitrotoluene	solid			can explode	(853)			X		
U107 Di-n-octylphthalate	liquid	425			15,400					
U108 1,4-Dioxane	liquid	54	356		11,590					
U109 1,2-Diphenylhydrazine	solid				---			X		
U110 Dipropylamine	liquid	63			---			X		
U111 Di-n-propylnitrosoamine					---			X		
U112 Ethylacetate	liquid	24	800		10,110					
U113 Ethylacrylate	liquid	60	721		11,880			X		
U114 Ethylenebisdithiocarbamate					---		X			
U115 Ethylene oxide	gas	<0	804		11,480		X			
U116 Ethylene thiourea	solid				---			X		
U117 Ethylether	liquid	-49	320		14,550					
U118 Ethylmethacrylate	liquid	68			---					
U119 Ethylmethanesulfonate					---					
U120 Fluoranthene	solid				---					
U121 Fluorotrichloromethane	liquid	--nonflammable--			(1341)	X				HF,Cl2
U122 Formaldehyde	gas		806		---					
U123 Formic acid	liquid	156	1114		7,045					
U124 Furan	liquid	<32			---					
U125 Furfural	liquid	140	600		10,490					
U126 Glycidylaldehyde	liquid				---					
U127 Hexachlorobenzene	solid				(509)	X				Cl2
U128 Hexachlorobutadiene	liquid		1130		---	X				Cl2

(a) Combustion values in parentheses are estimates.

Contaminant	Physical State	Flash Point °F	Auto Ignition Temperature °C	Explosive Carries Own Oxygen	Heat of Combustion Btu/lb (kg cal/gm)(a)	HCT/COCl2	SOx	NOx	Metal	Other
U129 Hexachlorocyclohexane	solid				---	X				
U130 Hexachlorocyclopentadiene	liquid				---	X				Cl2
U131 Hexachloroethane	solid				(110)	X				Cl2
U132 Hexachlorophene	solid				---	X				Cl2
U133 Hydrazine	liquid	100	518		8,345			X		NH3
U134 Hydrofluoric acid	liquid	--nonflammable--			---			X		HF
U135 Hydrogen sulfide	gas	--nonflammable--	500		---					
U136 Hydroxydimethylarsine oxide	solid				---					
U137 Indeno(1,2,3-cd)pyrene	liquid				---				X	
U138 Iodomethane	solid				---					I2
U139 Irondextran	solid				---				X	
U140 Isobutylalcohol	liquid	82	800		14,220					
U141 Isosafrole	liquid				(1234)					
U142 Kepone	solid	--nonflammable--			---					
U143 Lasiocarpine	solid				---	X				
U144 Lead acetate	solid	--nonflammable--			---			X	X	
U145 Lead phosphate	solid	--nonflammable--			---				X	POx
U146 Lead Subacetate	solid	--nonflammable--			---				X	
U147 Maleic anhydride	solid	215	890		3,340					
U148 Maleic hydrazide	solid				---			X		
U149 Malononitrile	solid				---			X		
U150 Melphalan	solid				---			X		
U151 Mercury	liquid	--nonflammable--			---	X			X	
U152 Methacrylonitrile	liquid	55			---			X		
U153 Methanethiol	gas	0			---		X			
U154 Methanol	liquid	52	725		8,419		X			
U155 Methapyrilene	liquid				---			X		
U156 Methyl chlorocarbonate	liquid				---	X				
U157 3-Methylcholanthrene	solid				---					
U158 4,4'-Methylene-bis-(2-chloroaniline)	solid				---	X		X		
U159 Methylethylketone	liquid	21	960		13,480					
U160 Methylethylketoneperoxide	liquid			may explode	---					
U161 Methylisobutylketone	liquid	73	860		10,400					
U162 Methylmethacrylate	liquid	50	790		11,400			X		
U163 N-Methyl-N'-nitro-N-nitrosoguanidine	solid				---					
U164 Methylthiouracil	solid				---		X	X		
U165 Naphthalene	solid	174	979		16,770					
U166 1,4-Naphthoquinone	solid				---					
U167 1-Naphthylamine	solid	315			(1264)			X		
U168 2-Naphthylamine	solid				(1264)			X		
U169 Nitrobenzene	liquid	190	900		10,420			X		
U170 4-Nitrophenol	solid				(689)			X		
U171 Nitropropane	liquid	120	789		(478)			X		
U172 N-Nitrosodi-n-butylamine	liquids				---			X		

(a) Combustion values in parentheses are estimates.

Contaminant	Physical State	Flash Point °F	Auto Ignition Temperature °C	Explosive Carries Own Oxygen	Heat of Combustion Btu/lb (kg cal/gm)(a)	HCT/CO/Cl2	SOx	NOx	Metal	Other
N-Nitrosodiethanolamine					---			x		
N-Nitrosodiethylamine					---			x		
N-Nitrosodi-n-propylamine					---			x		
N-Nitroso-n-ethylurea	solid				---			x		
N-Nitroso-n-methylurea	solid				---			x		
N-Nitroso-n-methylurethane	solid				---			x		
N-Nitrosopiperidine					---			x		
N-nitrosopyrrolidine					---			x		
5-Nitro-0-toluidine	solid	315			---			x		
Paraldehyde	liquid	96	460		---					
Pentachlorobenzene					---	x				Cl2
Pentachloroethane	liquid				---	x				Cl2
Pentachloronitrobenzene	solid				---	x		x		Cl2
1,3-Pentadiene					---					
Phenacetin	solid				---			x		
Phenol	solid	179	1319		13,400					
Phosphorous Sulfide	solid		500-554		---		x			POx
Phthalic anhydride	solid	305	1058		9,473					
2-Picoline	liquid	102	1000		---			x		
Pronamide					---	x		x		
1,3-Propane sulfone	liquid		604		(558)		x			
n-Propylamine	liquid	-35	604		14,390			x		
Pyridine	liquid	68	900		(657)			x		
Quinone	solid	559			---					
Reserpine	solid				---			x		
Resorcinol	solid	261	608		---					
Saccharin	solid		1126		---		x			
Safrole	liquid				(1244)					
Seleniousacid	solid	--nonflammable--			---					Se cmpds
Selenium sulfide	solid	--nonflammable--			---		x			Se cmpds
Streptozotocin		311			---			x		
1,2,4,5-tetrachlorobenzene	solid				---	x				Cl2
1,1,1,2-tetrachloroethane	liquid				---	x				Cl2
1,1,2,2-tetrachloroethane	liquid				---	x				Cl2
Tetrachloroethene	liquid	--nonflammable--			(37.3)	x				Cl2
Tetrachloromethane (carbon tetrachloride)	liquid				---	x				Cl2
2,3,4,6-Tetrachlorophenol	solid				---	x				Cl2
Tetrahydrofuran	liquid	6	610		14,990					
Thallium acetate	solid	--nonflammable--			---					
Thallium carbonate	solid	--nonflammable--			---				x	
Thallium chloride	solid	--nonflammable--			---	x			x	
Thallium nitrate	solid	--nonflammable--			---				x	
Thioacetamide	solid				---		x	x		
Thiouren	solid				---		x	x		
Toluene	liquid	40	896		17,430					
Toluene diamine	solid				---			x		

(a) Combustion values in parentheses are estimates.

Contaminant	Physical State	Flash Point °F	Auto Ignition Temperature °C	Explosive Carries Own Oxygen	Heat of Combustion Btu/lb (kg cal/gm)(a)	Off Gas Problems HCl/COCl2	SOx	NOx	Metal	Other
U222 o-Toluidine hydrochloride	solid	270	>300		---	X		X		
U223 Toluene diisocyanate	liquid				10,300			X		
U224 Toxaphene	solid	--nonflammable--			---	X				
U225 Tribromomethane	liquid	--nonflammable--			---	X				HBr
U226 1,1,1-Trichloroethane	liquid	109.4	932		4,700	X				
U227 1,1,2-Trichloroethane	liquid	90	770		4,700	X				
U228 Trichloroethene	liquid	--nonflammable--			---	X				Cl2
U229 Trichlorofluoromethane	liquid	--nonflammable--			---	X				HF,Cl2
U230 2,4,5-Trichlorophenol	solid	--nonflammable--			---	X				Cl2
U231 2,4,6-Trichlorophenol	solid	--nonflammable--			---	X				
U232 2,4,5-T	solid	--nonflammable--			---	X				
U233 2,4,5-Trichlorophenoxy-propionic acid	solid	--nonflammable--			---	X				
U234 Trinitrobenzene	solid			Can explode	(664)			X		
U235 Tris(2,3-dibromopropyl) phosphate	solid				---					POx,HBr
U236 Trypan blue	solid				---		X	X		
U237 Uracil mustard	solid				(397)	X		X		
U238 Urethane	solid							X		
U239 Xylene	liquid	81-90	869-986		17,554					

(a) Combustion values in parentheses are estimates.

Index

Academic public, 173
Acetic acid, 75
Acetylene, 203
Acoustic holography, 249–254
Acoustic reflection, 247–249, 250
Acquisition, 189
Acrolein, 367
Acrylonitrile, 266
Activated sludge, 337–338, 359
Additive effects, 56
Adsorption, 152, 259, 342–345
Adsorption isotherm, 343
Africa, 40
Agricultural Runoff Management (ARM)
  model, 257
Air, 8, 27, 40, 50, 59, 120
Air pollution, 9, 141, 239
Air quality regions, 13
Air quality standards, 30, 44
Alabama, 113
Alaska, 114, 120
Alkyl lead, 3
Alkyl mercury, 4, 53
Allergens, 69
Alpha particles, 66
Alum, 328–331
American Insurance Association, 182
American Petroleum Institute, 119
Ammonia, 334
Anaerobic digestion, 341
Aniline, 421
Antagonistic effects, 56
Aquatic toxicity, 63, 81, 86, 90, 95
Aquifer, 9, 29, 151, 165, 265, 272
Argentina, 37
Arkansas, 4
Arrhenius expression, 373
ARRRG model, 155
Arsenic, 2, 53, 328
Asbestos, 62, 284
Ascaris, 65
Ash, 10, 103
Asia, 40
Asphalt, 412
Association of the German Chemical Industry
  (UCI), 208

ASTM, 73–77
Atmospheric emissions, 9, 30
Atomic Energy Act of 1954, 2, 10, 65
Atomic Energy Commission, 78
Australia, 41
Austria, 267
Autoignition temperature, 68
Autopolymerization, 64

Baden-Wurtemberg, 164
Baghouse dust, 75
BA model, 99
Bangladesh, 41
Bans, 20, 21, 24, 67, 101, 388
Barge disposal, 420
Barrel reconditioners, 286–289
Basalt, 137
Batteries, 114, 120
Bedded salt, 21
Belgium, 40, 88–89
Beryllium, 89
Beta particles, 66
Bioassays, 58, 71, 87
Biochemical Oxygen Demand (BOD), 359
Bioconcentration, 4, 47, 62, 70, 83, 101,
  261
Biodegradation, 267
Biological treatment, 309, 335–342, 359
Blood elements, 65
Bond energies, 367–369
Bonding, 181–182
Boussinesq approximation model, 155
Brazil, 37
Brines, 8, 103, 209–211
Britain, 38, 51, 53, 103, 109, 119, 127–129,
  164, 208, 232, 305
Brokers, 204
Bromacil, 21
Bromates, 64
By-products, 10

Cadmium, 53, 56, 70, 89, 101, 324, 328,
  348, 412, 419
Calcium carbonate, 322
Calcium fluoride, 12

California, 20, 50, 53, 57–58, 75, 77, 81–86, 91, 100, 109, 118, 120, 162, 262, 292, 295, 305, 388, Appendix C
California Hazardous Wastes Control Act of 1973, 53, 57
California Hazardous Waste Working Group, 71
Calumet Container–Steel Container Corporation, 286
Camphor, 67
Canada, 37, 286
Caps, 268, 289
Carbamates, 21
Carbide process, 203
Carbon adsorption, 211, 342–345, 359
Carbon monoxide, 369
Carbon tetrachloride, 70, 369
Carcinogen, 1, 21, 71–72, 86, 418, 419
Carcinogenicity, 46, 71–73, 81, 86
Casualty radius, 63
Catalytic incineration, 381
Cation Exchange Capacity (CEC), 139, 163
Caves, 21
Cellular data structures, 149
Cementation, 411
Cement kiln dust, 103
Cement kilns, 21
Central America, 37
Centralized facility, 136
Centrifugation, 319, 357
Cesium, 334
Chelating agents, 75
Chemical Control Corporation, 199
Chemical Manufacturer's Association, 34, 230
Chemical Migration Risk Assessment (CMRA) method, 258–261
Chemical Recovery Association, 207
Chemical treatment, 319–335, 359
Chemical Wastes Act of 1977, 38
China, 41
Chlorates, 64
Chlorinated aliphatics, 21, 369
Chlorinated aromatics, 21, 369
Chlorine, 267, 333, 369
Chlorophenol, 267
Chromium, 53, 211, 324, 332–333, 348
Citrate, 75
Clarifiers, 312, 314, 359
Clean Air Act, 23, 30, 44, 45
Clean Water Act, 23, 44
Clinch River, 3

Closure, 7, 23, 180, 202, 388
Closure bonds, 221
Coal mining, 11
Coal slurry pipeline effluent, 21
Collection drains, 396
Colorado, 418
Columbia, 37
Combustible wastes, 211
Combustion, 8, 67, 234, 308
Combustion rate, 372–375, Appendix L
Committee On Interstate and Foreign Commerce, 230
Compensation, 34
Complexation, 75
Comprehensive Environmental Response Compensation and Liability Act of 1980 (CERCLA), 20–29, 44, 183, 234, 266
Connecticut, 235
Consulting, 284
Contained gases, 10
Contaminant transport, 154, 255–260
Contingency plans, 17, 25, 388
Control of Pollution Act, 38, 51
Convention on Prevention of Marine Pollution, 419
Copper, 328
Corps of Engineers, 12
Corrective Action, 21
Corrosivity, 47, 49, 52, 68–69, 79, 86
Council on Environmental Quality, 160
Council on Wage and Price Stability, 93
Cover, 158
Criteria:
   site, 135–147, 157–159, 200, 262, Appendix F, Appendix H
   waste, 19, 43, 57–78, 86–88, Appendix G
Culm banks, 12
Cyanide, 53, 79, 209, 325, 332

DACRIN model, 156
Dangerous wastes, 87, 91, 208
Davis FE model, 155
DDT, 4, 71
Decision model, 61
Definition, 9, 10, 15, 34, 43–105
Degradation, 151, 260, 266
Degree of hazard, 10, 64
Delaney clause, 73
Delaware, 295
Delisting, 21, 81, 194
Denmark, 38, 53, 208, 305

Department of Agriculture (USDA), 54
Department of Commerce (DOC), 233, 287
Department of Defense (DOD), 25
Department of Energy (DOE), 54, 152, 172, 239
Department of Health Services (DHS), 25
Department of Justice (DOJ), 20, 233
Department of State, 419
Department of Transportation (DOT), 30, 69, 82–84, 86, 233, 289, 292, 295
Department of Treasury, 23, 27, 28, 183
Deposition, 260
Dermal toxicity, 4, 62, 81, 86
Dermatitis, 69
Design standards, 5, 19, 33, 158
Desorption, 260
Detonation, 63, 67
Detoxification, 70
Detroit, 133
Dibenzofurans, 21
Dimethyl hydrazine, 21
Dioxin, 3, 21, 102, 175
Discovery, 26, 234
Dispersion, 155, 164, 179, 260
Disposal, 387–435
Dissemination strategy, 170
Distillation, 188
District of Columbia, 32
DNA, 71
Domestic sewage, 10, 21
Dose, 151–156
Dose models, 155–156
Dose-response, 73, 258
Double liners, 21, 395
Drilling muds, 52, 103
Drinking water contamination, 2
Drinking water standards, 29, 59, 80, 262
Drum reconditioners, 286–289
Dry subzones, 429
Dye and pigment wastes, 21, 71

Earthquakes, 138
Echardt survey, 231
Economically concerned public, 173
Economic dislocation, 4
Edema, 70
Electrical and electronics, 113
Electrical resistivity (ER), 245–247
Electromagnetic induction (EMI), 254
Electroplating and metal finishing, 119
Elimination approach to siting, 148

Emergency response, 24–27, 34, 302
Emergency Response Team (ERT), 25
Eminent domain, 175
Emulsified asphalt, 268
Emulsifiers, 266
Encapsulation, 387, 411–414
Endangerment assessment, 239
Energy recovery, 8, 179, 307
Engineered barriers, 6, 133, 157, 163, 405, 408, 410
Environmental Contamination Act, 37
Environmental Impact Statement (EIS), 159–162, 165
Environmental Protection Act of 1973, 38
Environmental Protection Agency, 2, 14, 15, 23–25, 34, 44, 49, 52, 57, 59, 66, 78–83, 91–97, 113, 118–120, 133, 162, 171, 174, 179, 182, 189, 200, 203, 227–231, 258, 264, 267, 268, 269, 284, 286, 289, 291, 292, 295, 298, 302, 307, 325, 382, 390, 417, 418, 419, 420
Epidemiology, 23
Erosion, 141
Erythema, 70
Ethyl parathion, 4
Etiological wastes, 10, 47
Europe, 38–40, 50, 53, 89, 101, 127, 179, 207–208, 233, 286, 288, 303
European Community (EC), 38, 50, 127, 207, 303
Evaporation, 346–348
Excavation, 263–266
Excreta, 65
Exemption quantities, 21, 102
Exhumation, 7, 263
EXPLORE model, 259
Explosiveness, 47, 59, 63–64, 234
Explosives, 3, 87
Export, 22
Exposure assessment, 21, 156
Exposure reduction, 272
Extraction procedures, 73–81, 328
Extraction of wastes, 263–266
Extremely hazardous wastes, 54, 87, 91, 295

Failure prediction model, 152
Faults, 136, 163
Feasibility study (FS), 262
Federal Emergency Management Agency (FEMA), 25

Federal Insecticide, Fungacide and Rodenticide Act (FIFRA), 30
Federal Water Pollution Control Act of 1972, 10, 23, 24, 29, 30, 43, 45, 81, 87, 165, 173
Fees, 20, 34
Ferric chloride, 318, 324
Ferric hydroxide, 324, 420
Field capacity, 151
Financial responsibility, 5, 6, 17–22, 28, 180–183, 202, 221
Finite difference method, 155
Finite element method, 155, 259
Fire point, 67
Fixation, 74, 207, 267, 389, 411–414
Flammability, 3, 11, 47, 52, 67–68, 79–88
Flannery decision, 44
Flash point, 57, 68, 78–79, 94
Flooding, 141
Floodplains, 138, 140, 158
Florida, 113, 114
Flotation, 316
Flow velocity, 151
Flue gas desulfurization, 103, 411
Fluidized bed incinerator, 381
Fluorine, 12
Fly ash, 103, 312
Fomites, 65
Food chain poisoning, 4
Food and Drug Administration (FDA), 4, 70
FOOD model, 155
Food Protection Committee, 46
Foreign regulation, 37–41, 303–305
Forests, 144
Formaldehyde, 267, 413
France, 46, 50, 101, 127, 179, 208
FRANCO routine, 260–261
Freundlich isotherm, 343
Fuel, 8, 21
Fuel oil, 266

Gamma radiation, 66
Garbage, 10
Gasoline, 264
General Accounting Office, 230
Geologic barriers, 136, 405
Geophysical surveys, 239–254
Germany, 40, 50, 53, 122, 164, 207, 233, 254, 286, 303
GETOUT model, 152
Glacial outwash, 137

Glass fixation, 413
Gob piles, 11
Go, no-go criteria, 144, 148, 149
Gradient, 151
Granular media filter, 316
Greater London Council, 53, 91, 97
Green Party, 179
Ground penetrating radar, 242–245, 250
Ground water, 8, 21, 59, 86, 99, 138–140, 147–156, 159, 162–165, 234, 257, 286, 308, 396, 401, 423, 428
Grout, 263, 267, 268, 272

Hanford Atomic Energy Reservation, 239–251, 389
Hazard assessments, 25
Hazardous air pollutants, 30, 44, 45
Hazardous Materials Transportation Act, 16, 30, 289
Hazardous substances, 29, 30, 44, 45, 78, 81, 143
Hazardous Waste Management Act of 1973, 14
Hazard Ranking System (HRS), 234
Heat of combustion, 367–369
Helminths, 65
Herfa-Neurode, 179, 207
History of regulation, 11–15
Hong Kong, 41
How clean is clean, 5, 262
HSSWDS, 396, 405
Human intrusion, 410
Hurricane, 141
Hydraulic gradient, 151
Hydrochloric acid, 322, 369
Hydrofracing, 269
Hydrogen, 12
Hydrogen peroxide, 333
Hydrogen sulfide, 3, 12, 79
Hydrologic flow, 152
Hydrometallurgy, 265

Iberville, 3
Idaho, 111, 120, 122, 175
Igneous rock, 137
Ignition, 67
Illinois, 120, 134, 163, 302
Immediate removal, 26
Imminent hazard, 20, 28, 45, 228
Immobilization, 7, 56, 262, 267–272
Impoundments, 21, 29, 122, 135, 188

Incentives, 14, 20, 32, 204
Incineration, 8, 10, 81, 126, 190, 200, 207,
    221, 266, 365–385, 387
Incineration chemistry, 366–375
Incinerators, 21, 30, 136, 188, 202, 211,
    375–383
India, 40
Indiana, 120, 286
Industry:
    development, 188–199
    European, 207–208
    finances, 196–199
    structural, 192–199
Infectiousness, 64
Infiltration, 139, 151, 268, 389, 392, 395
Infrared imaging, 247
Inhalation toxicity, 59, 81–86
Injection tubing, 425
Injection well, 8, 21, 29, 126, 136, 192, 199,
    208, 268, 418, 428–431, 433
Ink wastes, 75
Inorganic chemicals, 115, 120
Inorganic wastes, 8, 21, 213
In-place degradation, 7, 259, 266–267, 272
In-situ degradation, 7, 259, 266–267, 272
In-situ vitrification, 269
Inspection, 18
Institution of Chemical Engineers, 89
Interim status, 17, 21
International Committee On Radiation
    Protection Task Group On Lung Dynamics,
    156
Inventory, 6, 109
Invert wells, 265
Ion exchange, 334, 359
Ireland, 40, 208, 305
Iron, 328
Irritation, 47, 69, 86
Isolation, 172, 269
Italy, 40

James River, 4, 165, 171, 227, 266, 267
Japan, 41, 53, 268, 288, 412, 422

Kentucky, 279
Kepone, 4, 165, 171, 175, 179, 207, 227,
    266, 267
Kerfing, 269
Kill radius, 63, 67
Kin-Buc landfill, 3
Kommunekemi, 208

Kriging, 257
KRONIC model, 155

Labeling, 16, 30, 389
Laboratory analysis, 284
Lagoons, *see* Impoundments
Landfarm, 135, 266, 341–342
Landfill, 8, 20, 24, 40, 63, 68, 77, 80, 86,
    91, 126, 133, 140, 141, 151, 165, 173,
    176, 188, 190, 202, 207, 208, 211, 233,
    266, 387–414
Landslides, 138
Landspreading, 126, 192, 202, 341
LC50, 63, 81, 86, 90, 91, 260
LD50, 58, 63, 82–87
Leachate, 20, 21, 59, 73–78, 83, 141, 151,
    163–165, 265, 390, 401
Leaching, 67, 262, 401
Lead, 53, 70, 325, 412
Leather tanning and finishing, 113
Liability, 5, 6, 8, 12, 27, 180, 202, 272, 279,
    302, 365
Lime, 203, 215, 320–322, 359, 412–414
Limestone, 137, 141, 320
Liner compatibility, 405
Liner installation, 407–408
Liners, 158, 396–401, 405–408
Linuron, 21
Liquid injection incinerators, 375–378, 383
Liquids-solids separation, 310–319, 357
Listing, 15, 52–57, 88, Appendix B,
    Appendix C
Lithium batteries, 21
Long term storage, 9
Louisiana, 3
Love Canal, 1, 7, 20, 22, 127, 165, 199, 227,
    233, 249, 254, 255, 268, 272, 387
Luxembourg, 40, 305

Magnetometry, 240–242, 250
Maine, 295
Malignancy, 66
Manganese, 329
Manifest, 7, 16, 21, 33, 289–305
Manufacturing Chemists Association, 91, 100
Marine Protection, Research, and Sanctuary
    Act, 29, 417, 418
Marches, 144
Martin v. *Reynolds Metals*, 12
Maryland, 20, 176, 295
Massachusetts, 111, 229, 286, 295

Mathematical models, 149–156, 257, 389, 396
Maximum allowable toxicant concentration, 260
Maximum Permissible Concentration (MPC), 67
Meiosis, 71
Mercury, 4, 53, 70, 89, 101, 329, 417, 422
Metal detection, 240, 250
Metal finishing wastes, 309, 359, 411
Metamorphic rock, 137
Methane, 366, 394
Methyl mercury, 4
Methyl parathion, 4
Mexico, 37
Mexico City, 37
Michigan, 33, 120, 175
Microorganisms, 65
Middle East, 40
Mines, 21, 179
Minimata disease, 422
Mining, 10, 21, 103, 157
Minneapolis-St. Paul, 118
Minnesota, 2, 3, 84, 86, 87, 104–110, 113–119, 126, 133, 302, Appendix D
Minnesota Association of Commerce and Industry, 110, 119
Missouri, 3
Mitosis, 71
Mixed Liquor Suspended Solids (MLSS), 337, 359
MMT model, 155
Molten salt incineration, 382
Monitoring, 17, 20, 387, 408
Monochlorobenzene, 267
Montmorillonite clay, 140
Multiple-hearth furnace, 380
Mutagenicity, 46, 71–73
Mutations, 66

Naphthalene, 67
National Academy of Sciences (NAS), 46, 64, 68
National Advisory Committee on Oceans and Atmosphere, 417
National Association of Corrosion Engineers (NACE), 69
National Association of Counties, 14
National Barrel and Drum Association (NABADA), 286–288
National Conference of State Legislatures, 34
National Contingency Plan (NCP), 20, 24–27

National Disposal Sites (NDS), 2, 13, 166–169
National Environmental Policy Act (NEPA), 160, 165
National Fire Protection Association (NFPA), 64, 68, 86
National Institute of Occupational Safety and Health (NIOSH), 28, 81, 86
National Pollution Discharge Elimination System (NPDES), 12, 112
National Priority List (NPL), 26, 231, 234, Appendix I
National Response Team (NRT), 25
National Solid Waste Management Association (NSWMA), 49, 175, 176, 182
National Strike Force (NSF), 25
Necrosis, 70
Negligence, 12
Netherlands, 38, 51, 53, 91, 101, 208, 233, 305
Neutralization, 56, 188, 210, 267, 309, 320–323, 359
Neutrons, 66
New Hampshire, 255
New Jersey, 3, 120, 176, 188, 295, 433
New Mexico, 4
New York, 1, 120, 188, 202, 228, 278, 284, 295, 417, 433
Niagara Falls, 1
Nickel, 325–331, 360, 412
Nitrogen oxides, 370
Nitrates, 64
Noise, 50
North Carolina, 266, 271
Norway, 38, 207
Nuclear Regulatory Commission (NRC), 66
Nuclear Waste Repositories, 152, 162, 171
Nuisance, 12

Occupational Safety and Health Act of 1970, 3, 30
Occupational Safety and Health Administration (OSHA), 28, 31, 67, 80
Ocean disposal, 8, 29, 199, 209, 417–420, 433
Ocean incineration, 382
Octanol-water partition coefficient, 71
Oder, 12, 50
Ohio, 109–110, 119–120, 194
Ohio Manufacturers' Association, 119

Ohio Petroleum Council, 119
On-Scene Commander (OSC), 25
On-site management, 194, 215
Ontario, 37, 303
Oral toxicity, 62, 81, 92
Ordnance, 120
Ore benefiation, 265
Oregon, 12, 111–120, 151, 175, 181
Organic chemicals, pesticides, and explosives,
    119–122
Organic still bottoms, 75
Organic wastes, 8, 75, 328, 337, 365, 395
Organized public, 173
Organobromines, 21
Organohalogens, 89, 369
Organometallic compounds, 372
Organophosphates, 4, 53
Outcrops, 136
Oxidation, 210, 309, 331–335, 359
Oxidation ponds, 339
Oxidizers, 78, 84, 86
Ozone, 267, 333

Paints and allied products, 119, 231
Paint sludges, 4, 21, 119, 194
Parasites, 65
Particulate transport, 260
Pathogenic wastes, 10, 65
PATHS model, 154
PBBs, 267
PCBs, 30, 53, 70, 133–134, 136, 175, 224,
    266, 267, 272, 284, Appendix G
Peerless cement, 133
Pennsylvania, 129, 173, 286, 295
Perchlorates, 64
Percolation, 272, 396
Performance standards, 5, 156, 390–393
Permanganates, 64
Permeability, 140, 151, 158, 162–163, 257,
    269, 404
Permits, 12, 17, 21, 28, 33, 38, 136, 157,
    176, 183, 199, 200, 271, 292, 302, 419
Peroxides, 64
Perpetual care, 181
Pesticide containers, 21
Pesticides, 21, 32, 38, 56, 67, 89, 258, 288
Petroleum refining, 114, 119, 342
Pharmaceuticals, 114
Pharmaceutical wastes, 75
Phenol, 338, 341, 361
Phenolphthalein alkalinity, 322

Philippines, 41
PHL model, 94
Phosgene, 369
Phosphine, 371
Phosphoric acid, 323
Phosphorous oxides, 371
Photodegradation, 267
Phytotoxicity, 49, 62
Picatinny arsenal scale, 63
Pickle liquor sludges, 194, 309
Planned removal, 26
Plasticizers, 67
Plastics, 114, 231
Plating wastes, 81, 119, 332
Poisonous waste act, 38, 51, 53, 305
Poland, 128
Polygon data structure, 149
Polymer fixation, 413
Porosity, 151, 257
Post-closure, 27, 180
Post-closure insurance, 27
Post-closure Liability Trust Fund (PCLTF),
    22–24, 27, 100, 153, 183
Potency, 71
Pozzolans, 412
Precipitation/coagulation/flocculation, 323–331
Preliminary assessment, 234
Pretreatment standards, 29
Primary metals smelting and refining, 113, 119
Prioritization, 5, 27, 91–103, 176
Priority pollutants, 44
Profiles, 113, 167
Property damage, 3, 12, 47
Property value loss, 169, 173
Protection casing, 425
Provisional code of practice for disposal of
    wastes, 89
Public acceptance, 6, 165–180, 265
Public Affairs Assist Team (PAAT), 25
Public health, 12, 14, 20, 21, 43, 44, 47, 48,
    50, 81, 89, 91, 101, 162, 227, 228, 234,
    261, 387, 390, 423
Public Information Assist Team (PIAT), 25
Publicly Owned Treatment Works (POWT),
    21, 29, 75, 102
Puerto Rico, 32
Pure compound approach, 54–56, 73, 81
Pyrolysis, 382

Questionnaires, 108–114, 166, 167, Appendix
    D, Appendix E

Radioactive waste, 2, 6, 32, 47, 334, 411, 419, 423
Radioactivity, 65–66
Radon, 268
Random walk method, 155
Ranking, 97, 149, 234
Reactivity, 3, 47, 52, 63, 78
Recharge, 138
Record keeping, 16, 19, 289, 302
Recycle, 9, 126, 203, 284
Refining wastes, 21
Refractory substances, 336
Refuse, 9, 64, 80
Refuse Act, 12
Regional Processing Facility (RPF), 167–170
Regional Response Team (RRT), 25
Regulation, 11–41, 289–305
Regulatory Analysis Review Group (RARG), 93–94, 102
Relative value criteria, 144–149
Release scenario, 152
Relocation, 25
Remedial action, 24–27, 40, 172, 227–275, Appendix K
Remedial investigation (RI), 239
Report to Congress, 2, 14, 19, 27, 32, 49, 54, 59, 78, 96, 109, 113, 119, 135, 137, 143, 149, 166, 171, 181, 187, 208, 215, 227, 287, 289, Appendix A
Reporting, 17, 32, 33
Representative public, 173
Resource Conservation and Recovery Act (RCRA), 9, 15–22, 23, 26, 43, 48, 59, 94, 100, 102, 118, 122, 127, 133, 158, 175, 182, 199, 200, 204, 224, 227, 266, 279, 285, 289, 309, 387, Appendix B
Resource damage, 23
Resource recovery, 41, 189, 192, 202
Resource Recovery Act of 1970, 2, 10, 13
Resource restoration, 23
Restoration goals, 262
Resuspension, 262
Retrievability, 8, 10
Reverse osmosis (RO), 348–352, 361
Rhode Island, 295
Right-to-know, 34, 36
Risk, 21, 23, 38, 50, 136, 138, 141, 142, 150, 158, 169, 174, 176, 181, 183, 202, 234, 258, 266, 303, 410, 422
Risk assessment, 78, 101, 156, 183, 258, 260–262, 410
Rivers and Harbors Act of 1899, 12, 29

Rocky flats, 265
Rocky Mountain arsenal, 418
Rotary kiln, 265, 378–380, 383
Rubber, 114
Rubbish, 11

Safe Drinking Water Act, 29, 423
Safety assessment, 152
Safety factor, 58, 81
Safety and health standards, 32
St. Louis Regional Commerce and Growth Association, 204
SAKAB, 38, 134, 207
Salt disposal, 21, 431
Salt domes, 21
Sampling, 18
Sanitary Landfill Model (SLM-1), 151
Scientific Support Coordinator (SSC), 25
Scouring, 259
Scraping soil, 263
Screening wastes, 312
Scrubber sludge, 103
Sealants, 268, 272
Secretions, 65
Sedimentary rock, 137
Sedimentation, 211, 312–316, 359
Sediment transport, 259
Seismic risk, 138
Selenium, 56, 402
Self-oxidation, 64
SEMA, 37
Sensitization, 47, 69
SERATRA model, 259
Service radius, 216
Shakopee, Minnesota, 3
Sheet piling, 268
Shoreline management, 137, 143
Silicon tetrafluoride, 12
Selrisem, 199, 228
Silver, 208
Sinkholes, 138
Sinking fund, 181
Site assessment, 7, 234–261
Site characterization, 7
Site evaluation, 135, 144, 149–156
Site mitigation, 262–278
Site qualification, 135, 144, 156–162
Site remediation, 7, 239, 262–278
Site restoration, 7, 34
Site screening, 26, 135, 147–149, 157
Siting, 6, 127, 133–185, 189, 198, 202, 258, 387, 427, Appendix F, Appendix H

Siting board, 176
Size reduction, 74
Sludge, 9, 21, 30, 44, 64, 67, 211, 284
Small quantity generators, 20
Soda ash, 320–323
Sodium acetate, 75
Sodium bicarbonate, 320–323
Sodium hydroxide, 320–323
Soil, 40, 136–139, 151, 284
Soil Conservation Service, 140
Soil Protection Act, 40
Soil sorption capacity, 139
Soil washing, 264
Solidification, 411–414
Solid Waste Disposal Act of 1965, 12, 14, 15, 44
Solid wastes, 6, 9, 11, 12, 15, 20, 43, 48, 103, 136, 194, 197, 209
Solvent extraction, 352–354
Solvent recovery, 188, 307
Solvents, 4, 6, 21, 56, 67, 71, 89, 188, 189, 204, 207, 264, 284
Source term characterization, 239–254
South Africa, 40
South America, 37
South Dakota, 32
Spain, 40
Special machinery, 114
Special wastes, 22, 37, 50, 65, 102
Stabilization, 267–272, 411–414
Stabilization ponds, 339–341
Stagnant subzones, 429
Standard Industrial Classification (SIC), 111, 114
Standard Metropolitan Statistical Area (SMSA), 143
State programs, 17, 32–35
Steam distillation, 209
Steam stripping, 354–356, 361
Sterility, 66
Storage, 8, 10, 13, 16, 23, 27, 30, 33, 48, 68, 91, 144, 190, 207, 214, 284
Strategic petroleum reserve, 431
Structural integrity test, 74
Styrene, 413
SUBDOSA model, 156
Subzone of lethargic flow, 429
Sulfides, 324–331, 370
Sulfite, 332, 370
Sulfur dioxide, 12, 332, 370
Sulfuric acid, 320
Superfund, 20–23, 34, 387, Appendix I

Surface casing, 425
Surface filters, 317–319
Surface removal, 263
Surface water, 8, 27, 86, 140, 154–156, 162–165, 233, 308
Surfactants, 263
Surveys, 13, 107–119, 136, Appendix D, Appendix E
Synthetic membrane liners, 405–408
Swamps, 137, 144
Sweden, 38, 127, 134, 207, 234
Switzerland, 40, 208, 286
Synergistic effects, 56
Synthetic garbage juice, 75

Taiwan, 41
Tennessee, 115, 120
Teratogenicity, 46, 71–73
Terrain, 142
Testing, 52–57, 71, 81
Texas, 100, 109, 118, 120, 157, 292, Appendix H
Textile dying and finishing, 114, 126
Thallium, 89
Thermal infrared imaging, 240, 247
Thermoplastic fixation, 412
Three Mile Island, 171
Time-dose relations, 258
Times Beach, Missouri, 272
Titanium dioxide, 422
Titanium pigment wastes, 422
Tornadoes, 141
Toulene, 361
Toulene diisocyanate (TDI), 21
Toxaphene, 4
Toxicity, 47, 52, 62–63, 78, 84, 164
Toxic pollutants, 23, 29, 32, 43, 45
Toxic substances, 15, 27, 30, 78
Toxic Substances Control Act (TSCA), 23, 30, 44, 45, 224, 227
Toxic vapors, 3, 11
Training, 13
Transportation, 7, 10, 16, 30, 33, 38, 48, 63, 67, 79, 136, 141, 169, 187, 196, 197, 211, 221, 279–306
Travel time, 151
Treatment, 7, 10, 17, 23, 27, 29, 30, 33, 37, 38, 48, 53, 91, 136, 187, 211, 284, 307–364
Treatment costs, 356–361
Trespass, 12

Trickling filter, 338
Trustees, 26

UMTRCA, 21
Uncontained liquids, 21, 37
Uncontrolled landfills, 7, 26, 34, 227–278, 387, Appendix I
Underground Injection Control (UIC) program, 423–431
Unified soil classification system, 163
Unionization of waste service industry, 196
U.S. Coast Guard, 25
Uranium Mill Tailings, 21, 103, 268, 395
Urea, 413

Vacuum filtration, 210, 317, 359
Valley of the Drums, 22, 199, 223
Vapor density, 68
Vaporization, 20, 389
Vegetation, 144
Venezuela, 37
Vermont, 285
Versailles Borough v. *McKeesport Coal and Coke,* 11
Vertical integration of industry, 199, 286
Vinyl ester, 413
Virginia, 4, 266
Volatilization, 67, 255, 262
VTT model, 155
Vulcanus, 382

Warchak v. *Moffat,* 12
Warfare agents, 65, 69, 419

Washington (state), 57, 74, 75, 87–88, 91, 102, 107, 113–114, 120, 295
Waste Disposal Act, 40
Waste exchange, 7, 203–207, 307
Waste Generation Factor (WGF), 115
Waste Haulers, 279–286
Waste Management Advisory Council, 38
Waste munitions, 120
Waste oil refining, 114
Waste oils, 4, 6, 9, 21, 38, 67, 87, 101, 188, 204, 207, 284, 308, 419
Waste reduction, 196, 307
Waste separation, 307
Waste stabilization, 411–414
Wastewater, 9, 308–361
Wastewater treatment, 309–361
Watershed value, 144–152
Waybills, 291
Weighting factors, 99, 149, 234
Well points, 263, 268
Wet oxidation, 333
Wilsonville, Illinois, 13, 133, 171
Wind erosion, 255, 389
Wind rose, 141
Working party on toxic and dangerous waste, 303

Zeolites, 334
Zero discharge, 165
Zero risk, 5, 165
Zinc, 56, 329, 412
Zone of delayed circulation, 429
Zone of rapid circulation, 429
Zoning, 133, 136, 157, 175, 232